Student Solutions Manual:

Advanced Engineering Mathematics
Second Edition

Warren S. Wright
Loyola Marymount University

Carol D. Wright

JONES AND BARTLETT PUBLISHERS
Sudbury, Massachusetts
BOSTON   TORONTO   LONDON   SINGAPORE

World Headquarters
Jones and Bartlett Publishers
40 Tall Pine Drive
Sudbury, MA 01776
978-443-5000
info@jbpub.com
www.jbpub.com

Jones and Bartlett Publishers Canada
2100 Bloor St. West
Suite 6-272
Toronto, ON M6S 5A5
CANADA

Jones and Bartlett Publishers International
Barb House, Barb Mews
London W6 7PA
UK

ISBN: 0-7637-1285-X

Printed in the United States of America
04 03 02 01   10 9 8 7 6 5 4 3

Table of Contents

1 Introduction to Differential Equations

Exercises 1.1

3. The differential equation is first-order. Writing it in the form $x(dy/dx) + y^2 = 1$, we see that it is nonlinear in y because of y^2. However, writing it in the form $(y^2 - 1)(dx/dy) + x = 0$, we see that it is linear in x.

6. Second-order; nonlinear because of $\cos(r + u)$

9. Third-order; linear

12. From $y = \frac{6}{5} - \frac{6}{5}e^{-20t}$ we obtain $dy/dt = 24e^{-20t}$, so that

$$\frac{dy}{dt} + 20y = 24e^{-20t} + 20\left(\frac{6}{5} - \frac{6}{5}e^{-20t}\right) = 24.$$

15. Writing $\ln(2X - 1) - \ln(X - 1) = t$ and differentiating implicitly we obtain

$$\frac{2}{2X - 1}\frac{dX}{dt} - \frac{1}{X - 1}\frac{dX}{dt} = 1$$

$$\left(\frac{2}{2X - 1} - \frac{1}{X - 1}\right)\frac{dX}{dt} = 1$$

$$\frac{2X - 2 - 2X + 1}{(2X - 1)(X - 1)}\frac{dX}{dt} = 1$$

$$\frac{dX}{dt} = -(2X - 1)(X - 1) = (X - 1)(1 - 2X).$$

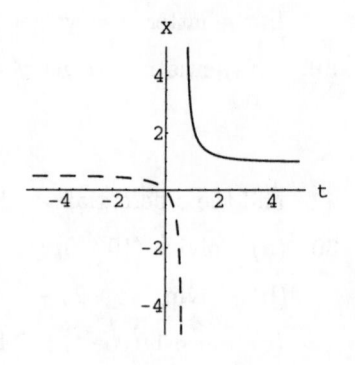

Exponentiating both sides of the implicit solution we obtain

$$\frac{2X - 1}{X - 1} = e^t \implies 2X - 1 = Xe^t - e^t \implies (e^t - 1) = (e^t - 2)X \implies X = \frac{e^t - 1}{e^t - 2}.$$

Solving $e^t - 2 = 0$ we get $t = \ln 2$. Thus, the solution is defined on $(-\infty, \ln 2)$ or on $(\ln 2, \infty)$. The graph of the solution defined on $(-\infty, \ln 2)$ is dashed, and the graph of the solution defined on $(\ln 2, \infty)$ is solid.

18. Differentiating $y = e^{-x^2}\int_0^x e^{t^2}\,dt + c_1 e^{-x^2}$ we obtain

$$y' = e^{-x^2}e^{x^2} - 2xe^{-x^2}\int_0^x e^{t^2}\,dt - 2c_1 xe^{-x^2} = 1 - 2xe^{-x^2}\int_0^x e^{t^2}\,dt - 2c_1 xe^{-x^2}.$$

Substituting into the differential equation, we have

$$y' + 2xy = 1 - 2xe^{-x^2}\int_0^x e^{t^2}\,dt - 2c_1 xe^{-x^2} + 2xe^{-x^2}\int_0^x e^{t^2}\,dt + 2c_1 xe^{-x^2} = 1.$$

21. From $y = \begin{cases} -x^2, & x < 0 \\ x^2, & x \geq 0 \end{cases}$ we obtain $y' = \begin{cases} -2x, & x < 0 \\ 2x, & x \geq 0 \end{cases}$ so that $xy' - 2y = 0$.

24. From $x = \cos 2t + \sin 2t + \frac{1}{5}e^t$ and $y = -\cos 2t - \sin 2t - \frac{1}{5}e^t$ we obtain

$$\frac{dx}{dt} = -2\sin 2t + 2\cos 2t + \frac{1}{5}e^t \quad \text{and} \quad \frac{dy}{dt} = 2\sin 2t - 2\cos 2t - \frac{1}{5}e^t$$

and

$$\frac{d^2x}{dt^2} = -4\cos 2t - 4\sin 2t + \frac{1}{5}e^t \quad \text{and} \quad \frac{d^2y}{dt^2} = 4\cos 2t + 4\sin 2t - \frac{1}{5}e^t.$$

Then

$$4y + e^t = 4(-\cos 2t - \sin 2t - \frac{1}{5}e^t) + e^t$$

$$= -4\cos 2t - 4\sin 2t + \frac{1}{5}e^t = \frac{d^2x}{dt^2}$$

and

$$4x - e^t = 4(\cos 2t + \sin 2t + \frac{1}{5}e^t) - e^t$$

$$= 4\cos 2t + 4\sin 2t - \frac{1}{5}e^t = \frac{d^2y}{dt^2}.$$

27. $(y')^2 + 1 = 0$ has no real solution.

30. Any function of the form $y = ce^t$ or $y = ce^{-t}$ is its own second derivative. The corresponding differential equation is $y'' - y = 0$. Functions of the form $y = c\sin t$ or $y = c\cos t$ have second derivatives that are the negatives of themselves. The differential equation is $y'' + y = 0$.

33. For the first-order differential equation integrate $f(x)$. For the second-order differential equation integrate twice. In the latter case we get $y = \int(\int f(t)dt)dt + c_1 t + c_2$.

36. Differentiating we get $y' = c_1 + 3c_2 t^2$ and $y'' = 6c_2 t$. Then $c_2 = y''/6t$ and $c_1 = y' - ty''/2$, so

$$y = \left(y' - \frac{ty''}{2}\right)t + \left(\frac{y''}{6t}\right)t^3 = ty' - \frac{1}{3}t^2 y''$$

and the differential equation is $t^2 y'' - 3ty' + 3y = 0$.

39. (a) Solving $(10 - 5y)/3x = 0$ we see that $y = 2$ is a constant solution.

(b) Solving $y^2 + 2y - 3 = (y + 3)(y - 1) = 0$ we see that $y = -3$ and $y = 1$ are constant solutions.

(c) Since $1/(y - 1) = 0$ has no solutions, the differential equation has no constant solutions.

(d) Setting $y' = 0$ we have $y'' = 0$ and $6y = 10$. Thus $y = 5/3$ is a constant solution.

42. One solution, with domain approximately $(-\infty, 1.6)$ is the portion of the graph in the second quadrant together with the lower part of the graph in the first quadrant. A second solution, with domain approximately $(0, 1.6)$ is the upper part of the graph in the first quadrant. The third solution, with domain $(0, \infty)$, is the part of the graph in the fourth quadrant.

45. Since $\phi'(x) > 0$ for all x in I, $\phi(x)$ is an increasing function on I. Hence, it can have no relative extrema on I.

Exercises 1.2

3. Using $x' = -c_1 \sin t + c_2 \cos t$ we obtain $c_1 = -1$ and $c_2 = 8$. The solution is $x = -\cos t + 8\sin t$.

6. Using $x' = -c_1 \sin t + c_2 \cos t$ we obtain

$$\frac{\sqrt{2}}{2} c_1 + \frac{\sqrt{2}}{2} c_2 = \sqrt{2}$$

$$-\frac{\sqrt{2}}{2} c_1 + \frac{\sqrt{2}}{2} c_2 = 2\sqrt{2}.$$

Solving we find $c_1 = -1$ and $c_2 = 3$. The solution is $x = -\cos t + 3\sin t$.

9. From the initial conditions we obtain

$$c_1 e^{-1} + c_2 e = 5$$

$$c_1 e^{-1} - c_2 e = -5.$$

Solving we get $c_1 = 0$ and $c_2 = 5e^{-1}$. A solution of the initial-value problem is $y = 5e^{-x-1}$.

12. Two solutions are $y = 0$ and $y = x^2$. (Also, any constant multiple of x^2 is a solution.)

15. For $f(x, y) = \dfrac{y}{x}$ we have $\dfrac{\partial f}{\partial y} = \dfrac{1}{x}$. Thus the differential equation will have a unique solution in any region where $x \neq 0$.

18. For $f(x, y) = \dfrac{x^2}{1 + y^3}$ we have $\dfrac{\partial f}{\partial y} = \dfrac{-3x^2 y^2}{(1 + y^3)^2}$. Thus the differential equation will have a unique solution in any region where $y \neq -1$.

21. The differential equation has a unique solution at $(1, 4)$.

24. The differential equation is not guaranteed to have a unique solution at $(-1, 1)$.

27. (a) Since $\dfrac{d}{dt}\left(-\dfrac{1}{t + c}\right) = \dfrac{1}{(t + c)^2} = y^2$, we see that $y = -\dfrac{1}{t + c}$ is a solution of the differential equation.

(b) Solving $y(0) = -1/c = 1$ we obtain $c = -1$ and $y = 1/(1 - t)$. Solving $y(0) = -1/c = -1$ we obtain $c = 1$ and $y = -1/(1 + t)$. Being sure to include $t = 0$, we see that the interval of existence of $y = 1/(1 - t)$ is $(-\infty, 1)$, while the interval of existence of $y = -1/(1 + t)$ is $(-1, \infty)$.

(c) Solving $y(0) = -1/c = y_0$ we obtain $c = -1/y_0$ and
$$y = -\dfrac{1}{-1/y_0 + t} = \dfrac{y_0}{1 - y_0 t}, \quad y_0 \neq 0.$$

Since we must have $-1/y_0 + t \neq 0$, the largest interval of existence (which must contain 0) is either $(-\infty, 1/y_0)$ when $y_0 > 0$ or $(1/y_0, \infty)$ when $y_0 < 0$.

(d) By inspection we see that $y = 0$ is a solution on $(-\infty, \infty)$.

30. If the solution is tangent to the x-axis at $(x_0, 0)$, then $y' = 0$ when $x = x_0$ and $y = 0$. Substituting these values into $y' + 2y = 3x - 6$ we get $0 + 0 = 3x_0 - 6$ or $x_0 = 2$.

33. The antiderivative of $y' = 8e^{2x} + 6x$ is $y = 4e^{2x} + 3x^2 + c$. Setting $x = 0$ and $y = 9$ we get $9 = 4 + c$, so $c = 5$ and $y = 4e^{2x} + 3x^2 + 5$.

Exercises 1.3

3. Let b be the rate of births and d the rate of deaths. Then $b = k_1 P$ and $d = k_2 P^2$. Since $dP/dt = b - d$, the differential equation is $dP/dt = k_1 P - k_2 P^2$.

6. By inspecting the graph we take T_m to be $T_m(t) = 80 - 30 \cos \pi t/12$. Then the temperature of the body at time t is determined by the differential equation
$$\frac{dT}{dt} = k\left[T - \left(80 - 30 \cos \frac{\pi}{12}t\right)\right], \quad t > 0.$$

9. The rate at which salt is leaving the tank is
$$(3 \text{ gal/min}) \cdot \left(\frac{A}{300} \text{ lb/gal}\right) = \frac{A}{100} \text{ lb/min.}$$

Thus $dA/dt = A/100$.

12. The volume of water in the tank at time t is $V = \frac{1}{3}\pi r^2 h = \frac{1}{3}A_w h$. Using the formula from Problem 11 for the volume of water leaving the tank we see that the differential equation is
$$\frac{dh}{dt} = \frac{3}{A_w}\frac{dV}{dt} = \frac{3}{A_w}(-cA_h\sqrt{2gh}) = -\frac{3cA_h}{A_w}\sqrt{2gh}.$$

Using $A_h = \pi(2/12)^2 = \pi/36$, $g = 32$, and $c = 0.6$, this becomes

$$\frac{dh}{dt} = -\frac{3(0.6)\pi/36}{A_w}\sqrt{64h} = -\frac{0.4\pi}{A_w}h^{1/2}.$$

To find A_w we let r be the radius of the top of the water. Then $r/h = 8/20$, so $r = 2h/5$ and $A_w = \pi(2h/5)^2 = 4\pi h^2/25$. Thus

$$\frac{dh}{dt} = -\frac{0.4\pi}{4\pi h^2/25}h^{1/2} = -2.5h^{-3/2}.$$

15. From Newton's second law we obtain $m\dfrac{dv}{dt} = -kv^2 + mg$.

18. From Problem 17, without a damping force, the differential equation is $m\,d^2x/dt^2 = -kx$. With a damping force proportional to velocity the differential equation becomes

$$m\frac{d^2x}{dt^2} = -kx - \beta\frac{dx}{dt} \qquad \text{or} \qquad m\frac{d^2x}{dt^2} + \beta\frac{dx}{dt} + kx = 0.$$

21. From $g = k/R^2$ we find $k = gR^2$. Using $a = d^2r/dt^2$ and the fact that the positive direction is upward we get

$$\frac{d^2r}{dt^2} = -a = -\frac{k}{r^2} = -\frac{gR^2}{r^2} \qquad \text{or} \qquad \frac{d^2r}{dt^2} + \frac{gR^2}{r^2} = 0.$$

24. The differential equation is $\dfrac{dA}{dt} = k_1(M - A) - k_2 A$.

27. We see from the figure that $2\theta + \alpha = \pi$. Thus

$$\frac{y}{-x} = \tan\alpha = \tan(\pi - 2\theta) = -\tan 2\theta = -\frac{2\tan\theta}{1 - \tan^2\theta}.$$

Since the slope of the tangent line is $y' = \tan\theta$ we have $y/x = 2y'[1 - (y')^2]$ or $y - y(y')^2 = 2xy'$, which is the quadratic equation $y(y')^2 + 2xy' - y = 0$ in y'. Using the quadratic formula we get

$$y' = \frac{-2x \pm \sqrt{4x^2 + 4y^2}}{2y} = \frac{-x \pm \sqrt{x^2 + y^2}}{y}.$$

Since $dy/dx > 0$, the differential equation is

$$\frac{dy}{dx} = \frac{-x + \sqrt{x^2 + y^2}}{y} \qquad \text{or} \qquad y\frac{dy}{dx} - \sqrt{x^2 + y^2} + x = 0.$$

30. The differential equation in (8) is $dA/dt = 6 - A/100$. If $A(t)$ attains a maximum, then $dA/dt = 0$ at this time and $A = 600$. If $A(t)$ continues to increase without reaching a maximum then $A'(t) > 0$ for $t > 0$ and A cannot exceed 600. In this case, if $A'(t)$ approaches 0 as t increases to infinity, we see that $A(t)$ approaches 600 as t increases to infinity.

33. From Problem 21, $d^2r/dt^2 = -gR^2/r^2$. Since R is a constant, if $r = R + s$, then $d^2r/dt^2 = d^2s/dt^2$ and, using a Taylor series, we get

$$\frac{d^2s}{dt^2} = -g\frac{R^2}{(R + s)^2} = -gR^2(R + s)^{-2} \approx -gR^2[R^{-2} - 2sR^{-3} + \cdots] = -g + \frac{2gs}{R^3} + \cdots.$$

Thus, for R much larger than s, the differential equation is approximated by $d^2s/dt^2 = -g$.

Chapter 1 Review Exercises

3. $\dfrac{d}{dx}(c_1 \cos kx + c_2 \sin kx) = -kc_1 \sin kx + kc_2 \cos kx;$

$\dfrac{d^2}{dx^2}(c_1 \cos kx + c_2 \sin kx) = -k^2 c_1 \cos kx - k^2 c_2 \sin kx = -k^2(c_1 \cos kx + c_2 \sin kx);$

$\dfrac{d^2 y}{dx^2} = -k^2 y$ or $\dfrac{d^2 y}{dx^2} + k^2 y = 0$

6. $y' = -c_1 e^x \sin x + c_1 e^x \cos x + c_2 e^x \cos x + c_2 e^x \sin x;$

$y'' = -c_1 e^x \cos x - c_1 e^x \sin x - c_1 e^x \sin x + c_1 e^x \cos x - c_2 e^x \sin x + c_2 e^x \cos x + c_2 e^x \cos x + c_2 e^x \sin x$

$\quad = -2c_1 e^x \sin x + 2c_2 e^x \cos x;$

$y'' - 2y' = -2c_1 e^x \cos x - 2c_2 e^x \sin x = -2y; \qquad y'' - 2y' + 2y = 0$

9. b

12. a,b,d

15. The slope of the tangent line at (x, y) is y', so the differential equation is $y' = x^2 + y^2$.

18. (a) Differentiating $y^2 - 2y = x^2 - x + c$ we obtain $2yy' - 2y' = 2x - 1$ or $(2y - 2)y' = 2x - 1$.

(b) Setting $x = 0$ and $y = 1$ in the solution we have $1 - 2 = 0 - 0 + c$ or $c = -1$. Thus, a solution of the initial-value problem is $y^2 - 2y = x^2 - x - 1$.

(c) Using the quadratic formula to solve $y^2 - 2y - (x^2 - x - 1) = 0$ we get $y = (2 \pm \sqrt{4 + 4(x^2 - x - 1)})/2$ $= 1 \pm \sqrt{x^2 - x} = 1 \pm \sqrt{x(x - 1)}$. Since $x(x-1) \geq 0$ for $x \leq 0$ or $x \geq 1$, we see that neither $y = 1 + \sqrt{x(x-1)}$ nor $y = 1 - \sqrt{x(x-1)}$ is differentiable at $x = 0$. Thus, both functions are solutions of the differential equation, but neither is a solution of the initial-value problem.

21. Differentiating $y = \sin(\ln x)$ we obtain $y' = \cos(\ln x)/x$ and $y'' = -[\sin(\ln x) + \cos(\ln x)]/x^2$. Then

$$x^2 y'' + xy' + y = x^2 \left(-\frac{\sin(\ln x) + \cos(\ln x)}{x^2} \right) + x \frac{\cos(\ln x)}{x} + \sin(\ln x) = 0.$$

24. The differential equation is

$$\frac{dh}{dt} = -\frac{cA_0}{A_w} \sqrt{2gh} .$$

Using $A_0 = \pi(1/24)^2 = \pi/576$, $A_w = \pi(2)^2 = 4\pi$, and $g = 32$, this becomes

$$\frac{dh}{dt} = -\frac{c\pi/576}{4\pi} \sqrt{64h} = \frac{c}{288} \sqrt{h} .$$

2 First-Order Differential Equations

_____ **Exercises 2.1** _____

3.

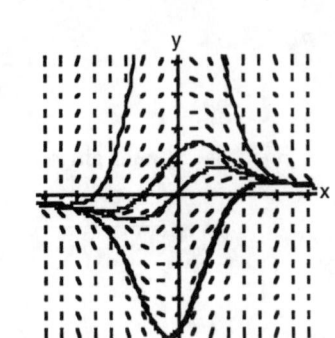

15. Writing the differential equation in the form $dy/dx = y(1-y)(1+y)$ we see that critical points are located at $y = -1$, $y = 0$, and $y = 1$. The phase portrait is shown below.

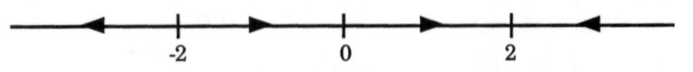

(a)

(b)

(c)

(d)

18. Solving $y^2 - y^3 = y^2(1-y) = 0$ we obtain the critical points 0 and 1.

From the phase portrait we see that 1 is asymptotically stable and 0 is semi-stable.

21. Solving $y^2(4-y^2) = y^2(2-y)(2+y) = 0$ we obtain the critical points -2, 0, and 2.

From the phase portrait we see that 2 is asymptotically stable, 0 is semi-stable, and -2 is unstable.

24. Solving $ye^y - 9y = y(e^y - 9) = 0$ we obtain the critical points 0 and $\ln 9$.

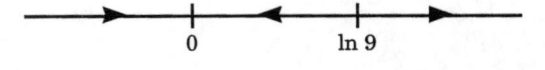

6

From the phase portrait we see that 0 is asymptotically stable and $\ln 9$ is unstable.

27. Critical points are $y = 0$ and $y = c$.

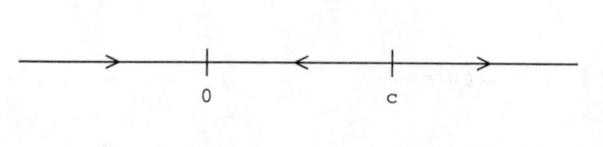

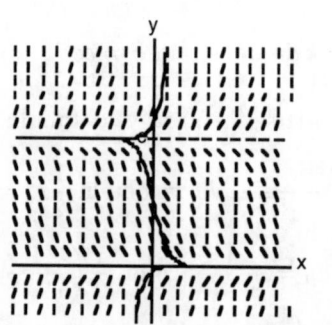

30. When $y < \frac{1}{2}x^2$, $y' = x^2 - 2y$ is positive and the portions of solution curves "inside" the nullcline parabola are increasing. When $y > \frac{1}{2}x^2$, $y' = x^2 - 2y$ is negative and the portions of the solution curves "outside" the nullcline parabola are decreasing.

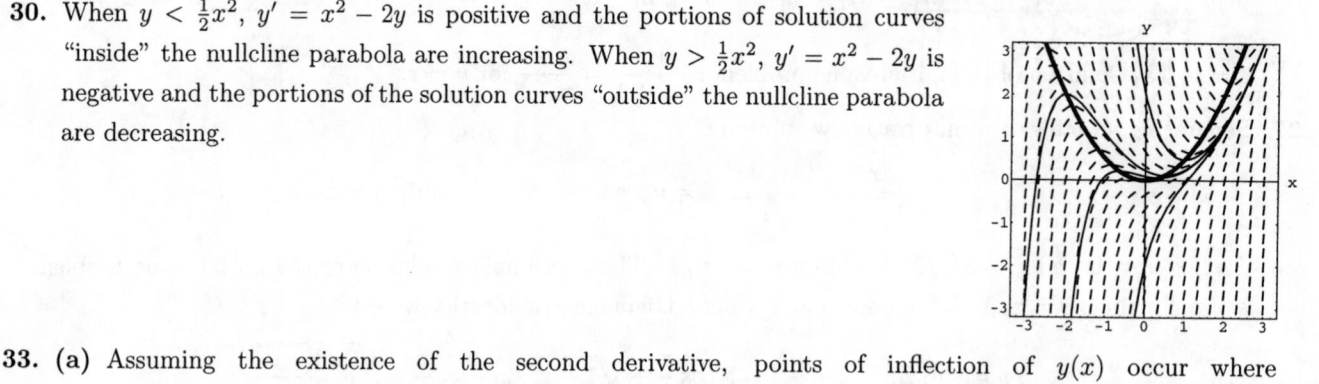

33. (a) Assuming the existence of the second derivative, points of inflection of $y(x)$ occur where $y''(x) = 0$. From $dy/dx = g(y)$ we have $d^2y/dx^2 = g'(y)\,dy/dx$. Thus, the y-coordinate of a point of inflection can be located by solving $g'(y) = 0$. (Points where $dy/dx = 0$ correspond to constant solutions of the differential equation.)

(b) Solving $y^2 - y - 6 = (y - 3)(y + 2) = 0$ we see that 3 and -2 are critical points. Now $d^2y/dx^2 = (2y - 1)\,dy/dx = (2y - 1)(y - 3)(y + 2)$, so the only possible point of inflection is at $y = \frac{1}{2}$, although the concavity of solutions can be different on either side of $y = -2$ and $y = 3$. Since $y''(x) < 0$ for $y < -2$ and $\frac{1}{2} < y < 3$, and $y''(x) > 0$ for $-2 < y < \frac{1}{2}$ and $y > 3$, we see that solution curves are concave down for $y < -2$ and $\frac{1}{2} < y < 3$ and concave up for $-2 < y < \frac{1}{2}$ and $y > 3$. Points of inflection of solutions of autonomous differential equations will have the same y-coordinates because between critical points they are horizontal translates of each other.

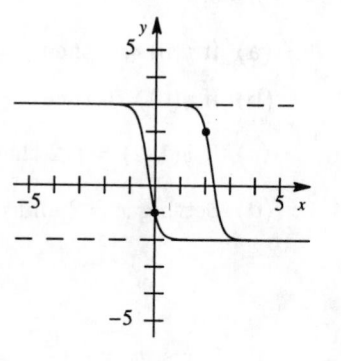

Exercises 2.2

In many of the following problems we will encounter an expression of the form $\ln |g(y)| = f(x) + c$. To solve for $g(y)$ we exponentiate both sides of the equation. This yields $|g(y)| = e^{f(x)+c} = e^c e^{f(x)}$ which implies $g(y) = \pm e^c e^{f(x)}$. Letting $c_1 = \pm e^c$ we obtain $g(y) = c_1 e^{f(x)}$.

3. From $dy = -e^{-3x}\,dx$ we obtain $y = \frac{1}{3}e^{-3x} + c$.

6. From $\frac{1}{y}\,dy = -2x\,dx$ we obtain $\ln |y| = -x^2 + c$ or $y = c_1 e^{-x^2}$.

9. From $\left(y + 2 + \frac{1}{y}\right)dy = x^2 \ln x\,dx$ we obtain $\frac{y^2}{2} + 2y + \ln |y| = \frac{x^3}{3}\ln |x| - \frac{1}{9}x^3 + c$.

12. From $2y\,dy = -\dfrac{\sin 3x}{\cos^3 3x}\,dx = -\tan 3x \sec^2 3x\,dx$ we obtain $y^2 = -\dfrac{1}{6}\sec^2 3x + c.$

15. From $\dfrac{1}{S}\,dS = k\,dr$ we obtain $S = ce^{kr}.$

18. From $\dfrac{1}{N}\,dN = \left(te^{t+2} - 1\right)dt$ we obtain $\ln|N| = te^{t+2} - e^{t+2} - t + c.$

21. From $x\,dx = \dfrac{1}{\sqrt{1-y^2}}\,dy$ we obtain $\dfrac{1}{2}x^2 = \sin^{-1}y + c$ or $y = \sin\left(\dfrac{x^2}{2} + c_1\right).$

24. From $\dfrac{1}{y^2-1}\,dy = \dfrac{1}{x^2-1}\,dx$ or $\dfrac{1}{2}\left(\dfrac{1}{y-1} - \dfrac{1}{y+1}\right)dy = \dfrac{1}{2}\left(\dfrac{1}{x-1} - \dfrac{1}{x+1}\right)dx$ we obtain

$\ln|y-1| - \ln|y+1| = \ln|x-1| - \ln|x+1| + \ln c$ or $\dfrac{y-1}{y+1} = \dfrac{c(x-1)}{x+1}.$ Using $y(2) = 2$ we find

$c = 1.$ The solution of the initial-value problem is $\dfrac{y-1}{y+1} = \dfrac{x-1}{x+1}$ or $y = x.$

27. Separating variables and integrating we obtain

$$\frac{dx}{\sqrt{1-x^2}} - \frac{dy}{\sqrt{1-y^2}} = 0 \quad \text{and} \quad \sin^{-1}x - \sin^{-1}y = c.$$

Setting $x = 0$ and $y = \sqrt{3}/2$ we obtain $c = -\pi/3.$ Thus, an implicit solution of the initial-value problem is $\sin^{-1}x - \sin^{-1}y = \pi/3.$ Solving for y and using a trigonometric identity we get

$$y = \sin\left(\sin^{-1}x + \frac{\pi}{3}\right) = x\cos\frac{\pi}{3} + \sqrt{1-x^2}\,\sin\frac{\pi}{3} = \frac{x}{2} + \frac{\sqrt{3}\sqrt{1-x^2}}{2}\,.$$

30. From $\left(\dfrac{1}{y-1} + \dfrac{-1}{y}\right)dy = \dfrac{1}{x}\,dx$ we obtain $\ln|y-1| - \ln|y| = \ln|x| + c$ or $y = \dfrac{1}{1-c_1 x}.$ Another solution is $y = 0.$

   (a) If $y(0) = 1$ then $y = 1.$

   (b) If $y(0) = 0$ then $y = 0.$

   (c) If $y(1/2) = 1/2$ then $y = \dfrac{1}{1+2x}\,.$

   (d) Setting $x = 2$ and $y = \frac{1}{4}$ we obtain

$$\frac{1}{4} = \frac{1}{1 - c_1(2)}, \quad 1 - 2c_1 = 4, \quad \text{and} \quad c_1 = -\frac{3}{2}.$$

   Thus, $y = \dfrac{1}{1 + \frac{3}{2}x} = \dfrac{2}{2 + 3x}\,.$

33. The singular solution $y = 1$ satisfies the initial-value problem.

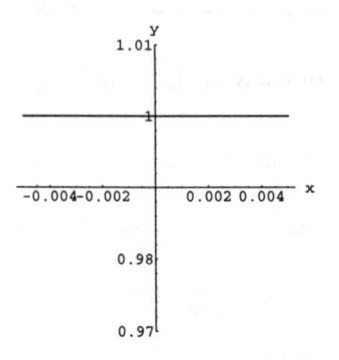

36. Separating variables we obtain $\dfrac{dy}{(y-1)^2 - 0.01} = dx$. Then

$$5 \ln \left| \frac{10y - 11}{10y - 9} \right| = x + c.$$

Setting $x = 0$ and $y = 1$ we obtain $c = 5 \ln 1 = 0$. The solution is

$$5 \ln \left| \frac{10y - 11}{10y - 9} \right| = x.$$

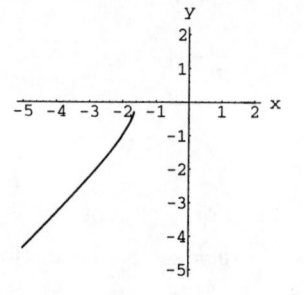

39. (a) Separating variables we have $2y\, dy = (2x + 1)dx$. Integrating gives $y^2 = x^2 + x + c$. When $y(-2) = -1$ we find $c = -1$, so $y^2 = x^2 + x - 1$ and $y = -\sqrt{x^2 + x - 1}$. The negative square root is chosen because of the initial condition.

(b) The interval of definition appears to be approximately $(-\infty, -1.65)$.

(c) Solving $x^2 + x - 1 = 0$ we get $x = -\frac{1}{2} \pm \frac{1}{2}\sqrt{5}$, so the exact interval of definition is $\left(-\infty, -\frac{1}{2} - \frac{1}{2}\sqrt{5}\right)$.

42. (a) Separating variables and integrating we obtain $x^2 - y^2 = c$. For $c \neq 0$ the graph is a square hyperbola centered at the origin. All four initial conditions imply $c = 0$ and $y = \pm x$. Since the differential equation is not defined for $y = 0$, solutions are $y = \pm x$, $x < 0$ and $y = \pm x$, $x > 0$. The solution for $y(a) = a$ is $y = x$, $x > 0$; for $y(a) = -a$ is $y = -x$; for $y(-a) = a$ is $y = -x$, $x < 0$; and for $y(-a) = -a$ is $y = x$, $x < 0$.

(b) Since x/y is not defined when $y = 0$, the initial-value problem has no solution.

(c) Setting $x = 1$ and $y = 2$ in $x^2 - y^2 = c$ we get $c = -3$, so $y^2 = x^2 + 3$ and $y(x) = \sqrt{x^2 + 3}$, where the positive square root is chosen because of the initial condition. The domain is all real numbers since $x^2 + 3 > 0$ for all x.

45. We are looking for a function $y(x)$ such that

$$y^2 + \left(\frac{dy}{dx}\right)^2 = 1.$$

Using the positive square root gives

$$\frac{dy}{dx} = \sqrt{1 - y^2} \implies \frac{dy}{\sqrt{1 - y^2}} = dx \implies \sin^{-1} y = x + c.$$

Thus a solution is $y = \sin(x + c)$. If we use the negative square root we obtain

$$y = \sin(c - x) = -\sin(x - c) = -\sin(x + c_1).$$

Note also that $y = 1$ and $y = -1$ are solutions.

Exercises 2.3

3. For $y' + y = e^{3x}$ an integrating factor is $e^{\int dx} = e^x$ so that $\dfrac{d}{dx}[e^x y] = e^{4x}$ and $y = \frac{1}{4}e^{3x} + ce^{-x}$ for $-\infty < x < \infty$. The transient term is ce^{-x}.

6. For $y' + 2xy = x^3$ an integrating factor is $e^{\int 2x\,dx} = e^{x^2}$ so that $\dfrac{d}{dx}\left[e^{x^2} y\right] = x^3 e^{x^2}$ and $y = \frac{1}{2}x^2 - \frac{1}{2} + ce^{-x^2}$ for $-\infty < x < \infty$. The transient term is ce^{-x^2}.

9. For $y' - \dfrac{1}{x}y = x\sin x$ an integrating factor is $e^{-\int (1/x)dx} = \dfrac{1}{x}$ so that $\dfrac{d}{dx}\left[\dfrac{1}{x}y\right] = \sin x$ and $y = cx - x\cos x$ for $0 < x < \infty$.

12. For $y' - \dfrac{x}{(1+x)}y = x$ an integrating factor is $e^{-\int [x/(1+x)]dx} = (x+1)e^{-x}$ so that $\dfrac{d}{dx}\left[(x+1)e^{-x}y\right] = x(x+1)e^{-x}$ and $y = -x - \dfrac{2x+3}{x+1} + \dfrac{ce^x}{x+1}$ for $-1 < x < \infty$.

15. For $\dfrac{dx}{dy} - \dfrac{4}{y}x = 4y^5$ an integrating factor is $e^{-\int (4/y)dy} = y^{-4}$ so that $\dfrac{d}{dy}\left[y^{-4}x\right] = 4y$ and $x = 2y^6 + cy^4$ for $0 < y < \infty$.

18. For $y' + (\cot x)y = \sec^2 x \csc x$ an integrating factor is $e^{\int \cot x\,dx} = \sin x$ so that $\dfrac{d}{dx}[(\sin x)\,y] = \sec^2 x$ and $y = \sec x + c\csc x$ for $0 < x < \pi/2$.

21. For $\dfrac{dr}{d\theta} + r\sec\theta = \cos\theta$ an integrating factor is $e^{\int \sec\theta\,d\theta} = \sec\theta + \tan\theta$ so that $\dfrac{d}{d\theta}[r(\sec\theta + \tan\theta)] = 1 + \sin\theta$ and $r(\sec\theta + \tan\theta) = \theta - \cos\theta + c$ for $-\pi/2 < \theta < \pi/2$.

24. For $y' + \dfrac{2}{x^2-1}y = \dfrac{x+1}{x-1}$ an integrating factor is $e^{\int [2/(x^2-1)]dx} = \dfrac{x-1}{x+1}$ so that $\dfrac{d}{dx}\left[\dfrac{x-1}{x+1}y\right] = 1$ and $(x-1)y = x(x+1) + c(x+1)$ for $-1 < x < 1$.

27. For $\dfrac{di}{dt} + \dfrac{R}{L}i = \dfrac{E}{L}$ an integrating factor is $e^{\int (R/L)\,dt} = e^{Rt/L}$ so that $\dfrac{d}{dt}\left[ie^{Rt/L}\right] = \dfrac{E}{L}e^{Rt/L}$ and $i = \dfrac{E}{R} + ce^{-Rt/L}$ for $-\infty < t < \infty$. If $i(0) = i_0$ then $c = i_0 - E/R$ and $i = \dfrac{E}{R} + \left(i_0 - \dfrac{E}{R}\right)e^{-Rt/L}$.

30. For $y' + (\tan x)y = \cos^2 x$ an integrating factor is $e^{\int \tan x\,dx} = \sec x$ so that $\dfrac{d}{dx}[(\sec x)\,y] = \cos x$ and $y = \sin x\cos x + c\cos x$ for $-\pi/2 < x < \pi/2$. If $y(0) = -1$ then $c = -1$ and $y = \sin x\cos x - \cos x$.

33. For $y' + 2xy = f(x)$ an integrating factor is e^{x^2} so that

$$ye^{x^2} = \begin{cases} \frac{1}{2}e^{x^2} + c_1, & 0 \le x \le 1; \\ c_2, & x > 1. \end{cases}$$

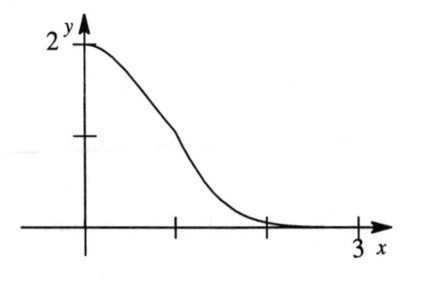

If $y(0) = 2$ then $c_1 = 3/2$ and for continuity we must have $c_2 = \frac{1}{2}e + \frac{3}{2}$ so that

$$y = \begin{cases} \frac{1}{2} + \frac{3}{2}e^{-x^2}, & 0 \le x \le 1; \\ \left(\frac{1}{2}e + \frac{3}{2}\right)e^{-x^2}, & x > 1. \end{cases}$$

42. We want 4 to be a critical point, so use $y' = 4 - y$.

45. The left-hand derivative of the function at $x = 1$ is $1/e$ and the right-hand derivative at $x = 1$ is $1 - 1/e$. Thus, y is not differentiable at $x = 1$.

48. We see by inspection that $y = 0$ is a solution.

_____ **Exercises 2.4** _____

3. Let $M = 5x + 4y$ and $N = 4x - 8y^3$ so that $M_y = 4 = N_x$. From $f_x = 5x + 4y$ we obtain $f = \frac{5}{2}x^2 + 4xy + h(y)$, $h'(y) = -8y^3$, and $h(y) = -2y^4$. The solution is $\frac{5}{2}x^2 + 4xy - 2y^4 = c$.

6. Let $M = 4x^3 - 3y\sin 3x - y/x^2$ and $N = 2y - 1/x + \cos 3x$ so that $M_y = -3\sin 3x - 1/x^2$ and $N_x = 1/x^2 - 3\sin 3x$. The equation is not exact.

9. Let $M = y^3 - y^2\sin x - x$ and $N = 3xy^2 + 2y\cos x$ so that $M_y = 3y^2 - 2y\sin x = N_x$. From $f_x = y^3 - y^2\sin x - x$ we obtain $f = xy^3 + y^2\cos x - \frac{1}{2}x^2 + h(y)$, $h'(y) = 0$, and $h(y) = 0$. The solution is $xy^3 + y^2\cos x - \frac{1}{2}x^2 = c$.

12. Let $M = 3x^2y + e^y$ and $N = x^3 + xe^y - 2y$ so that $M_y = 3x^2 + e^y = N_x$. From $f_x = 3x^2y + e^y$ we obtain $f = x^3y + xe^y + h(y)$, $h'(y) = -2y$, and $h(y) = -y^2$. The solution is $x^3y + xe^y - y^2 = c$.

15. Let $M = x^2y^3 - 1/(1 + 9x^2)$ and $N = x^3y^2$ so that $M_y = 3x^2y^2 = N_x$. From $f_x = x^2y^3 - 1/(1 + 9x^2)$ we obtain $f = \frac{1}{3}x^3y^3 - \frac{1}{3}\arctan(3x) + h(y)$, $h'(y) = 0$, and $h(y) = 0$. The solution is $x^3y^3 - \arctan(3x) = c$.

18. Let $M = 2y\sin x\cos x - y + 2y^2e^{xy^2}$ and $N = -x + \sin^2 x + 4xye^{xy^2}$ so that

$$M_y = 2\sin x\cos x - 1 + 4xy^3e^{xy^2} + 4ye^{xy^2} = N_x.$$

From $f_x = 2y\sin x\cos x - y + 2y^2e^{xy^2}$ we obtain $f = y\sin^2 x - xy + 2e^{xy^2} + h(y)$, $h'(y) = 0$, and $h(y) = 0$. The solution is $y\sin^2 x - xy + 2e^{xy^2} = c$.

21. Let $M = x^2 + 2xy + y^2$ and $N = 2xy + x^2 - 1$ so that $M_y = 2(x + y) = N_x$. From $f_x = x^2 + 2xy + y^2$ we obtain $f = \frac{1}{3}x^3 + x^2y + xy^2 + h(y)$, $h'(y) = -1$, and $h(y) = -y$. The general solution is $\frac{1}{3}x^3 + x^2y + xy^2 - y = c$. If $y(1) = 1$ then $c = \frac{4}{3}$ and the solution of the initial-value problem is $\frac{1}{3}x^3 + x^2y + xy^2 - y = \frac{4}{3}$.

24. Let $M = t/2y^4$ and $N = (3y^2 - t^2)/y^5$ so that $M_y = -2t/y^5 = N_t$. From $f_t = t/2y^4$ we obtain $f = \dfrac{t^2}{4y^4} + h(y)$, $h'(y) = \dfrac{3}{y^3}$, and $h(y) = -\dfrac{3}{2y^2}$. The general solution is $\dfrac{t^2}{4y^4} - \dfrac{3}{2y^2} = c$. If $y(1) = 1$ then $c = -5/4$ and the solution of the initial-value problem is $\dfrac{t^2}{4y^4} - \dfrac{3}{2y^2} = -\dfrac{5}{4}$.

27. Equating $M_y = 3y^2 + 4kxy^3$ and $N_x = 3y^2 + 40xy^3$ we obtain $k = 10$.

30. Let $M = (x^2 + 2xy - y^2)/(x^2 + 2xy + y^2)$ and $N = (y^2 + 2xy - x^2)/(y^2 + 2xy + x^2)$ so that $M_y = -4xy/(x + y)^3 = N_x$. From $f_x = (x^2 + 2xy + y^2 - 2y^2)/(x + y)^2$ we obtain $f = x + \dfrac{2y^2}{x + y} + h(y)$, $h'(y) = -1$, and $h(y) = -y$. The solution of the differential equation is $x^2 + y^2 = c(x + y)$.

33. We note that $(N_x - M_y)/M = 2/y$, so an integrating factor is $e^{\int 2dy/y} = y^2$. Let $M = 6xy^3$ and $N = 4y^3 + 9x^2y^2$ so that $M_y = 18xy^2 = N_x$. From $f_x = 6xy^3$ we obtain $f = 3x^2y^3 + h(y)$, $h'(y) = 4y^3$, and $h(y) = y^4$. The solution of the differential equation is $3x^2y^3 + y^4 = c$.

36. (a) Implicitly differentiating $x^3 + 2x^2y + y^2 = c$ and solving for dy/dx we obtain

$$3x^2 + 2x^2\frac{dy}{dx} + 4xy + 2y\frac{dy}{dx} = 0 \quad \text{and} \quad \frac{dy}{dx} = -\frac{3x^2 + 4xy}{2x^2 + 2y}.$$

Separating variables we get $(4xy + 3x^2)dx + (2y + 2x^2)dy = 0$.

(b) Setting $x = 0$ and $y = -2$ in $x^3 + 2x^2y + y^2 = c$ we find $c = 4$, and setting $x = y = 1$ we also find $c = 4$. Thus, both initial conditions determine the same implicit solution.

(c) Solving $x^3 + 2x^2y + y^2 = 4$ for y we get

$$y_1(x) = -x^2 - \sqrt{4 - x^3 + x^4} \quad \text{and} \quad y_2(x) = -x^2 + \sqrt{4 - x^3 + x^4}.$$

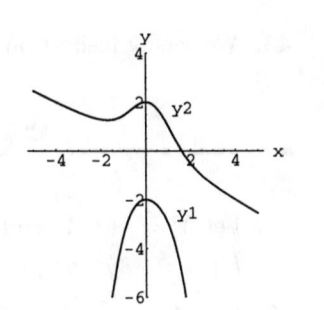

39. (a) Since $f_y = N(x, y) = xe^{xy} + 2xy + 1/x$ we obtain $f = e^{xy} + xy^2 + \dfrac{y}{x} + h(x)$ so that $f_x = ye^{xy} + y^2 - \dfrac{y}{x^2} + h'(x)$.

Let $M(x, y) = ye^{xy} + y^2 - \dfrac{y}{x^2}$.

(b) Since $f_x = M(x, y) = y^{1/2}x^{-1/2} + x\left(x^2 + y\right)^{-1}$ we obtain $f = 2y^{1/2}x^{1/2} + \dfrac{1}{2}\ln\left|x^2 + y\right| + g(y)$ so that

$f_y = y^{-1/2}x^{1/2} + \dfrac{1}{2}\left(x^2 + y\right)^{-1} + g'(x)$. Let $N(x, y) = y^{-1/2}x^{1/2} + \dfrac{1}{2}\left(x^2 + y\right)^{-1}$.

Exercises 2.5

3. Letting $x = vy$ we have

$$vy(v\,dy + y\,dv) + (y - 2vy)\,dy = 0$$

$$vy\,dv + \left(v^2 - 2v + 1\right)dy = 0$$

$$\frac{v\,dv}{(v - 1)^2} + \frac{dy}{y} = 0$$

$$\ln|v - 1| - \frac{1}{v - 1} + \ln|y| = c$$

$$\ln\left|\frac{x}{y} - 1\right| - \frac{1}{x/y - 1} + \ln y = c$$

$$(x - y)\ln|x - y| - y = c(x - y).$$

6. Letting $y = ux$ we have

$$\left(u^2x^2 + ux^2\right)dx + x^2(u\,dx + x\,du) = 0$$

$$\left(u^2 + 2u\right)dx + x\,du = 0$$

$$\frac{dx}{x} + \frac{du}{u(u + 2)} = 0$$

$$\ln|x| + \frac{1}{2}\ln|u| - \frac{1}{2}\ln|u + 2| = c$$

$$\frac{x^2u}{u + 2} = c_1$$

$$x^2\frac{y}{x} = c_1\left(\frac{y}{x} + 2\right)$$

$$x^2y = c_1(y + 2x).$$

9. Letting $y = ux$ we have

$$-ux\,dx + (x + \sqrt{u}\,x)(u\,dx + x\,du) = 0$$

$$(x + x\sqrt{u})\,du + u^{3/2}\,dx = 0$$

$$\left(u^{-3/2} + \frac{1}{u}\right)du + \frac{dx}{x} = 0$$

$$-2u^{-1/2} + \ln|u| + \ln|x| = c$$

$$\ln|y/x| + \ln|x| = 2\sqrt{x/y} + c$$

$$y(\ln|y| - c)^2 = 4x.$$

12. Letting $y = ux$ we have

$$(x^2 + 2u^2x^2)\,dx - ux^2(u\,dx + x\,du) = 0$$

$$(1 + u^2)\,dx - ux\,du = 0$$

$$\frac{dx}{x} - \frac{u\,du}{1 + u^2} = 0$$

$$\ln|x| - \frac{1}{2}\ln\left(1 + u^2\right) = c$$

$$\frac{x^2}{1 + u^2} = c_1$$

$$x^4 = c_1\left(y^2 + x^2\right).$$

Using $y(-1) = 1$ we find $c_1 = 1/2$. The solution of the initial-value problem is $2x^4 = y^2 + x^2$.

15. From $y' + \dfrac{1}{x}y = \dfrac{1}{x}y^{-2}$ and $w = y^3$ we obtain $\dfrac{dw}{dx} + \dfrac{3}{x}w = \dfrac{3}{x}$. An integrating factor is x^3 so that $x^3w = x^3 + c$ or $y^3 = 1 + cx^{-3}$.

18. From $y' - \left(1 + \dfrac{1}{x}\right)y = y^2$ and $w = y^{-1}$ we obtain $\dfrac{dw}{dx} + \left(1 + \dfrac{1}{x}\right)w = -1$. An integrating factor is xe^x so that $xe^xw = -xe^x + e^x + c$ or $y^{-1} = -1 + \dfrac{1}{x} + \dfrac{c}{x}e^{-x}$.

21. From $y' - \dfrac{2}{x}y = \dfrac{3}{x^2}y^4$ and $w = y^{-3}$ we obtain $\dfrac{dw}{dx} + \dfrac{6}{x}w = -\dfrac{9}{x^2}$. An integrating factor is x^6 so that $x^6w = -\frac{9}{5}x^5 + c$ or $y^{-3} = -\frac{9}{5}x^{-1} + cx^{-6}$. If $y(1) = \frac{1}{2}$ then $c = \frac{49}{5}$ and $y^{-3} = -\frac{9}{5}x^{-1} + \frac{49}{5}x^{-6}$.

24. Let $u = x + y$ so that $du/dx = 1 + dy/dx$. Then $\dfrac{du}{dx} - 1 = \dfrac{1 - u}{u}$ or $u\,du = dx$. Thus $\frac{1}{2}u^2 = x + c$ or $u^2 = 2x + c_1$, and $(x + y)^2 = 2x + c_1$.

27. Let $u = y - 2x + 3$ so that $du/dx = dy/dx - 2$. Then $\dfrac{du}{dx} + 2 = 2 + \sqrt{u}$ or $\dfrac{1}{\sqrt{u}}\,du = dx$. Thus $2\sqrt{u} = x + c$ and $2\sqrt{y - 2x + 3} = x + c$.

30. Let $u = 3x + 2y$ so that $du/dx = 3 + 2\,dy/dx$. Then $\dfrac{du}{dx} = 3 + \dfrac{2u}{u + 2} = \dfrac{5u + 6}{u + 2}$ and $\dfrac{u + 2}{5u + 6}\,du = dx$. Now

$$\frac{u + 2}{5u + 6} = \frac{1}{5} + \frac{4}{25u + 30}$$

so we have

$$\int\left(\frac{1}{5} + \frac{4}{25u + 30}\right)du = dx$$

and $\frac{1}{5}u + \frac{4}{25}\ln|25u + 30| = x + c$. Thus

$$\frac{1}{5}(3x + 2y) + \frac{4}{25}\ln|75x + 50y + 30| = x + c.$$

Setting $x = -1$ and $y = -1$ we obtain $c = \frac{4}{5}\ln 95$. The solution is

$$\frac{1}{5}(3x + 2y) + \frac{4}{25}\ln|75x + 50y + 30| = x + \frac{4}{5}\ln 95$$

or

$$5y - 5x + 2\ln|75x + 50y + 30| = 10\ln 95.$$

33. (a) The substitutions $y = y_1 + u$ and $\dfrac{dy}{dx} = \dfrac{dy_1}{dx} + \dfrac{du}{dx}$ lead to

$$\frac{dy_1}{dx} + \frac{du}{dx} = P + Q(y_1 + u) + R(y_1 + u)^2 = P + Qy_1 + Ry_1^2 + Qu + 2y_1 Ru + Ru^2$$

or

$$\frac{du}{dx} - (Q + 2y_1 R)u = Ru^2.$$

This is a Bernoulli equation with $n = 2$ which can be reduced to the linear equation

$$\frac{dw}{dx} + (Q + 2y_1 R)w = -R$$

by the substitution $w = u^{-1}$.

(b) Identify $P(x) = -4/x^2$, $Q(x) = -1/x$, and $R(x) = 1$. Then $\dfrac{dw}{dx} + \left(-\dfrac{1}{x} + \dfrac{4}{x}\right)w = -1$. An integrating

factor is x^3 so that $x^3 w = -\frac{1}{4}x^4 + c$ or $u = \left[-\frac{1}{4}x + cx^{-3}\right]^{-1}$. Thus, $y = \dfrac{2}{x} + u$.

Exercises 2.6

---

3. Separating variables and integrating, we have

$$\frac{dy}{y} = dx \quad \text{and} \quad \ln|y| = x + c.$$

Thus $y = c_1 e^x$ and, using $y(0) = 1$, we find $c = 1$, so $y = e^x$ is the solution of the initial-value problem.

h=0.1

x_n	y_n	True Value	Abs. Error	% Rel. Error
0.00	1.0000	1.0000	0.0000	0.00
0.10	1.1000	1.1052	0.0052	0.47
0.20	1.2100	1.2214	0.0114	0.93
0.30	1.3310	1.3499	0.0189	1.40
0.40	1.4641	1.4918	0.0277	1.86
0.50	1.6105	1.6487	0.0382	2.32
0.60	1.7716	1.8221	0.0506	2.77
0.70	1.9487	2.0138	0.0650	3.23
0.80	2.1436	2.2255	0.0820	3.68
0.90	2.3579	2.4596	0.1017	4.13
1.00	2.5937	2.7183	0.1245	4.58

h=0.05

x_n	y_n	True Value	Abs. Error	% Rel. Error
0.00	1.0000	1.0000	0.0000	0.00
0.05	1.0500	1.0513	0.0013	0.12
0.10	1.1025	1.1052	0.0027	0.24
0.15	1.1576	1.1618	0.0042	0.36
0.20	1.2155	1.2214	0.0059	0.48
0.25	1.2763	1.2840	0.0077	0.60
0.30	1.3401	1.3499	0.0098	0.72
0.35	1.4071	1.4191	0.0120	0.84
0.40	1.4775	1.4918	0.0144	0.96
0.45	1.5513	1.5683	0.0170	1.08
0.50	1.6289	1.6487	0.0198	1.20
0.55	1.7103	1.7333	0.0229	1.32
0.60	1.7959	1.8221	0.0263	1.44
0.65	1.8856	1.9155	0.0299	1.56
0.70	1.9799	2.0138	0.0338	1.68
0.75	2.0789	2.1170	0.0381	1.80
0.80	2.1829	2.2255	0.0427	1.92
0.85	2.2920	2.3396	0.0476	2.04
0.90	2.4066	2.4596	0.0530	2.15
0.95	2.5270	2.5857	0.0588	2.27
1.00	2.6533	2.7183	0.0650	2.39

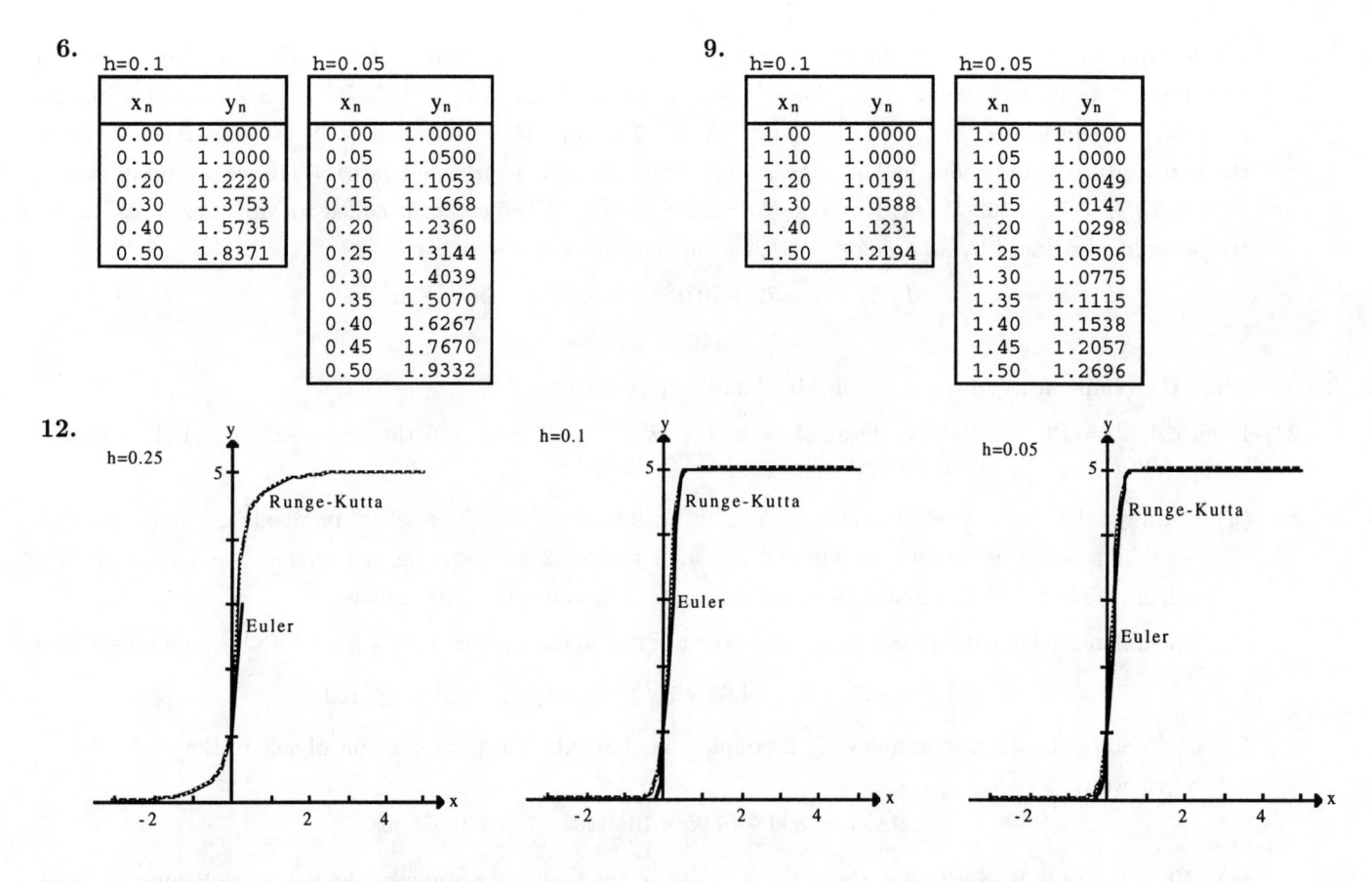

6.

h=0.1	
x_n	y_n
0.00	1.0000
0.10	1.1000
0.20	1.2220
0.30	1.3753
0.40	1.5735
0.50	1.8371

h=0.05	
x_n	y_n
0.00	1.0000
0.05	1.0500
0.10	1.1053
0.15	1.1668
0.20	1.2360
0.25	1.3144
0.30	1.4039
0.35	1.5070
0.40	1.6267
0.45	1.7670
0.50	1.9332

9.

h=0.1	
x_n	y_n
1.00	1.0000
1.10	1.0000
1.20	1.0191
1.30	1.0588
1.40	1.1231
1.50	1.2194

h=0.05	
x_n	y_n
1.00	1.0000
1.05	1.0000
1.10	1.0049
1.15	1.0147
1.20	1.0298
1.25	1.0506
1.30	1.0775
1.35	1.1115
1.40	1.1538
1.45	1.2057
1.50	1.2696

12.

Exercises 2.7

3. Let $P = P(t)$ be the population at time t. From $dP/dt = kt$ and $P(0) = P_0 = 500$ we obtain $P = 500e^{kt}$. Using $P(10) = 575$ we find $k = \frac{1}{10} \ln 1.15$. Then $P(30) = 500e^{3 \ln 1.15} \approx 760$ years.

6. From $dS/dt = rS$ we obtain $S = S_0 e^{rt}$ where $S(0) = S_0$.
   (a) If $S_0 = \$5000$ and $r = 5.75\%$ then $S(5) = \$6665.45$.
   (b) If $S(t) = \$10,000$ then $t = 12$ years.
   (c) $S \approx \$6651.82$

9. In Problem 8 the amount at time t is shown to be $N(t) = 100e^{kt}$. Setting $N(t) = 50$ we obtain

$$50 = 100e^{kt} \implies kt = \ln\frac{1}{2} \implies t = \frac{\ln 1/2}{(1/6)\ln 0.97} \approx 136.5 \text{ hours.}$$

12. From Example 3, the amount of carbon present at time t is $A(t) = A_0 e^{-0.00012378t}$. Letting $t = 660$ and solving for A_0 we have $A(660) = A_0 e^{-0.0001237(660)} = 0.921553A_0$. Thus, approximately 92% of the original amount of C-14 remained in the cloth as of 1988.

15. Assume that $dT/dt = k(T - 100)$ so that $T = 100 + ce^{kt}$. If $T(0) = 20°$ and $T(1) = 22°$ then $c = -80$ and $k = \ln(39/40)$ so that $T(t) = 90°$ implies $t = 82.1$ seconds. If $T(t) = 98°$ then $t = 145.7$ seconds.

18. We will assume that the temperature of both the room and the cream is 72°F, and that the temperature of the coffee when it is first put on the table is 175°F. If we let $T_1(t)$ represent the temperature of the coffee in Mr. Jones' cup at time t, then

$$\frac{dT_1}{dt} = k(T_1 - 72),$$

which implies $T_1 = 72 + c_1 e^{kt}$. At time $t = 0$ Mr. Jones adds cream to his coffee which immediately reduces its temperature by an amount α, so that $T_1(0) = 175 - \alpha$. Thus $175 - \alpha = T_1(0) = 72 + c_1$, which implies $c_1 = 103 - \alpha$, so that $T_1(t) = 72 + (103 - \alpha)e^{kt}$. At $t = 5$, $T_1(5) = 72 + (103 - \alpha)e^{5k}$. Now we let $T_2(t)$ represent the temperature of the coffee in Mrs. Jones' cup. From $T_2 = 72 + c_2 e^{kt}$ and $T_2(0) = 175$ we obtain $c_2 = 103$, so that $T_2(t) = 72 + 103e^{kt}$. At $t = 5$, $T_2(5) = 72 + 103e^{5k}$. When cream is added to Mrs. Jones' coffee the temperature is reduced by an amount α. Using the fact that $k < 0$ we have

$$T_2(5) - \alpha = 72 + 103e^{5k} - \alpha < 72 + 103e^{5k} - \alpha e^{5k}$$
$$= 72 + (103 - \alpha)e^{5k} = T_1(5).$$

Thus, the temperature of the coffee in Mr. Jones' cup is hotter.

21. From $dA/dt = 10 - A/100$ we obtain $A = 1000 + ce^{-t/100}$. If $A(0) = 0$ then $c = -1000$ and $A = 1000 - 1000e^{-t/100}$. At $t = 5$, $A(5) \approx 48.77$ pounds.

24. (a) Initially the tank contains 300 gallons of solution. Since brine is pumped in at a rate of 3 gal/min and the solution is pumped out at a rate of 2 gal/min, the net change is an increase of 1 gal/min. Thus, in 100 minutes the tank will contain its capacity of 400 gallons.

(b) The differential equation describing the amount of salt in the tank is $A'(t) = 6 - 2A/(300+t)$ with solution

$$A(t) = 600 + 2t - (4.95 \times 10^7)(300 + t)^{-2}, \qquad 0 \le t \le 100,$$

as noted in the discussion following Example 5 in the text. Thus, the amount of salt in the tank when it overflows is

$$A(100) = 800 - (4.95 \times 10^7)(400)^{-2} = 490.625 \text{ lbs.}$$

(c) When the tank is overflowing the amount of salt in the tank is governed by the differential equation

$$\frac{dA}{dt} = (3 \text{ gal/min})(2 \text{ lb/gal}) - (\frac{A}{400} \text{ lb/gal})(3 \text{ gal/min})$$
$$= 6 - \frac{3A}{400}, \qquad A(100) = 490.625.$$

Solving the equation we obtain $A(t) = 800 + ce^{-3t/400}$. The initial condition yields $c = -654.947$, so that

$$A(t) = 800 - 654.947e^{-3t/400}.$$

When $t = 150$, $A(150) = 587.37$ lbs.

(d) As $t \to \infty$, the amount of salt is 800 lbs, which is to be expected since (400 gal)(2 lbs/gal)= 800 lbs.

(e)

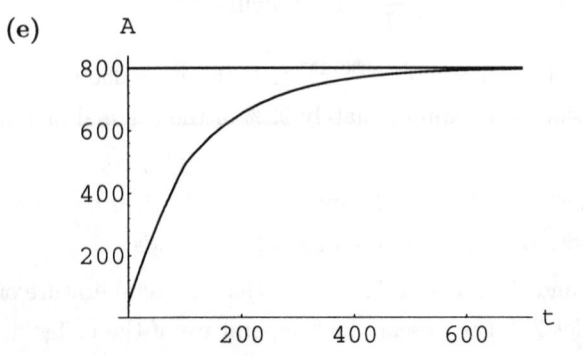

27. Assume $R\,dq/dt + (1/c)q = E(t)$, $R = 200$, $C = 10^{-4}$, and $E(t) = 100$ so that $q = 1/100 + ce^{-50t}$. If $q(0) = 0$ then $c = -1/100$ and $i = \frac{1}{2}e^{-50t}$.

30. Separating variables we obtain

$$\frac{dq}{E_0 - q/C} = \frac{dt}{k_1 + k_2 t} \implies -C \ln \left| E_0 - \frac{q}{C} \right| = \frac{1}{k_2} \ln |k_1 + k_2 t| + c_1 \implies \frac{(E_0 - q/C)^{-C}}{(k_1 + k_2 t)^{1/k_2}} = c_2.$$

Setting $q(0) = q_0$ we find $c_2 = \dfrac{(E_0 - q_0/C)^{-C}}{k_1^{1/k_2}}$, so

$$\frac{(E_0 - q/C)^{-C}}{(k_1 + k_2 t)^{1/k_2}} = \frac{(E_0 - q_0/C)^{-C}}{k_1^{1/k_2}} \implies \left(E_0 - \frac{q}{C} \right)^{-C} = \left(E_0 - \frac{q_0}{C} \right)^{-C} \left(\frac{k_1}{k + k_2 t} \right)^{-1/k_2}$$

$$\implies E_0 - \frac{q}{C} = \left(E_0 - \frac{q_0}{C} \right) \left(\frac{k_1}{k + k_2 t} \right)^{1/Ck_2}$$

$$\implies q = E_0 C + (q_0 - E_0 C) \left(\frac{k_1}{k + k_2 t} \right)^{1/Ck_2}$$

36. Separating variables we obtain

$$\frac{dP}{P} = k \cos t \, dt \implies \ln |P| = k \sin t + c \implies P = c_1 e^{k \sin t}.$$

If $P(0) = P_0$ then $c_1 = P_0$ and $P = P_0 e^{k \sin t}$.

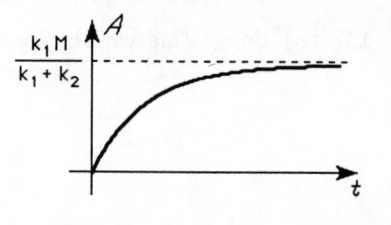

39. **(a)** Solving $k_1(M - A) - k_2 A = 0$ for A we find the equilibrium solution $A = k_1 M/(k_1 + k_2)$. From the phase portrait we see that $\lim_{t \to \infty} A(t) = k_1 M/(k_1 + k_2)$.

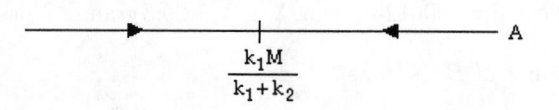

Since $k_2 > 0$, the material will never be completely memorized and the larger k_2 is, the less the amount of material will be memorized over time.

(b) Write the differential equation in the form $dA/dt + (k_1 + k_2)A = k_1 M$. Then an integrating factor is $e^{(k_1 + k_2)t}$, and

$$\frac{d}{dt} \left[e^{(k_1 + k_2)t} A \right] = k_1 M e^{(k_1 + k_2)t}$$

$$\implies e^{(k_1 + k_2)t} A = \frac{k_1 M}{k_1 + k_2} e^{(k_1 + k_2)t} + c$$

$$\implies A = \frac{k_1 M}{k_1 + k_2} + c e^{-(k_1 + k_2)t}.$$

Using $A(0) = 0$ we find $c = -\dfrac{k_1 M}{k_1 + k_2}$ and $A = \dfrac{k_1 M}{k_1 + k_2} \left(1 - e^{-(k_1 + k_2)t} \right)$. As $t \to \infty$, $A \to \dfrac{k_1 M}{k_1 + k_2}$.

_____ **Exercises 2.8** _____

3. From $\dfrac{dP}{dt} = P\left(10^{-1} - 10^{-7}P\right)$ and $P(0) = 5000$ we obtain $P = \dfrac{500}{0.0005 + 0.0995e^{-0.1t}}$ so that $P \to 1,000,000$

as $t \to \infty$. If $P(t) = 500,000$ then $t = 52.9$ months.

6. (a) Solving $P(5 - P) - \frac{25}{4} = 0$ for P we obtain the equilibrium solution $P = \frac{5}{2}$. For $P \neq \frac{5}{2}$, $dP/dt < 0$. Thus, if $P_0 < \frac{5}{2}$, the population becomes extinct (otherwise there would be another equilibrium solution.) Using separation of variables to solve the initial-value problem we get $P(t) = [4P_0 + (10P_0 - 25)t]/[4 + (4P_0 - 10)t]$. To find when the population becomes extinct for $P_0 < \frac{5}{2}$ we solve $P(t) = 0$ for t. We see that the time of extinction is $t = 4P_0/5(5 - 2P_0)$.

(b) Solving $P(5 - P) - 7 = 0$ for P we obtain complex roots, so there are no equilibrium solutions. Since $dP/dt < 0$ for all values of P, the population becomes extinct for any initial condition. Using separation of variables to solve the initial-value problem we get

$$P(t) = \frac{5}{2} + \frac{\sqrt{3}}{2}\tan\left[\tan^{-1}\left(\frac{2P_0 - 5}{\sqrt{3}}\right) - \frac{\sqrt{3}}{2}t\right].$$

Solving $P(t) = 0$ for t we see that the time of extinction is

$$t = \frac{2}{3}\left(\sqrt{3}\tan^{-1}(5/\sqrt{3}) + \sqrt{3}\tan^{-1}\left[(2P_0 - 5)/\sqrt{3}\right]\right).$$

9. Let $X = X(t)$ be the amount of C at time t and $\dfrac{dX}{dt} = k(120 - 2X)(150 - X)$. If $X(0) = 0$ and $X(5) = 10$ then $X = \dfrac{150 - 150e^{180kt}}{1 - 2.5e^{180kt}}$ where $k = .0001259$, and $X(20) = 29.3$ grams. Now $X \to 60$ as $t \to \infty$, so that the amount of $A \to 0$ and the amount of $B \to 30$ as $t \to \infty$.

12. To obtain the solution of this differential equation we use $h(t)$ from part (a) in Problem 11 with A_h replaced by cA_h. Then $h(t) = (A_w\sqrt{H} - 4cA_ht)^2/A_w^2$. Solving $h(t) = 0$ with $c = 0.6$ and the values from Problem 11 we see that the tank empties in 3035.79 seconds or 50.6 minutes.

15. (a) Separating variables we obtain

$$\frac{m\,dv}{mg - kv^2} = dt$$

$$\frac{1}{g}\frac{dv}{1 - (kv/mg)^2} = dt$$

$$\frac{\sqrt{mg}}{\sqrt{k}\,g}\frac{\sqrt{k/mg}\,dv}{1 - (\sqrt{k}\,v/\sqrt{mg})^2} = dt$$

$$\sqrt{\frac{m}{kg}}\tanh^{-1}\frac{\sqrt{k}\,v}{\sqrt{mg}} = t + c$$

$$\tanh^{-1}\frac{\sqrt{k}\,v}{\sqrt{mg}} = \sqrt{\frac{kg}{m}}\,t + c_1.$$

Thus the velocity at time t is

$$v(t) = \sqrt{\frac{mg}{k}}\tanh\left(\sqrt{\frac{kg}{m}}t + c_1\right).$$

Setting $t = 0$ and $v = v_0$ we find $c_1 = \tanh^{-1}(\sqrt{k}\,v_0/\sqrt{mg})$.

(b) Since $\tanh t \to 1$ as $t \to \infty$, we have $v \to \sqrt{mg/k}$ as $t \to \infty$.

(c) Integrating the expression for $v(t)$ in part (a) we obtain

$$s(t) = \sqrt{\frac{mg}{k}} \int \tanh\left(\sqrt{\frac{kg}{m}}t + c_1\right) dt = \frac{m}{k}\ln\left[\cosh\left(\sqrt{\frac{kg}{m}}t + c_1\right)\right] + c_2.$$

Setting $t = 0$ and $s = s_0$ we find $c_2 = s_0 - \frac{m}{k}\ln(\cosh c_1)$.

21. (a) Writing the equation in the form $(x - \sqrt{x^2 + y^2})dx + y\,dy$ we identify $M = x - \sqrt{x^2 + y^2}$ and $N = y$. Since M and N are both homogeneous of degree 1 we use the substitution $y = ux$. It follows that

$$\left(x - \sqrt{x^2 + u^2 x^2}\right) dx + ux(u\,dx + x\,du) = 0$$

$$x\left[\left(1 - \sqrt{1 + u^2}\right) + u^2\right] dx + x^2 u\,du = 0$$

$$-\frac{u\,du}{1 + u^2 - \sqrt{1 + u^2}} = \frac{dx}{x}$$

$$\frac{u\,du}{\sqrt{1 + u^2}\left(1 - \sqrt{1 + u^2}\right)} = \frac{dx}{x}.$$

Letting $w = 1 - \sqrt{1 + u^2}$ we have $dw = -u\,du/\sqrt{1 + u^2}$ so that

$$-\ln\left(1 - \sqrt{1 + u^2}\right) = \ln x + c$$

$$\frac{1}{1 - \sqrt{1 + u^2}} = c_1 x$$

$$1 - \sqrt{1 + u^2} = -\frac{c_2}{x} \qquad (-c_2 = 1/c_1)$$

$$1 + \frac{c_2}{x} = \sqrt{1 + \frac{y^2}{x^2}}$$

$$1 + \frac{2c_2}{x} + \frac{c_2^2}{x^2} = 1 + \frac{y^2}{x^2}.$$

Solving for y^2 we have

$$y^2 = 2c_2 x + c_2^2 = 4\left(\frac{c_2}{2}\right)\left(x + \frac{c_2}{2}\right)$$

which is a family of parabolas symmetric with respect to the x-axis with vertex at $(-c_2/2, 0)$ and focus at the origin.

(b) Let $u = x^2 + y^2$ so that

$$\frac{du}{dx} = 2x + 2y\frac{dy}{dx}.$$

Then

$$y\frac{dy}{dx} = \frac{1}{2}\frac{du}{dx} - x$$

and the differential equation can be written in the form

$$\frac{1}{2}\frac{du}{dx} - x = -x + \sqrt{u} \quad \text{or} \quad \frac{1}{2}\frac{du}{dx} = \sqrt{u}.$$

19

Separating variables and integrating we have

$$\frac{du}{2\sqrt{u}} = dx$$

$$\sqrt{u} = x + c$$

$$u = x^2 + 2cx + c^2$$

$$x^2 + y^2 = x^2 + 2cx + c^2$$

$$y^2 = 2cx + c^2.$$

Exercises 2.9

3. The amounts of x and y are the same at about $t = 5$ days. The amounts of x and z are the same at about $t = 20$ days. The amounts of y and z are the same at about $t = 147$ days. The time when y and z are the same makes sense because most of A and half of B are gone, so half of C should have been formed.

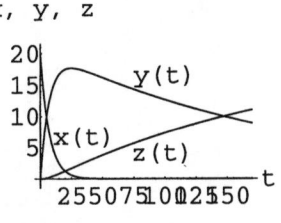

6. Let x_1, x_2, and x_3 be the amounts of salt in tanks A, B, and C, respectively, so that

$$x_1' = \frac{1}{100}x_2 \cdot 2 - \frac{1}{100}x_1 \cdot 6 = \frac{1}{50}x_2 - \frac{3}{50}x_1$$

$$x_2' = \frac{1}{100}x_1 \cdot 6 + \frac{1}{100}x_3 - \frac{1}{100}x_2 \cdot 2 - \frac{1}{100}x_2 \cdot 5 = \frac{3}{50}x_1 - \frac{7}{100}x_2 + \frac{1}{100}x_3$$

$$x_3' = \frac{1}{100}x_2 \cdot 5 - \frac{1}{100}x_3 - \frac{1}{100}x_3 \cdot 4 = \frac{1}{20}x_2 - \frac{1}{20}x_3.$$

9. From the graph we see that the populations are first equal at about $t = 5.6$. The approximate periods of x and y are both 45.

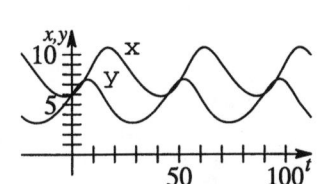

12. By Kirchoff's first law we have $i_1 = i_2 + i_3$. By Kirchoff's second law, on each loop we have $E(t) = Li_1' + R_1i_2$ and $E(t) = Li_1' + R_2i_3 + \frac{1}{C}q$ so that $q = CR_1i_2 - CR_2i_3$. Then $i_3 = q' = CR_1i_2' - CR_2i_3'$ so that the system is

$$Li_2' + Li_3' + R_1i_2 = E(t)$$

$$-R_1i_2' + R_2i_3' + \frac{1}{C}i_3 = 0.$$

15. We first note that $s(t) + i(t) + r(t) = n$. Now the rate of change of the number of susceptible persons, $s(t)$, is proportional to the number of contacts between the number of people infected and the number who are susceptible; that is, $ds/dt = -k_1si$. We use $-k_1$ because $s(t)$ is decreasing. Next, the rate of change of the number of persons who have recovered is proportional to the number infected; that is, $dr/dt = k_2i$ where k_2 is positive since r is increasing. Finally, to obtain di/dt we use

$$\frac{d}{dt}(s + i + r) = \frac{d}{dt}n = 0.$$

This gives

$$\frac{di}{dt} = -\frac{dr}{dt} - \frac{ds}{dt} = -k_2i + k_1si.$$

The system of equations is then

$$\frac{ds}{dt} = -k_1 s i$$

$$\frac{di}{dt} = -k_2 i + k_1 s i$$

$$\frac{dr}{dt} = k_2 i.$$

A reasonable set of initial conditions is $i(0) = i_0$, the number of infected people at time 0, $s(0) = n - i_0$, and $r(0) = 0$.

Chapter 2 Review Exercises

3. $\dfrac{dy}{dx} = (y-1)^2(y-3)^2$

6. The zero of f occurs at approximately 1.3. Since $P'(t) = f(P) > 0$ for $P < 1.3$ and $P'(t) = f(P) > 0$ for $P > 1.3$, $\lim_{t \to \infty} P(t) = 1.3$.

9. Separating variables we obtain

$$\cos^2 x \, dx = \frac{y}{y^2+1} \, dy \implies \frac{1}{2}x + \frac{1}{4}\sin 2x = \frac{1}{2}\ln(y^2+1) + c \implies 2x + \sin 2x = 2\ln(y^2+1) + c.$$

12. Write the differential equation in the form $(3y^2 + 2x)dx + (4y^2 + 6xy)dy = 0$. Letting $M = 3y^2 + 2x$ and $N = 4y^2 + 6xy$ we see that $M_y = 6y = N_x$ so the differential equation is exact. From $f_x = 3y^2 + 2x$ we obtain $f = 3xy^2 + x^2 + h(y)$. Then $f_y = 6xy + h'(y) = 4y^2 + 6xy$ and $h'(y) = 4y^2$ so $h(y) = \frac{4}{3}y^3$. The general solution is

$$3xy^2 + x^2 + \frac{4}{3}y^3 = c.$$

15. Write the equation in the form $\dfrac{dy}{dx} + \dfrac{8x}{x^2+4}y = \dfrac{2x}{x^2+4}$. An integrating factor is $(x^2+4)^4$, so

$$\frac{d}{dx}\left[(x^2+4)^4 y\right] = 2x(x^2+4)^3 \implies (x^2+4)^4 y = \frac{1}{4}(x^2+4)^4 + c \implies y = \frac{1}{4} + c(x^2+4)^{-4}.$$

18. Separating variables and integrating we have

$$\frac{dy}{y^2} = -2(t+1)\, dt$$

$$-\frac{1}{y} = -(t+1)^2 + c$$

$$y = \frac{1}{(t+1)^2 + c}.$$

The initial condition implies $c = -9$, so the solution of the initial-value problem is

$$y = \frac{1}{t^2 + 2t - 8} \quad \text{where} \quad -4 < t < 2.$$

21. The graph of $y_1(x)$ is the portion of the closed black curve lying in the fourth quadrant. Its interval of definition is approximately $(0.7, 4.3)$. The graph of $y_2(x)$ is the portion of the left-hand black curve lying in the third quadrant. Its interval of definition is $(-\infty, 0)$.

24. Let $A = A(t)$ be the volume of CO_2 at time t. From $\dfrac{dA}{dt} = 1.2 - \dfrac{A}{4}$ and $A(0) = 16\,\text{ft}^3$ we obtain $A = 4.8 + 11.2e^{-t/4}$. Since $A(10) = 5.7\,\text{ft}^3$, the concentration is 0.017%. As $t \to \infty$ we have $A \to 4.8\,\text{ft}^3$ or 0.06%.

27. (a) The differential equation is

$$\frac{dT}{dt} = k[T - T_2 - B(T_1 - T)] = k[(1 + B)T - (BT_1 + T_2)].$$

Separating variables we obtain $\dfrac{dT}{(1 + B)T - (BT_1 + T_2)} = k\,dt$. Then

$$\frac{1}{1 + B}\ln|(1 + B)T - (BT_1 + T_2)| = kt + c \quad\text{and}\quad T(t) = \frac{BT_1 + T_2}{1 + B} + c_3 e^{k(1+B)t}.$$

Since $T(0) = T_1$ we must have $c_3 = \dfrac{T_1 - T_2}{1 + B}$ and so

$$T(t) = \frac{BT_1 + T_2}{1 + B} + \frac{T_1 - T_2}{1 + B}e^{k(1+B)t}.$$

(b) Since $k < 0$, $\displaystyle\lim_{t\to\infty} e^{k(1+B)t} = 0$ and $\displaystyle\lim_{t\to\infty} T(t) = \frac{BT_1 + T_2}{1 + B}$.

(c) Since $T_s = T_2 + B(T_1 - T)$, $\displaystyle\lim_{t\to\infty} T_s = T_2 + BT_1 - B\left(\frac{BT_1 + T_2}{1 + B}\right) = \frac{BT_1 + T_2}{1 + B}$.

30. (a) From $y = -x - 1 + c_1 e^x$ we obtain $y' = y + x$ so that the differential equation of the orthogonal family is $\dfrac{dy}{dx} = -\dfrac{1}{y + x}$ or $\dfrac{dx}{dy} + x = -y$. An integrating factor is e^y, so

$$\frac{d}{dy}[e^y x] = -ye^y \implies e^y x = -ye^y + e^y + c \implies x = -y + 1 + ce^{-y}.$$

(b) Differentiating the family of curves, we have

$$y' = -\frac{1}{(x + c_1)^2} = -\frac{1}{y^2}.$$

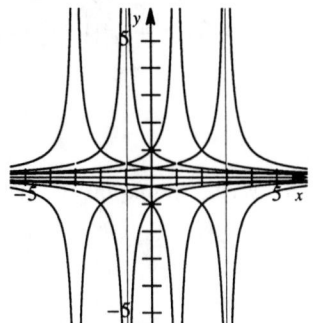

The differential equation for the family of orthogonal trajectories is then $y' = y^2$. Separating variables and integrating we get

$$\frac{dy}{y^2} = dx$$

$$-\frac{1}{y} = x + c_1$$

$$y = -\frac{1}{x + c_1}.$$

3 Higher-Order Differential Equations

———————— **Exercises 3.1** ————————————————

3. From $y = c_1 x + c_2 x \ln x$ we find $y' = c_1 + c_2(1 + \ln x)$. Then $y(1) = c_1 = 3$, $y'(1) = c_1 + c_2 = -1$ so that $c_1 = 3$ and $c_2 = -4$. The solution is $y = 3x - 4x \ln x$.

6. In this case we have $y(0) = c_1 = 0$, $y'(0) = 2c_2 \cdot 0 = 0$ so $c_1 = 0$ and c_2 is arbitrary. Two solutions are $y = x^2$ and $y = 2x^2$.

9. Since $a_2(x) = x - 2$ and $x_0 = 0$ the problem has a unique solution for $-\infty < x < 2$.

12. In this case we have $y(0) = c_1 = 1$, $y'(1) = 2c_2 = 6$ so that $c_1 = 1$ and $c_2 = 3$. The solution is $y = 1 + 3x^2$.

15. Since $(-4)x + (3)x^2 + (1)(4x - 3x^2) = 0$ the functions are linearly dependent.

18. Since $(1)\cos 2x + (1)1 + (-2)\cos^2 x = 0$ the functions are linearly dependent.

21. The functions are linearly independent since $W\left(1 + x, x, x^2\right) = \begin{vmatrix} 1+x & x & x^2 \\ 1 & 1 & 2x \\ 0 & 0 & 2 \end{vmatrix} = 2 \neq 0$.

24. The functions satisfy the differential equation and are linearly independent since

$$W(\cosh 2x, \sinh 2x) = 2$$

for $-\infty < x < \infty$. The general solution is

$$y = c_1 \cosh 2x + c_2 \sinh 2x.$$

27. The functions satisfy the differential equation and are linearly independent since

$$W\left(x^3, x^4\right) = x^6 \neq 0$$

for $0 < x < \infty$. The general solution is

$$y = c_1 x^3 + c_2 x^4.$$

30. The functions satisfy the differential equation and are linearly independent since

$$W(1, x, \cos x, \sin x) = 1$$

for $-\infty < x < \infty$. The general solution is

$$y = c_1 + c_2 x + c_3 \cos x + c_4 \sin x.$$

33. Since $e^{x-3} = e^{-3}e^x = (e^{-5}e^2)e^x = e^{-5}e^{x+2}$, we see that e^{x-3} is a constant multiple of e^{x+2} and the functions are linearly dependent.

36. The functions $y_1 = \cos x$ and $y_2 = \sin x$ form a fundamental set of solutions of the homogeneous equation, and $y_p = x \sin x + (\cos x) \ln(\cos x)$ is a particular solution of the nonhomogeneous equation.

39. (a) We have $y'_{p_1} = 6e^{2x}$ and $y''_{p_1} = 12e^{2x}$, so

$$y''_{p_1} - 6y'_{p_1} + 5y_{p_1} = 12e^{2x} - 36e^{2x} + 15e^{2x} = -9e^{2x}.$$

Also, $y'_{p_2} = 2x + 3$ and $y''_{p_2} = 2$, so

$$y''_{p_2} - 6y'_{p_2} + 5y_{p_2} = 2 - 6(2x + 3) + 5(x^2 + 3x) = 5x^2 + 3x - 16.$$

(b) By the superposition principle for nonhomogeneous equations a particular solution of $y'' - 6y' + 5y = 5x^2 + 3x - 16 - 9e^{2x}$ is $y_p = x^2 + 3x + 3e^{2x}$. A particular solution of the second equation is

$$y_p = -2y_{p_2} - \frac{1}{9}y_{p_1} = -2x^2 - 6x - \frac{1}{3}e^{2x}.$$

Exercises 3.2

In Problems 3 and 6 we use reduction of order to find a second solution. In Problems 9-15 we use formula (5) from the text.

3. Define $y = u(x)\cos 4x$ so

$$y' = -4u\sin 4x + u'\cos 4x, \quad y'' = u''\cos 4x - 8u'\sin 4x - 16u\cos 4x$$

and

$$y'' + 16y = (\cos 4x)u'' - 8(\sin 4x)u' = 0 \quad \text{or} \quad u'' - 8(\tan 4x)u' = 0.$$

If $w = u'$ we obtain the first-order equation $w' - 8(\tan 4x)w = 0$ which has the integrating factor $e^{-8\int \tan 4x\, dx} = \cos^2 4x$. Now

$$\frac{d}{dx}[(\cos^2 4x)w] = 0 \quad \text{gives} \quad (\cos^2 4x)w = c.$$

Therefore $w = u' = c\sec^2 4x$ and $u = c_1\tan 4x$. A second solution is $y_2 = \tan 4x\cos 4x = \sin 4x$.

6. Define $y = u(x)e^{5x}$ so

$$y' = 5e^{5x}u + e^{5x}u', \quad y'' = e^{5x}u'' + 10e^{5x}u' + 25e^{5x}u$$

and

$$y'' - 25y = e^{5x}(u'' + 10u') = 0 \quad \text{or} \quad u'' + 10u' = 0.$$

If $w = u'$ we obtain the first-order equation $w' + 10w = 0$ which has the integrating factor $e^{10\int dx} = e^{10x}$. Now

$$\frac{d}{dx}[e^{10x}w] = 0 \quad \text{gives} \quad e^{10x}w = c.$$

Therefore $w = u' = ce^{-10x}$ and $u = c_1e^{-10x}$. A second solution is $y_2 = e^{-10x}e^{5x} = e^{-5x}$.

9. Identifying $P(x) = -7/x$ we have

$$y_2 = x^4 \int \frac{e^{-\int -(7/x)\, dx}}{x^8}\, dx = x^4 \int \frac{1}{x}\, dx = x^4 \ln|x|.$$

A second solution is $y_2 = x^4 \ln|x|$.

12. Identifying $P(x) = 0$ we have

$$y_2 = x^{1/2}\ln x \int \frac{e^{-\int 0\, dx}}{x(\ln x)^2} = x^{1/2}\ln x\left(-\frac{1}{\ln x}\right) = -x^{1/2}.$$

A second solution is $y_2 = x^{1/2}$.

15. Identifying $P(x) = 2(1+x)/\left(1 - 2x - x^2\right)$ we have

$$y_2 = (x+1) \int \frac{e^{-\int 2(1+x)dx/\left(1-2x-x^2\right)}}{(x+1)^2}\, dx = (x+1) \int \frac{e^{\ln\left(1-2x-x^2\right)}}{(x+1)^2}\, dx$$

$$= (x+1) \int \frac{1 - 2x - x^2}{(x+1)^2}\, dx = (x+1) \int \left[\frac{2}{(x+1)^2} - 1\right] dx$$

$$= (x+1) \left[-\frac{2}{x+1} - x\right] = -2 - x^2 - x.$$

A second solution is $y_2 = x^2 + x + 2$.

18. Define $y = u(x) \cdot 1$ so

$$y' = u', \quad y'' = u'' \quad \text{and} \quad y'' + y' = u'' + u' = 0.$$

If $w = u'$ we obtain the first order equation $w' + w = 0$ which has the integrating factor $e^{\int dx} = e^x$. Now

$$\frac{d}{dx}[e^x w] = 0 \quad \text{gives} \quad e^x w = c.$$

Therefore $w = u' = ce^{-x}$ and $u = c_1 e^{-x}$. A second solution is $y_2 = 1 \cdot e^{-x} = e^{-x}$. We see by observation that a particular solution is $y_p = x$. The general solution is

$$y = c_1 + c_2 e^{-x} + x.$$

21. (a) For m_1 constant, let $y_1 = e^{m_1 x}$. Then $y_1' = m_1 e^{m_1 x}$ and $y_1'' = m_1^2 e^{m_1 x}$. Substituting into the differential equation we obtain

$$ay_1'' + by_1' + cy_1 = am_1^2 e^{m_1 x} + bm_1 e^{m_1 x} + ce^{m_1 x}$$

$$= e^{m_1 x}(am_1^2 + bm_1 + c) = 0.$$

Thus, $y_1 = e^{m_1 x}$ will be a solution of the differential equation whenever $am_1^2 + bm_1 + c = 0$. Since a quadratic equation always has at least one real or complex root, the differential equation must have a solution of the form $y_1 = e^{m_1 x}$.

(b) Write the differential equation in the form

$$y'' + \frac{b}{a}y' + \frac{c}{a}y = 0,$$

and let $y_1 = e^{m_1 x}$ be a solution. Then a second solution is given by

$$y_2 = e^{m_1 x} \int \frac{e^{-bx/a}}{e^{2m_1 x}}\, dx$$

$$= e^{m_1 x} \int e^{-(b/a + 2m_1)x}\, dx$$

$$= -\frac{1}{b/a + 2m_1} e^{m_1 x} e^{-(b/a + 2m_1)} \qquad (m_1 \neq -b/2a)$$

$$= -\frac{1}{b/a + 2m_1} e^{-(b/a + m_1)}.$$

Thus, when $m_1 \neq -b/2a$, a second solution is given by $y_2 = e^{m_2 x}$ where $m_2 = -b/a - m_1$. When $m_1 = -b/2a$ a second solution is given by

$$y_2 = e^{m_1 x} \int dx = xe^{m_1 x}.$$

(c) The functions

$$\sin x = \frac{1}{2i}(e^{ix} - e^{-ix}) \qquad \cos x = \frac{1}{2}(e^{ix} + e^{-ix})$$

$$\sinh x = \frac{1}{2}(e^x - e^{-x}) \qquad \cosh x = \frac{1}{2}(e^x + e^{-x})$$

are all expressible in terms of exponential functions.

_____ Exercises 3.3 _____

3. From $m^2 - m - 6 = 0$ we obtain $m = 3$ and $m = -2$ so that $y = c_1 e^{3x} + c_2 e^{-2x}$.

6. From $m^2 - 10m + 25 = 0$ we obtain $m = 5$ and $m = 5$ so that $y = c_1 e^{5x} + c_2 x e^{5x}$.

9. From $m^2 + 9 = 0$ we obtain $m = 3i$ and $m = -3i$ so that $y = c_1 \cos 3x + c_2 \sin 3x$.

12. From $2m^2 + 2m + 1 = 0$ we obtain $m = -1/2 \pm i/2$ so that

$$y = e^{-x/2}(c_1 \cos x/2 + c_2 \sin x/2).$$

15. From $m^3 - 4m^2 - 5m = 0$ we obtain $m = 0$, $m = 5$, and $m = -1$ so that

$$y = c_1 + c_2 e^{5x} + c_3 e^{-x}.$$

18. From $m^3 + 3m^2 - 4m - 12 = 0$ we obtain $m = -2$, $m = 2$, and $m = -3$ so that

$$y = c_1 e^{-2x} + c_2 e^{2x} + c_3 e^{-3x}.$$

21. From $m^3 + 3m^2 + 3m + 1 = 0$ we obtain $m = -1$, $m = -1$, and $m = -1$ so that

$$y = c_1 e^{-x} + c_2 x e^{-x} + c_3 x^2 e^{-x}.$$

24. From $m^4 - 2m^2 + 1 = 0$ we obtain $m = 1$, $m = 1$, $m = -1$, and $m = -1$ so that

$$y = c_1 e^x + c_2 x e^x + c_3 e^{-x} + c_4 x e^{-x}.$$

27. From $m^5 + 5m^4 - 2m^3 - 10m^2 + m + 5 = 0$ we obtain $m = -1$, $m = -1$, $m = 1$, and $m = 1$, and $m = -5$ so that

$$u = c_1 e^{-r} + c_2 r e^{-r} + c_3 e^r + c_4 r e^r + c_5 e^{-5r}.$$

30. From $m^2 + 1 = 0$ we obtain $m = \pm i$ so that $y = c_1 \cos \theta + c_2 \sin \theta$. If $y(\pi/3) = 0$ and $y'(\pi/3) = 2$ then

$$\frac{1}{2}c_1 + \frac{\sqrt{3}}{2}c_2 = 0, \quad -\frac{\sqrt{3}}{2}c_1 + \frac{1}{2}c_2 = 2, \text{ so } c_1 = -\sqrt{3}, \ c_2 = 1, \text{ and } y = -\sqrt{3}\cos\theta + \sin\theta.$$

33. From $m^2 + m + 2 = 0$ we obtain $m = -1/2 \pm \sqrt{7}i/2$ so that $y = e^{-x/2}\left(c_1 \cos \sqrt{7}x/2 + c_2 \sin \sqrt{7}x/2\right)$. If $y(0) = 0$ and $y'(0) = 0$ then $c_1 = 0$ and $c_2 = 0$ so that $y = 0$.

36. From $m^3 + 2m^2 - 5m - 6 = 0$ we obtain $m = -1$, $m = 2$, and $m = -3$ so that

$$y = c_1 e^{-x} + c_2 e^{2x} + c_3 e^{-3x}.$$

If $y(0) = 0$, $y'(0) = 0$, and $y''(0) = 1$ then

$$c_1 + c_2 + c_3 = 0, \quad -c_1 + 2c_2 - 3c_3 = 0, \quad c_1 + 4c_2 + 9c_3 = 1,$$

so $c_1 = -1/6$, $c_2 = 1/15$, $c_3 = 1/10$, and

$$y = -\frac{1}{6}e^{-x} + \frac{1}{15}e^{2x} + \frac{1}{10}e^{-3x}.$$

39. From $m^2 - 10m + 25 = 0$ we obtain $m = 5$ and $m = 5$ so that $y = c_1 e^{5x} + c_2 x e^{5x}$. If $y(0) = 1$ and $y(1) = 0$ then $c_1 = 1$, $c_1 e^5 + c_2 e^5 = 0$, so $c_1 = 1$, $c_2 = -1$, and $y = e^{5x} - x e^{5x}$.

42. From $m^2 - 2m + 2 = 0$ we obtain $m = 1 \pm i$ so that $y = e^x (c_1 \cos x + c_2 \sin x)$. If $y(0) = 1$ and $y(\pi) = 1$ then $c_1 = 1$ and $y(\pi) = e^\pi \cos \pi = -e^\pi$. Since $-e^\pi \neq 1$, the boundary-value problem has no solution.

51. The auxiliary equation should have a pair of complex roots $a \pm bi$ where $a < 0$, so that the solution has the form $e^{ax}(c_1 \cos bx + c_2 \sin bx)$. Thus, the differential equation is (e).

54. The differential equation should have the form $y'' + k^2 y = 0$ where $k = 2$ so that the period of the solution is π. Thus, the differential equation is (b).

57. Since $(m-4)(m+5)^2 = m^3 + 6m^2 - 15m - 100$ the differential equation is $y''' + 6y'' - 15y' - 100y = 0$. The differential equation is not unique since any constant multiple of the left-hand side of the differential equation would lead to the auxiliary roots.

60. Factoring the difference of two squares we obtain

$$m^4 + 1 = (m^2 + 1)^2 - 2m^2 = (m^2 + 1 - \sqrt{2}\,m)(m^2 + 1 + \sqrt{2}\,m) = 0.$$

Using the quadratic formula on each factor we get $m = \pm\sqrt{2}/2 \pm \sqrt{2}\,i/2$. The solution of the differential equation is

$$y(x) = e^{\sqrt{2}\,x/2}\left(c_1 \cos \frac{\sqrt{2}}{2}x + c_2 \sin \frac{\sqrt{2}}{2}x\right) + e^{-\sqrt{2}\,x/2}\left(c_3 \cos \frac{\sqrt{2}}{2}x + c_4 \sin \frac{\sqrt{2}}{2}x\right).$$

Exercises 3.4

3. From $m^2 - 10m + 25 = 0$ we find $m_1 = m_2 = 5$. Then $y_c = c_1 e^{5x} + c_2 x e^{5x}$ and we assume $y_p = Ax + B$. Substituting into the differential equation we obtain $25A = 30$ and $-10A + 25B = 3$. Then $A = \frac{6}{5}$, $B = \frac{6}{5}$, $y_p = \frac{6}{5}x + \frac{6}{5}$, and

$$y = c_1 e^{5x} + c_2 x e^{5x} + \frac{6}{5}x + \frac{6}{5}.$$

6. From $m^2 - 8m + 20 = 0$ we find $m_1 = 2 + 4i$ and $m_2 = 2 - 4i$. Then $y_c = e^{2x}(c_1 \cos 4x + c_2 \sin 4x)$ and we assume $y_p = Ax^2 + Bx + C + (Dx + E)e^x$. Substituting into the differential equation we obtain

$$2A - 8B + 20C = 0$$
$$-6D + 13E = 0$$
$$-16A + 20B = 0$$
$$13D = -26$$
$$20A = 100.$$

Then $A = 5$, $B = 4$, $C = \frac{11}{10}$, $D = -2$, $E = -\frac{12}{13}$, $y_p = 5x^2 + 4x + \frac{11}{10} + \left(-2x - \frac{12}{13}\right)e^x$ and

$$y = e^{2x}(c_1 \cos 4x + c_2 \sin 4x) + 5x^2 + 4x + \frac{11}{10} + \left(-2x - \frac{12}{13}\right)e^x.$$

9. From $m^2 - m = 0$ we find $m_1 = 1$ and $m_2 = 0$. Then $y_c = c_1 e^x + c_2$ and we assume $y_p = Ax$. Substituting into the differential equation we obtain $-A = -3$. Then $A = 3$, $y_p = 3x$ and $y = c_1 e^x + c_2 + 3x$.

12. From $m^2 - 16 = 0$ we find $m_1 = 4$ and $m_2 = -4$. Then $y_c = c_1 e^{4x} + c_2 e^{-4x}$ and we assume $y_p = Axe^{4x}$. Substituting into the differential equation we obtain $8A = 2$. Then $A = \frac{1}{4}$, $y_p = \frac{1}{4} xe^{4x}$ and

$$y = c_1 e^{4x} + c_2 e^{-4x} + \frac{1}{4} xe^{4x}.$$

15. From $m^2 + 1 = 0$ we find $m_1 = i$ and $m_2 = -i$. Then $y_c = c_1 \cos x + c_2 \sin x$ and we assume $y_p = (Ax^2 + Bx) \cos x + (Cx^2 + Dx) \sin x$. Substituting into the differential equation we obtain $4C = 0$, $2A + 2D = 0$, $-4A = 2$, and $-2B + 2C = 0$. Then $A = -\frac{1}{2}$, $B = 0$, $C = 0$, $D = \frac{1}{2}$, $y_p = -\frac{1}{2}x^2 \cos x + \frac{1}{2}x \sin x$, and

$$y = c_1 \cos x + c_2 \sin x - \frac{1}{2}x^2 \cos x + \frac{1}{2}x \sin x.$$

18. From $m^2 - 2m + 2 = 0$ we find $m_1 = 1 + i$ and $m_2 = 1 - i$. Then $y_c = e^x(c_1 \cos x + c_2 \sin x)$ and we assume $y_p = Ae^{2x} \cos x + Be^{2x} \sin x$. Substituting into the differential equation we obtain $A + 2B = 1$ and $-2A + B = -3$. Then $A = \frac{7}{5}$, $B = -\frac{1}{5}$, $y_p = \frac{7}{5}e^{2x} \cos x - \frac{1}{5}e^{2x} \sin x$ and

$$y = e^x(c_1 \cos x + c_2 \sin x) + \frac{7}{5}e^{2x} \cos x - \frac{1}{5}e^{2x} \sin x.$$

21. From $m^3 - 6m^2 = 0$ we find $m_1 = m_2 = 0$ and $m_3 = 6$. Then $y_c = c_1 + c_2 x + c_3 e^{6x}$ and we assume $y_p = Ax^2 + B \cos x + C \sin x$. Substituting into the differential equation we obtain $-12A = 3$, $6B - C = -1$, and $B + 6C = 0$. Then $A = -\frac{1}{4}$, $B = -\frac{6}{37}$, $C = \frac{1}{37}$, $y_p = -\frac{1}{4}x^2 - \frac{6}{37} \cos x + \frac{1}{37} \sin x$, and

$$y = c_1 + c_2 x + c_3 e^{6x} - \frac{1}{4}x^2 - \frac{6}{37} \cos x + \frac{1}{37} \sin x.$$

24. From $m^3 - m^2 - 4m + 4 = 0$ we find $m_1 = 1$, $m_2 = 2$, and $m_3 = -2$. Then $y_c = c_1 e^x + c_2 e^{2x} + c_3 e^{-2x}$ and we assume $y_p = A + Bxe^x + Cxe^{2x}$. Substituting into the differential equation we obtain $4A = 5$, $-3B = -1$, and $4C = 1$. Then $A = \frac{5}{4}$, $B = \frac{1}{3}$, $C = \frac{1}{4}$, $y_p = \frac{5}{4} + \frac{1}{3}xe^x + \frac{1}{4}xe^{2x}$, and

$$y = c_1 e^x + c_2 e^{2x} + c_3 e^{-2x} + \frac{5}{4} + \frac{1}{3}xe^x + \frac{1}{4}xe^{2x}.$$

27. We have $y_c = c_1 \cos 2x + c_2 \sin 2x$ and we assume $y_p = A$. Substituting into the differential equation we find $A = -\frac{1}{2}$. Thus $y = c_1 \cos 2x + c_2 \sin 2x - \frac{1}{2}$. From the initial conditions we obtain $c_1 = 0$ and $c_2 = \sqrt{2}$, so $y = \sqrt{2} \sin 2x - \frac{1}{2}$.

30. We have $y_c = c_1 e^{-2x} + c_2 xe^{-2x}$ and we assume $y_p = (Ax^3 + Bx^2)e^{-2x}$. Substituting into the differential equation we find $A = \frac{1}{6}$ and $B = \frac{3}{2}$. Thus $y = c_1 e^{-2x} + c_2 xe^{-2x} + \left(\frac{1}{6}x^3 + \frac{3}{2}x^2\right)e^{-2x}$. From the initial conditions we obtain $c_1 = 2$ and $c_2 = 9$, so

$$y = 2e^{-2x} + 9xe^{-2x} + \left(\frac{1}{6}x^3 + \frac{3}{2}x^2\right)e^{-2x}.$$

33. We have $x_c = c_1 \cos \omega t + c_2 \sin \omega t$ and we assume $x_p = At \cos \omega t + Bt \sin \omega t$. Substituting into the differential equation we find $A = -F_0/2\omega$ and $B = 0$. Thus $x = c_1 \cos \omega t + c_2 \sin \omega t - (F_0/2\omega)t \cos \omega t$. From the initial conditions we obtain $c_1 = 0$ and $c_2 = F_0/2\omega^2$, so

$$x = (F_0/2\omega^2) \sin \omega t - (F_0/2\omega)t \cos \omega t.$$

36. We have $y_c = c_1 e^{-2x} + e^x(c_2 \cos \sqrt{3}\, x + c_3 \sin \sqrt{3}\, x)$ and we assume $y_p = Ax + B + Cxe^{-2x}$. Substituting into the differential equation we find $A = \frac{1}{4}$, $B = -\frac{5}{8}$, and $C = \frac{2}{3}$. Thus

$$y = c_1 e^{-2x} + e^x(c_2 \cos \sqrt{3}\, x + c_3 \sin \sqrt{3}\, x) + \frac{1}{4}x - \frac{5}{8} + \frac{2}{3}xe^{-2x}.$$

From the initial conditions we obtain $c_1 = -\frac{23}{12}$, $c_2 = -\frac{59}{24}$, and $c_3 = \frac{17}{72}\sqrt{3}$, so

$$y = -\frac{23}{12}e^{-2x} + e^x\left(-\frac{59}{24}\cos\sqrt{3}\,x + \frac{17}{72}\sqrt{3}\sin\sqrt{3}\,x\right) + \frac{1}{4}x - \frac{5}{8} + \frac{2}{3}xe^{-2x}.$$

39. We have $y_c = c_1\cos 2x + c_2\sin 2x$ and we assume $y_p = A\cos x + B\sin x$ on $[0, \pi/2]$. Substituting into the differential equation we find $A = 0$ and $B = \frac{1}{3}$. Thus $y = c_1\cos 2x + c_2\sin 2x + \frac{1}{3}\sin x$ on $[0, \pi/2]$. On $(\pi/2, \infty)$ we have $y = c_3\cos 2x + c_4\sin 2x$. From $y(0) = 1$ and $y'(0) = 2$ we obtain

$$c_1 = 1$$

$$\frac{1}{3} + 2c_2 = 2.$$

Solving this system we find $c_1 = 1$ and $c_2 = \frac{5}{6}$. Thus $y = \cos 2x + \frac{5}{6}\sin 2x + \frac{1}{3}\sin x$ on $[0, \pi/2]$. Now continuity of y at $x = \pi/2$ implies

$$\cos\pi + \frac{5}{6}\sin\pi + \frac{1}{3}\sin\frac{\pi}{2} = c_3\cos\pi + c_4\sin\pi$$

or $-1 + \frac{1}{3} = -c_3$. Hence $c_3 = \frac{2}{3}$. Continuity of y' at $x = \pi/2$ implies

$$-2\sin\pi + \frac{5}{3}\cos\pi + \frac{1}{3}\cos\frac{\pi}{2} = -2c_3\sin\pi + 2c_4\cos\pi$$

or $-\frac{5}{3} = -2c_4$. Then $c_4 = \frac{5}{6}$ and the solution of the boundary-value problem is

$$y(x) = \begin{cases} \cos 2x + \frac{5}{6}\sin 2x + \frac{1}{3}\sin x, & 0 \le x \le \pi/2 \\ \frac{2}{3}\cos 2x + \frac{5}{6}\sin 2x, & x > \pi/2. \end{cases}$$

42. $f(t) = e^{-t}$. We see that $y_p \to \infty$ as $t \to \infty$ and $y_p \to \infty$ as $t \to -\infty$.

Exercises 3.5

The particular solution, $y_p = u_1 y_1 + u_2 y_2$, in the following problems can take on a variety of forms, especially where trigonometric functions are involved. The validity of a particular form can best be checked by substituting it back into the differential equation.

3. The auxiliary equation is $m^2 + 1 = 0$, so $y_c = c_1\cos x + c_2\sin x$ and

$$W = \begin{vmatrix} \cos x & \sin x \\ -\sin x & \cos x \end{vmatrix} = 1.$$

Identifying $f(x) = \sin x$ we obtain

$$u_1' = -\sin^2 x$$

$$u_2' = \cos x \sin x.$$

Then

$$u_1 = \frac{1}{4}\sin 2x - \frac{1}{2}x = \frac{1}{2}\sin x\cos x - \frac{1}{2}x$$

$$u_2 = -\frac{1}{2}\cos^2 x.$$

and

$$y = c_1\cos x + c_2\sin x + \frac{1}{2}\sin x\cos^2 x - \frac{1}{2}x\cos x - \frac{1}{2}\cos^2 x\sin x$$

$$= c_1\cos x + c_2\sin x - \frac{1}{2}x\cos x.$$

6. The auxiliary equation is $m^2 + 1 = 0$, so $y_c = c_1 \cos x + c_2 \sin x$ and

$$W = \begin{vmatrix} \cos x & \sin x \\ -\sin x & \cos x \end{vmatrix} = 1.$$

Identifying $f(x) = \sec^2 x$ we obtain

$$u_1' = -\frac{\sin x}{\cos^2 x}$$

$$u_2' = \sec x.$$

Then

$$u_1 = -\frac{1}{\cos x} = -\sec x$$

$$u_2 = \ln|\sec x + \tan x|$$

and

$$y = c_1 \cos x + c_2 \sin x - \cos x \sec x + \sin x \ln|\sec x + \tan x|$$

$$= c_1 \cos x + c_2 \sin x - 1 + \sin x \ln|\sec x + \tan x|.$$

9. The auxiliary equation is $m^2 - 4 = 0$, so $y_c = c_1 e^{2x} + c_2 e^{-2x}$ and

$$W = \begin{vmatrix} e^{2x} & e^{-2x} \\ 2e^{2x} & -2e^{-2x} \end{vmatrix} = -4.$$

Identifying $f(x) = e^{2x}/x$ we obtain $u_1' = 1/4x$ and $u_2' = -e^{4x}/4x$. Then

$$u_1 = \frac{1}{4} \ln|x|,$$

$$u_2 = -\frac{1}{4} \int_{x_0}^x \frac{e^{4t}}{t}\, dt$$

and

$$y = c_1 e^{2x} + c_2 e^{-2x} + \frac{1}{4}\left(e^{2x} \ln|x| - e^{-2x} \int_{x_0}^x \frac{e^{4t}}{t}\, dt \right), \qquad x_0 > 0.$$

12. The auxiliary equation is $m^2 - 2m + 1 = (m-1)^2 = 0$, so $y_c = c_1 e^x + c_2 x e^x$ and

$$W = \begin{vmatrix} e^x & x e^x \\ e^x & x e^x + e^x \end{vmatrix} = e^{2x}.$$

Identifying $f(x) = e^x/(1 + x^2)$ we obtain

$$u_1' = -\frac{x e^x e^x}{e^{2x}(1+x^2)} = -\frac{x}{1+x^2}$$

$$u_2' = \frac{e^x e^x}{e^{2x}(1+x^2)} = \frac{1}{1+x^2}.$$

Then $u_1 = -\frac{1}{2} \ln(1+x^2)$, $u_2 = \tan^{-1} x$, and

$$y = c_1 e^x + c_2 x e^x - \frac{1}{2} e^x \ln(1+x^2) + x e^x \tan^{-1} x.$$

15. The auxiliary equation is $m^2 + 2m + 1 = (m+1)^2 = 0$, so $y_c = c_1 e^{-t} + c_2 t e^{-t}$ and

$$W = \begin{vmatrix} e^{-t} & t e^{-t} \\ -e^{-t} & -t e^{-t} + e^{-t} \end{vmatrix} = e^{-2t}.$$

Identifying $f(t) = e^{-t} \ln t$ we obtain

$$u_1' = -\frac{te^{-t}e^{-t} \ln t}{e^{-2t}} = -t \ln t$$

$$u_2' = \frac{e^{-t}e^{-t} \ln t}{e^{-2t}} = \ln t.$$

Then

$$u_1 = -\frac{1}{2}t^2 \ln t + \frac{1}{4}t^2$$

$$u_2 = t \ln t - t$$

and

$$y = c_1 e^{-t} + c_2 t e^{-t} - \frac{1}{2}t^2 e^{-t} \ln t + \frac{1}{4}t^2 e^{-t} + t^2 e^{-t} \ln t - t^2 e^{-t}$$

$$= c_1 e^{-t} + c_2 t e^{-t} + \frac{1}{2}t^2 e^{-t} \ln t - \frac{3}{4}t^2 e^{-t}.$$

18. The auxiliary equation is $4m^2 - 4m + 1 = (2m - 1)^2 = 0$, so $y_c = c_1 e^{x/2} + c_2 x e^{x/2}$ and

$$W = \begin{vmatrix} e^{x/2} & xe^{x/2} \\ \frac{1}{2}e^{x/2} & \frac{1}{2}xe^{x/2} + e^{x/2} \end{vmatrix} = e^x.$$

Identifying $f(x) = \frac{1}{4}e^{x/2}\sqrt{1 - x^2}$ we obtain

$$u_1' = -\frac{xe^{x/2}e^{x/2}\sqrt{1 - x^2}}{4e^x} = -\frac{1}{4}x\sqrt{1 - x^2}$$

$$u_2' = \frac{e^{x/2}e^{x/2}\sqrt{1 - x^2}}{4e^x} = \frac{1}{4}\sqrt{1 - x^2}.$$

Then

$$u_1 = \frac{1}{12}\left(1 - x^2\right)^{3/2}$$

$$u_2 = \frac{x}{8}\sqrt{1 - x^2} + \frac{1}{8}\sin^{-1} x$$

and

$$y = c_1 e^{x/2} + c_2 x e^{x/2} + \frac{1}{12}e^{x/2}\left(1 - x^2\right)^{3/2} + \frac{1}{8}x^2 e^{x/2}\sqrt{1 - x^2} + \frac{1}{8}x e^{x/2} \sin^{-1} x.$$

21. The auxiliary equation is $m^2 + 2m - 8 = (m - 2)(m + 4) = 0$, so $y_c = c_1 e^{2x} + c_2 e^{-4x}$ and

$$W = \begin{vmatrix} e^{2x} & e^{-4x} \\ 2e^{2x} & -4e^{-4x} \end{vmatrix} = -6e^{-2x}.$$

Identifying $f(x) = 2e^{-2x} - e^{-x}$ we obtain

$$u_1' = \frac{1}{3}e^{-4x} - \frac{1}{6}e^{-3x}$$

$$u_2' = -\frac{1}{6}e^{3x} - \frac{1}{3}e^{2x}.$$

Then

$$u_1 = -\frac{1}{12}e^{-4x} + \frac{1}{18}e^{-3x}$$

$$u_2 = \frac{1}{18}e^{3x} - \frac{1}{6}e^{2x}.$$

Thus

$$y = c_1 e^{2x} + c_2 e^{-4x} - \frac{1}{12} e^{-2x} + \frac{1}{18} e^{-x} + \frac{1}{18} e^{-x} - \frac{1}{6} e^{-2x}$$

$$= c_1 e^{2x} + c_2 e^{-4x} - \frac{1}{4} e^{-2x} + \frac{1}{9} e^{-x}$$

and

$$y' = 2c_1 e^{2x} - 4c_2 e^{-4x} + \frac{1}{2} e^{-2x} - \frac{1}{9} e^{-x}.$$

The initial conditions imply

$$c_1 + c_2 - \frac{5}{36} = 1$$

$$2c_1 - 4c_2 + \frac{7}{18} = 0.$$

Thus $c_1 = 25/36$ and $c_2 = 4/9$, and

$$y = \frac{25}{36} e^{2x} + \frac{4}{9} e^{-4x} - \frac{1}{4} e^{-2x} + \frac{1}{9} e^{-x}.$$

24. Write the equation in the form

$$y'' + \frac{1}{x} y' + \frac{1}{x^2} y = \frac{\sec(\ln x)}{x^2}$$

and identify $f(x) = \sec(\ln x)/x^2$. From $y_1 = \cos(\ln x)$ and $y_2 = \sin(\ln x)$ we compute

$$W = \begin{vmatrix} \cos(\ln x) & \sin(\ln x) \\ -\dfrac{\sin(\ln x)}{x} & \dfrac{\cos(\ln x)}{x} \end{vmatrix} = \frac{1}{x}.$$

Now

$$u_1' = -\frac{\tan(\ln x)}{x} \quad \text{so} \quad u_1 = \ln|\cos(\ln x)|,$$

and

$$u_2' = \frac{1}{x} \quad \text{so} \quad u_2 = \ln x.$$

Thus, a particular solution is

$$y_p = \cos(\ln x) \ln|\cos(\ln x)| + (\ln x) \sin(\ln x).$$

27. The auxiliary equation is $3m^2 - 6m + 30 = 0$, which has roots $1 + 3i$, so $y_c = e^x(c_1 \cos 3x + c_2 \sin 3x)$. We consider first the differential equation $3y'' - 6y' + 30y = 15 \sin x$, which can be solved using undetermined coefficients. Letting $y_{p_1} = A \cos x + B \sin x$ and substituting into the differential equation we get

$$(27A - 6B) \cos x + (6a + 27b) \sin x = 15 \sin x.$$

Then

$$27A - 6B = 0 \quad \text{and} \quad 6a + 27b = 15,$$

so $A = \frac{2}{17}$ and $B = \frac{9}{17}$. Thus, $y_{p_1} = \frac{2}{17} \cos x + \frac{9}{17} \sin x$. Next, we consider the differential equation $3y'' - 6y' + 30y$, for which a particular solution y_{p_2} can be found using variation of parameters. The Wronskian is

$$W = \begin{vmatrix} e^x \cos 3x & e^x \sin 3x \\ e^x \cos 3x - 3e^x \sin 3x & 3e^x \cos 3x + e^x \sin 3x \end{vmatrix} = 3e^{2x}.$$

Identifying $f(x) = \frac{1}{3} e^x \tan x$ we obtain

$$u_1' = -\frac{1}{9} \sin 3x \tan 3x \quad \text{and} \quad u_2' = \frac{1}{9} \sin 3x.$$

Then

$$u_1 = \frac{1}{27}\sin 3x + \frac{1}{27}\left[\ln\left(\cos\frac{3x}{2} - \sin\frac{3x}{2}\right) - \ln\left(\cos\frac{3x}{2} + \sin\frac{3x}{2}\right)\right]$$

$$u_2 = -\frac{1}{27}\cos 3x.$$

Thus

$$y_{p_2} = \frac{1}{27}e^x\cos 3x\left[\ln\left(\cos\frac{3x}{2} - \sin\frac{3x}{2}\right) - \ln\left(\cos\frac{3x}{2} + \sin\frac{3x}{2}\right)\right]$$

and the general solution of the original differential equation is

$$y = e^x(c_1\cos 3x + c_2\sin 3x) + y_{p_1}(x) + y_{p_2}(x).$$

30. We are given that $y_1 = x^2$ is a solution of $x^4y'' + x^3y' - 4x^2y = 0$. To find a second solution we use reduction of order. Let $y = x^2u(x)$. Then the product rule gives

$$y' = x^2u' + 2xu \quad \text{and} \quad y'' = x^2u'' + 4xu' + 2u,$$

so

$$x^4y'' + x^3y' - 4x^2y = x^5(xu'' + 5u') = 0.$$

Letting $w = u'$, this becomes $xw' + 5w = 0$. Separating variables and integrating we have

$$\frac{dw}{w} = -\frac{5}{x}\,dx \quad \text{and} \quad \ln|w| = -5\ln x + c.$$

Thus, $w = x^{-5}$ and $u = -\frac{1}{4}x^{-4}$. A second solution is then $y_2 = x^2x^{-4} = 1/x^2$, and the general solution of the homogeneous differential equation is $y_c = c_1x^2 + c_2/x^2$. To find a particular solution, y_p, we use variation of parameters. The Wronskian is

$$W = \begin{vmatrix} x^2 & 1/x^2 \\ 2x & -2/x^3 \end{vmatrix} = -\frac{4}{x}.$$

Identifying $f(x) = 1/x^4$ we obtain $u_1' = \frac{1}{4}x^{-5}$ and $u_2' = -\frac{1}{4}x^{-1}$. Then $u_1 = -\frac{1}{16}x^{-4}$ and $u_2 = -\frac{1}{4}\ln x$, so

$$y_p = -\frac{1}{16}x^{-4}x^2 - \frac{1}{4}(\ln x)x^{-2} = -\frac{1}{16}x^{-2} - \frac{1}{4}x^{-2}\ln x.$$

The general solution is

$$y = c_1x^2 + \frac{c_2}{x^2} - \frac{1}{16x^2} - \frac{1}{4x^2}\ln x.$$

Exercises 3.6

3. The auxiliary equation is $m^2 = 0$ so that $y = c_1 + c_2\ln x$.

6. The auxiliary equation is $m^2 + 4m + 3 = (m+1)(m+3) = 0$ so that $y = c_1x^{-1} + c_2x^{-3}$.

9. The auxiliary equation is $25m^2 + 1 = 0$ so that $y = c_1\cos\left(\frac{1}{5}\ln x\right) + c_2\left(\frac{1}{5}\ln x\right)$.

12. The auxiliary equation is $m^2 + 7m + 6 = (m+1)(m+6) = 0$ so that $y = c_1x^{-1} + c_2x^{-6}$.

15. Assuming that $y = x^m$ and substituting into the differential equation we obtain

$$m(m-1)(m-2) - 6 = m^3 - 3m^2 + 2m - 6 = (m-3)(m^2+2) = 0.$$

Thus

$$y = c_1x^3 + c_2\cos\left(\sqrt{2}\ln x\right) + c_3\sin\left(\sqrt{2}\ln x\right).$$

18. Assuming that $y = x^m$ and substituting into the differential equation we obtain

$$m(m-1)(m-2)(m-3) + 6m(m-1)(m-2) + 9m(m-1) + 3m + 1 = m^4 + 2m^2 + 1 = (m^2+1)^2 = 0.$$

Thus

$$y = c_1 \cos(\ln x) + c_2 \sin(\ln x) + c_3 \ln x \cos(\ln x) + c_4 \ln x \sin(\ln x).$$

21. The auxiliary equation is $m^2 - 2m + 1 = (m-1)^2 = 0$ so that $y_c = c_1 x + c_2 x \ln x$ and

$$W(x, x \ln x) = \begin{vmatrix} x & x \ln x \\ 1 & 1 + \ln x \end{vmatrix} = x.$$

Identifying $f(x) = 2/x$ we obtain $u_1' = -2\ln x/x$ and $u_2' = 2/x$. Then $u_1 = -(\ln x)^2$, $u_2 = 2\ln x$, and

$$y = c_1 x + c_2 x \ln x - x(\ln x)^2 + 2x(\ln x)^2$$

$$= c_1 x + c_2 x \ln x + x(\ln x)^2.$$

24. The auxiliary equation is $m^2 - 6m + 8 = (m-2)(m-4) = 0$, so that

$$y = c_1 x^2 + c_2 x^4 \quad \text{and} \quad y' = 2c_1 x + 4c_2 x^3.$$

The initial conditions imply

$$4c_1 + 16c_2 = 32$$

$$4c_1 + 32c_2 = 0.$$

Thus, $c_1 = 16$, $c_2 = -2$, and $y = 16x^2 - 2x^4$.

27. The auxiliary equation is $m^2 = 0$ so that $y_c = c_1 + c_2 \ln x$ and

$$W(1, \ln x) = \begin{vmatrix} 1 & \ln x \\ 0 & 1/x \end{vmatrix} = \frac{1}{x}.$$

Identifying $f(x) = 1$ we obtain $u_1' = -x \ln x$ and $u_2' = x$. Then $u_1 = \frac{1}{4}x^2 - \frac{1}{2}x^2 \ln x$, $u_2 = \frac{1}{2}x^2$, and

$$y = c_1 + c_2 \ln x + \frac{1}{4}x^2 - \frac{1}{2}x^2 \ln x + \frac{1}{2}x^2 \ln x = c_1 + c_2 \ln x + \frac{1}{4}x^2.$$

The initial conditions imply $c_1 + \frac{1}{4} = 1$ and $c_2 + \frac{1}{2} = -\frac{1}{2}$. Thus, $c_1 = \frac{3}{4}$, $c_2 = -1$, and $y = \frac{3}{4} - \ln x + \frac{1}{4}x^2$.

30. Substituting into the differential equation we obtain

$$\frac{d^2 y}{dt^2} - 5\frac{dy}{dt} + 6y = 2t.$$

The auxiliary equation is $m^2 - 5m + 6 = (m-2)(m-3) = 0$ so that $y_c = c_1 e^{2t} + c_2 e^{3t}$. Using undetermined coefficients we try $y_p = At + B$. This leads to $(-5A + 6B) + 6At = 2t$, so that $A = 1/3$, $B = 5/18$, and

$$y = c_1 e^{2t} + c_2 e^{3t} + \frac{1}{3}t + \frac{5}{18} = c_1 x^2 + c_2 x^3 + \frac{1}{3}\ln x + \frac{5}{18}.$$

39. Letting $u = x + 2$ we obtain

$$\frac{dy}{dx} = \frac{dy}{du}$$

and, using the chain rule,

$$\frac{d^2 y}{dx^2} = \frac{d}{dx}\left(\frac{dy}{du}\right) = \frac{d^2 y}{du^2}\frac{du}{dx} = \frac{d^2 y}{du^2}(1) = \frac{d^2 y}{du^2}.$$

Substituting into the differential equation we obtain

$$u^2 \frac{d^2 y}{du^2} + u\frac{dy}{du} + y = 0.$$

The auxiliary equation is $m^2 + 1 = 0$ so that

$$y = c_1 \cos(\ln u) + c_2 \sin(\ln u) = c_1 \cos[\ln(x+2)] + c_2 \sin[\ln(x+2)].$$

42. The function $y(x) = -\sqrt{x}\cos(\ln x)$ is defined for $x > 0$ and has x-intercepts where $\ln x = \pi/2 + k\pi$ for k an integer or where $x = e^{\pi/2 + k\pi}$. Solving $\pi/2 + k\pi = 0.5$ we get $k \approx -0.34$, so $e^{\pi/2 + k\pi} < 0.5$ for all negative integers and the graph has infinitely many x-intercepts in $(0, 0.5)$.

Exercises 3.7

3. Let $u = y'$ so that $u' = y''$. The equation becomes $u' = -u - 1$ which is separable. Thus

$$\frac{du}{u^2 + 1} = -dx \implies \tan^{-1} u = -x + c_1 \implies y' = \tan(c_1 - x) \implies y = \ln|\cos(c_1 - x)| + c_2.$$

6. Let $u = y'$ so that $y'' = u\dfrac{du}{dy}$. The equation becomes $(y+1)u\dfrac{du}{dy} = u^2$. Separating variables we obtain

$$\frac{du}{u} = \frac{dy}{y+1} \implies \ln|u| = \ln|y+1| + \ln c_1 \implies u = c_1(y+1)$$

$$\implies \frac{dy}{dx} = c_1(y+1) \implies \frac{dy}{y+1} = c_1\,dx$$

$$\implies \ln|y+1| = c_1 x + c_2 \implies y + 1 = c_3 e^{c_1 x}.$$

12. Let $u = y'$ so that $u' = y''$. The equation becomes $u' - \dfrac{1}{x}u = u^2$, which is Bernoulli. Using the substitution $w = u^{-1}$ we obtain $\dfrac{dw}{dx} + \dfrac{1}{x}w = -1$. An integrating factor is x, so

$$\frac{d}{dx}[xw] = -x \implies w = -\frac{1}{2}x + \frac{1}{x}c \implies \frac{1}{u} = \frac{c_1 - x^2}{2x} \implies u = \frac{2x}{c_1 - x^2} \implies y = -\ln|c_1 - x^2| + c_2.$$

21. Letting $u = y''$, separating variables, and integrating we have

$$\frac{du}{dx} = \sqrt{1 + u^2}, \quad \frac{du}{\sqrt{1+u^2}} = dx, \quad \text{and} \quad \sinh^{-1} u = x + c_1.$$

Then

$$u = y'' = \sinh(x + c_1), \quad y' = \cosh(x + c_1) + c_2, \quad \text{and} \quad y = \sinh(x + c_1) + c_2 x + c_3.$$

Exercises 3.8

3. From $\frac{3}{4}x'' + 72x = 0$, $x(0) = -1/4$, and $x'(0) = 0$ we obtain $x = -\frac{1}{4}\cos 4\sqrt{6}\,t$.

6. From $50x'' + 200x = 0$, $x(0) = 0$, and $x'(0) = -10$ we obtain $x = -5\sin 2t$ and $x' = -10\cos 2t$.

9. From $\frac{1}{4}x'' + x = 0$, $x(0) = 1/2$, and $x'(0) = 3/2$ we obtain

$$x = \frac{1}{2}\cos 2t + \frac{3}{4}\sin 2t = \frac{\sqrt{13}}{4}\sin(2t + 0.588).$$

12. From $x' + 9x = 0$, $x(0) = -1$, and $x'(0) = -\sqrt{3}$ we obtain

$$x = -\cos 3t - \frac{\sqrt{3}}{3}\sin 3t = \frac{2}{\sqrt{3}}\sin\left(37 + \frac{4\pi}{3}\right)$$

and $x' = 2\sqrt{3}\cos(3t + 4\pi/3)$. If $x' = 3$ then $t = -7\pi/18 + 2n\pi/3$ and $t = -\pi/2 + 2n\pi/3$ for $n = 1, 2, 3, \ldots$.

15. For large values of t the differential equation is approximated by $x'' = 0$. The solution of this equation is the linear function $x = c_1 t + c_2$. Thus, for large time, the restoring force will have decayed to the point where the spring is incapable of returning the mass, and the spring will simply keep on stretching.

18. (a) below, since $x(0) > 0$.   **(b)** from rest, since $x'(0) = 0$.

21. From $\frac{1}{8}x'' + x' + 2x = 0$, $x(0) = -1$, and $x'(0) = 8$ we obtain $x = 4te^{-4t} - e^{-4t}$ and $x' = 8e^{-4t} - 16te^{-4t}$. If $x = 0$ then $t = 1/4$ second. If $x' = 0$ then $t = 1/2$ second and the extreme displacement is $x = e^{-2}$ feet.

24. (a) $x = \frac{1}{3}e^{-8t}\left(4e^{6t} - 1\right)$ is never zero; the extreme displacement is $x(0) = 1$ meter.

    (b) $x = \frac{1}{3}e^{-8t}\left(5 - 2e^{6t}\right) = 0$ when $t = \frac{1}{6}\ln\frac{5}{2} \approx 0.153$ second; if $x' = \frac{4}{3}e^{-8t}\left(e^{6t} - 10\right) = 0$ then $t = \frac{1}{6}\ln 10 \approx$ 0.384 second and the extreme displacement is $x = -0.232$ meter.

27. From $\frac{5}{16}x'' + \beta x' + 5x = 0$ we find that the roots of the auxiliary equation are $m = -\frac{8}{5}\beta \pm \frac{4}{5}\sqrt{4\beta^2 - 25}$.

    (a) If $4\beta^2 - 25 > 0$ then $\beta > 5/2$.

    (b) If $4\beta^2 - 25 = 0$ then $\beta = 5/2$.

    (c) If $4\beta^2 - 25 < 0$ then $0 < \beta < 5/2$.

30. (a) If $x'' + 2x' + 5x = 12\cos 2t + 3\sin 2t$, $x(0) = -1$, and $x'(0) = 5$ then $x_c = e^{-t}(c_1\cos 2t + c_2\sin 2t)$ and $x_p = 3\sin 2t$ so that the equation of motion is

$$x = e^{-t}\cos 2t + 3\sin 2t.$$

(b)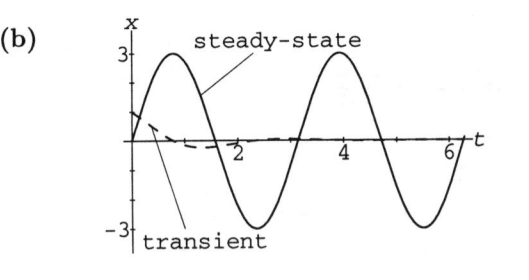

(c)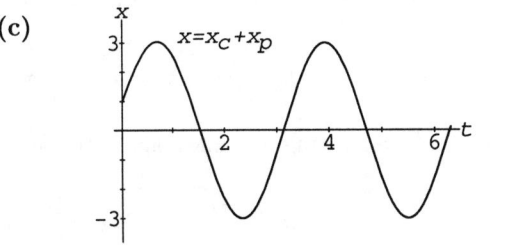

33. From $2x'' + 32x = 68e^{-2t}\cos 4t$, $x(0) = 0$, and $x'(0) = 0$ we obtain $x_c = c_1\cos 4t + c_2\sin 4t$ and $x_p = \frac{1}{2}e^{-2t}\cos 4t - 2e^{-2t}\sin 4t$ so that

$$x = -\frac{1}{2}\cos 4t + \frac{9}{4}\sin 4t + \frac{1}{2}e^{-2t}\cos 4t - 2e^{-2t}\sin 4t.$$

36. (a) From $100x'' + 1600x = 1600\sin 8t$, $x(0) = 0$, and $x'(0) = 0$ we obtain $x_c = c_1\cos 4t + c_2\sin 4t$ and $x_p = -\frac{1}{3}\sin 8t$ so that

$$x = \frac{2}{3}\sin 4t - \frac{1}{3}\sin 8t.$$

    (b) If $x = \frac{1}{3}\sin 4t(2 - 2\cos 4t) = 0$ then $t = n\pi/4$ for $n = 0, 1, 2, \ldots$.

    (c) If $x' = \frac{8}{3}\cos 4t - \frac{8}{3}\cos 8t = \frac{8}{3}(1 - \cos 4t)(1 + 2\cos 4t) = 0$ then $t = \pi/3 + n\pi/2$ and $t = \pi/6 + n\pi/2$ for $n = 0, 1, 2, \ldots$ at the extreme values. *Note:* There are many other values of t for which $x' = 0$.

    (d) $x(\pi/6 + n\pi/2) = \sqrt{3}/2$ cm. and $x(\pi/3 + n\pi/2) = -\sqrt{3}/2$ cm.

(e)

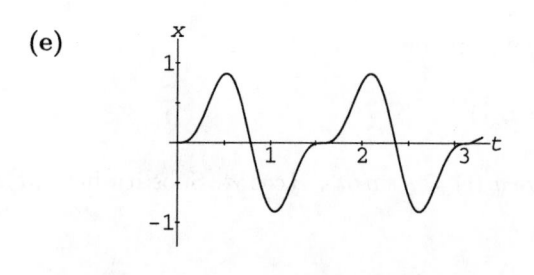

39. (a) From $x'' + \omega^2 x = F_0 \cos \gamma t$, $x(0) = 0$, and $x'(0) = 0$ we obtain $x_c = c_1 \cos \omega t + c_2 \sin \omega t$ and $x_p = (F_0 \cos \gamma t)/(\omega^2 - \gamma^2)$ so that

$$x = -\frac{F_0}{\omega^2 - \gamma^2} \cos \omega t + \frac{F_0}{\omega^2 - \gamma^2} \cos \gamma t.$$

(b) $\displaystyle \lim_{\gamma \to \omega} \frac{F_0}{\omega^2 - \gamma^2} (\cos \gamma t - \cos \omega t) = \lim_{\gamma \to \omega} \frac{-F_0 t \sin \gamma t}{-2\gamma} = \frac{F_0}{2\omega} t \sin \omega t.$

45. Solving $\frac{1}{20} q'' + 2q' + 100q = 0$ we obtain $q(t) = e^{-20t}(c_1 \cos 40t + c_2 \sin 40t)$. The initial conditions $q(0) = 5$ and $q'(0) = 0$ imply $c_1 = 5$ and $c_2 = 5/2$. Thus

$$q(t) = e^{-20t}\left(5\cos 40t + \frac{5}{2}\sin 40t\right) \approx \sqrt{25 + 25/4}\, e^{-20t} \sin(40t + 1.1071)$$

and $q(0.01) \approx 4.5676$ coulombs. The charge is zero for the first time when $40t + 0.4636 = \pi$ or $t \approx 0.0509$ second.

48. Solving $q'' + 100q' + 2500q = 30$ we obtain $q(t) = c_1 e^{-50t} + c_2 t e^{-50t} + 0.012$. The initial conditions $q(0) = 0$ and $q'(0) = 2$ imply $c_1 = -0.012$ and $c_2 = 1.4$. Thus

$$q(t) = -0.012e^{-50t} + 1.4te^{-50t} + 0.012 \quad \text{and} \quad i(t) = 2e^{-50t} - 70te^{-50t}.$$

Solving $i(t) = 0$ we see that the maximum charge occurs when $t = 1/35$ and $q(1/35) \approx 0.01871$.

51. The differential equation is $\frac{1}{2}q'' + 20q' + 1000q = 100\sin t$. To use Example 11 in the text we identify $E_0 = 100$ and $\gamma = 60$. Then

$$X = L\gamma - \frac{1}{c\gamma} = \frac{1}{2}(60) - \frac{1}{0.001(60)} \approx 13.3333,$$

$$Z = \sqrt{X^2 + R^2} = \sqrt{X^2 + 400} \approx 24.0370,$$

and

$$\frac{E_0}{Z} = \frac{100}{Z} \approx 4.1603.$$

From Problem 50, then

$$i_p(t) \approx 4.1603(60t + \phi)$$

where $\sin \phi = -X/Z$ and $\cos \phi = R/Z$. Thus $\tan \phi = -X/R \approx -0.6667$ and ϕ is a fourth quadrant angle. Now $\phi \approx -0.5880$ and

$$i_p(t) \approx 4.1603(60t - 0.5880).$$

54. By Problem 50 the amplitude of the steady-state current is E_0/Z, where $Z = \sqrt{X^2 + R^2}$ and $X = L\gamma - 1/C\gamma$. Since E_0 is constant the amplitude will be a maximum when Z is a minimum. Since R is constant, Z will be a minimum when $X = 0$. Solving $L\gamma - 1/C\gamma = 0$ for γ we obtain $\gamma = 1/\sqrt{LC}$. The maximum amplitude will be E_0/R.

57. In an *L-C* series circuit there is no resistor, so the differential equation is

$$L\frac{d^2q}{dt^2} + \frac{1}{C}q = E(t).$$

Then $q(t) = c_1 \cos\left(t/\sqrt{LC}\right) + c_2 \sin\left(t/\sqrt{LC}\right) + q_p(t)$ where $q_p(t) = A\sin\gamma t + B\cos\gamma t$. Substituting $q_p(t)$ into the differential equation we find

$$\left(\frac{1}{C} - L\gamma^2\right)A\sin\gamma t + \left(\frac{1}{C} - L\gamma^2\right)B\cos\gamma t = E_0\cos\gamma t.$$

Equating coefficients we obtain $A = 0$ and $B = E_0C/(1 - LC\gamma^2)$. Thus, the charge is

$$q(t) = c_1 \cos\frac{1}{\sqrt{LC}}t + c_2 \sin\frac{1}{\sqrt{LC}}t + \frac{E_0C}{1 - LC\gamma^2}\cos\gamma t.$$

The initial conditions $q(0) = q_0$ and $q'(0) = i_0$ imply $c_1 = q_0 - E_0C/(1 - LC\gamma^2)$ and $c_2 = i_0\sqrt{LC}$. The current is

$$i(t) = -\frac{c_1}{\sqrt{LC}}\sin\frac{1}{\sqrt{LC}}t + \frac{c_2}{\sqrt{LC}}\cos\frac{1}{\sqrt{LC}}t - \frac{E_0C\gamma}{1 - LC\gamma^2}\sin\gamma t$$

$$= i_0\cos\frac{1}{\sqrt{LC}}t - \frac{1}{\sqrt{LC}}\left(q_0 - \frac{E_0C}{1 - LC\gamma^2}\right)\sin\frac{1}{\sqrt{LC}}t - \frac{E_0C\gamma}{1 - LC\gamma^2}\sin\gamma t.$$

Exercises 3.9

3. (a) The general solution is

$$y(x) = c_1 + c_2 x + c_3 x^2 + c_4 x^3 + \frac{w_0}{24EI}x^4.$$

The boundary conditions are $y(0) = 0$, $y'(0) = 0$, $y(L) = 0$, $y''(L) = 0$. The first two conditions give $c_1 = 0$ and $c_2 = 0$. The conditions at $x = L$ give the system

$$c_3L^2 + c_4L^3 + \frac{w_0}{24EI}L^4 = 0$$

$$2c_3 + 6c_4L + \frac{w_0}{2EI}L^2 = 0.$$

Solving, we obtain $c_3 = w_0L^2/16EI$ and $c_4 = -5w_0L/48EI$. The deflection is

$$y(x) = \frac{w_0}{48EI}(3L^2x^2 - 5Lx^3 + 2x^4).$$

(b)

6. (a) The deflection of the beam in Problem 2 is

$$y(x) = \frac{w_0}{24EI}(L^3x - 2Lx^3 + x^4).$$

Since $y(x)$ is a differentiable function on $[0, L]$ we can find the maximum deflection by comparing the deflections at $x = 0$, $x = L$, and at any critical points. Setting

$$y' = \frac{w_0}{24EI}(4x^3 - 6Lx^2 + L^3) = 0,$$

we see that $x = L/2$ is the only critical point in $[0, L]$. Since $y(0) = y(L) = 0$, the maximum deflection is $y_{max} = y(L/2) = 5w_0 L^4/384EI$.

(b) The maximum deflection in Example 1 is $y(L/2) = (w_0/24EI)L^4/16 = w_0 L^4/384EI$, which is $1/5$ of the maximum displacement of the beam in Problem 2.

9. For $\lambda \leq 0$ the only solution of the boundary-value problem is $y = 0$. For $\lambda > 0$ we have

$$y = c_1 \cos \sqrt{\lambda}\, x + c_2 \sin \sqrt{\lambda}\, x.$$

Now $y(0) = 0$ implies $c_1 = 0$, so

$$y(\pi) = c_2 \sin \sqrt{\lambda}\, \pi = 0$$

gives

$$\sqrt{\lambda}\, \pi = n\pi \quad \text{or} \quad \lambda = n^2, \; n = 1, 2, 3, \ldots .$$

The eigenvalues n^2 correspond to the eigenfunctions $\sin nx$ for $n = 1, 2, 3, \ldots$.

12. For $\lambda \leq 0$ the only solution of the boundary-value problem is $y = 0$. For $\lambda > 0$ we have

$$y = c_1 \cos \sqrt{\lambda}\, x + c_2 \sin \sqrt{\lambda}\, x.$$

Now $y(0) = 0$ implies $c_1 = 0$, so

$$y'\left(\frac{\pi}{2}\right) = c_2 \sqrt{\lambda} \cos \sqrt{\lambda}\, \frac{\pi}{2} = 0$$

gives

$$\sqrt{\lambda}\, \frac{\pi}{2} = \frac{(2n-1)\pi}{2} \quad \text{or} \quad \lambda = (2n-1)^2, \; n = 1, 2, 3, \ldots .$$

The eigenvalues $(2n-1)^2$ correspond to the eigenfunctions $\sin(2n-1)x$.

15. The auxiliary equation has solutions

$$m = \frac{1}{2}\left(-2 \pm \sqrt{4 - 4(\lambda + 1)}\right) = -1 \pm \sqrt{-\lambda}.$$

For $\lambda < 0$ we have

$$y = e^{-x}\left(c_1 \cosh \sqrt{-\lambda}\, x + c_2 \sinh \sqrt{-\lambda}\, x\right).$$

The boundary conditions imply

$$y(0) = c_1 = 0$$
$$y(5) = c_2 e^{-5} \sinh 5\sqrt{-\lambda} = 0$$

so $c_1 = c_2 = 0$ and the only solution of the boundary-value problem is $y = 0$.
For $\lambda = 0$ we have

$$y = c_1 e^{-x} + c_2 x e^{-x}$$

and the only solution of the boundary-value problem is $y = 0$.
For $\lambda > 0$ we have

$$y = e^{-x}\left(c_1 \cos \sqrt{\lambda}\, x + c_2 \sin \sqrt{\lambda}\, x\right).$$

Now $y(0) = 0$ implies $c_1 = 0$, so

$$y(5) = c_2 e^{-5} \sin 5\sqrt{\lambda} = 0$$

gives

$$5\sqrt{\lambda} = n\pi \quad \text{or} \quad \lambda = \frac{n^2\pi^2}{25}, \quad n = 1, 2, 3, \dots .$$

The eigenvalues $n^2\pi^2/25$ correspond to the eigenfunctions $e^{-x}\sin\frac{n\pi}{5}x$ for $n = 1, 2, 3, \dots$.

18. For $\lambda = 0$ the only solution of the boundary-value problem is $y = 0$. For $\lambda \neq 0$ we have

$$y = c_1\cos\lambda x + c_2\sin\lambda x.$$

Now $y(0) = 0$ implies $c_1 = 0$, so

$$y'(3\pi) = c_2\lambda\cos 3\pi\lambda = 0$$

gives

$$3\pi\lambda = \frac{(2n-1)\pi}{2} \quad \text{or} \quad \lambda = \frac{2n-1}{6}, \quad n = 1, 2, 3, \dots .$$

The eigenvalues $(2n-1)/6$ correspond to the eigenfunctions $\sin\frac{2n-1}{6}x$ for $n = 1, 2, 3, \dots$.

21. For $\lambda = 0$ the general solution is $y = c_1 + c_2\ln x$. Now $y' = c_2/x$, so $y'(1) = c_2 = 0$ and $y = c_1$. Since $y'(e^2) = 0$ for any c_1 we see that $y(x) = 1$ is an eigenfunction corresponding to the eigenvalue $\lambda = 0$.

For $\lambda < 0$, $y = c_1 x^{-\sqrt{-\lambda}} + c_2 x^{\sqrt{-\lambda}}$. The initial conditions imply $c_1 = c_2 = 0$, so $y(x) = 0$.

For $\lambda > 0$, $y = c_1\cos(\sqrt{\lambda}\ln x) + c_2\sin(\sqrt{\lambda}\ln x)$. Now

$$y' = -c_1\frac{\sqrt{\lambda}}{x}\sin(\sqrt{\lambda}\ln x) + c_2\frac{\sqrt{\lambda}}{x}\cos(\sqrt{\lambda}\ln x),$$

and $y'(1) = c_2\sqrt{\lambda} = 0$ implies $c_2 = 0$. Finally, $y'(e^2) = -(c_1\sqrt{\lambda}/e^2)\sin(2\sqrt{\lambda}) = 0$ implies $\lambda = n^2\pi^2/4$ for $n = 1$, 2, 3, $\dots$. The corresponding eigenfunctions are

$$y = \cos\left(\frac{n\pi}{2}\ln x\right).$$

27. The general solution is

$$y = c_1\cos\sqrt{\frac{\rho\omega^2}{T}}\,x + c_2\sin\sqrt{\frac{\rho\omega^2}{T}}\,x.$$

From $y(0) = 0$ we obtain $c_1 = 0$. Setting $y(L) = 0$ we find $\sqrt{\rho\omega^2/T}\,L = n\pi$, $n = 1, 2, 3, \dots$. Thus, critical speeds are $\omega_n = n\pi\sqrt{T}/L\sqrt{\rho}$, $n = 1, 2, 3, \dots$. The corresponding deflection curves are

$$y(x) = c_2\sin\frac{n\pi}{L}x, \quad n = 1, 2, 3, \dots ,$$

where $c_2 \neq 0$.

30. The auxiliary equation is $m^2 = 0$ so that $u(r) = c_1 + c_2\ln r$. The boundary conditions $u(a) = u_0$ and $u(b) = u_1$ yield the system $c_1 + c_2\ln a = u_0$, $c_1 + c_2\ln b = u_1$. Solving gives

$$c_1 = \frac{u_1\ln a - u_0\ln b}{\ln(a/b)} \quad \text{and} \quad c_2 = \frac{u_0 - u_1}{\ln(a/b)}.$$

Thus

$$u(r) = \frac{u_1\ln a - u_0\ln b}{\ln(a/b)} + \frac{u_0 - u_1}{\ln(a/b)}\ln r = \frac{u_0\ln(r/b) - u_1\ln(r/a)}{\ln(a/b)}.$$

33. (a) A solution curve has the same y-coordinate at both ends of the interval $[-\pi, \pi]$ and the tangent lines at the endpoints of the interval are parallel.

(b) For $\lambda = 0$ the solution of $y'' = 0$ is $y = c_1 x + c_2$. From the first boundary condition we have

$$y(-\pi) = -c_1\pi + c_2 = y(\pi) = c_1\pi + c_2$$

or $2c_1\pi = 0$. Thus, $c_1 = 0$ and $y = c_2$. This constant solution is seen to satisfy the boundary-value problem. For $\lambda < 0$ we have $y = c_1 \cosh \lambda x + c_2 \sinh \lambda x$. In this case the first boundary condition gives

$$y(-\pi) = c_1 \cosh(-\lambda\pi) + c_2 \sinh(-\lambda\pi)$$
$$= c_1 \cosh \lambda\pi - c_2 \sinh \lambda\pi$$
$$= y(\pi) = c_1 \cosh \lambda\pi + c_2 \sinh \lambda\pi$$

or $2c_2 \sinh \lambda\pi = 0$. Thus $c_2 = 0$ and $y = c_1 \cosh \lambda x$. The second boundary condition implies in a similar fashion that $c_1 = 0$. Thus, for $\lambda < 0$, the only solution of the boundary-value problem is $y = 0$.

For $\lambda > 0$ we have $y = c_1 \cos \lambda x + c_2 \sin \lambda x$. The first boundary condition implies

$$y(-\pi) = c_1 \cos(-\lambda\pi) + c_2 \sin(-\lambda\pi)$$
$$= c_1 \cos \lambda\pi - c_2 \sin \lambda\pi$$
$$= y(\pi) = c_1 \cos \lambda\pi + c_2 \sin \lambda\pi$$

or $2c_2 \sin \lambda\pi = 0$. Similarly, the second boundary condition implies $2c_1\lambda \sin \lambda\pi = 0$. If $c_1 = c_2 = 0$ the solution is $y = 0$. However, if $c_1 \neq 0$ or $c_2 \neq 0$, then $\sin \lambda\pi = 0$, which implies that λ must be an integer, n. Therefore, for c_1 and c_2 not both 0, $y = c_1 \cos nx + c_2 \sin nx$ is a nontrivial solution of the boundary-value problem. Since $\cos(-nx) = \cos nx$ and $\sin(-nx) = -\sin nx$, we may assume without loss of generality that the eigenvalues are $\lambda_n = n$, for n a positive integer. The corresponding eigenfunctions are $y_n = \cos nx$ and $y_n = \sin nx$.

(c)

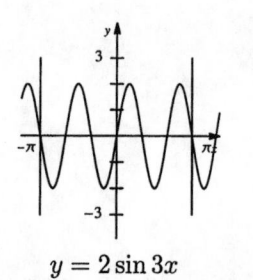

$y = 2\sin 3x$

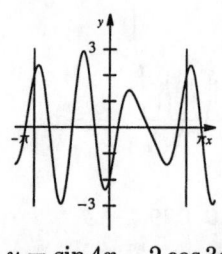

$y = \sin 4x - 2\cos 3x$

Exercises 3.10

3. The period corresponding to $x(0) = 1$, $x'(0) = 1$ is approximately 5.8. The second initial-value problem does not have a periodic solution.

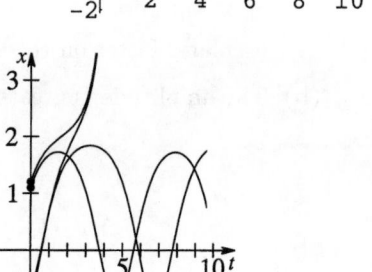

6. From the graphs we see that the interval is approximately $(-0.8, 1.1)$.

41

9. This is a damped hard spring, so all solutions should be oscillatory with $x \to 0$ as $t \to \infty$.

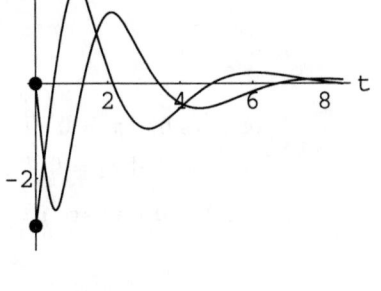

12. (a)

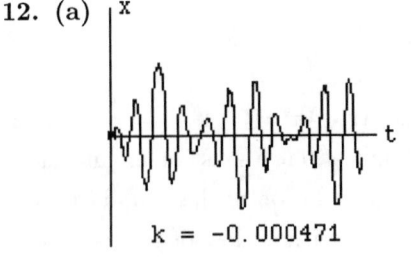

k = -0.000471

k = -0.000472

The system appears to be oscillatory for $-0.000471 \le k_1 < 0$ and nonoscillatory for $k_1 \le -0.000472$.

(b)

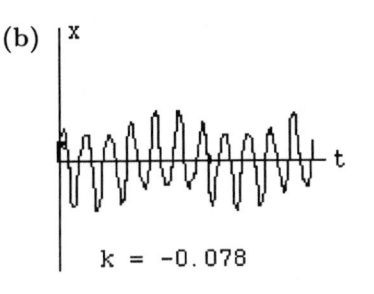

k = -0.078

k = -0.079

The system appears to be oscillatory for $-0.077 \le k_1 < 0$ and nonoscillatory for $k_1 \le 0.078$.

15. (a) Write the differential equation as

$$\frac{d^2\theta}{dt^2} + \omega^2 \sin\theta = 0,$$

where $\omega^2 = g/\ell$. To test for differences between the earth and the moon we take $\ell = 3$, $\theta(0) = 1$, and $\theta'(0) = 2$. Using $g = 32$ on the earth and $g = 5.5$ on the moon we obtain the graphs shown in the figure. Comparing the apparent periods of the graphs, we see that the pendulum oscillates faster on the earth than on the moon.

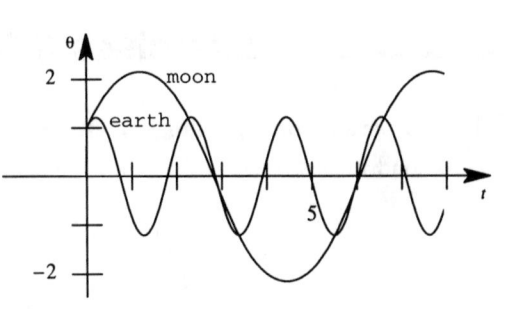

(b) The amplitude is greater on the moon than on the earth.

(c) The linear model is

$$\frac{d^2\theta}{dt^2} + \omega^2\theta = 0,$$

where $\omega^2 = g/\ell$. When $g = 32$, $\ell = 3$, $\theta(0) = 1$, and $\theta'(0) = 2$, the general solution is

$$\theta(t) = \cos 3.266t + 0.612 \sin 3.266t.$$

When $g = 5.5$ the general solution is

$$\theta(t) = \cos 1.354t + 1.477 \sin 1.354t.$$

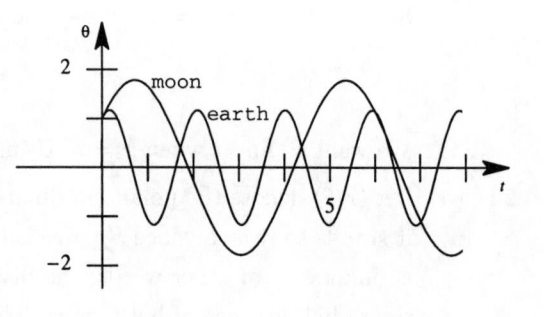

As in the nonlinear case, the pendulum oscillates faster on the earth than on the moon and still has greater amplitude on the moon.

18. (a) Setting $dy/dt = v$, the differential equation in (13) becomes $dv/dt = -gR^2/y^2$. But, by the chain rule, $dv/dt = (dv/dy)(dy/dt) = v\,dv/dt$, so $v\,dv/dy = -gR^2/y^2$. Separating variables and integrating we obtain

$$v\,dv = -gR^2\frac{dy}{y^2} \quad \text{and} \quad \frac{1}{2}v^2 = \frac{gR^2}{y} + c.$$

Setting $v = v_0$ and $y = R$ we find $c = -gR + \frac{1}{2}v_0^2$ and

$$v^2 = 2g\frac{R^2}{y} - 2gR + v_0^2.$$

(b) As $y \to \infty$ we assume that $v \to 0^+$. Then $v_0^2 = 2gR$ and $v_0 = \sqrt{2gR}$.

(c) Using $g = 32$ ft/s and $R = 4000(5280)$ ft we find

$$v_0 = \sqrt{2(32)(4000)(5280)} \approx 36765.2 \text{ ft/s} \approx 25067 \text{ mi/hr}.$$

(d) $v_0 = \sqrt{2(0.165)(32)(1080)} \approx 7760 \text{ ft/s} \approx 5291 \text{ mi/hr}$

21. (a) The weight of x feet of the chain is $2x$, so the corresponding mass is $m = 2x/32 = x/16$. The only force acting on the chain is the weight of the portion of the chain hanging over the edge of the platform. Thus, by Newton's second law,

$$\frac{d}{dt}(mv) = \frac{d}{dt}\left(\frac{x}{16}v\right) = \frac{1}{16}\left(x\frac{dv}{dt} + v\frac{dx}{dt}\right) = \frac{1}{16}\left(x\frac{dv}{dt} + v^2\right) = 2x$$

and $x\,dv/dt + v^2 = 32x$. Now, by the chain rule, $dv/dt = (dv/dx)(dx/dt) = v\,dv/dx$, so $xv\,dv/dx + v^2 = 32x$.

(b) We separate variables and write the differential equation as $(v^2 - 32x)\,dx + xv\,dv = 0$. This is not an exact form, but $\mu(x) = x$ is an integrating factor. Multiplying by x we get $(xv^2 - 32x^2)\,dx + x^2v\,dv = 0$. This form is the total differential of $u = \frac{1}{2}x^2v^2 - \frac{32}{3}x^3$, so an implicit solution is $\frac{1}{2}x^2v^2 - \frac{32}{3}x^3 = c$. Letting $x = 3$ and $v = 0$ we find $c = -288$. Solving for v we get

$$\frac{dx}{dt} = v = \frac{8\sqrt{x^3 - 27}}{\sqrt{3}x}, \quad 3 \le x \le 8.$$

(c) Separating variables and integrating we obtain

$$\frac{x}{\sqrt{x^3 - 27}}\,dx = \frac{8}{\sqrt{3}}\,dt \quad \text{and} \quad \int_3^x \frac{s}{\sqrt{s^3 - 27}}\,ds = \frac{8}{\sqrt{3}}t + c.$$

Since $x = 3$ when $t = 0$, we see that $c = 0$ and

$$t = \frac{\sqrt{3}}{8} \int_3^x \frac{s}{\sqrt{s^3 - 27}} \, ds.$$

We want to find t when $x = 7$. Using a CAS we find $t(7) = 0.576$ seconds.

24. **(a)** Let (r, θ) denote the polar coordinates of the destroyer S_1. When S_1 travels the 6 miles from $(9, 0)$ to $(3, 0)$ it stands to reason, since S_2 travels half as fast as S_1, that the polar coordinates of S_2 are $(3, \theta_2)$, where θ_2 is unknown. In other words, the distances of the ships from $(0, 0)$ are the same and $r(t) = 15t$ then gives the radial distance of both ships. This is necessary if S_1 is to intercept S_2.

(b) The differential of arc length in polar coordinates is $(ds)^2 = (r \, d\theta)^2 + (dr)^2$, so that

$$\left(\frac{ds}{dt} \right)^2 = r^2 \left(\frac{d\theta}{dt} \right)^2 + \left(\frac{dr}{dt} \right)^2 .$$

Using $ds/dt = 30$ and $dr/dt = 15$ then gives

$$900 = 225t^2 \left(\frac{d\theta}{dt} \right)^2 + 225$$

$$675 = 225t^2 \left(\frac{d\theta}{dt} \right)^2$$

$$\frac{d\theta}{dt} = \frac{\sqrt{3}}{t}$$

$$\theta(t) = \sqrt{3} \ln t + c = \sqrt{3} \ln \frac{r}{15} + c.$$

When $r = 3$, $\theta = 0$, so $c = -\sqrt{3} \ln(1/5)$ and

$$\theta(t) = \sqrt{3} \left(\ln \frac{r}{15} - \ln \frac{1}{5} \right) = \sqrt{3} \ln \frac{r}{3} .$$

Thus $r = 3e^{\theta/\sqrt{3}}$, whose graph is a logarithmic spiral.

(c) The time for S_1 to go from $(9, 0)$ to $(3, 0) = \frac{1}{5}$ hour. Now S_1 must intercept the path of S_2 for some angle β, where $0 < \beta < 2\pi$. At the time of interception t_2 we have $15t_2 = 3e^{\beta/\sqrt{3}}$ or $t = e^{\beta/\sqrt{3}}/5$. The total time is then

$$t = \frac{1}{5} + \frac{1}{5} e^{\beta/\sqrt{3}} < \frac{1}{5} (1 + e^{2\pi/\sqrt{3}}).$$

Exercises 3.11

3. From $Dx = -y + t$ and $Dy = x - t$ we obtain $y = t - Dx$, $Dy = 1 - D^2 x$, and $(D^2 + 1)x = 1 + t$. Then

$$x = c_1 \cos t + c_2 \sin t + 1 + t$$

and

$$y = c_1 \sin t - c_2 \cos t + t - 1.$$

6. From $(D + 1)x + (D - 1)y = 2$ and $3x + (D + 2)y = -1$ we obtain $x = -\frac{1}{3} - \frac{1}{3}(D + 2)y$, $Dx = -\frac{1}{3}(D^2 + 2D)y$, and $(D^2 + 5)y = -7$. Then

$$y = c_1 \cos \sqrt{5}\, t + c_2 \sin \sqrt{5}\, t - \frac{7}{5}$$

and

$$x = \left(-\frac{2}{3}c_1 - \frac{\sqrt{5}}{3}c_2\right)\cos\sqrt{5}\,t + \left(\frac{\sqrt{5}}{3}c_1 - \frac{2}{3}c_2\right)\sin\sqrt{5}\,t + \frac{3}{5}.$$

9. From $Dx + D^2y = e^{3t}$ and $(D+1)x + (D-1)y = 4e^{3t}$ we obtain $D(D^2+1)x = 34e^{3t}$ and $D(D^2+1)y = -8e^{3t}$. Then

$$y = c_1 + c_2\sin t + c_3\cos t - \frac{4}{15}e^{3t}$$

and

$$x = c_4 + c_5\sin t + c_6\cos t + \frac{17}{15}e^{3t}.$$

Substituting into $(D+1)x + (D-1)y = 4e^{3t}$ gives

$$(c_4 - c_1) + (c_5 - c_6 - c_3 - c_2)\sin t + (c_6 + c_5 + c_2 - c_3)\cos t = 0$$

so that $c_4 = c_1$, $c_5 = c_3$, $c_6 = -c_2$, and

$$x = c_1 - c_2\cos t + c_3\sin t + \frac{17}{15}e^{3t}.$$

12. From $(2D^2 - D - 1)x - (2D+1)y = 1$ and $(D-1)x + Dy = -1$ we obtain $(2D+1)(D-1)(D+1)x = -1$ and $(2D+1)(D+1)y = -2$. Then

$$x = c_1e^{-t/2} + c_2e^{-t} + c_3e^t + 1$$

and

$$y = c_4e^{-t/2} + c_5e^{-t} - 2.$$

Substituting into $(D-1)x + Dy = -1$ gives

$$\left(-\frac{3}{2}c_1 - \frac{1}{2}c_4\right)e^{-t/2} + (-2c_2 - c_5)e^{-t} = 0$$

so that $c_4 = -3c_1$, $c_5 = -2c_2$, and

$$y = -3c_1e^{-t/2} - 2c_2e^{-t} - 2.$$

15. Multiplying the first equation by $D+1$ and the second equation by D^2+1 and subtracting we obtain $(D^4 - D^2)x = 1$. Then

$$x = c_1 + c_2t + c_3e^t + c_4e^{-t} - \frac{1}{2}t^2.$$

Multiplying the first equation by $D+1$ and subtracting we obtain $D^2(D+1)y = 1$. Then

$$y = c_5 + c_6t + c_7e^{-t} - \frac{1}{2}t^2.$$

Substituting into $(D-1)x + (D^2+1)y = 1$ gives

$$(-c_1 + c_2 + c_5 - 1) + (-2c_4 + 2c_7)e^{-t} + (-1 - c_2 + c_6)t = 1$$

so that $c_5 = c_1 - c_2 + 2$, $c_6 = c_2 + 1$, and $c_7 = c_4$. The solution of the system is

$$x = c_1 + c_2t + c_3e^t + c_4e^{-t} - \frac{1}{2}t^2$$
$$y = (c_1 - c_2 + 2) + (c_2 + 1)t + c_4e^{-t} - \frac{1}{2}t^2.$$

45

18. From $Dx + z = e^t$, $(D-1)x + Dy + Dz = 0$, and $x + 2y + Dz = e^t$ we obtain $z = -Dx + e^t$, $Dz = -D^2x + e^t$, and the system $(-D^2 + D - 1)x + Dy = -e^t$ and $(-D^2 + 1)x + 2y = 0$. Then $y = \frac{1}{2}(D^2 - 1)x$, $Dy = \frac{1}{2}D(D^2 - 1)x$, and $(D-2)(D^2 + 1)x = -2e^t$ so that

$$x = c_1 e^{2t} + c_2 \cos t + c_3 \sin t + e^t,$$

$$y = \frac{3}{2}c_1 e^{2t} - c_2 \cos t - c_3 \sin t,$$

and

$$z = -2c_1 e^{2t} - c_3 \cos t + c_2 \sin t.$$

21. From $(D + 5)x + y = 0$ and $4x - (D + 1)y = 0$ we obtain $y = -(D + 5)x$ so that $Dy = -(D^2 + 5D)x$. Then $4x + (D^2 + 5D)x + (D + 5)x = 0$ and $(D + 3)^2 x = 0$. Thus

$$x = c_1 e^{-3t} + c_2 t e^{-3t}$$

and

$$y = -(2c_1 + c_2)e^{-3t} - 2c_2 t e^{-3t}.$$

Using $x(1) = 0$ and $y(1) = 1$ we obtain

$$c_1 e^{-3} + c_2 e^{-3} = 0$$
$$-(2c_1 + c_2)e^{-3} - 2c_2 e^{-3} = 1$$

or

$$c_1 + c_2 = 0$$
$$2c_1 + 3c_2 = -e^3.$$

Thus $c_1 = e^3$ and $c_2 = -e^3$. The solution of the initial value problem is

$$x = e^{-3t+3} - te^{-3t+3}$$
$$y = -e^{-3t+3} + 2te^{-3t+3}.$$

24. Write the system as

$$0.04i_2' + 30i_2 + 10i_3 = 50 \qquad \qquad i_2' + 750i_2 + 250i_3 = 1250$$
$$\text{or}$$
$$0.02i_3' + 10i_2 + 10i_3 = 50 \qquad \qquad i_3' + 500i_2 + 500i_3 = 2500.$$

In operator notation this becomes

$$(D + 750)i_2 + 250i_3 = 1250$$
$$500i_2 + (D + 500)i_3 = 2500.$$

Eliminating i_3 we get

$$\left[(D + 500)(D + 750) - 250(500)\right]i_2 = (D + 250)(D + 1000)i_2 = 0.$$

Thus

$$i_2(t) = c_1 e^{-250t} + c_2 e^{-1000t}.$$

Since $i_3 = \frac{1}{250}[1250 - (D + 750)i_2]$, we have

$$i_3(t) = -2c_1 e^{-250t} + c_2 e^{-1000t} + 5.$$

The initial conditions imply

$$i_2(0) = c_1 + c_2 = 0$$
$$i_3(0) = -2c_1 + c_2 + 5 = 0,$$

so $c_1 = \frac{5}{3}$ and $c_2 = -\frac{5}{3}$. Thus

$$i_2(t) = \frac{5}{3}e^{-250t} - \frac{5}{3}e^{-1000t}$$

$$i_3(t) = -\frac{10}{3}e^{-250t} - \frac{5}{3}e^{-1000t} + 5.$$

Chapter 3 Review Exercises

3. True

6. $2\pi/5$, since $\frac{1}{4}x'' + 6.25x = 0$.

9. They are linearly independent over $(-\infty, \infty)$ and linearly dependent over $(0, \infty)$.

12. **(a)** The auxiliary equation is $am(m-1) + bm + c = am^2 + (b-a)m + c = 0$. If the roots are 3 and -1, then we want $(m-3)(m+1) = m^2 - 2m - 3 = 0$. Thus, let $a = 1$, $b = -1$, and $c = -3$, so that the differential equation is $x^2y'' - xy' - 3y = 0$.

(b) In this case we want the auxiliary equation to be $m^2 + 1 = 0$, so let $a = 1$, $b = 1$, and $c = 1$. Then the differential equation is $x^2y'' + xy' + y = 0$.

15. From $m^3 + 10m^2 + 25m = 0$ we obtain $m = 0$, $m = -5$, and $m = -5$ so that

$$y = c_1 + c_2e^{-5x} + c_3xe^{-5x}.$$

18. From $2m^4 + 3m^3 + 2m^2 + 6m - 4 = 0$ we obtain $m = 1/2$, $m = -2$, and $m = \pm\sqrt{2}\,i$ so that

$$y = c_1e^{x/2} + c_2e^{-2x} + c_3\cos\sqrt{2}\,x + c_4\sin\sqrt{2}\,x.$$

21. Applying $D(D^2 + 1)$ to the differential equation we obtain

$$D(D^2 + 1)(D^3 - 5D^2 + 6D) = D^2(D^2 + 1)(D - 2)(D - 3) = 0.$$

Then

$$y = \underbrace{c_1 + c_2e^{2x} + c_3e^{3x}}_{y_c} + c_4x + c_5\cos x + c_6\sin x$$

and $y_p = Ax + B\cos x + C\sin x$. Substituting y_p into the differential equation yields

$$6A + (5B + 5C)\cos x + (-5B + 5C)\sin x = 8 + 2\sin x.$$

Equating coefficients gives $A = 4/3$, $B = -1/5$, and $C = 1/5$. The general solution is

$$y = c_1 + c_2e^{2x} + c_3e^{3x} + \frac{4}{3}x - \frac{1}{5}\cos x + \frac{1}{5}\sin x.$$

24. The auxiliary equation is $m^2 - 1 = 0$, so $y_c = c_1e^x + c_2e^{-x}$ and

$$W = \begin{vmatrix} e^x & e^{-x} \\ e^x & -e^{-x} \end{vmatrix} = -2.$$

Identifying $f(x) = 2e^x/(e^x + e^{-x})$ we obtain

$$u_1' = \frac{1}{e^x + e^{-x}} = \frac{e^x}{1 + e^{2x}}$$

$$u_2' = -\frac{e^{2x}}{e^x + e^{-x}} = -\frac{e^{3x}}{1 + e^{2x}} = -e^x + \frac{e^x}{1 + e^{2x}}.$$

Then $u_1 = \tan^{-1} e^x$, $u_2 = -e^x + \tan^{-1} e^x$, and

$$y = c_1 e^x + c_2 e^{-x} + e^x \tan^{-1} e^x - 1 + e^{-x} \tan^{-1} e^x.$$

27. The auxiliary equation is $m^2 - 5m + 6 = (m-2)(m-3) = 0$ and a particular solution is $y_p = x^4 - x^2 \ln x$ so that

$$y = c_1 x^2 + c_2 x^3 + x^4 - x^2 \ln x.$$

30. (a) If $y = \sin x$ is a solution then so is $y = \cos x$ and $m^2 + 1$ is a factor of the auxiliary equation $m^4 + 2m^3 + 11m^2 + 2m + 10 = 0$. Dividing by $m^2 + 1$ we get $m^2 + 2m + 10$, which has roots $-1 \pm 3i$. The general solution of the differential equation is

$$y = c_1 \cos x + c_2 \sin x + e^{-x}(c_3 \cos 3x + c_4 \sin 3x).$$

(b) The auxiliary equation is $m(m+1) = m^2 + m = 0$, so the associated homogeneous differential equation is $y'' + y' = 0$. Letting $y = c_1 + c_2 e^{-x} + \frac{1}{2}x^2 - x$ and computing $y'' + y'$ we get x. Thus, the differential equation is $y'' + y' = x$.

33. The auxiliary equation is $m^2 - 2m + 2 = 0$ so that $m = 1 \pm i$ and $y = e^x(c_1 \cos x + c_2 \sin x)$. Setting $y(\pi/2) = 0$ and $y(\pi) = -1$ we obtain $c_1 = e^{-\pi}$ and $c_2 = 0$. Thus, $y = e^{x-\pi} \cos x$.

36. The auxiliary equation is $m^2 + 1 = 0$, so $y_c = c_1 \cos x + c_2 \sin x$ and

$$W = \begin{vmatrix} \cos x & \sin x \\ -\sin x & \cos x \end{vmatrix} = 1.$$

Identifying $f(x) = \sec^3 x$ we obtain

$$u_1' = -\sin x \sec^3 x = -\frac{\sin x}{\cos^3 x}$$

$$u_2' = \cos x \sec^3 x = \sec^2 x.$$

Then

$$u_1 = -\frac{1}{2} \frac{1}{\cos^2 x} = -\frac{1}{2} \sec^2 x$$

$$u_2 = \tan x.$$

Thus

$$y = c_1 \cos x + c_2 \sin x - \frac{1}{2} \cos x \sec^2 x + \sin x \tan x$$

$$= c_1 \cos x + c_2 \sin x - \frac{1}{2} \sec x + \frac{1 - \cos^2 x}{\cos x}$$

$$= c_3 \cos x + c_2 \sin x + \frac{1}{2} \sec x.$$

and

$$y_p' = -c_3 \sin x + c_2 \cos x + \frac{1}{2} \sec x \tan x.$$

The initial conditions imply

$$c_3 + \frac{1}{2} = 1$$

$$c_2 = \frac{1}{2}.$$

Thus $c_3 = c_2 = 1/2$ and

$$y = \frac{1}{2} \cos x + \frac{1}{2} \sin x + \frac{1}{2} \sec x.$$

42. From $(D-2)x - y = t-2$ and $-3x + (D-4)y = -4t$ we obtain $(D-1)(D-5)x = 9-8t$. Then

$$x = c_1 e^t + c_2 e^{5t} - \frac{8}{5}t - \frac{3}{25}$$

and

$$y = (D-2)x - t + 2 = -c_1 e^t + 3c_2 e^{5t} + \frac{16}{25} + \frac{11}{25}t.$$

45. The period of a spring mass system is given by $T = 2\pi/\omega$ where $\omega^2 = k/m = kg/W$, where k is the spring constant, W is the weight of the mass attached to the spring, and g is the acceleration due to gravity. Thus, the period of oscillation is $T = (2\pi/\sqrt{kg})\sqrt{W}$. If the weight of the original mass is W, then $(2\pi/\sqrt{kg})\sqrt{W} = 3$ and $(2\pi/\sqrt{kg})\sqrt{W-8} = 2$. Dividing, we get $\sqrt{W}/\sqrt{W-8} = 3/2$ or $W = \frac{9}{4}(W-8)$. Solving for W we find that the weight of the original mass was 14.4 pounds.

48. From $x'' + \beta x' + 64x = 0$ we see that oscillatory motion results if $\beta^2 - 256 < 0$ or $0 \le |\beta| < 16$.

51. For $\lambda > 0$ the general solution is $y = c_1 \cos \sqrt{\lambda}\, x + c_2 \sin \sqrt{\lambda}\, x$. Now $y(0) = c_1$ and $y(2\pi) = c_1 \cos 2\pi\sqrt{\lambda} + c_2 \sin 2\pi\sqrt{\lambda}$, so the condition $y(0) = y(2\pi)$ implies

$$c_1 = c_1 \cos 2\pi\sqrt{\lambda} + c_2 \sin 2\pi\sqrt{\lambda}$$

which is true when $\sqrt{\lambda} = n$ or $\lambda = n^2$ for $n = 1, 2, 3, \ldots$. Since $y' = -\sqrt{\lambda}\, c_1 \sin \sqrt{\lambda}\, x + \sqrt{\lambda}\, c_2 \cos \sqrt{\lambda}\, x = -nc_1 \sin nx + nc_2 \cos nx$, we see that $y'(0) = nc_2 = y'(2\pi)$ for $n = 1, 2, 3, \ldots$. Thus, the eigenvalues are n^2 for $n = 1, 2, 3, \ldots$, with corresponding eigenfunctions $\cos nx$ and $\sin nx$. When $\lambda = 0$, the general solution is $y = c_1 x + c_2$ and the corresponding eigenfunction is $y = 1$. For $\lambda < 0$ the general solution is $y = c_1 \cosh \sqrt{-\lambda}\, x + c_2 \sinh \sqrt{-\lambda}\, x$. In this case $y(0) = c_1$ and $y(2\pi) = c_1 \cosh 2\pi\sqrt{-\lambda} + c_2 \sinh 2\pi\sqrt{-\lambda}$, so $y(0) = y(2\pi)$ can only be valid for $\lambda = 0$. Thus, there are no eigenvalues corresponding to $\lambda < 0$.

4 The Laplace Transform

3. $\mathcal{L}\{f(t)\} = \displaystyle\int_0^1 te^{-st}dt + \int_1^\infty e^{-st}dt = \left(-\frac{1}{s}te^{-st} - \frac{1}{s^2}e^{-st}\right)\Big|_0^1 - \frac{1}{s}e^{-st}\Big|_1^\infty$

$\qquad = \left(-\dfrac{1}{s}e^{-s} - \dfrac{1}{s^2}e^{-s}\right) - \left(0 - \dfrac{1}{s^2}\right) - \dfrac{1}{s}(0 - e^{-s}) = \dfrac{1}{s^2}(1 - e^{-s}), \quad s > 0$

6. $\mathcal{L}\{f(t)\} = \displaystyle\int_{\pi/2}^\infty (\cos t)e^{-st}dt = \left(-\frac{s}{s^2 + 1}e^{-st}\cos t + \frac{1}{s^2 + 1}e^{-st}\sin t\right)\Big|_{\pi/2}^\infty$

$\qquad = 0 - \left(0 + \dfrac{1}{s^2 + 1}e^{-\pi s/2}\right) = -\dfrac{1}{s^2 + 1}e^{-\pi s/2}, \quad s > 0$

9. $f(t) = \begin{cases} 1 - t, & 0 < t < 1 \\ 0, & t > 0 \end{cases}$

$\qquad \mathcal{L}\{f(t)\} = \displaystyle\int_0^1 (1 - t)e^{-st}\,dt = \left(-\frac{1}{s}(1 - t)e^{-st} + \frac{1}{s^2}e^{-st}\right)\Big|_0^1 = \frac{1}{s^2}e^{-s} + \frac{1}{s} - \frac{1}{s^2}, \quad s > 0$

12. $\mathcal{L}\{f(t)\} = \displaystyle\int_0^\infty e^{-2t-5}e^{-st}dt = e^{-5}\int_0^\infty e^{-(s+2)t}dt = -\frac{e^{-5}}{s + 2}e^{-(s+2)t}\Big|_0^\infty = \frac{e^{-5}}{s + 2}, \quad s > -2$

15. $\mathcal{L}\{f(t)\} = \displaystyle\int_0^\infty e^{-t}(\sin t)e^{-st}dt = \int_0^\infty (\sin t)e^{-(s+1)t}dt$

$\qquad = \left(\dfrac{-(s + 1)}{(s + 1)^2 + 1}e^{-(s+1)t}\sin t - \dfrac{1}{(s + 1)^2 + 1}e^{-(s+1)t}\cos t\right)\Big|_0^\infty$

$\qquad = \dfrac{1}{(s + 1)^2 + 1} = \dfrac{1}{s^2 + 2s + 2}, \quad s > -1$

18. $\mathcal{L}\{f(t)\} = \displaystyle\int_0^\infty t(\sin t)e^{-st}dt = \left[\left(-\frac{t}{s^2 + 1} - \frac{2s}{(s^2 + 1)^2}\right)(\cos t)e^{-st} - \left(\frac{st}{s^2 + 1} + \frac{s^2 - 1}{(s^2 + 1)^2}\right)(\sin t)e^{-st}\right]_0^\infty$

$\qquad = \dfrac{2s}{(s^2 + 1)^2}, \quad s > 0$

21. $\mathcal{L}\{4t - 10\} = \dfrac{4}{s^2} - \dfrac{10}{s}$

24. $\mathcal{L}\{-4t^2 + 16t + 9\} = -4\dfrac{2}{s^3} + \dfrac{16}{s^2} + \dfrac{9}{s}$

27. $\mathcal{L}\{1 + e^{4t}\} = \dfrac{1}{s} + \dfrac{1}{s - 4}$

30. $\mathcal{L}\{e^{2t} - 2 + e^{-2t}\} = \dfrac{1}{s - 2} - \dfrac{2}{s} + \dfrac{1}{s + 2}$

33. $\mathcal{L}\{\sinh kt\} = \dfrac{k}{s^2 - k^2}$

36. $\mathcal{L}\{e^{-t}\cosh t\} = \mathcal{L}\left\{e^{-t}\dfrac{e^t + e^{-t}}{2}\right\} = \mathcal{L}\left\{\dfrac{1}{2} + \dfrac{1}{2}e^{-2t}\right\} = \dfrac{1}{2s} + \dfrac{1}{2(s + 2)}$

39. (a) Using integration by parts for $\alpha > 0$,

$$\Gamma(\alpha + 1) = \int_0^\infty t^\alpha e^{-t}\,dt = -t^\alpha e^{-t}\Big|_0^\infty + \alpha\int_0^\infty t^{\alpha-1}e^{-t}\,dt = \alpha\Gamma(\alpha).$$

(b) Let $u = st$ so that $du = s\,dt$. Then

$$\mathscr{L}\{t^\alpha\} = \int_0^\infty e^{-st}t^\alpha dt = \int_0^\infty e^{-u}\left(\frac{u}{s}\right)^\alpha \frac{1}{s}\,du = \frac{1}{s^{\alpha+1}}\Gamma(\alpha+1), \quad \alpha > -1.$$

42. Let $F(t) = t^{1/3}$. Then $F(t)$ is of exponential order, but $f(t) = F'(t) = \frac{1}{3}t^{-2/3}$ is unbounded near $t = 0$ and hence is not of exponential order.

Let $f(t) = 2te^{t^2}\cos e^{t^2} = \dfrac{d}{dt}\sin e^{t^2}$. This function is not of exponential order, but we can show that its Laplace transform exists. Using integration by parts we have

$$\mathscr{L}\{2te^{t^2}\cos e^{t^2}\} = \int_0^\infty e^{-st}\left(\frac{d}{dt}\sin e^{t^2}\right)dt = \lim_{a\to\infty}\left[e^{-st}\sin e^{t^2}\Big|_0^a + s\int_0^a e^{-st}\sin e^{t^2}\,dt\right]$$

$$= s\int_0^\infty e^{-st}\sin e^{t^2}\,dt = s\,\mathscr{L}\{\sin e^{t^2}\}.$$

Since $\sin e^{t^2}$ is continuous and of exponential order, $\mathscr{L}\{\sin e^{t^2}\}$ exists, and therefore $\mathscr{L}\{2te^{t^2}\cos e^{t^2}\}$ exists.

45. By part (c) of Theorem 4.1

$$\mathscr{L}\{e^{(a+ib)t}\} = \frac{1}{s-(a+ib)} = \frac{1}{(s-a)-ib}\frac{(s-a)+ib}{(s-a)+ib} = \frac{s-a+ib}{(s-a)^2+b^2}.$$

By Euler's formula, $e^{i\theta} = \cos\theta + i\sin\theta$, so

$$\mathscr{L}\{e^{(a+ib)t}\} = \mathscr{L}\{e^{at}e^{ibt}\}\,\mathscr{L}\{e^{at}(\cos bt + i\sin bt)\}$$

$$= \mathscr{L}\{e^{at}\cos bt\} + i\,\mathscr{L}\{e^{at}\sin bt\}$$

$$= \frac{s-a}{(s-a)^2+b^2} + i\frac{b}{(s-a)^2+b^2}.$$

Equating real and imaginary parts we get

$$\mathscr{L}\{e^{at}\cos bt\} = \frac{s-a}{(s-a)^2+b^2} \quad\text{and}\quad \mathscr{L}\{e^{at}\sin bt\} = \frac{b}{(s-a)^2+b^2}.$$

Exercises 4.2

3. $\mathscr{L}^{-1}\left\{\dfrac{1}{s^2} - \dfrac{48}{s^5}\right\} = \mathscr{L}^{-1}\left\{\dfrac{1}{s^2} - \dfrac{48}{24}\cdot\dfrac{4!}{s^5}\right\} = t - 2t^4$

6. $\mathscr{L}^{-1}\left\{\dfrac{(s+2)^2}{s^3}\right\} = \mathscr{L}^{-1}\left\{\dfrac{1}{s} + 4\cdot\dfrac{1}{s^2} + 2\cdot\dfrac{2}{s^3}\right\} = 1 + 4t + 2t^2$

9. $\mathscr{L}^{-1}\left\{\dfrac{1}{4s+1}\right\} = \mathscr{L}^{-1}\left\{\dfrac{1}{4}\cdot\dfrac{1}{s+1/4}\right\} = \dfrac{1}{4}e^{-t/4}$

12. $\mathscr{L}^{-1}\left\{\dfrac{10s}{s^2+16}\right\} = 10\cos 4t$

15. $\mathscr{L}^{-1}\left\{\dfrac{2s-6}{s^2+9}\right\} = \mathscr{L}^{-1}\left\{2\cdot\dfrac{s}{s^2+9} - 2\cdot\dfrac{3}{s^2+9}\right\} = 2\cos 3t - 2\sin 3t$

18. $\mathscr{L}^{-1}\left\{\dfrac{s+1}{s^2-4s}\right\} = \mathscr{L}^{-1}\left\{-\dfrac{1}{4}\cdot\dfrac{1}{s} + \dfrac{5}{4}\cdot\dfrac{1}{s-4}\right\} = -\dfrac{1}{4} + \dfrac{5}{4}e^{4t}$

21. $\mathscr{L}^{-1}\left\{\dfrac{0.9s}{(s-0.1)(s+0.2)}\right\} = \mathscr{L}^{-1}\left\{(0.3)\cdot\dfrac{1}{s-0.1} + (0.6)\cdot\dfrac{1}{s+0.2}\right\} = 0.3e^{0.1t} + 0.6e^{-0.2t}$

24. $\mathscr{L}^{-1}\left\{\dfrac{s^2+1}{s(s-1)(s+1)(s-2)}\right\} = \mathscr{L}^{-1}\left\{\dfrac{1}{2}\cdot\dfrac{1}{s} - \dfrac{1}{s-1} - \dfrac{1}{3}\cdot\dfrac{1}{s+1} + \dfrac{5}{6}\cdot\dfrac{1}{s-2}\right\}$

$$= \dfrac{1}{2} - e^t - \dfrac{1}{3}e^{-t} + \dfrac{5}{6}e^{2t}$$

27. $\mathscr{L}^{-1}\left\{\dfrac{2s-4}{(s^2+s)(s^2+1)}\right\} = \mathscr{L}^{-1}\left\{\dfrac{2s-4}{s(s^2+1)^2}\right\} = \mathscr{L}^{-1}\left\{-\dfrac{4}{s} + \dfrac{3}{s+1} + \dfrac{s}{s^2+1} + \dfrac{3}{s^2+1}\right\}$

$$= -4 + 3e^{-t} + \cos t + 3\sin t$$

30. $\mathscr{L}^{-1}\left\{\dfrac{6s+3}{(s^2+1)(s^2+4)}\right\} = \mathscr{L}^{-1}\left\{2\cdot\dfrac{s}{s^2+1} + \dfrac{1}{s^2+1} - 2\cdot\dfrac{s}{s^2+4} - \dfrac{1}{2}\cdot\dfrac{2}{s^2+4}\right\}$

$$= 2\cos t + \sin t - 2\cos 2t - \dfrac{1}{2}\sin 2t$$

33. The Laplace transform of the differential equation is

$$s\mathscr{L}\{y\} - y(0) + 6\mathscr{L}\{y\} = \dfrac{1}{s-4}.$$

Solving for $\mathscr{L}\{y\}$ we obtain

$$\mathscr{L}\{y\} = \dfrac{1}{(s-4)(s+6)} + \dfrac{2}{s+6} = \dfrac{1}{10}\cdot\dfrac{1}{s-4} + \dfrac{19}{10}\cdot\dfrac{1}{s+6}.$$

Thus

$$y = \dfrac{1}{10}e^{4t} + \dfrac{19}{10}e^{-6t}.$$

36. The Laplace transform of the differential equation is

$$s^2\mathscr{L}\{y\} - sy(0) - y'(0) - 4\left[s\mathscr{L}\{y\} - y(0)\right] = \dfrac{6}{s-3} - \dfrac{3}{s+1}.$$

Solving for $\mathscr{L}\{y\}$ we obtain

$$\mathscr{L}\{y\} = \dfrac{6}{(s-3)(s^2-4s)} - \dfrac{3}{(s+1)(s^2-4s)} + \dfrac{s-5}{s^2-4s}$$

$$= \dfrac{5}{2}\cdot\dfrac{1}{s} - \dfrac{2}{s-3} - \dfrac{3}{5}\cdot\dfrac{1}{s+1} + \dfrac{11}{10}\cdot\dfrac{1}{s-4}.$$

Thus

$$y = \dfrac{5}{2} - 2e^{3t} - \dfrac{3}{5}e^{-t} + \dfrac{11}{10}e^{4t}.$$

39. The Laplace transform of the differential equation is

$$2\left[s^3\mathscr{L}\{y\} - s^2(0) - sy'(0) - y''(0)\right] + 3\left[s^2\mathscr{L}\{y\} - sy(0) - y'(0)\right] - 3[s\mathscr{L}\{y\} - y(0)] - 2\mathscr{L}\{y\} = \dfrac{1}{s+1}.$$

Solving for $\mathscr{L}\{y\}$ we obtain

$$\mathscr{L}\{y\} = \dfrac{2s+3}{(s+1)(s-1)(2s+1)(s+2)} = \dfrac{1}{2}\dfrac{1}{s+1} + \dfrac{5}{18}\dfrac{1}{s-1} - \dfrac{8}{9}\dfrac{1}{s+1/2} + \dfrac{1}{9}\dfrac{1}{s+2}.$$

Thus

$$y = \dfrac{1}{2}e^{-t} + \dfrac{5}{18}e^t - \dfrac{8}{9}e^{-t/2} + \dfrac{1}{9}e^{-2t}.$$

42. For $y'' - 4y' = 6e^{3t} - 3e^{-t}$ the transfer function is $W(s) = 1/(s^2-4s)$. The zero-input response is

$$y_0(t) = \mathscr{L}^{-1}\left\{\dfrac{s-5}{s^2-4s}\right\} = \mathscr{L}^{-1}\left\{\dfrac{5}{4}\cdot\dfrac{1}{s} - \dfrac{1}{4}\cdot\dfrac{1}{s-4}\right\} = \dfrac{5}{4} - \dfrac{1}{4}e^{4t},$$

and the zero-state response is

$$y_1(t) = \mathscr{L}^{-1}\left\{\frac{6}{(s-3)(s^2-4s)} - \frac{3}{(s+1)(s^2-4s)}\right\}$$

$$= \mathscr{L}^{-1}\left\{\frac{27}{20}\cdot\frac{1}{s-4} - \frac{2}{s-3} + \frac{5}{4}\cdot\frac{1}{s} - \frac{3}{5}\cdot\frac{1}{s+1}\right\}$$

$$= \frac{27}{20}e^{4t} - 2e^{3t} + \frac{5}{4} - \frac{3}{5}e^{-t}.$$

Exercises 4.3

3. $\mathscr{L}\{t^3 e^{-2t}\} = \dfrac{3!}{(s+2)^4}$

6. $\mathscr{L}\{e^{2t}(t-1)^2\} = \mathscr{L}\{t^2 e^{2t} - 2te^{2t} + e^{2t}\} = \dfrac{2}{(s-2)^3} - \dfrac{2}{(s-2)^2} + \dfrac{1}{s-2}$

9. $\mathscr{L}\{(1 - e^t + 3e^{-4t})\cos 5t\} = \mathscr{L}\{\cos 5t - e^t\cos 5t + 3e^{-4t}\cos 5t\} = \dfrac{s}{s^2+25} - \dfrac{s-1}{(s-1)^2+25} + \dfrac{3(s+4)}{(s+4)^2+25}$

12. $\mathscr{L}^{-1}\left\{\dfrac{1}{(s-1)^4}\right\} = \mathscr{L}^{-1}\left\{\dfrac{1}{6}\dfrac{3!}{(s-1)^4}\right\} = \dfrac{1}{6}t^3 e^t$

15. $\mathscr{L}^{-1}\left\{\dfrac{s}{s^2+4s+5}\right\} = \mathscr{L}^{-1}\left\{\dfrac{(s+2)}{(s+2)^2+1^2} - 2\dfrac{1}{(s+2)^2+1^2}\right\} = e^{-2t}\cos t - 2e^{-2t}\sin t$

18. $\mathscr{L}^{-1}\left\{\dfrac{5s}{(s-2)^2}\right\} = \mathscr{L}^{-1}\left\{\dfrac{5(s-2)+10}{(s-2)^2}\right\} = \mathscr{L}^{-1}\left\{\dfrac{5}{s-2} + \dfrac{10}{(s-2)^2}\right\} = 5e^{2t} + 10te^{2t}$

21. The Laplace transform of the differential equation is

$$s\,\mathscr{L}\{y\} - y(0) + 4\,\mathscr{L}\{y\} = \frac{1}{s+4}.$$

Solving for $\mathscr{L}\{y\}$ we obtain $\mathscr{L}\{y\} = \dfrac{1}{(s+4)^2} + \dfrac{2}{s+4}$. Thus

$$y = te^{-4t} + 2e^{-4t}.$$

24. The Laplace transform of the differential equation is

$$s^2\,\mathscr{L}\{y\} - sy(0) - y'(0) - 4\,[s\,\mathscr{L}\{y\} - y(0)] + 4\,\mathscr{L}\{y\} = \frac{6}{(s-2)^4}.$$

Solving for $\mathscr{L}\{y\}$ we obtain $\mathscr{L}\{y\} = \dfrac{1}{20}\dfrac{5!}{(s-2)^6}$. Thus, $y = \dfrac{1}{20}t^5 e^{2t}$.

27. The Laplace transform of the differential equation is

$$s^2\,\mathscr{L}\{y\} - sy(0) - y'(0) - 6\,[s\,\mathscr{L}\{y\} - y(0)] + 13\,\mathscr{L}\{y\} = 0.$$

Solving for $\mathscr{L}\{y\}$ we obtain

$$\mathscr{L}\{y\} = -\frac{3}{s^2-6s+13} = -\frac{3}{2}\frac{2}{(s-3)^2+2^2}.$$

Thus

$$y = -\frac{3}{2}e^{3t}\sin 2t.$$

30. The Laplace transform of the differential equation is

$$s^2 \mathcal{L}\{y\} - sy(0) - y'(0) - 2\left[s\,\mathcal{L}\{y\} - y(0)\right] + 5\,\mathcal{L}\{y\} = \frac{1}{s} + \frac{1}{s^2}.$$

Solving for $\mathcal{L}\{y\}$ we obtain

$$\mathcal{L}\{y\} = \frac{4s^2 + s + 1}{s^2(s^2 - 2s + 5)} = \frac{7}{25}\frac{1}{s} + \frac{1}{5}\frac{1}{s^2} + \frac{-7s/25 + 109/25}{s^2 - 2s + 5}$$

$$= \frac{7}{25}\frac{1}{s} + \frac{1}{5}\frac{1}{s^2} - \frac{7}{25}\frac{s-1}{(s-1)^2 + 2^2} + \frac{51}{25}\frac{2}{(s-1)^2 + 2^2}.$$

Thus

$$y = \frac{7}{25} + \frac{1}{5}t - \frac{7}{25}e^t \cos 2t + \frac{51}{25}e^t \sin 2t.$$

33. Recall from Chapter 3 that $mx'' = -kx - \beta x'$. Now $m = W/g = 4/32 = \frac{1}{8}$ slug, and $4 = 2k$ so that $k = 2$ lb/ft. Thus, the differential equation is $x'' + 7x' + 16x = 0$. The initial conditions are $x(0) = -3/2$ and $x'(0) = 0$. The Laplace transform of the differential equation is

$$s^2 \mathcal{L}\{x\} + \frac{3}{2}s + 7s\mathcal{L}\{x\} + \frac{21}{2} + 16\mathcal{L}\{x\} = 0.$$

Solving for $\mathcal{L}\{x\}$ we obtain

$$\mathcal{L}\{x\} = \frac{-3s/2 - 21/2}{s^2 + 7s + 16} = -\frac{3}{2}\frac{s + 7/2}{(s + 7/2)^2 + (\sqrt{15}/2)^2} - \frac{7\sqrt{15}}{10}\frac{\sqrt{15}/2}{(s + 7/2)^2 + (\sqrt{15}/2)^2}.$$

Thus

$$x = -\frac{3}{2}e^{-7t/2}\cos\frac{\sqrt{15}}{2}t - \frac{7\sqrt{15}}{10}e^{-7t/2}\sin\frac{\sqrt{15}}{2}t.$$

36. The differential equation is

$$R\frac{dq}{dt} + \frac{1}{C}q = E_0 e^{-kt}, \quad q(0) = 0.$$

The Laplace transform of this equation is

$$R\,\mathcal{L}\{q\} + \frac{1}{C}\,\mathcal{L}\{q\} = E_0 \frac{1}{s + k}.$$

Solving for $\mathcal{L}\{q\}$ we obtain

$$\mathcal{L}\{q\} = \frac{E_0 C}{(s + k)(RC_s + 1)} = \frac{E_0/R}{(s + k)(s + 1/RC)}.$$

When $1/RC \neq k$ we have by partial fractions

$$\mathcal{L}\{q\} = \frac{E_0}{R}\left(\frac{1/(1/RC - k)}{s + k} - \frac{1/(1/RC - k)}{s + 1/RC}\right) = \frac{E_0}{R}\frac{1}{1/RC - k}\left(\frac{1}{s + k} - \frac{1}{s + 1/RC}\right).$$

Thus

$$q(t) = \frac{E_0 C}{1 - kRC}\left(e^{-kt} - e^{-t/RC}\right).$$

When $1/RC = k$ we have

$$\mathcal{L}\{q\} = \frac{E_0}{R}\frac{1}{(s + k)^2}.$$

Thus

$$q(t) = \frac{E_0}{R}te^{-kt} = \frac{E_0}{R}te^{-t/RC}.$$

39. $\mathcal{L}\{t\,\mathcal{U}(t-2)\} = \mathcal{L}\{(t-2)\,\mathcal{U}(t-2) + 2\,\mathcal{U}(t-2)\} = \dfrac{e^{-2s}}{s^2} + \dfrac{2e^{-2s}}{s}$

42. $\mathcal{L}\left\{\sin t \,\mathcal{U}\left(t - \dfrac{\pi}{2}\right)\right\} = \mathcal{L}\left\{\cos\left(t - \dfrac{\pi}{2}\right)\mathcal{U}\left(t - \dfrac{\pi}{2}\right)\right\} = \dfrac{se^{-\pi s}}{s^2 + 1}$

45. $\mathcal{L}^{-1}\left\{\dfrac{e^{-\pi s}}{s^2 + 1}\right\} = \sin(t - \pi)\,\mathcal{U}(t - \pi)$

48. $\mathcal{L}^{-1}\left\{\dfrac{e^{-2s}}{s^2(s-1)}\right\} = \mathcal{L}^{-1}\left\{-\dfrac{e^{-2s}}{s} - \dfrac{e^{-2s}}{s^2} + \dfrac{e^{-2s}}{s-1}\right\} = -\mathcal{U}(t-2) - (t-2)\,\mathcal{U}(t-2) + e^{t-2}\,\mathcal{U}(t-2)$

51. (f)

54. (d)

57. $\mathcal{L}\{t^2\,\mathcal{U}(t-1)\} = \mathcal{L}\left\{\left[(t-1)^2 + 2t - 1\right]\mathcal{U}(t-1)\right\} = \mathcal{L}\left\{\left[(t-1)^2 + 2(t-1) - 1\right]\mathcal{U}(t-1)\right\}$

$$= \left(\dfrac{2}{s^3} + \dfrac{2}{s^2} + \dfrac{1}{s}\right)e^{-s}$$

60. $\mathcal{L}\{\sin t - \sin t\,\mathcal{U}(t-2\pi)\} = \mathcal{L}\{\sin t - \sin(t-2\pi)\,\mathcal{U}(t-2\pi)\} = \dfrac{1}{s^2 + 1} - \dfrac{e^{-2\pi s}}{s^2 + 1}$

63. The Laplace transform of the differential equation is

$$s\,\mathcal{L}\{y\} - y(0) + \mathcal{L}\{y\} = \dfrac{5}{s}e^{-s}.$$

Solving for $\mathcal{L}\{y\}$ we obtain

$$\mathcal{L}\{y\} = \dfrac{5e^{-s}}{s(s+1)} = 5e^{-s}\left[\dfrac{1}{s} - \dfrac{1}{s+1}\right].$$

Thus

$$y = 5\,\mathcal{U}(t-1) - 5e^{-(t-1)}\,\mathcal{U}(t-1).$$

66. The Laplace transform of the differential equation is

$$s^2\,\mathcal{L}\{y\} - sy(0) - y'(0) + 4\,\mathcal{L}\{y\} = \dfrac{1}{s} - \dfrac{e^{-s}}{s}.$$

Solving for $\mathcal{L}\{y\}$ we obtain

$$\mathcal{L}\{y\} = \dfrac{1-s}{s(s^2+4)} - e^{-s}\dfrac{1}{s(s^2+4)} = \dfrac{1}{4}\dfrac{1}{s} - \dfrac{1}{4}\dfrac{s}{s^2+4} - \dfrac{1}{2}\dfrac{2}{s^2+4} - e^{-s}\left[\dfrac{1}{4}\dfrac{1}{s} - \dfrac{1}{4}\dfrac{s}{s^2+4}\right].$$

Thus

$$y = \dfrac{1}{4} - \dfrac{1}{4}\cos 2t - \dfrac{1}{2}\sin 2t - \left[\dfrac{1}{4} - \dfrac{1}{4}\cos 2(t-1)\right]\mathcal{U}(t-1).$$

69. The Laplace transform of the differential equation is

$$s^2\,\mathcal{L}\{y\} - sy(0) - y'(0) + \mathcal{L}\{y\} = \dfrac{e^{-\pi s}}{s} - \dfrac{e^{-2\pi s}}{s}.$$

Solving for $\mathcal{L}\{y\}$ we obtain

$$\mathcal{L}\{y\} = e^{-\pi s}\left[\dfrac{1}{s} - \dfrac{s}{s^2+1}\right] - e^{-2\pi s}\left[\dfrac{1}{s} - \dfrac{s}{s^2+1}\right] + \dfrac{1}{s^2+1}.$$

Thus

$$y = [1 - \cos(t-\pi)]\mathcal{U}(t-\pi) - [1 - \cos(t-2\pi)]\mathcal{U}(t-2\pi) + \sin t.$$

72. Recall from Chapter 3 that $mx'' = -kx + f(t)$. Now $m = W/g = 32/32 = 1$ slug, and $32 = 2k$ so that $k = 16$ lb/ft. Thus, the differential equation is $x'' + 16x = f(t)$. The initial conditions are $x(0) = 0$, $x'(0) = 0$. Also, since

$$f(t) = \begin{cases} \sin t, & 0 \le t < 2\pi \\ 0, & t \ge 2\pi \end{cases}$$

55

and $\sin t = \sin(t - 2\pi)$ we can write

$$f(t) = \sin t - \sin(t - 2\pi)\,\mathcal{U}(t - 2\pi).$$

The Laplace transform of the differential equation is

$$s^2 \mathcal{L}\{x\} + 16\,\mathcal{L}\{x\} = \frac{1}{s^2 + 1} - \frac{1}{s^2 + 1}e^{-2\pi s}.$$

Solving for $\mathcal{L}\{x\}$ we obtain

$$\mathcal{L}\{x\} = \frac{1}{(s^2 + 16)(s^2 + 1)} - \frac{1}{(s^2 + 16)(s^2 + 1)}e^{-2\pi s}$$

$$= \frac{-1/15}{s^2 + 16} + \frac{1/15}{s^2 + 1} - \left[\frac{-1/15}{s^2 + 16} + \frac{1/15}{s^2 + 1}\right]e^{-2\pi s}.$$

Thus

$$x(t) = -\frac{1}{60}\sin 4t + \frac{1}{15}\sin t + \frac{1}{60}\sin 4(t - 2\pi)\,\mathcal{U}(t - 2\pi) - \frac{1}{15}\sin(t - 2\pi)\,\mathcal{U}(t - 2\pi)$$

$$= \begin{cases} -\frac{1}{60}\sin 4t + \frac{1}{15}\sin t, & 0 \le t < 2\pi \\ 0, & t \ge 2\pi. \end{cases}$$

75. (a) The differential equation is

$$\frac{di}{dt} + 10i = \sin t + \cos\left(t - \frac{3\pi}{2}\right)\mathcal{U}\left(t - \frac{3\pi}{2}\right), \quad i(0) = 0.$$

The Laplace transform of this equation is

$$s\,\mathcal{L}\{i\} + 10\,\mathcal{L}\{i\} = \frac{1}{s^2 + 1} + \frac{se^{-3\pi s/2}}{s^2 + 1}.$$

Solving for $\mathcal{L}\{i\}$ we obtain

$$\mathcal{L}\{i\} = \frac{1}{(s^2 + 1)(s + 10)} + \frac{s}{(s^2 + 1)(s + 10)}e^{-3\pi s/2}$$

$$= \frac{1}{101}\left(\frac{1}{s + 10} - \frac{s}{s^2 + 1} + \frac{10}{s^2 + 1}\right) + \frac{1}{101}\left(\frac{-10}{s + 10} + \frac{10s}{s^2 + 1} + \frac{1}{s^2 + 1}\right)e^{-3\pi s/2}.$$

Thus

$$i(t) = \frac{1}{101}\left(e^{-10t} - \cos t + 10\sin t\right)$$

$$+ \frac{1}{101}\left[-10e^{-10(t - 3\pi/2)} + 10\cos\left(t - \frac{3\pi}{2}\right) + \sin\left(t - \frac{3\pi}{2}\right)\right]\mathcal{U}\left(t - \frac{3\pi}{2}\right).$$

(b)

The maximum value of $i(t)$ is approximately 0.1 at $t = 1.7$, the minimum is approximately -0.1 at 4.7.

78. The differential equation is

$$EI\frac{d^4y}{dx^4} = w_0[\,\mathcal{U}(x - L/3) - \mathcal{U}(x - 2L/3)].$$

Taking the Laplace transform of both sides and using $y(0) = y'(0) = 0$ we obtain

$$s^4 \mathscr{L}\{y\} - sy''(0) - y'''(0) = \frac{w_0}{EI} \frac{1}{s} \left(e^{-Ls/3} - e^{-2Ls/3}\right).$$

Letting $y''(0) = c_1$ and $y'''(0) = c_2$ we have

$$\mathscr{L}\{y\} = \frac{c_1}{s^3} + \frac{c_2}{s^4} + \frac{w_0}{EI} \frac{1}{s^5} \left(e^{-Ls/3} - e^{-2Ls/3}\right)$$

so that

$$y(x) = \frac{1}{2}c_1 x^2 + \frac{1}{6}c_2 x^3 + \frac{1}{24} \frac{w_0}{EI} \left[\left(x - \frac{L}{3}\right)^4 \mathscr{U}\left(x - \frac{L}{3}\right) - \left(x - \frac{2L}{3}\right)^4 \mathscr{U}\left(x - \frac{2L}{3}\right)\right].$$

To find c_1 and c_2 we compute

$$y''(x) = c_1 + c_2 x + \frac{1}{2}\frac{w_0}{EI}\left[\left(x - \frac{L}{3}\right)^2 \mathscr{U}\left(x - \frac{L}{3}\right) - \left(x - \frac{2L}{3}\right)^2 \mathscr{U}\left(x - \frac{2L}{3}\right)\right]$$

and

$$y'''(x) = c_2 + \frac{w_0}{EI}\left[\left(x - \frac{L}{3}\right)\mathscr{U}\left(x - \frac{L}{3}\right) - \left(x - \frac{2L}{3}\right)\mathscr{U}\left(x - \frac{2L}{3}\right)\right].$$

Then $y''(L) = y'''(L) = 0$ yields the system

$$c_1 + c_2 L + \frac{1}{2}\frac{w_0}{EI}\left[\left(\frac{2L}{3}\right)^2 - \left(\frac{L}{3}\right)^2\right] = c_1 + c_2 L + \frac{1}{6}\frac{w_0 L^2}{EI} = 0$$

$$c_2 + \frac{w_0}{EI}\left[\frac{2L}{3} - \frac{L}{3}\right] = c_2 + \frac{1}{3}\frac{w_0 L}{EI} = 0.$$

Solving for c_1 and c_2 we obtain $c_1 = \frac{1}{6}w_0 L^2 / EI$ and $c_2 = -\frac{1}{3}w_0 L / EI$. Thus

$$y(x) = \frac{w_0}{EI}\left(\frac{1}{12}L^2 x^2 - \frac{1}{18}Lx^3 + \frac{1}{24}\left[\left(x - \frac{L}{3}\right)^4 \mathscr{U}\left(x - \frac{L}{3}\right) - \left(x - \frac{2L}{3}\right)^4 \mathscr{U}\left(x - \frac{2L}{3}\right)\right]\right).$$

81. From Theorem 4.6 we have $\mathscr{L}\{te^{kti}\} = 1/(s - ki)^2$. Then, using Euler's formula,

$$\mathscr{L}\{te^{kti}\} = \mathscr{L}\{t\cos kt + it\sin kt\} = \mathscr{L}\{t\cos kt\} + i\,\mathscr{L}\{t\sin kt\}$$

$$= \frac{1}{(s - ki)^2} = \frac{(s + ki)^2}{(s^2 + k^2)^2} = \frac{s^2 - k^2}{(s^2 + k^2)^2} + i\frac{2ks}{(s^2 + k^2)^2}.$$

Equating real and imaginary parts we have

$$\mathscr{L}\{t\cos kt\} = \frac{s^2 - k^2}{(s^2 + k^2)^2} \quad \text{and} \quad \mathscr{L}\{t\sin kt\} = \frac{2ks}{(s^2 + k^2)^2}.$$

Exercises 4.4

3. $\mathcal{L}\{t^2 \sinh t\} = \dfrac{d^2}{ds^2}\left(\dfrac{1}{s^2 - 1}\right) = \dfrac{6s^2 + 2}{(s^2 - 1)^3}$

6. $\mathcal{L}\{te^{-3t}\cos 3t\} = -\dfrac{d}{ds}\left(\dfrac{s + 3}{(s + 3)^2 + 9}\right) = \dfrac{(s + 3)^2 - 9}{[(s + 3)^2 + 9]^2}$

9. $\mathcal{L}\{e^{-t} * e^t \cos t\} = \dfrac{s - 1}{(s + 1)\left[(s - 1)^2 + 1\right]}$

12. $\mathcal{L}\left\{\displaystyle\int_0^t \cos \tau \, d\tau\right\} = \dfrac{1}{s}\mathcal{L}\{\cos t\} = \dfrac{s}{s(s^2 + 1)} = \dfrac{1}{s^2 + 1}$

15. $\mathcal{L}\left\{\displaystyle\int_0^t \tau e^{t - \tau} \, d\tau\right\} = \mathcal{L}\{t\}\,\mathcal{L}\{e^t\} = \dfrac{1}{s^2(s - 1)}$

18. $\mathcal{L}\left\{t\displaystyle\int_0^t \tau e^{-\tau} \, d\tau\right\} = -\dfrac{d}{ds}\mathcal{L}\left\{\displaystyle\int_0^t \tau e^{-\tau} \, d\tau\right\} = -\dfrac{d}{ds}\left(\dfrac{1}{s}\dfrac{1}{(s + 1)^2}\right) = \dfrac{3s + 1}{s^2(s + 1)^3}$

21. $f(t) = -\dfrac{1}{t}\mathcal{L}^{-1}\left\{\dfrac{d}{ds}[\ln(s - 3) - \ln(s + 1)]\right\} = -\dfrac{1}{t}\mathcal{L}^{-1}\left\{\dfrac{1}{s - 3} - \dfrac{1}{s + 1}\right\} = -\dfrac{1}{t}\left(e^{3t} - e^{-t}\right)$

24. $\mathcal{L}\{f(t)\} = \dfrac{1}{1 - e^{-2as}}\displaystyle\int_0^a e^{-st}\,dt = \dfrac{1}{s(1 + e^{-as})}$

27. $\mathcal{L}\{f(t)\} = \dfrac{1}{1 - e^{-\pi s}}\displaystyle\int_0^\pi e^{-st}\sin t\,dt = \dfrac{1}{s^2 + 1}\cdot\dfrac{e^{\pi s/2} + e^{-\pi s/2}}{e^{\pi s/2} - e^{-\pi s/2}} = \dfrac{1}{s^2 + 1}\coth\dfrac{\pi s}{2}$

30. The Laplace transform of the differential equation is

$$s\,\mathcal{L}\{y\} - \mathcal{L}\{y\} = \dfrac{2(s - 1)}{((s^2 - 1)^2 + 1)^2}.$$

Solving for $\mathcal{L}\{y\}$ we obtain

$$\mathcal{L}\{y\} = \dfrac{2}{((s - 1)^2 + 1)^2}.$$

Thus

$$y = e^t \sin t - te^t \cos t.$$

33. The Laplace transform of the differential equation is

$$s^2 \mathcal{L}\{y\} - sy(0) - y'(0) + 16\mathcal{L}\{y\} = \mathcal{L}\{\cos 4t - \cos 4t\,\mathcal{U}(t - \pi)\}$$

or

$$(s^2 + 16)\mathcal{L}\{y\} = 1 + \dfrac{s}{s^2 + 16} - e^{-\pi s}\mathcal{L}\{\cos 4(t + \pi)\}$$

$$= 1 + \dfrac{s}{s^2 + 16} - e^{-\pi s}\mathcal{L}\{\cos 4t\}$$

$$= 1 + \dfrac{s}{s^2 + 16} - \dfrac{s}{s^2 + 16}e^{-\pi s}.$$

Thus

$$\mathcal{L}\{y\} = \dfrac{1}{s^2 + 16} + \dfrac{s}{(s^2 + 16)^2} - \dfrac{s}{(s^2 + 16)^2}e^{-\pi s}$$

and

$$y = \dfrac{1}{4}\sin 4t + \dfrac{1}{8}t\sin 4t - \dfrac{1}{8}(t - \pi)\sin 4(t - \pi)\,\mathcal{U}(t - \pi).$$

36. (a)

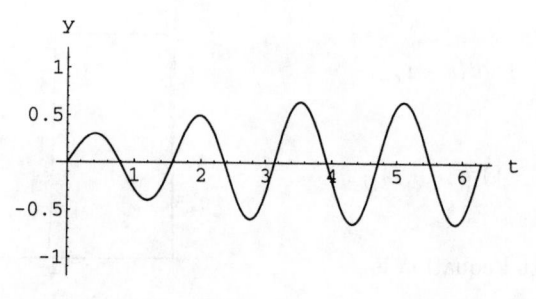

(b)

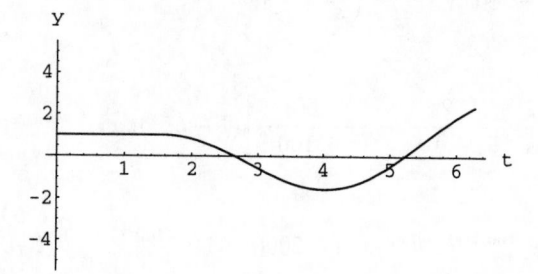

39. The Laplace transform of the given equation is

$$\mathscr{L}\{f\} = \mathscr{L}\{te^t\} + \mathscr{L}\{t\}\,\mathscr{L}\{f\}.$$

Solving for $\mathscr{L}\{f\}$ we obtain

$$\mathscr{L}\{f\} = \frac{s^2}{(s-1)^3(s+1)} = \frac{1}{8}\frac{1}{s-1} + \frac{3}{4}\frac{1}{(s-1)^2} + \frac{1}{4}\frac{2}{(s-1)^3} - \frac{1}{8}\frac{1}{s+1}.$$

Thus

$$f(t) = \frac{1}{8}e^t + \frac{3}{4}te^t + \frac{1}{4}t^2e^t - \frac{1}{8}e^{-t}$$

42. The Laplace transform of the given equation is

$$\mathscr{L}\{f\} = \mathscr{L}\{\cos t\} + \mathscr{L}\{e^{-t}\}\,\mathscr{L}\{f\}.$$

Solving for $\mathscr{L}\{f\}$ we obtain

$$\mathscr{L}\{f\} = \frac{s}{s^2+1} + \frac{1}{s^2+1}.$$

Thus

$$f(t) = \cos t + \sin t.$$

45. The Laplace transform of the given equation is

$$s\,\mathscr{L}\{y\} - y(0) = \mathscr{L}\{1\} - \mathscr{L}\{\sin t\} - \mathscr{L}\{1\}\,\mathscr{L}\{y\}.$$

Solving for $\mathscr{L}\{f\}$ we obtain

$$\mathscr{L}\{y\} = \frac{s^3 - s^2 + s}{s(s^2+1)^2} = \frac{1}{s^2+1} + \frac{1}{2}\frac{2s}{(s^2+1)^2}.$$

Thus

$$y = \sin t - \frac{1}{2}t\sin t.$$

48. The differential equation is

$$0.005\frac{di}{dt} + i + \frac{1}{0.02}\int_0^t i(\tau)d\tau = 100\big[t - (t-1)\,\mathcal{U}(t-1)\big]$$

or

$$\frac{di}{st} + 200i + 10{,}000\int_0^t i(\tau)d\tau = 20{,}000\big[t - (t-1)\,\mathcal{U}(t-1)\big],$$

where $i(0) = 0$. The Laplace transform of the differential equation is

$$s\,\mathcal{L}\{i\} + 200\,\mathcal{L}\{i\} + \frac{10{,}000}{s}\mathcal{L}\{i\} = 20{,}000\left(\frac{1}{s^2} - \frac{1}{s^2}e^{-s}\right).$$

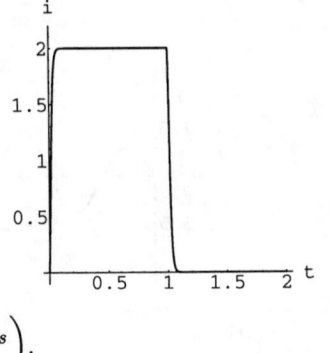

Solving for $\mathcal{L}\{i\}$ we obtain

$$\mathcal{L}\{i\} = \frac{20{,}000}{s(s+100)^2}(1-e^{-s}) = \left[\frac{2}{s} - \frac{2}{s+100} - \frac{200}{(s+100)^2}\right](1-e^{-s}).$$

Thus

$$i(t) = 2 - 2e^{-100t} - 200te^{-100t} - 2\,\mathcal{U}(t-1) + 2e^{-100(t-1)}\,\mathcal{U}(t-1) + 200(t-1)e^{-100(t-1)}\,\mathcal{U}(t-1).$$

51. The differential equation is $x'' + 2x' + 10x = 20f(t)$, where $f(t)$ is the meander function with $a = \pi$. Using the initial conditions $x(0) = x'(0) = 0$ and taking the Laplace transform we obtain

$$(s^2 + 2s + 10)\,\mathcal{L}\{x(t)\} = \frac{20}{s}(1 - e^{-\pi s})\frac{1}{1 + e^{-\pi s}}$$

$$= \frac{20}{s}(1 - e^{-\pi s})(1 - e^{-\pi s} + e^{-2\pi s} - e^{-3\pi s} + \cdots)$$

$$= \frac{20}{s}(1 - 2e^{-\pi s} + 2e^{-2\pi s} - 2e^{-3\pi s} + \cdots)$$

$$= \frac{20}{s} + \frac{40}{s}\sum_{n=1}^{\infty}(-1)^n e^{-n\pi s}.$$

Then

$$\mathcal{L}\{x(t)\} = \frac{20}{s(s^2 + 2s + 10)} + \frac{40}{s(s^2 + 2s + 10)}\sum_{n=1}^{\infty}(-1)^n e^{-n\pi s}$$

$$= \frac{2}{s} - \frac{2s+4}{s^2 + 2s + 10} + \sum_{n=1}^{\infty}(-1)^n\left[\frac{4}{s} - \frac{4s+8}{s^2 + 2s + 10}\right]e^{-n\pi s}$$

$$= \frac{2}{s} - \frac{2(s+1)+2}{(s+1)^2 + 9} + 4\sum_{n=1}^{\infty}(-1)^n\left[\frac{1}{s} - \frac{(s+1)+1}{(s+1)^2 + 9}\right]e^{-n\pi s}$$

and

$$x(t) = 2\left(1 - e^{-t}\cos 3t - \frac{1}{3}e^{-t}\sin 3t\right) + 4\sum_{n=1}^{\infty}(-1)^n\left[1 - e^{-(t-n\pi)}\cos 3(t - n\pi)\right.$$

$$\left. - \frac{1}{3}e^{-(t-n\pi)}\sin 3(t - n\pi)\right]\mathcal{U}(t - n\pi).$$

The graph of $x(t)$ on the interval $[0, 2\pi)$ is shown.

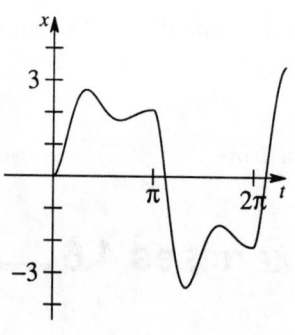

54. $f * (g + h) = \displaystyle\int_0^t f(\tau)[g(t - \tau) + h(t - \tau)]\, d\tau = \int_0^t f(\tau)g(t - \tau)\, d\tau + \int_0^t f(\tau)h(t - \tau)\, d\tau$

$\qquad = \displaystyle\int_0^t f(\tau)[g(t - \tau) + h(t - \tau)]\, d\tau = f * g + f * h$

Exercises 4.5

3. The Laplace transform of the differential equation yields

$$\mathcal{L}\{y\} = \frac{1}{s^2 + 1}\left(1 + e^{-2\pi s}\right)$$

so that

$$y = \sin t + \sin t\, \mathcal{U}(t - 2\pi).$$

6. The Laplace transform of the differential equation yields

$$\mathcal{L}\{y\} = \frac{s}{s^2 + 1} + \frac{1}{s^2 + 1}(e^{-2\pi s} + e^{-4\pi s})$$

so that

$$y = \cos t + \sin t[\, \mathcal{U}(t - 2\pi) + \mathcal{U}(t - 4\pi)].$$

9. The Laplace transform of the differential equation yields

$$\mathcal{L}\{y\} = \frac{1}{(s + 2)^2 + 1}e^{-2\pi s}$$

so that

$$y = e^{-2(t - 2\pi)}\sin t\, \mathcal{U}(t - 2\pi).$$

12. The Laplace transform of the differential equation yields

$$\mathcal{L}\{y\} = \frac{1}{(s - 1)^2(s - 6)} + \frac{e^{-2s} + e^{-4s}}{(s - 1)(s - 6)}$$

$$= -\frac{1}{25}\frac{1}{s - 1} - \frac{1}{5}\frac{1}{(s - 1)^2} + \frac{1}{25}\frac{1}{s - 6} + \left[-\frac{1}{5}\frac{1}{s - 1} + \frac{1}{5}\frac{1}{s - 6}\right](e^{-2s} + e^{-4s})$$

so that

$$y = -\frac{1}{25}e^t - \frac{1}{5}te^t + \frac{1}{25}e^{6t} + \left[-\frac{1}{5}e^{t - 2} + \frac{1}{5}e^{6(t - 2)}\right]\mathcal{U}(t - 2)$$

$$+ \left[-\frac{1}{5}e^{t - 4} + \frac{1}{5}e^{6(t - 4)}\right]\mathcal{U}(t - 4).$$

15. The Laplace transform of the differential equation yields

$$\mathcal{L}\{y\} = \frac{1}{s^2 + \omega^2}$$

so that $y(t) = \sin \omega t$. Note that $y'(0) = 1$, even though the initial condition was $y'(0) = 0$.

—————— Exercises 4.6 ——————————————————

3. Taking the Laplace transform of the system gives

$$s\,\mathcal{L}\{x\} + 1 = \mathcal{L}\{x\} - 2\mathcal{L}\{y\}$$
$$s\,\mathcal{L}\{y\} - 2 = 5\mathcal{L}\{x\} - \mathcal{L}\{y\}$$

so that

$$\mathcal{L}\{x\} = \frac{-s-5}{s^2+9} = -\frac{s}{s^2+9} - \frac{5}{3}\frac{3}{s^2+9}$$

and

$$x = -\cos 3t - \frac{5}{3}\sin 3t.$$

Then

$$y = \frac{1}{2}x - \frac{1}{2}x' = 2\cos 3t - \frac{7}{3}\sin 3t.$$

6. Taking the Laplace transform of the system gives

$$(s+1)\,\mathcal{L}\{x\} - (s-1)\mathcal{L}\{y\} = -1$$
$$s\,\mathcal{L}\{x\} + (s+2)\,\mathcal{L}\{y\} = 1$$

so that

$$\mathcal{L}\{y\} = \frac{s+1/2}{s^2+s+1} = \frac{s+1/2}{(s+1/2)^2 + (\sqrt{3}/2)^2}$$

and

$$\mathcal{L}\{x\} = \frac{-3/2}{s^2+s+1} = \frac{-3/2}{(s+1/2)^2 + (\sqrt{3}/2)^2}.$$

Then

$$y = e^{-t/2}\cos\frac{\sqrt{3}}{2}t \quad \text{and} \quad x = e^{-t/2}\sin\frac{\sqrt{3}}{2}t.$$

9. Adding the equations and then subtracting them gives

$$\frac{d^2x}{dt^2} = \frac{1}{2}t^2 + 2t$$

$$\frac{d^2y}{dt^2} = \frac{1}{2}t^2 - 2t.$$

Taking the Laplace transform of the system gives

$$\mathcal{L}\{x\} = 8\frac{1}{s} + \frac{1}{24}\frac{4!}{s^5} + \frac{1}{3}\frac{3!}{s^4}$$

and

$$\mathcal{L}\{y\} = \frac{1}{24}\frac{4!}{s^5} - \frac{1}{3}\frac{3!}{s^4}$$

so that

$$x = 8 + \frac{1}{24}t^4 + \frac{1}{3}t^3 \quad \text{and} \quad y = \frac{1}{24}t^4 - \frac{1}{3}t^3.$$

12. Taking the Laplace transform of the system gives

$$(s-4)\,\mathcal{L}\{x\} + 2\mathcal{L}\{y\} = \frac{2e^{-s}}{s}$$

$$-3\,\mathcal{L}\{x\} + (s+1)\,\mathcal{L}\{y\} = \frac{1}{2} + \frac{e^{-s}}{s}$$

so that

$$\mathcal{L}\{x\} = \frac{-1/2}{(s-1)(s-2)} + e^{-s}\frac{1}{(s-1)(s-2)}$$

$$= \left[\frac{1}{2}\frac{1}{s-1} - \frac{1}{2}\frac{1}{s-2}\right] + e^{-s}\left[-\frac{1}{s-1} + \frac{1}{s-2}\right]$$

and

$$\mathcal{L}\{y\} = \frac{e^{-s}}{s} + \frac{s/4 - 1}{(s-1)(s-2)} + e^{-s}\frac{-s/2 + 2}{(s-1)(s-2)}$$

$$= \frac{3}{4}\frac{1}{s-1} - \frac{1}{2}\frac{1}{s-2} + e^{-s}\left[\frac{1}{s} - \frac{3}{2}\frac{1}{s-1} + \frac{1}{s-2}\right].$$

Then

$$x = \frac{1}{2}e^t - \frac{1}{2}e^{2t} + \left[-e^{t-1} + e^{2(t-1)}\right]\mathcal{U}(t-1)$$

and

$$y = \frac{3}{4}e^t - \frac{1}{2}e^{2t} + \left[1 - \frac{3}{2}e^{t-1} + e^{2(t-1)}\right]\mathcal{U}(t-1).$$

15. (a) By Kirchoff's first law we have $i_1 = i_2 + i_3$. By Kirchoff's second law, on each loop we have $E(t) = Ri_1 + L_1 i_2'$ and $E(t) = Ri_1 + L_2 i_3'$ or $L_1 i_2' + Ri_2 + Ri_3 = E(t)$ and $L_2 i_3' + Ri_2 + Ri_3 = E(t)$.

(b) Taking the Laplace transform of the system

$$0.01 i_2' + 5i_2 + 5i_3 = 100$$

$$0.0125 i_3' + 5i_2 + 5i_3 = 100$$

gives

$$(s+500)\,\mathcal{L}\{i_2\} + 500\mathcal{L}\{i_3\} = \frac{10{,}000}{s}$$

$$400\mathcal{L}\{i_2\} + (s+400)\,\mathcal{L}\{i_3\} = \frac{8{,}000}{s}$$

so that

$$\mathcal{L}\{i_3\} = \frac{8{,}000}{s^2 + 900s} = \frac{80}{9}\frac{1}{s} - \frac{80}{9}\frac{1}{s+900}.$$

63

Then

$$i_3 = \frac{80}{9} - \frac{80}{9}e^{-900t}$$

and

$$i_2 = 20 - 0.0025i_3' - i_3 = \frac{100}{9} - \frac{100}{9}e^{-900t}.$$

(c) $i_1 = i_2 + i_3 = 20 - 20e^{-900t}$

18. Taking the Laplace transform of the system

$$0.5i_1' + 50i_2 = 60$$

$$0.005i_2' + i_2 - i_1 = 0$$

gives

$$s\mathcal{L}\{i_1\} + 100\mathcal{L}\{i_2\} = \frac{120}{s}$$

$$-200\mathcal{L}\{i_1\} + (s+200)\mathcal{L}\{i_2\} = 0$$

so that

$$\mathcal{L}\{i_2\} = \frac{24{,}000}{s(s^2 + 200s + 20{,}000)} = \frac{6}{5}\frac{1}{s} - \frac{6}{5}\frac{s+100}{(s+100)^2 + 100^2} - \frac{6}{5}\frac{100}{(s+100)^2 + 100^2}.$$

Then

$$i_2 = \frac{6}{5} - \frac{6}{5}e^{-100t}\cos 100t - \frac{6}{5}e^{-100t}\sin 100t$$

and

$$i_1 = 0.005i_2' + i_2 = \frac{6}{5} - \frac{6}{5}e^{-100t}\cos 100t.$$

Chapter 4 Review Exercises

3. False; consider $f(t) = t^{-1/2}$.

6. False; consider $f(t) = 1$ and $g(t) = 1$.

9. $\mathcal{L}\{\sin 2t\} = \dfrac{2}{s^2 + 4}$

12. $\mathcal{L}\{\sin 2t\,\mathcal{U}(t - \pi)\} = \mathcal{L}\{\sin 2(t - \pi)\mathcal{U}(t - \pi)\} = \dfrac{2}{s^2 + 4}e^{-\pi s}$

15. $\mathcal{L}^{-1}\left\{\dfrac{1}{(s-5)^3}\right\} = \mathcal{L}^{-1}\left\{\dfrac{1}{2}\dfrac{2}{(s-5)^3}\right\} = \dfrac{1}{2}t^2 e^{5t}$

18. $\mathcal{L}^{-1}\left\{\dfrac{1}{s^2}e^{-5s}\right\} = (t - 5)\mathcal{U}(t - 5)$

21. $\mathcal{L}\{e^{-5t}\}$ exists for $s > -5$.

24. $\mathcal{L}\left\{\displaystyle\int_0^t e^{a\tau} f(\tau)\,d\tau\right\} = \dfrac{1}{s}\mathcal{L}\{e^{at} f(t)\} = \dfrac{F(s - a)}{s}$, whereas

$$\mathscr{L}\left\{e^{at}\int_0^t f(\tau)\,d\tau\right\} = \mathscr{L}\left\{\int_0^t f(\tau)\,d\tau\right\}\bigg|_{s\to s-a} = \frac{F(s)}{s}\bigg|_{s\to s-a} = \frac{F(s-a)}{s-a}.$$

27. $f(t-t_0)\,\mathscr{U}(t-t_0)$

30. $f(t) = \sin t\,\mathscr{U}(t-\pi) - \sin t\,\mathscr{U}(t-3\pi) = -\sin(t-\pi)\,\mathscr{U}(t-\pi) + \sin(t-3\pi)\,\mathscr{U}(t-3\pi)$

$$\mathscr{L}\{f(t)\} = -\frac{1}{s^2+1}e^{-\pi s} + \frac{1}{s^2+1}e^{-3\pi s}$$

$$\mathscr{L}\{e^t f(t)\} = -\frac{1}{(s-1)^2+1}e^{-\pi(s-1)} + \frac{1}{(s-1)^2+1}e^{-3\pi(s-1)}$$

33. Taking the Laplace transform of the differential equation we obtain

$$\mathscr{L}\{y\} = \frac{5}{(s-1)^2} + \frac{1}{2}\frac{2}{(s-1)^3}$$

so that

$$y = 5te^t + \frac{1}{2}t^2 e^t.$$

36. Taking the Laplace transform of the differential equation we obtain

$$\mathscr{L}\{y\} = \frac{s^3+2}{s^3(s-5)} - \frac{2+2s+s^2}{s^3(s-5)}e^{-s}$$

$$= -\frac{2}{125}\frac{1}{s} - \frac{2}{25}\frac{1}{s^2} - \frac{1}{5}\frac{2}{s^3} + \frac{127}{125}\frac{1}{s-5} - \left[-\frac{37}{125}\frac{1}{s} - \frac{12}{25}\frac{1}{s^2} - \frac{1}{5}\frac{2}{s^3} + \frac{37}{125}\frac{1}{s-5}\right]e^{-s}$$

so that

$$y = -\frac{2}{125} - \frac{2}{25}t - \frac{1}{5}t^2 + \frac{127}{125}e^{5t} - \left[-\frac{37}{125} - \frac{12}{25}(t-1) - \frac{1}{5}(t-1)^2 + \frac{37}{125}e^{5(t-1)}\right]\mathscr{U}(t-1).$$

39. Taking the Laplace transform of the system gives

$$s\mathscr{L}\{x\} + \mathscr{L}\{y\} = \frac{1}{s^2} + 1$$

$$4\mathscr{L}\{x\} + s\mathscr{L}\{y\} = 2$$

so that

$$\mathscr{L}\{x\} = \frac{s^2-2s+1}{s(s-2)(s+2)} = -\frac{1}{4}\frac{1}{s} + \frac{1}{8}\frac{1}{s-2} + \frac{9}{8}\frac{1}{s+2}.$$

Then

$$x = -\frac{1}{4} + \frac{1}{8}e^{2t} + \frac{9}{8}e^{-2t} \quad\text{and}\quad y = -x' + t = \frac{9}{4}e^{-2t} - \frac{1}{4}e^{2t} + t.$$

42. The differential equation is

$$\frac{1}{2}\frac{d^2q}{dt^2} + 10\frac{dq}{dt} + 100q = 10 - 10\,\mathscr{U}(t-5).$$

Taking the Laplace transform we obtain

$$\mathscr{L}\{q\} = \frac{20}{2(s^2+20s+200)}\left(1 - e^{-5s}\right)$$

$$= \left[\frac{1}{10}\frac{1}{s} - \frac{1}{10}\frac{s+10}{(s+10)^2+10^2} - \frac{1}{10}\frac{10}{(s+10)^2+10^2}\right]\left(1 - e^{-5s}\right)$$

so that

$$q(t) = \frac{1}{10} - \frac{1}{10}e^{-10t}\cos 10t - \frac{1}{10}e^{-10t}\sin 10t$$

$$- \left[\frac{1}{10} - \frac{1}{10}e^{-10(t-5)}\cos 10(t-5) - \frac{1}{10}e^{-10(t-5)}\sin 10(t-5)\right]\mathscr{U}(t-5).$$

5 Series Solutions of Linear Equations

Exercises 5.1
---

3. $\lim_{k\to\infty}\left|\dfrac{a_{k+1}}{a_k}\right| = \lim_{k\to\infty}\left|\dfrac{(x-5)^{k+1}/10^{k+1}}{(x-5)^k/10^k}\right| = \lim_{k\to\infty}\dfrac{1}{10}|x-5| = \dfrac{1}{10}|x-5|$

The series is absolutely convergent for $\frac{1}{10}|x-5| < 1$, $|x-5| < 10$, or on $(-5,15)$. At $x = -5$, the series

$\displaystyle\sum_{k=1}^{\infty}\dfrac{(-1)^k(-10)^k}{10^k} = \sum_{k=1}^{\infty}1$ diverges by the k-th term test. At $x = 15$, the series $\displaystyle\sum_{k=1}^{\infty}\dfrac{(-1)^k10^k}{10^k} = \sum_{k=1}^{\infty}(-1)^k$

diverges by the k-th term test. Thus, the series converges on $(-5,15)$.

6. $e^{-x}\cos x = \left(1 - x + \dfrac{x^2}{2} - \dfrac{x^3}{6} + \dfrac{x^4}{24} - \cdots\right)\left(1 - \dfrac{x^2}{2} + \dfrac{x^4}{24} - \cdots\right) = 1 - x + \dfrac{x^3}{3} - \dfrac{x^4}{6} + \cdots$

9. $\displaystyle\sum_{n=1}^{\infty}2nc_nx^{n-1} + \sum_{n=0}^{\infty}6c_nx^{n+1} = 2\cdot 1\cdot c_1x^0 + \underbrace{\sum_{n=2}^{\infty}2nc_nx^{n-1}}_{k=n-1} + \underbrace{\sum_{n=0}^{\infty}6c_nx^{n+1}}_{k=n+1}$

$$= 2c_1 + \sum_{k=1}^{\infty}2(k+1)c_{k+1}x^k + \sum_{k=1}^{\infty}6c_{k-1}x^k$$

$$= 2c_1 + \sum_{k=1}^{\infty}[2(k+1)c_{k+1} + 6c_{k-1}]x^k$$

12. $y' = \displaystyle\sum_{n=1}^{\infty}\dfrac{(-1)^n2n}{2^{2n}(n!)^2}x^{2n-1}, \qquad y'' = \sum_{n=1}^{\infty}\dfrac{(-1)^n2n(2n-1)}{2^{2n}(n!)^2}x^{2n-2}$

$xy'' + y' + xy = \underbrace{\displaystyle\sum_{n=1}^{\infty}\dfrac{(-1)^n2n(2n-1)}{2^{2n}(n!)^2}x^{2n-1}}_{k=n} + \underbrace{\sum_{n=1}^{\infty}\dfrac{(-1)^n2n}{2^{2n}(n!)^2}x^{2n-1}}_{k=n} + \underbrace{\sum_{n=0}^{\infty}\dfrac{(-1)^n}{2^{2n}(n!)^2}x^{2n+1}}_{k=n+1}$

$$= \sum_{k=1}^{\infty}\left[\dfrac{(-1)^k2k(2k-1)}{2^{2k}(k!)^2} + \dfrac{(-1)^k2k}{2^{2k}(k!)^2} + \dfrac{(-1)^{k-1}}{2^{2k-2}[(k-1)!]^2}\right]x^{2k-1}$$

$$= \sum_{k=1}^{\infty}\left[\dfrac{(-1)^k(2k)^2}{2^{2k}(k!)^2} - \dfrac{(-1)^k}{2^{2k-2}[(k-1)!]^2}\right]x^{2k-1} = \sum_{k=1}^{\infty}(-1)^k\left[\dfrac{(2k)^2 - 2^2k^2}{2^{2k}(k!)^2}\right]x^{2k-1} = 0$$

15. Substituting $y = \sum_{n=0}^{\infty}c_nx^n$ into the differential equation we have

$$y'' - 2xy' + y = \underbrace{\sum_{n=2}^{\infty}n(n-1)c_nx^{n-2}}_{k=n-2} - 2\underbrace{\sum_{n=1}^{\infty}nc_nx^n}_{k=n} + \underbrace{\sum_{n=0}^{\infty}c_nx^n}_{k=n}$$

$$= \sum_{k=0}^{\infty}(k+2)(k+1)c_{k+2}x^k - 2\sum_{k=1}^{\infty}kc_kx^k + \sum_{k=0}^{\infty}c_kx^k$$

$$= 2c_2 + c_0 + \sum_{k=1}^{\infty}[(k+2)(k+1)c_{k+2} - (2k-1)c_k]x^k = 0.$$

Thus

$$2c_2 + c_0 = 0$$

$$(k+2)(k+1)c_{k+2} - (2k-1)c_k = 0$$

and

$$c_2 = -\frac{1}{2}c_0$$

$$c_{k+2} = \frac{2k-1}{(k+2)(k+1)}\,c_k, \quad k = 1, 2, 3, \ldots.$$

Choosing $c_0 = 1$ and $c_1 = 0$ we find

$$c_2 = -\frac{1}{2}$$

$$c_3 = c_5 = c_7 = \cdots = 0$$

$$c_4 = -\frac{1}{8}$$

$$c_6 = -\frac{7}{336}$$

and so on. For $c_0 = 0$ and $c_1 = 1$ we obtain

$$c_2 = c_4 = c_6 = \cdots = 0$$

$$c_3 = \frac{1}{6}$$

$$c_5 = \frac{1}{24}$$

$$c_7 = \frac{1}{112}$$

and so on. Thus, two solutions are

$$y_1 = 1 - \frac{1}{2}x^2 - \frac{1}{8}x^4 - \frac{7}{336}x^6 - \cdots \qquad \text{and} \qquad y_2 = x + \frac{1}{6}x^3 + \frac{1}{24}x^5 + \frac{1}{112}x^7 + \cdots.$$

18. Substituting $y = \sum_{n=0}^{\infty} c_n x^n$ into the differential equation we have

$$y'' + 2xy' + 2y = \underbrace{\sum_{n=2}^{\infty} n(n-1)c_n x^{n-2}}_{k=n-2} + 2\underbrace{\sum_{n=1}^{\infty} nc_n x^n}_{k=n} + 2\underbrace{\sum_{n=0}^{\infty} c_n x^n}_{k=n}$$

$$= \sum_{k=0}^{\infty}(k+2)(k+1)c_{k+2}x^k + 2\sum_{k=1}^{\infty} kc_k x^k + 2\sum_{k=0}^{\infty} c_k x^k$$

$$= 2c_2 + 2c_0 + \sum_{k=1}^{\infty}[(k+2)(k+1)c_{k+2} + 2(k+1)c_k]x^k = 0.$$

Thus

$$2c_2 + 2c_0 = 0$$

$$(k+2)(k+1)c_{k+2} + 2(k+1)c_k = 0$$

and

$$c_2 = -c_0$$

$$c_{k+2} = -\frac{2}{k+2}\,c_k, \quad k = 1, 2, 3, \ldots.$$

67

Choosing $c_0 = 1$ and $c_1 = 0$ we find

$$c_2 = -1$$

$$c_3 = c_5 = c_7 = \cdots = 0$$

$$c_4 = \frac{1}{2}$$

$$c_6 = -\frac{1}{6}$$

and so on. For $c_0 = 0$ and $c_1 = 1$ we obtain

$$c_2 = c_4 = c_6 = \cdots = 0$$

$$c_3 = -\frac{2}{3}$$

$$c_5 = \frac{4}{15}$$

$$c_7 = -\frac{8}{105}$$

and so on. Thus, two solutions are

$$y_1 = 1 - x^2 + \frac{1}{2}x^4 - \frac{1}{6}x^6 + \cdots \quad \text{and} \quad y_2 = x - \frac{2}{3}x^3 + \frac{4}{15}x^5 - \frac{8}{105}x^7 + \cdots .$$

21. Substituting $y = \sum_{n=0}^{\infty} c_n x^n$ into the differential equation we have

$$y'' - (x+1)y' - y = \underbrace{\sum_{n=2}^{\infty} n(n-1)c_n x^{n-2}}_{k=n-2} - \underbrace{\sum_{n=1}^{\infty} nc_n x^n}_{k=n} - \underbrace{\sum_{n=1}^{\infty} nc_n x^{n-1}}_{k=n-1} - \underbrace{\sum_{n=0}^{\infty} c_n x^n}_{k=n}$$

$$= \sum_{k=0}^{\infty}(k+2)(k+1)c_{k+2}x^k - \sum_{k=1}^{\infty} kc_k x^k - \sum_{k=0}^{\infty}(k+1)c_{k+1}x^k - \sum_{k=0}^{\infty} c_k x^k$$

$$= 2c_2 - c_1 - c_0 + \sum_{k=1}^{\infty}[(k+2)(k+1)c_{k+2} - (k+1)c_{k+1} - (k+1)c_k]x^k = 0.$$

Thus

$$2c_2 - c_1 - c_0 = 0$$

$$(k+2)(k+1)c_{k+2} - (k-1)(c_{k+1} + c_k) = 0$$

and

$$c_2 = \frac{c_1 + c_0}{2}$$

$$c_{k+2} = \frac{c_{k+1} + c_k}{k+2}c_k, \quad k = 2, 3, 4, \ldots .$$

Choosing $c_0 = 1$ and $c_1 = 0$ we find

$$c_2 = \frac{1}{2}, \qquad c_3 = \frac{1}{6}, \qquad c_4 = \frac{1}{6}$$

and so on. For $c_0 = 0$ and $c_1 = 1$ we obtain

$$c_2 = \frac{1}{2}, \qquad c_3 = \frac{1}{2}, \qquad c_4 = \frac{1}{4}$$

and so on. Thus, two solutions are

$$y_1 = 1 + \frac{1}{2}x^2 + \frac{1}{6}x^3 + \frac{1}{6}x^4 + \cdots \qquad \text{and} \qquad y_2 = x + \frac{1}{2}x^2 + \frac{1}{2}x^3 + \frac{1}{4}x^4 + \cdots .$$

24. Substituting $y = \sum_{n=0}^{\infty} c_n x^n$ into the differential equation we have

$$\left(x^2 - 1\right) y'' + xy' - y = \underbrace{\sum_{n=2}^{\infty} n(n-1)c_n x^n}_{k=n} - \underbrace{\sum_{n=2}^{\infty} n(n-1)c_n x^{n-2}}_{k=n-2} + \underbrace{\sum_{n=1}^{\infty} nc_n x^n}_{k=n} - \underbrace{\sum_{n=0}^{\infty} c_n x^n}_{k=n}$$

$$= \sum_{k=2}^{\infty} k(k-1)c_k x^k - \sum_{k=0}^{\infty} (k+2)(k+1)c_{k+2} x^k + \sum_{k=1}^{\infty} kc_k x^k - \sum_{k=0}^{\infty} c_k x^k$$

$$= (-2c_2 - c_0) - 6c_3 x + \sum_{k=2}^{\infty} \left[-(k+2)(k+1)c_{k+2} + \left(k^2 - 1\right) c_k\right] x^k = 0.$$

Thus

$$-2c_2 - c_0 = 0$$

$$-6c_3 = 0$$

$$-(k+2)(k+1)c_{k+2} + (k-1)(k+1)c_k = 0$$

and

$$c_2 = -\frac{1}{2}c_0$$

$$c_3 = 0$$

$$c_{k+2} = \frac{k-1}{k+2} c_k, \quad k = 2, 3, 4, \dots.$$

Choosing $c_0 = 1$ and $c_1 = 0$ we find

$$c_2 = -\frac{1}{2}$$

$$c_3 = c_5 = c_7 = \dots = 0$$

$$c_4 = -\frac{1}{8}$$

and so on. For $c_0 = 0$ and $c_1 = 1$ we obtain

$$c_2 = c_4 = c_6 = \dots = 0$$

$$c_3 = c_5 = c_7 = \dots = 0.$$

Thus, two solutions are

$$y_1 = 1 - \frac{1}{2}x^2 - \frac{1}{8}x^4 - \dots \quad \text{and} \quad y_2 = x.$$

27. Substituting $y = \sum_{n=0}^{\infty} c_n x^n$ into the differential equation we have

$$y'' - 2xy' + 8y = \underbrace{\sum_{n=2}^{\infty} n(n-1)c_n x^{n-2}}_{k=n-2} - 2\underbrace{\sum_{n=1}^{\infty} nc_n x^n}_{k=n} + 8\underbrace{\sum_{n=0}^{\infty} c_n x^n}_{k=n}$$

$$= \sum_{k=0}^{\infty} (k+2)(k+1)c_{k+2} x^k - 2\sum_{k=1}^{\infty} kc_k x^k + 8\sum_{k=0}^{\infty} c_k x^k$$

$$= 2c_2 + 8c_0 + \sum_{k=1}^{\infty} [(k+2)(k+1)c_{k+2} + (8 - 2k)c_k]x^k = 0.$$

Thus

$$2c_2 + 8c_0 = 0$$

$$(k+2)(k+1)c_{k+2} + (8-2k)c_k = 0$$

and

$$c_2 = -4c_0$$

$$c_{k+2} = \frac{2k-8}{(k+2)(k+1)}c_k, \quad k = 1,2,3,\ldots .$$

Choosing $c_0 = 1$ and $c_1 = 0$ we find

$$c_2 = -4$$

$$c_3 = c_5 = c_7 = \cdots = 0$$

$$c_4 = \frac{4}{3}$$

$$c_6 = c_8 = c_{10} = \cdots = 0.$$

For $c_0 = 0$ and $c_1 = 1$ we obtain

$$c_2 = c_4 = c_6 = \cdots = 0$$

$$c_3 = -1$$

$$c_5 = \frac{1}{10}$$

and so on. Thus,

$$y = C_1\left(1 - 4x^2 + \frac{4}{3}x^4\right) + C_2\left(x - x^3 + \frac{1}{10}x^5 + \cdots\right)$$

and

$$y' = C_1\left(-8x + \frac{16}{3}x^3\right) + C_2\left(1 - 3x^2 + \frac{1}{2}x^4 + \cdots\right).$$

The initial conditions imply $C_1 = 3$ and $C_2 = 0$, so

$$y = 3\left(1 - 4x^2 + \frac{4}{3}x^4\right) = 3 - 12x^2 + 4x^4.$$

30. Substituting $y = \sum_{n=0}^{\infty} c_n x^n$ into the differential equation we have

$$y'' + e^x y' - y = \sum_{n=2}^{\infty} n(n-1)c_n x^{n-2}$$

$$+ \left(1 + x + \frac{1}{2}x^2 + \frac{1}{6}x^3 + \cdots\right)\left(c_1 + 2c_2 x + 3c_3 x^2 + 4c_4 x^3 + \cdots\right) - \sum_{n=0}^{\infty} c_n x^n$$

$$= \left[2c_2 + 6c_3 x + 12c_4 x^2 + 20c_5 x^3 + \cdots\right]$$

$$+ \left[c_1 + (2c_2 + c_1)x + \left(3c_3 + 2c_2 + \frac{1}{2}c_1\right)x^2 + \cdots\right] - \left[c_0 + c_1 x + c_2 x^2 + \cdots\right]$$

$$= (2c_2 + c_1 - c_0) + (6c_3 + 2c_2)x + \left(12c_4 + 3c_3 + c_2 + \frac{1}{2}c_1\right)x^2 + \cdots = 0.$$

Thus

$$2c_2 + c_1 - c_0 = 0$$

$$6c_3 + 2c_2 = 0$$

$$12c_4 + 3c_3 + c_2 + \frac{1}{2}c_1 = 0$$

and

$$c_2 = \frac{1}{2}c_0 - \frac{1}{2}c_1$$

$$c_3 = -\frac{1}{3}c_2$$

$$c_4 = -\frac{1}{4}c_3 + \frac{1}{12}c_2 - \frac{1}{24}c_1.$$

Choosing $c_0 = 1$ and $c_1 = 0$ we find

$$c_2 = \frac{1}{2}, \qquad c_3 = -\frac{1}{6}, \qquad c_4 = 0$$

and so on. For $c_0 = 0$ and $c_1 = 1$ we obtain

$$c_2 = -\frac{1}{2}, \qquad c_3 = \frac{1}{6}, \qquad c_4 = -\frac{1}{24}$$

and so on. Thus, two solutions are

$$y_1 = 1 + \frac{1}{2}x^2 - \frac{1}{6}x^3 + \cdots \qquad \text{and} \qquad y_2 = x - \frac{1}{2}x^2 + \frac{1}{6}x^3 - \frac{1}{24}x^4 + \cdots.$$

36. If $x > 0$ and $y > 0$, then $y'' = -xy < 0$ and the graph of a solution curve is concave down. Thus, whatever portion of a solution curve lies in the first quadrant is concave down. When $x > 0$ and $y < 0$, $y'' = -xy > 0$, so whatever portion of a solution curve lies in the fourth quadrant is concave up.

Exercises 5.2

3. Irregular singular point: $x = 3$; regular singular point: $x = -3$

6. Irregular singular point: $x = 5$; regular singular point: $x = 0$

9. Irregular singular point: $x = 0$; regular singular points: $x = 2, \pm 5$

12. Writing the differential equation in the form

$$y'' + \frac{x+3}{x}y' + 7xy = 0$$

we see that $x_0 = 0$ is a regular singular point. Multiplying by x^2, the differential equation can be put in the form

$$x^2 y'' + x(x+3)y' + 7x^3 y = 0.$$

We identify $p(x) = x + 3$ and $q(x) = 7x^3$.

15. Substituting $y = \sum_{n=0}^{\infty} c_n x^{n+r}$ into the differential equation and collecting terms, we obtain

$$2xy'' - y' + 2y = \left(2r^2 - 3r\right)c_0 x^{r-1} + \sum_{k=1}^{\infty}[2(k+r-1)(k+r)c_k - (k+r)c_k + 2c_{k-1}]x^{k+r-1} = 0,$$

which implies

$$2r^2 - 3r = r(2r - 3) = 0$$

and

$$(k+r)(2k + 2r - 3)c_k + 2c_{k-1} = 0.$$

The indicial roots are $r = 0$ and $r = 3/2$. For $r = 0$ the recurrence relation is

$$c_k = -\frac{2c_{k-1}}{k(2k - 3)}, \qquad k = 1, 2, 3, \ldots,$$

71

and

$$c_1 = 2c_0, \qquad c_2 = -2c_0, \qquad c_3 = \frac{4}{9}c_0.$$

For $r = 3/2$ the recurrence relation is

$$c_k = -\frac{2c_{k-1}}{(2k+3)k}, \qquad k = 1, 2, 3, \ldots,$$

and

$$c_1 = -\frac{2}{5}c_0, \qquad c_2 = \frac{2}{35}c_0, \qquad c_3 = -\frac{4}{945}c_0.$$

The general solution on $(0, \infty)$ is

$$y = C_1 \left(1 + 2x - 2x^2 + \frac{4}{9}x^3 + \cdots \right) + C_2 x^{3/2} \left(1 - \frac{2}{5}x + \frac{2}{35}x^2 - \frac{4}{945}x^3 + \cdots \right).$$

18. Substituting $y = \sum_{n=0}^{\infty} c_n x^{n+r}$ into the differential equation and collecting terms, we obtain

$$2x^2 y'' - xy' + (x^2 + 1)y = (2r^2 - 3r + 1)c_0 x^r + (2r^2 + r)c_1 x^{r+1}$$

$$+ \sum_{k=2}^{\infty} [2(k+r)(k+r-1)c_k - (k+r)c_k + c_k + c_{k-2}]x^{k+r}$$

$$= 0,$$

which implies

$$2r^2 - 3r + 1 = (2r-1)(r-1) = 0,$$

$$(2r^2 + r)c_1 = 0,$$

and

$$[(k+r)(2k+2r-3)+1]c_k + c_{k-2} = 0.$$

The indicial roots are $r = 1/2$ and $r = 1$, so $c_1 = 0$. For $r = 1/2$ the recurrence relation is

$$c_k = -\frac{c_{k-2}}{k(2k-1)}, \qquad k = 2, 3, 4, \ldots,$$

and

$$c_2 = -\frac{1}{6}c_0, \qquad c_3 = 0, \qquad c_4 = \frac{1}{168}c_0.$$

For $r = 1$ the recurrence relation is

$$c_k = -\frac{c_{k-2}}{k(2k+1)}, \qquad k = 2, 3, 4, \ldots,$$

and

$$c_2 = -\frac{1}{10}c_0, \qquad c_3 = 0, \qquad c_4 = \frac{1}{360}c_0.$$

The general solution on $(0, \infty)$ is

$$y = C_1 x^{1/2} \left(1 - \frac{1}{6}x^2 + \frac{1}{168}x^4 + \cdots \right) + C_2 x \left(1 - \frac{1}{10}x^2 + \frac{1}{360}x^4 + \cdots \right).$$

21. Substituting $y = \sum_{n=0}^{\infty} c_n x^{n+r}$ into the differential equation and collecting terms, we obtain

$$2xy'' - (3+2x)y' + y = (2r^2 - 5r)c_0 x^{r-1} + \sum_{k=1}^{\infty} [2(k+r)(k+r-1)c_k$$

$$- 3(k+r)c_k - 2(k+r-1)c_{k-1} + c_{k-1}]x^{k+r-1}$$

$$= 0,$$

which implies

$$2r^2 - 5r = r(2r - 5) = 0$$

and

$$(k + r)(2k + 2r - 5)c_k - (2k + 2r - 3)c_{k-1} = 0.$$

The indicial roots are $r = 0$ and $r = 5/2$. For $r = 0$ the recurrence relation is

$$c_k = \frac{(2k - 3)c_{k-1}}{k(2k - 5)}, \quad k = 1, 2, 3, \ldots,$$

and

$$c_1 = \frac{1}{3}c_0, \quad c_2 = -\frac{1}{6}c_0, \quad c_3 = -\frac{1}{6}c_0.$$

For $r = 5/2$ the recurrence relation is

$$c_k = \frac{2(k + 1)c_{k-1}}{k(2k + 5)}, \quad k = 1, 2, 3, \ldots,$$

and

$$c_1 = \frac{4}{7}c_0, \quad c_2 = \frac{4}{21}c_0, \quad c_3 = \frac{32}{693}c_0.$$

The general solution on $(0, \infty)$ is

$$y = C_1 \left(1 + \frac{1}{3}x - \frac{1}{6}x^2 - \frac{1}{6}x^3 + \cdots \right) + C_2 x^{5/2} \left(1 + \frac{4}{7}x + \frac{4}{21}x^2 + \frac{32}{693}x^3 + \cdots \right).$$

24. Substituting $y = \sum_{n=0}^{\infty} c_n x^{n+r}$ into the differential equation and collecting terms, we obtain

$$2x^2 y'' + 3xy' + (2x - 1)y = \left(2r^2 + r - 1 \right) c_0 x^r$$

$$+ \sum_{k=1}^{\infty} [2(k + r)(k + r - 1)c_k + 3(k + r)c_k - c_k + 2c_{k-1}]x^{k+r}$$

$$= 0,$$

which implies

$$2r^2 + r - 1 = (2r - 1)(r + 1) = 0$$

and

$$[(k + r)(2k + 2r + 1) - 1]c_k + 2c_{k-1} = 0.$$

The indicial roots are $r = -1$ and $r = 1/2$. For $r = -1$ the recurrence relation is

$$c_k = -\frac{2c_{k-1}}{k(2k - 3)}, \quad k = 1, 2, 3, \ldots,$$

and

$$c_1 = 2c_0, \quad c_2 = -2c_0, \quad c_3 = \frac{4}{9}c_0.$$

For $r = 1/2$ the recurrence relation is

$$c_k = -\frac{2c_{k-1}}{k(2k + 3)}, \quad k = 1, 2, 3, \ldots,$$

and

$$c_1 = -\frac{2}{5}c_0, \quad c_2 = \frac{2}{35}c_0, \quad c_3 = -\frac{4}{945}c_0.$$

The general solution on $(0, \infty)$ is

$$y = C_1 x^{-1} \left(1 + 2x - 2x^2 + \frac{4}{9}x^3 + \cdots \right) + C_2 x^{1/2} \left(1 - \frac{2}{5}x + \frac{2}{35}x^2 - \frac{4}{945}x^3 + \cdots \right).$$

27. Substituting $y = \sum_{n=0}^{\infty} c_n x^{n+r}$ into the differential equation and collecting terms, we obtain

$$xy'' - xy' + y = (r^2 - r)\, c_0 x^{r-1} + \sum_{k=0}^{\infty} [(k+r+1)(k+r)c_{k+1} - (k+r)c_k + c_k]x^{k+r} = 0$$

which implies

$$r^2 - r = r(r-1) = 0$$

and

$$(k+r+1)(k+r)c_{k+1} - (k+r-1)c_k = 0.$$

The indicial roots are $r_1 = 1$ and $r_2 = 0$. For $r_1 = 1$ the recurrence relation is

$$c_{k+1} = \frac{kc_k}{(k+2)(k+1)}, \quad k = 0,1,2,\dots,$$

and one solution is $y_1 = c_0 x$. A second solution is

$$y_2 = x \int \frac{e^{-\int -dx}}{x^2}\, dx = x \int \frac{e^x}{x^2}\, dx = x \int \frac{1}{x^2}\left(1 + x + \frac{1}{2}x^2 + \frac{1}{3!}x^3 + \cdots\right) dx$$

$$= x \int \left(\frac{1}{x^2} + \frac{1}{x} + \frac{1}{2} + \frac{1}{3!}x + \frac{1}{4!}x^2 + \cdots\right) dx = x\left[-\frac{1}{x} + \ln x + \frac{1}{2}x + \frac{1}{12}x^2 + \frac{1}{72}x^3 + \cdots\right]$$

$$= x \ln x - 1 + \frac{1}{2}x^2 + \frac{1}{12}x^3 + \frac{1}{72}x^4 + \cdots.$$

The general solution on $(0, \infty)$ is

$$y = C_1 x + C_2 y_2(x).$$

30. Substituting $y = \sum_{n=0}^{\infty} c_n x^{n+r}$ into the differential equation and collecting terms, we obtain

$$xy'' + y' + y = r^2 c_0 x^{r-1} + \sum_{k=1}^{\infty} [(k+r)(k+r-1)c_k + (k+r)c_k + c_{k-1}]x^{k+r-1} = 0$$

which implies $r^2 = 0$ and

$$(k+r)^2 c_k + c_{k-1} = 0.$$

The indicial roots are $r_1 = r_2 = 0$ and the recurrence relation is

$$c_k = -\frac{c_{k-1}}{k^2}, \quad k = 1,2,3,\dots.$$

One solution is

$$y_1 = c_0\left(1 - x + \frac{1}{2^2}x^2 - \frac{1}{(3!)^2}x^3 + \frac{1}{(4!)^2}x^4 - \cdots\right) = c_0 \sum_{n=0}^{\infty} \frac{(-1)^n}{(n!)^2}x^n.$$

A second solution is

$$y_2 = y_1 \int \frac{e^{-\int (1/x)dx}}{y_1^2} \, dx = y_1 \int \frac{dx}{x \left(1 - x + \frac{1}{4}x^2 - \frac{1}{36}x^3 + \cdots\right)^2}$$

$$= y_1 \int \frac{dx}{x \left(1 - 2x + \frac{3}{2}x^2 - \frac{5}{9}x^3 + \frac{35}{288}x^4 - \cdots\right)}$$

$$= y_1 \int \frac{1}{x} \left(1 + 2x + \frac{5}{2}x^2 + \frac{23}{9}x^3 + \frac{677}{288}x^4 + \cdots\right) dx$$

$$= y_1 \int \left(\frac{1}{x} + 2 + \frac{5}{2}x + \frac{23}{9}x^2 + \frac{677}{288}x^3 + \cdots\right) dx$$

$$= y_1 \left[\ln x + 2x + \frac{5}{4}x^2 + \frac{23}{27}x^3 + \frac{677}{1,152}x^4 + \cdots\right]$$

$$= y_1 \ln x + y_1 \left(2x + \frac{5}{4}x^2 + \frac{23}{27}x^3 + \frac{677}{1,152}x^4 + \cdots\right).$$

The general solution on $(0, \infty)$ is

$$y = C_1 y_1(x) + C_2 y_2(x).$$

33. (a) From $t = 1/x$ we have $dt/dx = -1/x^2 = -t^2$. Then

$$\frac{dy}{dx} = \frac{dy}{dt}\frac{dt}{dx} = -t^2 \frac{dy}{dt}$$

and

$$\frac{d^2y}{dx^2} = \frac{d}{dx}\left(\frac{dy}{dx}\right) = \frac{d}{dx}\left(-t^2\frac{dy}{dt}\right) = -t^2\frac{d^2y}{dt^2}\frac{dt}{dx} - \frac{dy}{dt}\left(2t\frac{dt}{dx}\right) = t^4\frac{d^2y}{dt^2} + 2t^3\frac{dy}{dt}.$$

Now

$$x^4\frac{d^2y}{dx^2} + \lambda y = \frac{1}{t^4}\left(t^4\frac{d^2y}{dt^2} + 2t^3\frac{dy}{dt}\right) + \lambda y = \frac{d^2y}{dt^2} + \frac{2}{t}\frac{dy}{dt} + \lambda y = 0$$

becomes

$$t\frac{d^2y}{dt^2} + 2\frac{dy}{dt} + \lambda ty = 0.$$

(b) Substituting $y = \sum_{n=0}^{\infty} c_n t^{n+r}$ into the differential equation and collecting terms, we obtain

$$t\frac{d^2y}{dt^2} + 2\frac{dy}{dt} + \lambda ty = (r^2 + r)c_0 t^{r-1} + (r^2 + 3r + 2)c_1 t^r$$

$$+ \sum_{k=2}^{\infty}[(k+r)(k+r-1)c_k + 2(k+r)c_k + \lambda c_{k-2}]t^{k+r-1}$$

$$= 0,$$

which implies

$$r^2 + r = r(r+1) = 0,$$
$$(r^2 + 3r + 2)\, c_1 = 0,$$

and

$$(k+r)(k+r+1)c_k + \lambda c_{k-2} = 0.$$

The indicial roots are $r_1 = 0$ and $r_2 = -1$, so $c_1 = 0$. For $r_1 = 0$ the recurrence relation is

$$c_k = -\frac{\lambda c_{k-2}}{k(k+1)}, \quad k = 2, 3, 4, \ldots,$$

and

$$c_2 = -\frac{\lambda}{3!} c_0$$

$$c_3 = c_5 = c_7 = \cdots = 0$$

$$c_4 = \frac{\lambda^2}{5!} c_0$$

$$c_{2n} = (-1)^n \frac{\lambda^n}{(2n+1)!} c_0.$$

For $r_2 = -1$ the recurrence relation is

$$c_k = -\frac{\lambda c_{k-2}}{k(k-1)}, \quad k = 2, 3, 4, \ldots,$$

and

$$c_2 = -\frac{\lambda}{2!} c_0$$

$$c_3 = c_5 = c_7 = \cdots = 0$$

$$c_4 = \frac{\lambda^2}{4!} c_0$$

$$c_{2n} = (-1)^n \frac{\lambda^n}{(2n)!} c_0.$$

The general solution on $(0, \infty)$ is

$$y(t) = C_1 \sum_{n=0}^{\infty} \frac{(-1)^n}{(2n+1)!} (\sqrt{\lambda}\,t)^{2n} + C_2 t^{-1} \sum_{n=0}^{\infty} \frac{(-1)^n}{(2n)!} (\sqrt{\lambda}\,t)^{2n}$$

$$= \frac{1}{t} \left[C_1 \sum_{n=0}^{\infty} \frac{(-1)^n}{(2n+1)!} (\sqrt{\lambda}\,t)^{2n+1} + C_2 \sum_{n=0}^{\infty} \frac{(-1)^n}{(2n)!} (\sqrt{\lambda}\,t)^{2n} \right]$$

$$= \frac{1}{t} [C_1 \sin \sqrt{\lambda}\,t + C_2 \cos \sqrt{\lambda}\,t].$$

(c) Using $t = 1/x$, the solution of the original equation is

$$y(x) C_1 x \sin \frac{\sqrt{\lambda}}{x} + C_2 x \cos \frac{\sqrt{\lambda}}{x}.$$

Exercises 5.3

3. Since $\nu^2 = 25/4$ the general solution is $y = c_1 J_{5/2}(x) + c_2 J_{-5/2}(x)$.

6. Since $\nu^2 = 4$ the general solution is $y = c_1 J_2(x) + c_2 Y_2(x)$.

9. If $y = x^{-1/2} v(x)$ then

$$y' = x^{-1/2} v'(x) - \frac{1}{2} x^{-3/2} v(x),$$

$$y'' = x^{-1/2} v''(x) - x^{-3/2} v'(x) + \frac{3}{4} x^{-5/2} v(x),$$

and

$$x^2 y'' + 2xy' + \lambda^2 x^2 y = x^{3/2} v'' + x^{1/2} v' + \left(\lambda^2 x^{3/2} - \frac{1}{4} x^{-1/2} \right) v.$$

Multiplying by $x^{1/2}$ we obtain

$$x^2 v'' + xv' + \left(\lambda^2 x^2 - \frac{1}{4} \right) v = 0,$$

whose solution is $v = c_1 J_{1/2}(\lambda x) + c_2 J_{-1/2}(\lambda x)$. Then $y = c_1 x^{-1/2} J_{1/2}(\lambda x) + c_2 x^{-1/2} J_{-1/2}(\lambda x)$.

12. From $y = \sqrt{x}\, J_\nu(\lambda x)$ we find

$$y' = \lambda \sqrt{x}\, J_\nu'(\lambda x) + \frac{1}{2} x^{-1/2} J_\nu(\lambda x)$$

and

$$y'' = \lambda^2 \sqrt{x}\, J_\nu''(\lambda x) + \lambda x^{-1/2} J_\nu'(\lambda x) - \frac{1}{4} x^{-3/2} J_\nu(\lambda x).$$

Substituting into the differential equation, we have

$$x^2 y'' + \left(\lambda^2 x^2 - \nu^2 + \frac{1}{4} \right) y = \sqrt{x}\, \left[\lambda^2 x^2 J_\nu''(\lambda x) + \lambda x J_\nu'(\lambda x) + \left(\lambda^2 x^2 - \nu^2 \right) J_\nu(\lambda x) \right]$$

$$= \sqrt{x} \cdot 0 \qquad \text{(since } J_n \text{ is a solution of Bessel's equation)}$$

$$= 0.$$

Therefore, $\sqrt{x}\, J_\nu(\lambda x)$ is a solution of the original equation.

15. From Problem 10 with $n = -1$ we find $y = x^{-1} J_{-1}(x)$. From Problem 11 with $n = 1$ we find $y = x^{-1} J_1(x) = -x^{-1} J_{-1}(x)$.

18. From Problem 10 with $n = 3$ we find $y = x^3 J_3(x)$. From Problem 11 with $n = -3$ we find $y = x^3 J_{-3}(x) = -x^3 J_3(x)$.

21. Letting $\nu = 1$ in (15) in the text we have

$$x J_0(x) = \frac{d}{dx}[x J_1(x)] \qquad \text{so} \qquad \int_0^x r J_0(r)\, dr = r J_1(r) \Big|_{r=0}^{r=x} = x J_1(x).$$

24. (a) By Problem 20, with $\nu = 1/2$, we obtain $J_{1/2}(x) = x J_{3/2}(x) + x J_{-1/2}(x)$ so that

$$J_{3/2}(x) = \sqrt{\frac{2}{\pi x}} \left(\frac{\sin x}{x} - \cos x \right);$$

with $\nu = -1/2$ we obtain $-J_{-1/2}(x) = x J_{1/2}(x) + x J_{-3/2}(x)$ so that

$$J_{-3/2}(x) = -\sqrt{\frac{2}{\pi x}} \left(\frac{\cos x}{x} + \sin x \right);$$

and with $\nu = 3/2$ we obtain $3 J_{3/2}(x) = x J_{5/2}(x) + x J_{1/2}(x)$ so that

$$J_{5/2}(x) = \sqrt{\frac{2}{\pi x}} \left(\frac{3 \sin x}{x^2} - \frac{3 \cos x}{x} - \sin x \right).$$

(b)

27. Letting

$$s = \frac{2}{\alpha}\sqrt{\frac{k}{m}}\,e^{-\alpha t/2},$$

we have

$$\frac{dx}{dt} = \frac{dx}{ds}\frac{ds}{dt} = \frac{dx}{dt}\left[\frac{2}{\alpha}\sqrt{\frac{k}{m}}\left(-\frac{\alpha}{2}\right)e^{-\alpha t/2}\right] = \frac{dx}{ds}\left(-\sqrt{\frac{k}{m}}\,e^{-\alpha t/2}\right)$$

and

$$\frac{d^2x}{dt^2} = \frac{d}{dt}\left(\frac{dx}{dt}\right) = \frac{dx}{ds}\left(\frac{\alpha}{2}\sqrt{\frac{k}{m}}\,e^{-\alpha t/2}\right) + \frac{d}{dt}\left(\frac{dx}{ds}\right)\left(-\sqrt{\frac{k}{m}}\,e^{-\alpha t/2}\right)$$

$$= \frac{dx}{ds}\left(\frac{\alpha}{2}\sqrt{\frac{k}{m}}\,e^{-\alpha t/2}\right) + \frac{d^2x}{ds^2}\frac{ds}{dt}\left(-\sqrt{\frac{k}{m}}\,e^{-\alpha t/2}\right)$$

$$= \frac{dx}{ds}\left(\frac{\alpha}{2}\sqrt{\frac{k}{m}}\,e^{-\alpha t/2}\right) + \frac{d^2x}{ds^2}\left(\frac{k}{m}\,e^{-\alpha t}\right).$$

Then

$$m\frac{d^2x}{dt^2} + ke^{-\alpha t}x = ke^{-\alpha t}\frac{d^2x}{ds^2} + \frac{m\alpha}{2}\sqrt{\frac{k}{m}}\,e^{-\alpha t/2}\frac{dx}{dt} + ke^{-\alpha t}x = 0.$$

Multiplying by $2^2/\alpha^2 m$ we have

$$\frac{2^2}{\alpha^2}\frac{k}{m}\,e^{-\alpha t}\frac{d^2x}{ds^2} + \frac{2}{\alpha}\sqrt{\frac{k}{m}}\,e^{-\alpha t/2}\frac{dx}{dt} + \frac{2}{\alpha^2}\frac{k}{m}\,e^{-\alpha t}x = 0$$

or, since $s = (2/\alpha)\sqrt{k/m}\,e^{-\alpha t/2}$,

$$s^2\frac{d^2x}{ds^2} + s\frac{dx}{ds} + s^2x = 0.$$

30. The general solution of Bessel's equation is

$$w(t) = c_1 J_{1/3}(t) + c_2 J_{-1/3}(t), \qquad t > 0.$$

Thus, the general solution of Airy's equation for $x > 0$ is

$$y = x^{1/2}w\left(\frac{2}{3}\alpha x^{3/2}\right) = c_1 x^{1/2}J_{1/3}\left(\frac{2}{3}\alpha x^{3/2}\right) + c_2 x^{1/2}J_{-1/3}\left(\frac{2}{3}\alpha x^{3/2}\right).$$

33. Setting $y = \sqrt{x}\,J_1(2\sqrt{x})$ and differentiating we obtain

$$y' = \sqrt{x}\,J_1'(2\sqrt{x})\frac{2}{2\sqrt{x}} + \frac{1}{2\sqrt{x}}\,J_1(2\sqrt{x}) = J_1'(2\sqrt{x}) + \frac{1}{2\sqrt{x}}\,J_1(2\sqrt{x})$$

and

$$y'' = J_1''(2\sqrt{x})\,\frac{2}{2\sqrt{x}} + \frac{1}{2\sqrt{x}}\,J_1'(2\sqrt{x})\,\frac{2}{2\sqrt{x}} - \frac{1}{4x^{3/2}}\,J_1(2\sqrt{x})$$

$$= \frac{1}{\sqrt{x}}\,J_1''(2\sqrt{x}) + \frac{1}{2x}\,J_1'(2\sqrt{x}) - \frac{1}{4x^{3/2}}\,J_1(2\sqrt{x}).$$

Substituting into the differential equation and letting $t = 2\sqrt{x}$ we have

$$xy'' + y = \sqrt{x}\,J_1''(2\sqrt{x}) + \frac{1}{2}J_1'(2\sqrt{x}) - \frac{1}{4\sqrt{x}}\,J_1(2\sqrt{x}) + \sqrt{x}\,J_1(2\sqrt{x})$$

$$= \frac{1}{\sqrt{x}}\left[xJ_1''(2\sqrt{x}) + \frac{\sqrt{x}}{2}\,J_1'(2\sqrt{x}) + \left(x - \frac{1}{4}\right)J_1(2\sqrt{x})\right]$$

$$= \frac{2}{t}\left[\frac{t^2}{4}J_1''(t) + \frac{t}{4}\,J_1'(t) + \left(\frac{t^2}{4} - \frac{1}{4}\right)J_1(t)\right]$$

$$= \frac{1}{2t}\left[t^2 J_1''(t) + tJ_1'(t) + (t^2 - 1)J_1(t)\right].$$

Since $J_1(t)$ is a solution of $t^2 y'' + ty' + (t^2 - 1)y = 0$, we see that the last expression above is 0 and $y = \sqrt{x}\,J_1(2\sqrt{x})$ is a solution of $xy'' + y = 0$.

Chapter 5 Review Exercises

3. Solving $x^2 - 2x + 10 = 0$ we obtain $x = 1 \pm \sqrt{11}$, which are singular points. Thus, the minimum radius of convergence is $|1 - \sqrt{11}| = \sqrt{11} - 1$.

6. The differential equation $(x - 1)(x + 3)y'' + y = 0$ has regular singular points at $x = 1$ and $x = -3$.

9. Substituting $y = \sum_{n=0}^{\infty} c_n x^n$ into the differential equation we obtain

$$(x - 1)y'' + 3y = (-2c_2 + 3c_0) + \sum_{k=3}^{\infty}(k - 1)(k - 2)c_{k-1} - k(k - 1)c_k + 3c_{k-2}]x^{k-2} = 0$$

which implies $c_2 = 3c_0/2$ and

$$c_k = \frac{(k - 1)(k - 2)c_{k-1} + 3c_{k-2}}{k(k - 1)}, \quad k = 3, 4, 5, \ldots .$$

Choosing $c_0 = 1$ and $c_1 = 0$ we find

$$c_2 = \frac{3}{2}, \qquad c_3 = \frac{1}{2}, \qquad c_4 = \frac{5}{8}$$

and so on. For $c_0 = 0$ and $c_1 = 1$ we obtain

$$c_2 = 0, \qquad c_3 = \frac{1}{2}, \qquad c_4 = \frac{1}{4}$$

and so on. Thus, two solutions are

$$y_1 = C_1\left(1 + \frac{3}{2}x^2 + \frac{1}{2}x^3 + \frac{5}{8}x^4 + \cdots\right)$$

and

$$y_2 = C_2\left(x + \frac{1}{2}x^3 + \frac{1}{4}x^4 + \cdots\right).$$

12. Substituting $y = \sum_{n=0}^{\infty} c_n x^n$ into the differential equation we have

$$(\cos x)y'' + y = \left(1 - \frac{1}{2}x^2 + \frac{1}{24}x^4 - \frac{1}{720}x^6 + \cdots\right)\left(2c_2 + 6c_3 x + 12c_4 x^2 + 20c_5 x^3 + 30c_6 x^4 + \cdots\right)$$

$$+ \sum_{n=0}^{\infty} c_n x^n$$

$$= \left[2c_2 + 6c_3 x + (12c_4 - c_2)x^2 + (20c_5 - 3c_3)x^3 + \left(30c_6 - 6c_4 + \frac{1}{12}c_2\right)x^4 + \cdots\right]$$

$$+ \left[c_0 + c_1 x + c_2 x^2 + c_3 x^3 + c_4 x^4 + \cdots\right]$$

$$= (c_0 + 2c_2) + (c_1 + 6c_3)x + 12c_4 x^2 + (20c_5 - 2c_3)x^3 + \left(30c_6 - 5c_4 + \frac{1}{12}c_2\right)x^4 + \cdots$$

$$= 0.$$

Thus

$$c_0 + 2c_2 = 0$$
$$c_1 + 6c_3 = 0$$
$$12c_4 = 0$$
$$20c_5 - 2c_3 = 0$$
$$30c_6 - 5c_4 + \frac{1}{12}c_2 = 0$$

and

$$c_2 = -\frac{1}{2}c_0$$
$$c_3 = -\frac{1}{6}c_1$$
$$c_4 = 0$$
$$c_5 = \frac{1}{10}c_3$$
$$c_6 = \frac{1}{6}c_4 - \frac{1}{360}c_2.$$

Choosing $c_0 = 1$ and $c_1 = 0$ we find

$$c_2 = -\frac{1}{2}, \quad c_3 = 0, \quad c_4 = 0, \quad c_5 = 0, \quad c_6 = \frac{1}{720}$$

and so on. For $c_0 = 0$ and $c_1 = 1$ we find

$$c_2 = 0, \quad c_3 = -\frac{1}{6}, \quad c_4 = 0, \quad c_5 = -\frac{1}{60}, \quad c_6 = 0$$

and so on. Thus, two solutions are

$$y_1 = 1 - \frac{1}{2}x^2 + \frac{1}{720}x^6 + \cdots \quad \text{and} \quad y_2 = x - \frac{1}{6}x^3 - \frac{1}{60}x^5 + \cdots.$$

15. Writing the differential equation in the form

$$y'' + \left(\frac{1 - \cos x}{x}\right)y' + xy = 0,$$

and noting that

$$\frac{1 - \cos x}{x} = \frac{x}{2} - \frac{x^3}{24} + \frac{x^5}{720} - \cdots$$

is analytic at $x = 0$, we conclude that $x = 0$ is an ordinary point of the differential equation.

18. (a) From $y = -\dfrac{1}{u}\dfrac{du}{dx}$ we obtain

$$\frac{dy}{dx} = -\frac{1}{u}\frac{d^2u}{dx^2} + \frac{1}{u^2}\left(\frac{du}{dx}\right)^2.$$

Then $dy/dx = x^2 + y^2$ becomes

$$-\frac{1}{u}\frac{d^2u}{dx^2} + \frac{1}{u^2}\left(\frac{du}{dx}\right)^2 = x^2 + \frac{1}{u^2}\left(\frac{du}{dx}\right)^2,$$

so $\quad \dfrac{d^2u}{dx^2} + x^2u = 0.$

(b) If $u = x^{1/2}w(\tfrac{1}{2}x^2)$ then

$$u' = x^{3/2}w'\left(\frac{1}{2}x^2\right) + \frac{1}{2}x^{-1/2}w\left(\frac{1}{2}x^2\right)$$

and

$$u'' = x^{5/2}w''\left(\frac{1}{2}x^2\right) + 2x^{1/2}w'\left(\frac{1}{2}x^2\right) - \frac{1}{4}x^{-3/2}w\left(\frac{1}{2}x^2\right),$$

so

$$u'' + x^2u = x^{1/2}\left[x^2w''\left(\frac{1}{2}x^2\right) + 2w'\left(\frac{1}{2}x^2\right) + \left(x^2 - \frac{1}{4}x^{-2}\right)w\left(\frac{1}{2}x^2\right)\right] = 0.$$

Letting $t = \tfrac{1}{2}x^2$ we have

$$\sqrt{2t}\left[2tw''(t) + 2w'(t) + \left(2t - \frac{1}{4\cdot 2t}\right)w(t)\right] = 0$$

or

$$t^2w''(t) + tw'(t) + \left(t^2 - \frac{1}{16}\right)w(t) = 0.$$

This is Bessel's equation with $\nu = 1/4$, so

$$w(t) = c_1 J_{1/4}(t) + c_2 J_{-1/4}(t).$$

(c) We have

$$
\begin{aligned}
y &= -\frac{1}{u}\frac{du}{dx} = -\frac{1}{x^{1/2}w(t)}\frac{d}{dx}x^{1/2}w(t)\\[2mm]
&= -\frac{1}{x^{1/2}w}\left[x^{1/2}\frac{dw}{dt}\frac{dt}{dx} + \frac{1}{2}x^{-1/2}w\right]\\[2mm]
&= -\frac{1}{x^{1/2}w}\left[x^{3/2}\frac{dw}{dt} + \frac{1}{2x^{1/2}}w\right]\\[2mm]
&= -\frac{1}{2xw}\left[2x^2\frac{dw}{dt} + w\right] = -\frac{1}{2xw}\left[4t\frac{dw}{dt} + w\right].
\end{aligned}
$$

Now

$$4t\frac{dw}{dt} + w = 4t\frac{d}{dt}[c_1 J_{1/4}(t) + c_2 J_{-1/4}(t)] + c_1 J_{1/4}(t) + c_2 J_{-1/4}(t)$$

$$= 4t\left[c_1\left(J_{-3/4}(t) - \frac{1}{4t}J_{1/4}(t)\right) + c_2\left(-\frac{1}{4t}J_{-1/4}(t) - J_{3/4}(t)\right)\right]$$

$$+ c_1 J_{1/4}(t) + c_2 J_{-1/4}(t)$$

$$= 4c_1 t J_{-3/4}(t) - 4c_2 t J_{3/4}(t)$$

$$= 2c_1 x^2 J_{-3/4}\left(\frac{1}{2}x^2\right) - 2c_2 x^2 J_{3/4}\left(\frac{1}{2}x^2\right),$$

so

$$y = -\frac{2c_1 x^2 J_{-3/4}(\frac{1}{2}x^2) - 2c_2 x^2 J_{3/4}(\frac{1}{2}x^2)}{2x[c_1 J_{1/4}(\frac{1}{2}x^2) + c_2 J_{-1/4}(\frac{1}{2}x^2)]}$$

$$= x\frac{-c_1 J_{-3/4}(\frac{1}{2}x^2) + c_2 J_{3/4}(\frac{1}{2}x^2)}{c_1 J_{1/4}(\frac{1}{2}x^2) + c_2 J_{-1/4}(\frac{1}{2}x^2)}.$$

Letting $c = c_1/c_2$ we have

$$y = x\frac{J_{3/4}(\frac{1}{2}x^2) - cJ_{-3/4}(\frac{1}{2}x^2)}{cJ_{1/4}(\frac{1}{2}x^2) + J_{-1/4}(\frac{1}{2}x^2)}.$$

6 Numerical Solutions of Ordinary Differential Equations

_____ **Exercises 6.1** _____

All tables in this chapter were constructed in a spreadsheet program which does not support subscripts. Consequently, x_n and y_n will be indicated as $x(n)$ and $y(n)$, respectively.

3.

$h = 0.1$		$h = 0.05$	
$x(n)$	$y(n)$	$x(n)$	$y(n)$
0.00	0.0000	0.00	0.0000
0.10	0.1005	0.05	0.0501
0.20	0.2030	0.10	0.1004
0.30	0.3098	0.15	0.1512
0.40	0.4234	0.20	0.2028
0.50	0.5470	0.25	0.2554
		0.30	0.3095
		0.35	0.3652
		0.40	0.4230
		0.45	0.4832
		0.50	0.5465

6.

$h = 0.1$		$h = 0.05$	
$x(n)$	$y(n)$	$x(n)$	$y(n)$
0.00	0.0000	0.00	0.0000
0.10	0.0050	0.05	0.0013
0.20	0.0200	0.10	0.0050
0.30	0.0451	0.15	0.0113
0.40	0.0805	0.20	0.0200
0.50	0.1266	0.25	0.0313
		0.30	0.0451
		0.35	0.0615
		0.40	0.0805
		0.45	0.1022
		0.50	0.1266

9.

$h = 0.1$		$h = 0.05$	
$x(n)$	$y(n)$	$x(n)$	$y(n)$
1.00	1.0000	1.00	1.0000
1.10	1.0095	1.05	1.0024
1.20	1.0404	1.10	1.0100
1.30	1.0967	1.15	1.0228
1.40	1.1866	1.20	1.0414
1.50	1.3260	1.25	1.0663
		1.30	1.0984
		1.35	1.1389
		1.40	1.1895
		1.45	1.2526
		1.50	1.3315

12. (a)

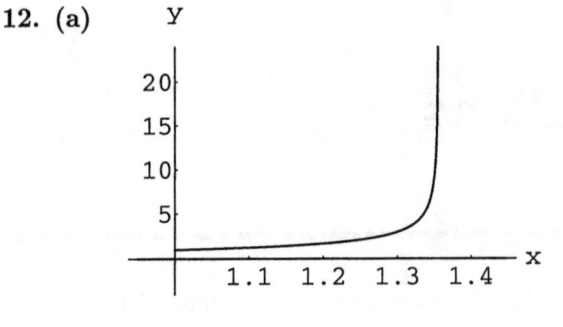

(b)

h=0.1	EULER	IMPROVED EULER
x(n)	y(n)	y(n)
1.00	1.0000	1.0000
1.10	1.2000	1.2469
1.20	1.4938	1.6668
1.30	1.9711	2.6427
1.40	2.9060	8.7988

15. (a) Using the Euler method we obtain $y(0.1) \approx y_1 = 0.8$.

(b) Using $y'' = 5e^{-2x}$ we see that the local truncation error is

$$5e^{-2c} \frac{(0.1)^2}{2} = 0.025e^{-2c}.$$

Since e^{-2x} is a decreasing function, $e^{-2c} \leq e^0 = 1$ for $0 \leq c \leq 0.1$. Thus an upper bound for the local truncation error is $0.025(1) = 0.025$.

(c) Since $y(0.1) = 0.8234$, the actual error is $y(0.1) - y_1 = 0.0234$, which is less than 0.025.

(d) Using the Euler method with $h = 0.05$ we obtain $y(0.1) \approx y_2 = 0.8125$.

(e) The error in (d) is $0.8234 - 0.8125 = 0.0109$. With global truncation error $O(h)$, when the step size is halved we expect the error for $h = 0.05$ to be one-half the error when $h = 0.1$. Comparing 0.0109 with 0.0234 we see that this is the case.

18. (a) Using $y''' = -114e^{-3(x-1)}$ we see that the local truncation error is

$$\left| y'''(c) \frac{h^3}{6} \right| = 114e^{-3(x-1)} \frac{h^3}{6} = 19h^3 e^{-3(c-1)}.$$

(b) Since $e^{-3(x-1)}$ is a decreasing function for $1 \leq x \leq 1.5$, $e^{-3(c-1)} \leq e^{-3(1-1)} = 1$ for $1 \leq c \leq 1.5$ and

$$\left| y'''(c) \frac{h^3}{6} \right| \leq 19(0.1)^3 (1) = 0.019.$$

(c) Using the improved Euler method with $h = 0.1$ we obtain $y(1.5) \approx 2.080108$. With $h = 0.05$ we obtain $y(1.5) \approx 2.059166$.

(d) Since $y(1.5) = 2.053216$, the error for $h = 0.1$ is $E_{0.1} = 0.026892$, while the error for $h = 0.05$ is $E_{0.05} = 0.005950$. With global truncation error $O(h^2)$ we expect $E_{0.1}/E_{0.05} \approx 4$. We actually have $E_{0.1}/E_{0.05} = 4.52$.

21. Because y_{n+1}^* depends on y_n and is used to determine y_{n+1}, all of the y_n^* cannot be computed at one time independently of the corresponding y_n values. For example, the computation of y_4^* involves the value of y_3.

Exercises 6.2

3.

x(n)	y(n)
1.00	5.0000
1.10	3.9724
1.20	3.2284
1.30	2.6945
1.40	2.3163
1.50	2.0533

6.

x(n)	y(n)
0.00	1.0000
0.10	1.1115
0.20	1.2530
0.30	1.4397
0.40	1.6961
0.50	2.0670

9.

x(n)	y(n)
0.00	0.5000
0.10	0.5213
0.20	0.5358
0.30	0.5443
0.40	0.5482
0.50	0.5493

12.

x(n)	y(n)
0.00	0.5000
0.10	0.5250
0.20	0.5498
0.30	0.5744
0.40	0.5987
0.50	0.6225

15. (a)

	h = 0.05	h = 0.1
x(n)	y(n)	y(n)
1.00	1.0000	1.0000
1.05	1.1112	
1.10	1.2511	1.2511
1.15	1.4348	
1.20	1.6934	1.6934
1.25	2.1047	
1.30	2.9560	2.9425
1.35	7.8981	
1.40	1.06E+15	903.0282

(b)

18. (a) Using $y^{(5)} = -1026e^{-3(x-1)}$ we see that the local truncation error is

$$\left| y^{(5)}(c)\frac{h^5}{120} \right| = 8.55h^5 e^{-3(c-1)}.$$

(b) Since $e^{-3(x-1)}$ is a decreasing function for $1 \le x \le 1.5$, $e^{-3(c-1)} \le e^{-3(1-1)} = 1$ for $1 \le c \le 1.5$ and

$$y^{(5)}(c)\frac{h^5}{120} \le 8.55(0.1)^5(1) = 0.0000855.$$

(c) Using the fourth-order Runge-Kutta method with $h = 0.1$ we obtain $y(1.5) \approx 2.053338827$. With $h = 0.05$ we obtain $y(1.5) \approx 2.053222989$.

Exercises 6.3

3.

x(n)	y(n)	
0.00	1.0000	initial condition
0.20	0.7328	Runge-Kutta
0.40	0.6461	Runge-Kutta
0.60	0.6585	Runge-Kutta
	0.7332	*predictor*
0.80	0.7232	corrector

6.

x(n)	y(n)	
0.00	1.0000	initial condition
0.20	1.4414	Runge-Kutta
0.40	1.9719	Runge-Kutta
0.60	2.6028	Runge-Kutta
	3.3483	*predictor*
0.80	3.3486	corrector
	4.2276	*predictor*
1.00	4.2280	corrector

x(n)	y(n)	
0.00	1.0000	initial condition
0.10	1.2102	Runge-Kutta
0.20	1.4414	Runge-Kutta
0.30	1.6949	Runge-Kutta
	1.9719	*predictor*
0.40	1.9719	corrector
	2.2740	*predictor*
0.50	2.2740	corrector
	2.6028	*predictor*
0.60	2.6028	corrector
	2.9603	*predictor*
0.70	2.9603	corrector
	3.3486	*predictor*
0.80	3.3486	corrector
	3.7703	*predictor*
0.90	3.7703	corrector
	4.2280	*predictor*
1.00	4.2280	corrector

Exercises 6.4

3. The substitution $y' = u$ leads to the system

$$y' = u, \qquad u' = 4u - 4y.$$

Using formula (4) in the text with x corresponding to t, y corresponding to x, and u corresponding to y, we obtain

Runge-Kutta method with h=0.2

m1	m2	m3	m4	k1	k2	k3	k4	x	y	u
								0.00	-2.0000	1.0000
0.2000	0.4400	0.5280	0.9072	2.4000	3.2800	3.5360	4.8064	0.20	-1.4928	4.4731

Runge-Kutta method with h=0.1

m1	m2	m3	m4	k1	k2	k3	k4	x	y	u
								0.00	-2.0000	1.0000
0.1000	0.1600	0.1710	0.2452	1.2000	1.4200	1.4520	1.7124	0.10	-1.8321	2.4427
0.2443	0.3298	0.3444	0.4487	1.7099	2.0031	2.0446	2.3900	0.20	-1.4919	4.4753

6.

Runge-Kutta method with h=0.1

m1	m2	m3	m4	k1	k2	k3	k4	t	i1	i2
								0.00	0.0000	0.0000
10.0000	0.0000	12.5000	-20.0000	0.0000	5.0000	-5.0000	22.5000	0.10	2.5000	3.7500
8.7500	-2.5000	13.4375	-28.7500	-5.0000	4.3750	-10.6250	29.6875	0.20	2.8125	5.7813
10.1563	-4.3750	17.0703	-40.0000	-8.7500	5.0781	-16.0156	40.3516	0.30	2.0703	7.4023
13.2617	-6.3672	22.9443	-55.1758	-12.7344	6.6309	-22.5488	55.3076	0.40	0.6104	9.1919
17.9712	-8.8867	31.3507	-75.9326	-17.7734	8.9856	-31.2024	75.9821	0.50	-1.5619	11.4877

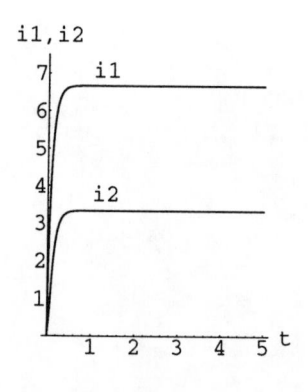

i1,i2

As $t \to \infty$ we see that $i_1(t) \to 6.75$ and $i_2(t) \to 3.4$.

9.

								Runge-Kutta method with h=0.2		
m1	m2	m3	m4	k1	k2	k3	k4	t	x	y
								0.00	-3.0000	5.0000
-1.0000	-0.9200	-0.9080	-0.8176	-0.6000	-0.7200	-0.7120	-0.8216	0.20	-3.9123	4.2857

								Runge-Kutta method with h=0.1		
m1	m2	m3	m4	k1	k2	k3	k4	t	x	y
								0.00	-3.0000	5.0000
-0.5000	-0.4800	-0.4785	-0.4571	-0.3000	-0.3300	-0.3290	-0.3579	0.10	-3.4790	4.6707
-0.4571	-0.4342	-0.4328	-0.4086	-0.3579	-0.3858	-0.3846	-0.4112	0.20	-3.9123	4.2857

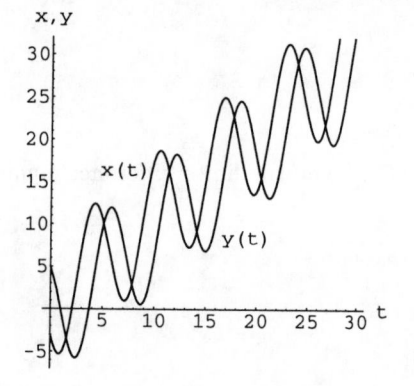

x,y

12. Solving for x' and y' we obtain the system

$$x' = \frac{1}{2}y - 3t^2 + 2t - 5$$
$$y' = -\frac{1}{2}y + 3t^2 + 2t + 5.$$

								Runge-Kutta method with h=0.2		
m1	m2	m3	m4	k1	k2	k3	k4	t	x	y
								0.00	3.0000	-1.0000
-1.1000	-1.0110	-1.0115	-0.9349	1.1000	1.0910	1.0915	1.0949	0.20	1.9867	0.0933

								Runge-Kutta method with h=0.1		
m1	m2	m3	m4	k1	k2	k3	k4	t	x	y
								0.00	3.0000	-1.0000
-0.5500	-0.5270	-0.5271	-0.5056	0.5500	0.5470	0.5471	0.5456	0.10	2.4727	-0.4527
-0.5056	-0.4857	-0.4857	-0.4673	0.5456	0.5457	0.5457	0.5473	0.20	1.9867	0.0933

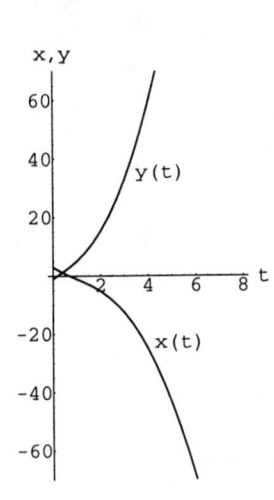

Exercises 6.5

3. We identify $P(x) = 2$, $Q(x) = 1$, $f(x) = 5x$, and $h = (1 - 0)/5 = 0.2$. Then the finite difference equation is

$$1.2y_{i+1} - 1.96y_i + 0.8y_{i-1} = 0.04(5x_i).$$

The solution of the corresponding linear system gives

x	0.0	0.2	0.4	0.6	0.8	1.0
y	0.0000	-0.2259	-0.3356	-0.3308	-0.2167	0.0000

6. We identify $P(x) = 5$, $Q(x) = 0$, $f(x) = 4\sqrt{x}$, and $h = (2 - 1)/6 = 0.1667$. Then the finite difference equation is

$$1.4167y_{i+1} - 2y_i + 0.5833y_{i-1} = 0.2778(4\sqrt{x_i}).$$

The solution of the corresponding linear system gives

x	1.0000	1.1667	1.3333	1.5000	1.6667	1.8333	2.0000
y	1.0000	-0.5918	-1.1626	-1.3070	-1.2704	-1.1541	-1.0000

9. We identify $P(x) = 1 - x$, $Q(x) = x$, $f(x) = x$, and $h = (1 - 0)/10 = 0.1$. Then the finite difference equation is

$$[1 + 0.05(1 - x_i)]y_{i+1} + [-2 + 0.01x_i]y_i + [1 - 0.05(1 - x_i)]y_{i-1} = 0.01x_i.$$

The solution of the corresponding linear system gives

x	0.0	0.1	0.2	0.3	0.4	0.5	0.6
y	0.0000	0.2660	0.5097	0.7357	0.9471	1.1465	1.3353

0.7	0.8	0.9	1.0
1.5149	1.6855	1.8474	2.0000

12. We identify $P(r) = 2/r$, $Q(r) = 0$, $f(r) = 0$, and $h = (4 - 1)/6 = 0.5$. Then the finite difference equation is

$$\left(1 + \frac{0.5}{r_i}\right)u_{i+1} - 2u_i + \left(1 - \frac{0.5}{r_i}\right)u_{i-1} = 0.$$

The solution of the corresponding linear system gives

r	1.0	1.5	2.0	2.5	3.0	3.5	4.0
u	50.0000	72.2222	83.3333	90.0000	94.4444	97.6190	100.0000

Chapter 6 Review Exercises

3.

h=0.1 x(n)	EULER	IMPROVED EULER	RUNGE KUTTA
0.50	0.5000	0.5000	0.5000
0.60	0.6000	0.6048	0.6049
0.70	0.7095	0.7191	0.7194
0.80	0.8283	0.8427	0.8431
0.90	0.9559	0.9752	0.9757
1.00	1.0921	1.1163	1.1169

h=0.05 x(n)	EULER	IMPROVED EULER	RUNGE KUTTA
0.50	0.5000	0.5000	0.5000
0.55	0.5500	0.5512	0.5512
0.60	0.6024	0.6049	0.6049
0.65	0.6573	0.6609	0.6610
0.70	0.7144	0.7193	0.7194
0.75	0.7739	0.7800	0.7801
0.80	0.8356	0.8430	0.8431
0.85	0.8996	0.9082	0.9083
0.90	0.9657	0.9755	0.9757
0.95	1.0340	1.0451	1.0452
1.00	1.1044	1.1168	1.1169

6.

x(n)	y(n)	
0.00	2.0000	initial condition
0.10	2.4734	Runge-Kutta
0.20	3.1781	Runge-Kutta
0.30	4.3925	Runge-Kutta
	6.7689	*predictor*
0.40	7.0783	corrector

7 Vectors

Exercises 7.1

3. (a) $\langle 12, 0 \rangle$ (b) $\langle 4, -5 \rangle$ (c) $\langle 4, 5 \rangle$ (d) $\sqrt{41}$ (e) $\sqrt{41}$

6. (a) $\langle 3, 9 \rangle$ (b) $\langle -4, -12 \rangle$ (c) $\langle 6, 18 \rangle$ (d) $4\sqrt{10}$ (e) $6\sqrt{10}$

9. (a) $\langle 4, -12 \rangle - \langle -2, 2 \rangle = \langle 6, -14 \rangle$ (b) $\langle -3, 9 \rangle - \langle -5, 5 \rangle = \langle 2, 4 \rangle$

12. (a) $\langle 8, 0 \rangle - \langle 0, -6 \rangle = \langle 8, 6 \rangle$ (b) $\langle -6, 0 \rangle - \langle 0, -15 \rangle = \langle -6, 15 \rangle$

15.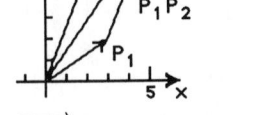

$\overrightarrow{P_1 P_2} = \langle 2, 5 \rangle$

18.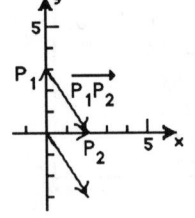

$\overrightarrow{P_1 P_2} = \langle 2, -3 \rangle$

21. $a(= -\mathbf{a})$, $b(= -\frac{1}{4}\mathbf{a})$, $c(= \frac{5}{2}\mathbf{a})$, $e(= 2\mathbf{a})$, and $f(= -\frac{1}{2}\mathbf{a})$ are parallel to $\mathbf{a}$.

24. $\langle 5, 2 \rangle$

27. $\|\mathbf{a}\| = 5$;   (a) $\mathbf{u} = \frac{1}{5}\langle 0, -5 \rangle = \langle 0, -1 \rangle$;   (b) $-\mathbf{u} = \langle 0, 1 \rangle$

30. $\|\mathbf{a} + \mathbf{b}\| = \|\langle -5, 4 \rangle\| = \sqrt{25 + 16} = \sqrt{41}$; $\mathbf{u} = \frac{1}{\sqrt{41}}\langle -5, 4 \rangle = \langle -\frac{5}{\sqrt{41}}, \frac{4}{\sqrt{41}} \rangle$

33. $-\frac{3}{4}\mathbf{a} = \langle -3, -15/2 \rangle$

36.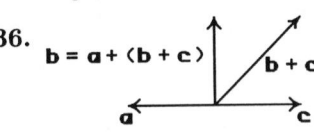

39.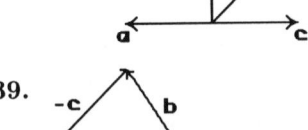

$\mathbf{b} = (-\mathbf{c}) - \mathbf{a}$; $(\mathbf{b} + \mathbf{c}) + \mathbf{a} = \mathbf{0}$; $\mathbf{a} + \mathbf{b} + \mathbf{c} = \mathbf{0}$

42. From $2\mathbf{i} + 3\mathbf{j} = k_1 \mathbf{b} + k_2 \mathbf{c} = k_1(-2\mathbf{i} + 4\mathbf{j}) + k_2(5\mathbf{i} + 7\mathbf{j}) = (-2k_1 + 5k_2)\mathbf{i} + (4k_1 + 7k_2)\mathbf{j}$ we obtain the system of equations $-2k_1 + 5k_2 = 2$, $4k_1 + 7k_2 = 3$. Solving, we find $k_1 = \frac{1}{34}$ and $k_2 = \frac{7}{17}$.

45. (a) Since $\mathbf{F}_f = -\mathbf{F}_g$, $\|\mathbf{F}_g\| = \|\mathbf{F}_f\| = \mu\|\mathbf{F}_n\|$ and $\tan\theta = \|\mathbf{F}_g\|/\|\mathbf{F}_n\| = \mu\|\mathbf{F}_n\|/\|\mathbf{F}_n\| = \mu$.

     (b) $\theta = \tan^{-1} 0.6 \approx 31°$.

48. Place one corner of the parallelogram at the origin and let two adjacent sides be $\overrightarrow{OP_1}$ and $\overrightarrow{OP_2}$. Let M be the midpoint of the diagonal connecting P_1 and P_2 and N be the midpoint of the other diagonal. 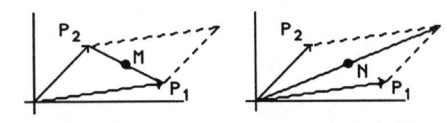 By Problem 37, $\overrightarrow{OM} = \frac{1}{2}(\overrightarrow{OP_1} + \overrightarrow{OP_2})$. Since $\overrightarrow{OP_1} + \overrightarrow{OP_2}$ is the main diagonal of the parallelogram and N is its midpoint, $\overrightarrow{ON} = \frac{1}{2}(\overrightarrow{OP_1} + \overrightarrow{OP_2})$. Thus, $\overrightarrow{OM} = \overrightarrow{ON}$ and the diagonals bisect each other.

Exercises 7.2

3.

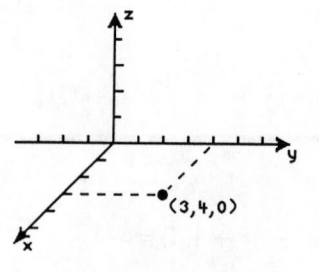

6.

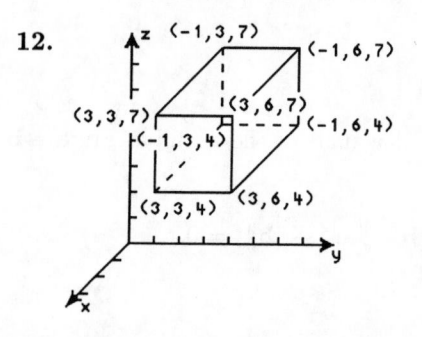

9. A line perpendicular to the xy-plane at $(2,3,0)$

12.

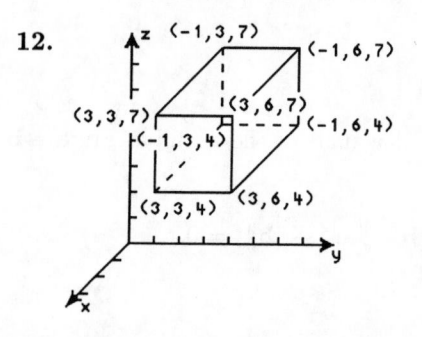

15. The union of the planes $x = 0$, $y = 0$, and $z = 0$

18. The union of the planes $x = 2$ and $z = 8$

21. $d = \sqrt{(3-6)^2 + (-1-4)^2 + (2-8)^2} = \sqrt{70}$

24. (a) 2; (b) $d = \sqrt{(-6)^2 + 2^2 + (-3)^2} = 7$

27. $d(P_1, P_2) = \sqrt{(4-1)^2 + (1-2)^2 + (3-3)^2} = \sqrt{10}$

$d(P_1, P_3) = \sqrt{(4-1)^2 + (6-2)^2 + (4-3)^2} = \sqrt{26}$

$d(P_2, P_3) = \sqrt{(4-4)^2 + (6-1)^2 + (4-3)^2} = \sqrt{26}$; The triangle is an isosceles triangle.

30. $d(P_1, P_2) = \sqrt{(1-2)^2 + (4-3)^2 + (4-2)^2} = \sqrt{6}$

$d(P_1, P_3) = \sqrt{(5-2)^2 + (0-3)^2 + (-4-2)^2} = 3\sqrt{6}$

$d(P_2, P_3) = \sqrt{(5-1)^2 + (0-4)^2 + (-4-4)^2} = 4\sqrt{6}$

Since $d(P_1, P_2) + d(P_1, P_3) = d(P_2, P_3)$, the points P_1, P_2, and P_3 are collinear.

33. $\left(\dfrac{1+7}{2}, \dfrac{3+(-2)}{2}, \dfrac{1/2 + 5/2}{2} \right) = (4, 1/2, 3/2)$

36. $(-3 + (-5))/2 = x_3 = -4$; $(4+8)/2 = y_3 = 6$; $(1+3)/2 = z_3 = 2$.

The coordinates of P_3 are $(-4, 6, 2)$.

(a) $\left(\dfrac{-3 + (-4)}{2}, \dfrac{4+6}{2}, \dfrac{1+2}{2} \right) = (-7/2, 5, 3/2)$

(b) $\left(\dfrac{-4 + (-5)}{2}, \dfrac{6+8}{2}, \dfrac{2+3}{2} \right) = (-9/2, 7, 5/2)$

39. $\overrightarrow{P_1 P_2} = \langle 2, 1, 1 \rangle$

42. $2\mathbf{a} - (\mathbf{b} - \mathbf{c}) = \langle 2, -6, 4 \rangle - \langle -3, -5, -8 \rangle = \langle 5, -1, 12 \rangle$

45. $\|\mathbf{a} + \mathbf{c}\| = \|\langle 3, 3, 11 \rangle\| = \sqrt{9 + 9 + 121} = \sqrt{139}$

48. $\|\mathbf{b}\|\mathbf{a} + \|\mathbf{a}\|\mathbf{b} = \sqrt{1 + 1 + 1}\,\langle 1, -3, 2 \rangle + \sqrt{1 + 9 + 4}\,\langle -1, 1, 1 \rangle = \langle \sqrt{3}, -3\sqrt{3}, 2\sqrt{3} \rangle + \langle -\sqrt{14}, \sqrt{14}, \sqrt{14} \rangle$

$\qquad = \langle \sqrt{3} - \sqrt{14}, -3\sqrt{3} + \sqrt{14}, 2\sqrt{3} + \sqrt{14} \rangle$

51. $\mathbf{b} = 4\mathbf{a} = 4\mathbf{i} - 4\mathbf{j} + 4\mathbf{k}$

Exercises 7.3

3. $\mathbf{a} \cdot \mathbf{b} = 2(-1) + (-3)2 + 4(5) = 12$

6. $\mathbf{a} \cdot (\mathbf{b} + \mathbf{c}) = 2(2) + (-3)8 + 4(4) = -4$

9. $\mathbf{a} \cdot \mathbf{a} = 2^2 + (-3)^2 + 4^2 = 29$

12. $(2\mathbf{a}) \cdot (\mathbf{a} - 2\mathbf{b}) = 4(4) + (-6)(-7) + 8(-6) = 10$

15. a and f, b and e, c and d

18. If $\mathbf{a}$ and $\mathbf{b}$ represent adjacent sides of the rhombus, then $\|\mathbf{a}\| = \|\mathbf{b}\|$, the diagonals of the rhombus are $\mathbf{a} + \mathbf{b}$ and $\mathbf{a} - \mathbf{b}$, and

$$(\mathbf{a} + \mathbf{b}) \cdot (\mathbf{a} - \mathbf{b}) = \mathbf{a} \cdot \mathbf{a} - \mathbf{a} \cdot \mathbf{b} + \mathbf{b} \cdot \mathbf{a} - \mathbf{b} \cdot \mathbf{b} = \mathbf{a} \cdot \mathbf{a} - \mathbf{b} \cdot \mathbf{b} = \|\mathbf{a}\|^2 - \|\mathbf{b}\|^2 = 0.$$

Thus, the diagonals are perpendicular.

21. $\mathbf{a} \cdot \mathbf{b} = 3(2) + (-1)2 = 4;\quad \|\mathbf{a}\| = \sqrt{10},\ \|\mathbf{b}\| = 2\sqrt{2}$

$\cos\theta = \dfrac{4}{(\sqrt{10})(2\sqrt{2})} = \dfrac{1}{\sqrt{5}} \implies \theta = \cos^{-1}\dfrac{1}{\sqrt{5}} \approx 1.11 \text{ rad} \approx 63.43°$

24. $\mathbf{a} \cdot \mathbf{b} = \frac{1}{2}(2) + \frac{1}{2}(-4) + \frac{3}{2}(6) = 8;\quad \|\mathbf{a}\| = \sqrt{11}/2,\ \|\mathbf{b}\| = 2\sqrt{14}$

$\cos\theta = \dfrac{8}{(\sqrt{11}/2)(2\sqrt{14})} = \dfrac{8}{\sqrt{154}} \implies \theta = \cos^{-1}(8/\sqrt{154}) \approx 0.87 \text{ rad} \approx 49.86°$

27. $\|\mathbf{a}\| = 2;\quad \cos\alpha = 1/2,\ \alpha = 60°;\quad \cos\beta = 0,\ \beta = 90°;\quad \cos\gamma = -\sqrt{3}/2,\ \gamma = 150°$

30. If $\mathbf{a}$ and $\mathbf{b}$ are orthogonal, then $\mathbf{a} \cdot \mathbf{b} = 0$ and

$$\cos\alpha_1\cos\alpha_2 + \cos\beta_1\cos\beta_2 + \cos\gamma_1\cos\gamma_2 = \frac{a_1}{\|\mathbf{a}\|}\frac{b_1}{\|\mathbf{b}\|} + \frac{a_2}{\|\mathbf{a}\|}\frac{b_2}{\|\mathbf{b}\|} + \frac{a_3}{\|\mathbf{a}\|}\frac{b_3}{\|\mathbf{b}\|}$$

$$= \frac{1}{\|\mathbf{a}\|\,\|\mathbf{b}\|}(a_1b_1 + a_2b_2 + a_3b_3) = \frac{1}{\|\mathbf{a}\|\,\|\mathbf{b}\|}(\mathbf{a} \cdot \mathbf{b}) = 0.$$

33. $\text{comp}_\mathbf{b}\mathbf{a} = \mathbf{a} \cdot \mathbf{b}/\|\mathbf{b}\| = \langle 1, -1, 3 \rangle \cdot \langle 2, 6, 3 \rangle / 7 = 5/7$

36. $\mathbf{a} + \mathbf{b} = \langle 3, 5, 6 \rangle;\quad 2\mathbf{b} = \langle 4, 12, 6 \rangle;\quad \text{comp}_{2\mathbf{b}}(\mathbf{a} + \mathbf{b}) \cdot 2\mathbf{b}/|2\mathbf{b}| = \langle 3, 5, 6 \rangle \cdot \langle 4, 12, 6 \rangle / 14 = 54/7$

39. (a) $\text{comp}_\mathbf{b}\mathbf{a} = \mathbf{a} \cdot \mathbf{b}/\|\mathbf{b}\| = (-5\mathbf{i} + 5\mathbf{j}) \cdot (-3\mathbf{i} + 4\mathbf{j})/5 = 7$

$\text{proj}_\mathbf{b}\mathbf{a} = (\text{comp}_\mathbf{b}\mathbf{a})\mathbf{b}/\|\mathbf{b}\| = 7(-3\mathbf{i} + 4\mathbf{j})/5 = -\frac{21}{5}\mathbf{i} + \frac{28}{5}\mathbf{j}$

(b) $\text{proj}_{\mathbf{b}\perp}\mathbf{a} = \mathbf{a} - \text{proj}_\mathbf{b}\mathbf{a} = (-5\mathbf{i} + 5\mathbf{j}) - (-\frac{21}{5}\mathbf{i} + \frac{28}{5}\mathbf{j}) = -\frac{4}{5}\mathbf{i} - \frac{3}{5}\mathbf{j}$

42. (a) $\text{comp}_\mathbf{b}\mathbf{a} = \mathbf{a} \cdot \mathbf{b}/\|\mathbf{b}\| = \langle 1, 1, 1 \rangle \cdot \langle -2, 2, -1 \rangle / 3 = -1/3$

$\text{proj}_\mathbf{b}\mathbf{a} = (\text{comp}_\mathbf{b}\mathbf{a})\mathbf{b}/\|\mathbf{b}\| = -\frac{1}{3}\langle -2, 2, -1 \rangle / 3 = \langle 2/9, -2/9, 1/9 \rangle$

(b) $\text{proj}_{\mathbf{b}\perp}\mathbf{a} = \mathbf{a} - \text{proj}_\mathbf{b}\mathbf{a} = \langle 1, 1, 1 \rangle - \langle 2/9, -2/9, 1/9 \rangle = \langle 7/9, 11/9, 8/9 \rangle$

45. We identify $\|\mathbf{F}\| = 20$, $\theta = 60°$ and $\|\mathbf{d}\| = 100$. Then $W = \|\mathbf{F}\|\,\|\mathbf{d}\|\cos\theta = 20(100)(\frac{1}{2}) = 1000$ ft-lb.

48. Using $\mathbf{d} = 6\mathbf{i} + 2\mathbf{j}$ and $\mathbf{F} = 3(\frac{3}{5}\mathbf{i} + \frac{4}{5}\mathbf{j})$, $W = \mathbf{F} \cdot \mathbf{d} = \langle \frac{9}{5}, \frac{12}{5} \rangle \cdot \langle 6, 2 \rangle = \frac{78}{5}$ ft-lb.

51. $\|\mathbf{a} + \mathbf{b}\|^2 = (\mathbf{a} + \mathbf{b}) \cdot (\mathbf{a} + \mathbf{b}) = \mathbf{a} \cdot \mathbf{a} + 2\mathbf{a} \cdot \mathbf{b} + \mathbf{b} \cdot \mathbf{b} = \|\mathbf{a}\|^2 + 2\mathbf{a} \cdot \mathbf{b} + \|\mathbf{b}\|^2$

$\qquad \leq \|\mathbf{a}\|^2 + 2|\mathbf{a} \cdot \mathbf{b}| + \|\mathbf{b}\|^2 \quad \boxed{\text{since } x \leq |x|}$

$\qquad \leq \|\mathbf{a}\|^2 + 2\|\mathbf{a}\|\,\|\mathbf{b}\| + \|\mathbf{b}\|^2 = (\|\mathbf{a}\| + \|\mathbf{b}\|)^2 \quad \boxed{\text{by Problem 50}}$

Thus, since $\|\mathbf{a} + \mathbf{b}\|$ and $\|\mathbf{a}\| + \|\mathbf{b}\|$ are positive, $\|\mathbf{a} + \mathbf{b}\| \leq \|\mathbf{a}\| + \|\mathbf{b}\|$.

Exercises 7.4

3. $\mathbf{a} \times \mathbf{b} = \begin{vmatrix} \mathbf{i} & \mathbf{j} & \mathbf{k} \\ 1 & -3 & 1 \\ 2 & 0 & 4 \end{vmatrix} = \begin{vmatrix} -3 & 1 \\ 0 & 4 \end{vmatrix} \mathbf{i} - \begin{vmatrix} 1 & 1 \\ 2 & 4 \end{vmatrix} \mathbf{j} + \begin{vmatrix} 1 & -3 \\ 2 & 0 \end{vmatrix} \mathbf{k} = \langle -12, -2, 6 \rangle$

6. $\mathbf{a} \times \mathbf{b} = \begin{vmatrix} \mathbf{i} & \mathbf{j} & \mathbf{k} \\ 4 & 1 & -5 \\ 2 & 3 & -1 \end{vmatrix} = \begin{vmatrix} 1 & -5 \\ 3 & -1 \end{vmatrix} \mathbf{i} - \begin{vmatrix} 4 & -5 \\ 2 & -1 \end{vmatrix} \mathbf{j} + \begin{vmatrix} 4 & 1 \\ 2 & 3 \end{vmatrix} \mathbf{k} = 14\mathbf{i} - 6\mathbf{j} + 10\mathbf{k}$

9. $\mathbf{a} \times \mathbf{b} = \begin{vmatrix} \mathbf{i} & \mathbf{j} & \mathbf{k} \\ 2 & 2 & -4 \\ -3 & -3 & 6 \end{vmatrix} = \begin{vmatrix} 2 & -4 \\ -3 & 6 \end{vmatrix} \mathbf{i} - \begin{vmatrix} 2 & -4 \\ -3 & 6 \end{vmatrix} \mathbf{j} + \begin{vmatrix} 2 & 2 \\ -3 & -3 \end{vmatrix} \mathbf{k} = \langle 0, 0, 0 \rangle$

12. $\overrightarrow{P_1P_2} = (0, 1, 1); \quad \overrightarrow{P_1P_3} = (1, 2, 2); \quad \overrightarrow{P_1P_2} \times \overrightarrow{P_1P_3} = \begin{vmatrix} \mathbf{i} & \mathbf{j} & \mathbf{k} \\ 0 & 1 & 1 \\ 1 & 2 & 2 \end{vmatrix} = \begin{vmatrix} 1 & 1 \\ 2 & 2 \end{vmatrix} \mathbf{i} - \begin{vmatrix} 0 & 1 \\ 1 & 2 \end{vmatrix} \mathbf{j} + \begin{vmatrix} 0 & 1 \\ 1 & 2 \end{vmatrix} \mathbf{k} = \mathbf{j} - \mathbf{k}$

15. $\mathbf{a} \times \mathbf{b} = \begin{vmatrix} \mathbf{i} & \mathbf{j} & \mathbf{k} \\ 5 & -2 & 1 \\ 2 & 0 & -7 \end{vmatrix} = \begin{vmatrix} -2 & 1 \\ 0 & -7 \end{vmatrix} \mathbf{i} - \begin{vmatrix} 5 & 1 \\ 2 & -7 \end{vmatrix} \mathbf{j} + \begin{vmatrix} 5 & -2 \\ 2 & 0 \end{vmatrix} \mathbf{k} = \langle 14, 37, 4 \rangle$

$\quad \mathbf{a} \cdot (\mathbf{a} \times \mathbf{b}) = \langle 5, -2, -1 \rangle \cdot \langle 14, 37, 4 \rangle = 70 - 74 + 4 = 0; \quad \mathbf{b} \cdot (\mathbf{a} \times \mathbf{b}) = \langle 2, 0, -7 \rangle \cdot \langle 14, 37, 4 \rangle = 28 + 0 - 28 = 0$

18. (a) $\mathbf{b} \times \mathbf{c} = \begin{vmatrix} \mathbf{i} & \mathbf{j} & \mathbf{k} \\ 1 & 2 & -1 \\ -1 & 5 & 8 \end{vmatrix} = \begin{vmatrix} 2 & -1 \\ 5 & 8 \end{vmatrix} \mathbf{i} - \begin{vmatrix} 1 & -1 \\ -1 & 8 \end{vmatrix} \mathbf{j} + \begin{vmatrix} 1 & 2 \\ -1 & 5 \end{vmatrix} \mathbf{k} = 21\mathbf{i} - 7\mathbf{j} + 7\mathbf{k}$

$\quad \mathbf{a} \times (\mathbf{b} \times \mathbf{c}) = \begin{vmatrix} \mathbf{i} & \mathbf{j} & \mathbf{k} \\ 3 & 0 & -4 \\ 21 & -7 & 7 \end{vmatrix} = \begin{vmatrix} 0 & -4 \\ -7 & 7 \end{vmatrix} \mathbf{i} - \begin{vmatrix} 3 & -4 \\ 21 & 7 \end{vmatrix} \mathbf{j} + \begin{vmatrix} 3 & 0 \\ 21 & -7 \end{vmatrix} \mathbf{k} = -28\mathbf{i} - 105\mathbf{j} - 21\mathbf{k}$

(b) $\mathbf{a} \cdot \mathbf{c} = (3\mathbf{i} - 4\mathbf{k}) \cdot (-\mathbf{i} + 5\mathbf{j} + 8\mathbf{k}) = -35; \quad (\mathbf{a} \cdot \mathbf{c})\mathbf{b} = -35(\mathbf{i} + 2\mathbf{j} - \mathbf{k}) = -35\mathbf{i} - 70\mathbf{j} + 35\mathbf{k}$

$\quad \mathbf{a} \cdot \mathbf{b} = (3\mathbf{i} - 4\mathbf{k}) \cdot (\mathbf{i} + 2\mathbf{j} - \mathbf{k}) = 7; \quad (\mathbf{a} \cdot \mathbf{b})\mathbf{c} = 7(-\mathbf{i} + 5\mathbf{j} + 8\mathbf{k}) = -7\mathbf{i} + 35\mathbf{j} + 56\mathbf{k}$

$\quad \mathbf{a} \times (\mathbf{b} \times \mathbf{c}) = (\mathbf{a} \cdot \mathbf{c})\mathbf{b} - (\mathbf{a} \cdot \mathbf{b})\mathbf{c} = (-35\mathbf{i} - 70\mathbf{j} + 35\mathbf{k}) - (-7\mathbf{i} + 35\mathbf{j} + 56\mathbf{k}) = -28\mathbf{i} - 105\mathbf{j} - 21\mathbf{k}$

21. $\mathbf{k} \times (2\mathbf{i} - \mathbf{j}) = \mathbf{k} \times (2\mathbf{i}) + \mathbf{k} \times (-\mathbf{j}) = 2(\mathbf{k} \times \mathbf{i}) - (\mathbf{k} \times \mathbf{j}) = 2\mathbf{j} - (-\mathbf{i}) = \mathbf{i} + 2\mathbf{j}$

24. $(2\mathbf{i} - \mathbf{j} + 5\mathbf{k}) \times \mathbf{i} = (2\mathbf{i} \times \mathbf{i}) + (-\mathbf{j} \times \mathbf{i}) + (5\mathbf{k} \times \mathbf{i}) = 2(\mathbf{i} \times \mathbf{i}) + (\mathbf{i} \times \mathbf{j}) + 5(\mathbf{k} \times \mathbf{i}) = 5\mathbf{j} + \mathbf{k}$

27. $\mathbf{k} \cdot (\mathbf{j} \times \mathbf{k}) = \mathbf{k} \cdot \mathbf{i} = 0$

30. $(\mathbf{i} \times \mathbf{j}) \cdot (3\mathbf{j} \times \mathbf{i}) = \mathbf{k} \cdot (-3\mathbf{k}) = -3(\mathbf{k} \cdot \mathbf{k}) = -3$

33. $(\mathbf{i} \times \mathbf{i}) \times \mathbf{j} = \mathbf{0} \times \mathbf{j} = \mathbf{0}$

36. $(\mathbf{i} \times \mathbf{k}) \times (\mathbf{j} \times \mathbf{i}) = (-\mathbf{j}) \times (-\mathbf{k}) = (-1)(-1)(\mathbf{j} \times \mathbf{k}) = \mathbf{j} \times \mathbf{k} = \mathbf{i}$

39. $(-\mathbf{a}) \times \mathbf{b} = -(\mathbf{a} \times \mathbf{b}) = -4\mathbf{i} + 3\mathbf{j} - 6\mathbf{k}$

42. $(\mathbf{a} \times \mathbf{b}) \cdot \mathbf{c} = 4(2) + (-3)4 + 6(-1) = -10$

45. (a) Let $A = (1,3,0)$, $B = (2,0,0)$, $C = (0,0,4)$, and $D = (1,-3,4)$. Then $\overrightarrow{AB} = \mathbf{i} - 3\mathbf{j}$, $\overrightarrow{AC} = -\mathbf{i} - 3\mathbf{j} + 4\mathbf{k}$, $\overrightarrow{CD} = \mathbf{i} - 3\mathbf{j}$, and $\overrightarrow{BD} = -\mathbf{i} - 3\mathbf{j} + 4\mathbf{k}$. Since $\overrightarrow{AB} = \overrightarrow{CD}$ and $\overrightarrow{AC} = \overrightarrow{BD}$, the quadrilateral is a parallelogram.

(b) Computing

$$\overrightarrow{AB} \times \overrightarrow{AC} = \begin{vmatrix} \mathbf{i} & \mathbf{j} & \mathbf{k} \\ 1 & -3 & 0 \\ -1 & -3 & 4 \end{vmatrix} = -12\mathbf{i} - 4\mathbf{j} - 6\mathbf{k}$$

we find that the area is $\| -12\mathbf{i} - 4\mathbf{j} - 6\mathbf{k}\| = \sqrt{144 + 16 + 36} = 14$.

48. $\overrightarrow{P_1P_2} = \mathbf{j} + 2\mathbf{k}$; $\overrightarrow{P_2P_3} = 2\mathbf{i} + \mathbf{j} - 2\mathbf{k}$

$$\overrightarrow{P_1P_2} \times \overrightarrow{P_2P_3} = \begin{vmatrix} \mathbf{i} & \mathbf{j} & \mathbf{k} \\ 0 & 1 & 2 \\ 2 & 1 & -2 \end{vmatrix} = \begin{vmatrix} 1 & 2 \\ 1 & -2 \end{vmatrix}\mathbf{i} - \begin{vmatrix} 0 & 2 \\ 2 & -2 \end{vmatrix}\mathbf{j} + \begin{vmatrix} 0 & 1 \\ 2 & 1 \end{vmatrix}\mathbf{k} = -4\mathbf{i} + 4\mathbf{j} - 2\mathbf{k}$$

$A = \frac{1}{2}\| -4\mathbf{i} + 4\mathbf{j} - 2\mathbf{k}\| = 3$ sq. units

51. $\mathbf{b} \times \mathbf{c} = \begin{vmatrix} \mathbf{i} & \mathbf{j} & \mathbf{k} \\ -1 & 4 & 0 \\ 2 & 2 & 2 \end{vmatrix} = \begin{vmatrix} 4 & 0 \\ 2 & 2 \end{vmatrix}\mathbf{i} - \begin{vmatrix} -1 & 0 \\ 2 & 2 \end{vmatrix}\mathbf{j} + \begin{vmatrix} -1 & 4 \\ 2 & 2 \end{vmatrix}\mathbf{k} = 8\mathbf{i} + 2\mathbf{j} - 10\mathbf{k}$

$\mathbf{v} = |\mathbf{a} \cdot (\mathbf{b} \times \mathbf{c})| = |(\mathbf{i} + \mathbf{j}) \cdot (8\mathbf{i} + 2\mathbf{j} - 10\mathbf{k})| = |8 + 2 + 0| = 10$ cu. units

54. The four points will be coplanar if the three vectors $\overrightarrow{P_1P_2} = \langle 3, -1, -1\rangle$, $\overrightarrow{P_2P_3} = \langle -3, -5, 13\rangle$, and $\overrightarrow{P_3P_4} = \langle -8, 7, -6\rangle$ are coplanar.

$$\overrightarrow{P_2P_3} \times \overrightarrow{P_3P_4} = \begin{vmatrix} \mathbf{i} & \mathbf{j} & \mathbf{k} \\ -3 & -5 & 13 \\ -8 & 7 & -6 \end{vmatrix} = \begin{vmatrix} -5 & 13 \\ 7 & -6 \end{vmatrix}\mathbf{i} - \begin{vmatrix} -3 & 13 \\ -8 & -6 \end{vmatrix}\mathbf{j} + \begin{vmatrix} -3 & -5 \\ -8 & 7 \end{vmatrix}\mathbf{k} = \langle -61, -122, -61\rangle$$

$\overrightarrow{P_1P_2} \cdot (\overrightarrow{P_2P_3} \times \overrightarrow{P_3P_4}) = \langle 3, -1, -1\rangle \cdot \langle -61, -122, -61\rangle = -183 + 122 + 61 = 0$

The four points are coplanar.

57. (a) We note first that $\mathbf{a} \times \mathbf{b} = \mathbf{k}$, $\mathbf{b} \times \mathbf{c} = \frac{1}{2}(\mathbf{i} - \mathbf{k})$, $\mathbf{c} \times \mathbf{a} = \frac{1}{2}(\mathbf{j} - \mathbf{k})$, $\mathbf{a} \cdot (\mathbf{b} \times \mathbf{c}) = \frac{1}{2}$, $\mathbf{b} \cdot (\mathbf{c} \times \mathbf{a}) = \frac{1}{2}$, and $\mathbf{c} \cdot (\mathbf{a} \times \mathbf{b}) = \frac{1}{2}$. Then

$$\mathbf{A} = \frac{\frac{1}{2}(\mathbf{i} - \mathbf{k})}{\frac{1}{2}} = \mathbf{i} - \mathbf{k}, \quad \mathbf{B} = \frac{\frac{1}{2}(\mathbf{j} - \mathbf{k})}{\frac{1}{2}} = \mathbf{j} - \mathbf{k}, \quad \text{and} \quad \mathbf{C} = \frac{\mathbf{k}}{\frac{1}{2}} = 2\mathbf{k}.$$

(b) We need to compute $\mathbf{A} \cdot (\mathbf{B} \times \mathbf{C})$. Using formula (10) in the text we have

$$\mathbf{B} \times \mathbf{C} = \frac{(\mathbf{c} \times \mathbf{a}) \times (\mathbf{a} \times \mathbf{b})}{[\mathbf{b} \cdot (\mathbf{c} \times \mathbf{a})][\mathbf{c} \cdot (\mathbf{a} \times \mathbf{b})]} = \frac{[(\mathbf{c} \times \mathbf{a}) \cdot \mathbf{b}]\mathbf{a} - [(\mathbf{c} \times \mathbf{a}) \cdot \mathbf{a}]\mathbf{b}}{[\mathbf{b} \cdot (\mathbf{c} \times \mathbf{a})][\mathbf{c} \cdot (\mathbf{a} \times \mathbf{b})]}$$

$$= \frac{\mathbf{a}}{\mathbf{c} \cdot (\mathbf{a} \times \mathbf{b})} \quad \boxed{\text{since } (\mathbf{c} \times \mathbf{a}) \cdot \mathbf{a} = 0.}$$

Then

$$\mathbf{A} \cdot (\mathbf{B} \times \mathbf{C}) = \frac{\mathbf{b} \times \mathbf{c}}{\mathbf{a} \cdot (\mathbf{b} \times \mathbf{c})} \cdot \frac{\mathbf{a}}{\mathbf{c} \cdot (\mathbf{a} \times \mathbf{b})} = \frac{1}{\mathbf{c} \cdot (\mathbf{a} \times \mathbf{b})}$$

and the volume of the unit cell of the reciprocal latrice is the reciprocal of the volume of the unit cell of the original lattice.

60. The statement is false since $\mathbf{i} \times (\mathbf{i} \times \mathbf{j}) = \mathbf{i} \times \mathbf{k} = -\mathbf{j}$ and $(\mathbf{i} \times \mathbf{i}) \times \mathbf{j} = \mathbf{0} \times \mathbf{j} = \mathbf{0}$.

63. Since
$$\|\mathbf{a} \times \mathbf{b}\|^2 = (a_2b_3 - a_3b_2)^2 + (a_1b_3 - a_3b_1)^2 + (a_1b_2 - a_2b_1)^2$$
$$= a_2^2b_3^2 - 2a_2b_3a_3b_2 + a_3^2b_2^2 + a_1^2b_3^2 - 2a_1b_3a_3b_1 + a_3^2b_1^2 + a_1^2b_2^2 - 2a_1b_2a_2b_1 + a_2^2b_1^2$$

and
$$\|\mathbf{a}\|^2\|\mathbf{b}\|^2 - (\mathbf{a}\cdot\mathbf{b})^2 = (a_1^2 + a_2^2 + a_3^2)(b_1^2 + b_2^2 + b_3^2) - (a_1b_1 + a_2b_2 + a_3b_3)^2$$
$$= a_1^2a_2^2 + a_1^2b_2^2 + a_1^2b_3^2 + a_2^2b_1^2 + a_2^2b_2^2 + a_2^2b_3^2 + a_3^2b_1^2 + a_3^2b_2^2 + a_3^2b_3^2$$
$$\quad - a_1^2b_1^2 - a_2^2b_2^2 - a_3^2b_3^2 - 2a_1b_1a_2b_2 - 2a_1b_1a_3b_3 - 2a_2b_2a_3b_3$$
$$= a_1^2b_2^2 + a_1^2b_3^2 + a_2^2b_1^2 + a_2^2b_3^2 + a_3^2b_1^2 + a_3^2b_2^2 - 2a_1a_2b_1b_2 - 2a_1a_3b_1b_3 - 2a_2a_3b_2b_3$$

we see that $\|\mathbf{a} \times \mathbf{b}\|^2 = \|\mathbf{a}\|^2\|\mathbf{b}\|^2 - (\mathbf{a}\cdot\mathbf{b})^2$.

Exercises 7.5

The equation of a line through P_1 and P_2 in 3-space with $\mathbf{r}_1 = \overrightarrow{OP_1}$ and $\mathbf{r}_2 = \overrightarrow{OP_2}$ can be expressed as $\mathbf{r} = \mathbf{r}_1 + t(k\mathbf{a})$ or $\mathbf{r} = \mathbf{r}_2 + t(k\mathbf{a})$ where $\mathbf{a} = \mathbf{r}_2 - \mathbf{r}_1$ and k is any non-zero scalar. Thus, the form of the equation of a line is not unique.

3. $\mathbf{a} = \langle 1/2 - (-3/2), -1/2 - 5/2, 1 - (-1/2) \rangle = \langle 2, -3, 3/2 \rangle$; $\langle x, y, z \rangle = \langle 1/2, -1/2, 1 \rangle + t\langle 2, -3, 3/2 \rangle$

6. $\mathbf{a} = \langle 3 - 5/2, 2 - 1, 1 - (-2) \rangle = \langle 1/2, 1, 3 \rangle$; $\langle x, y, z \rangle = \langle 3, 2, 1 \rangle + t\langle 1/2, 1, 3 \rangle$

9. $\mathbf{a} = \langle 1 - 3, 0 - (-2), 0 - (-7) \rangle = \langle -2, 2, 7 \rangle$; $x = 1 - 2t$, $y = 2t$, $z = 7t$

12. $\mathbf{a} = \langle -3 - 4, 7 - (-8), 9 - (-1) \rangle = \langle -7, 15, 10 \rangle$; $x = -3 - 7t$, $y = 7 + 15t$, $z = 9 + 10t$

15. $a_1 = -7 - 4 = -11$, $a_2 = 2 - 2 = 0$, $a_3 = 5 - 1 = 4$; $\dfrac{x+7}{-11} = \dfrac{z-5}{4}$, $y = 2$

18. $a_1 = 5/6 - 1/3 = 1/2$; $a_2 = -1/4 - 3/8 = -5/8$; $a_3 = 1/5 - 1/10 = 1/10$
$$\frac{x - 5/6}{1/2} = \frac{y + 1/4}{-5/8} = \frac{z - 1/5}{1/10}$$

21. parametric: $x = 5t$, $y = 9t$, $z = 4t$; symmetric: $\dfrac{x}{5} = \dfrac{y}{9} = \dfrac{z}{4}$

24. A direction vector is $\langle 5, 1/3, -2 \rangle$. Symmetric equations for the line are $(x-4)/5 = (y+11)/(1/3) = (z+7)/(-2)$.

27. Both lines go through the points $(0,0,0)$ and $(6,6,6)$. Since two points determine a line, the lines are the same.

30. The parametric equations for the line are $x = 1 + 2t$, $y = -2 + 3t$, $z = 4 + 2t$. In the xy-plane, $z = 4 + 2t = 0$ and $t = -2$. Then $x = 1 + 2(-2) = -3$ and $y = -2 + 3(-2) = -8$. The point is $(-3, -8, 0)$. In the xz-plane, $y = -2 + 3t = 0$ and $t = 2/3$. Then $x = 1 + 2(2/3) = 7/3$ and $z = 4 + 2(2/3) = 16/3$. The point is $(7/3, 0, 16/3)$. In the yz-plane, $x = 1 + 2t = 0$ and $t = -1/2$. Then $y = -2 + 3(-1/2) = -7/2$ and $z = 4 + 2(-1/2) = 3$. The point is $(0, -7/2, 3)$.

33. The system of equations $2 - t = 4 + s$, $3 + t = 1 + s$, $1 + t = 1 - s$, or $t + s = -2$, $t - s = -2$, $t + s = 0$ has no solution since $-2 \neq 0$. Thus, the lines do not intersect.

36. $\mathbf{a} = \langle 2, 7, -1 \rangle$, $\mathbf{b} = \langle -2, 1, 4 \rangle$, $\mathbf{a}\cdot\mathbf{b} = -1$, $\|\mathbf{a}\| = 3\sqrt{6}$, $\|\mathbf{b}\| = \sqrt{21}$;
$$\cos\theta = \frac{\mathbf{a}\cdot\mathbf{b}}{\|\mathbf{a}\|\,\|\mathbf{b}\|} = \frac{-1}{(3\sqrt{6})(\sqrt{21})} = -\frac{1}{9\sqrt{14}}; \quad \theta = \cos^{-1}\left(-\frac{1}{9\sqrt{14}}\right) \approx 91.70°$$

39. $2(x - 5) - 3(y - 1) + 4(z - 3) = 0$; $2x - 3y + 4z = 19$

42. $6x - y + 3z = 0$

45. From the points $(3, 5, 2)$ and $(2, 3, 1)$ we obtain the vector $\mathbf{u} = \mathbf{i} + 2\mathbf{j} + \mathbf{k}$. From the points $(2, 3, 1)$ and $(-1, -1, 4)$ we obtain the vector $\mathbf{v} = 3\mathbf{i} + 4\mathbf{j} - 3\mathbf{k}$. From the points $(-1, -1, 4)$ and (x, y, z) we obtain the vector $\mathbf{w} = (x + 1)\mathbf{i} + (y + 1)\mathbf{j} + (z - 4)\mathbf{k}$. Then, a normal vector is

$$\mathbf{u} \times \mathbf{v} = \begin{vmatrix} \mathbf{i} & \mathbf{j} & \mathbf{k} \\ 1 & 2 & 1 \\ 3 & 4 & -3 \end{vmatrix} = -10\mathbf{i} + 6\mathbf{j} - 2\mathbf{k}.$$

A vector equation of the plane is $-10(x + 1) + 6(y + 1) - 2(z - 4) = 0$ or $5x - 3y + z = 2$.

48. The three points are not colinear and all satisfy $x = 0$, which is the equation of the plane.

51. A normal vector to $x + y - 4z = 1$ is $\langle 1, 1, -4 \rangle$. The equation of the parallel plane is $(x - 2) + (y - 3) - 4(z + 5) = 0$ or $x + y - 4z = 25$.

54. A normal vector is $\langle 0, 1, 0 \rangle$. The equation of the plane is $y + 5 = 0$ or $y = -5$.

57. A direction vector for the two lines is $\langle 1, 2, 1 \rangle$. Points on the lines are $(1, 1, 3)$ and $(3, 0, -2)$. Thus, another vector parallel to the plane is $\langle 1 - 3, 1 - 0, 3 + 2 \rangle = \langle -2, 1, 5 \rangle$. A normal vector to the plane is $\langle 1, 2, 1 \rangle \times \langle -2, 1, 5 \rangle = \langle 9, -7, 5 \rangle$. Using the point $(3, 0, -2)$ in the plane, the equation of the plane is $9(x - 3) - 7(y - 0) + 5(z + 2) = 0$ or $9x - 7y + 5z = 17$.

60. A normal vector to the plane is $\langle 2 - 1, 6 - 0, -3 + 2 \rangle = \langle 1, 6, -1 \rangle$. The equation of the plane is $(x - 1) + 6(y - 1) - (z - 1) = 0$ or $x + 6y - z = 6$.

63. A direction vector of the line is $\langle -6, 9, 3 \rangle$, and the normal vectors of the planes are **(a)** $\langle 4, 1, 2 \rangle$, **(b)** $\langle 2, -3, 1 \rangle$, **(c)** $\langle 10, -15, -5 \rangle$, **(d)** $\langle -4, 6, 2 \rangle$. Vectors **(c)** and **(d)** are multiples of the direction vector and hence the corresponding planes are perpendicular to the line.

66. Letting $y = t$ in both equations and solving $x - z = 2 - 2t$, $3x + 2z = 1 + t$, we obtain $x = 1 - \frac{3}{5}t$, $y = t$, $z = -1 + \frac{7}{5}t$ or, letting $t = 5s$, $x = 1 - 3s$, $y = 5s$, $z = -1 + 7s$.

69. Substituting the parametric equations into the equation of the plane, we obtain $2(1 + 2t) - 3(2 - t) + 2(-3t) = -7$ or $t = -3$. Letting $t = -3$ in the equation of the line, we obtain the point of intersection $(-5, 5, 9)$.

72. Substituting the parametric equations into the equation of the plane, we obtain $4 + t - 3(2 + t) + 2(1 + 5t) = 0$ or $t = 0$. Letting $t = 0$ in the equation of the line, we obtain the point of intersection $(4, 2, 1)$.

75. The cross product of the direction vector of a line with the normal vector of a plane will be a normal vector to the desired plane. In this case a direction vector of the line is $\langle 3, -1, 5 \rangle$ and a normal vector to the given plane is $\langle 1, 1, 1 \rangle$. A normal vector to the desired plane is $\langle 3, -1, 5 \rangle \times \langle 1, 1, 1 \rangle = \langle -6, 2, 4 \rangle$. A point on the line, and hence in the plane, is $(4, 0, 1)$. The equation of the plane is $-6(x - 4) + 2(y - 0) + 4(z - 1) = 0$ or $3x - y - 2z = 10$.

78.

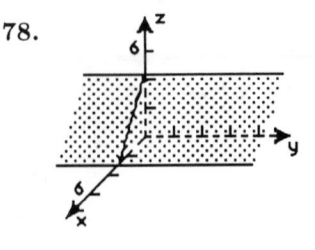

81.

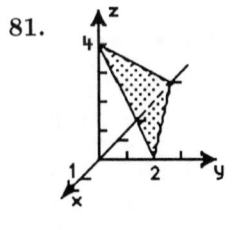

Exercises 7.6

3. Not a vector space. Axiom (**x**) is not satisfied.

6. A vector space

9. A vector space

12. Not a subspace. Axiom (**i**) is not satisfied.

15. A subspace

18. A subspace

21. Let (x_1, y_1, z_1) and (x_2, y_2, z_2) be in S. Then

$$(x_1, y_1, z_1) + (x_2, y_2, z_2) = (at_1, bt_1, ct_1) + (at_2, bt_2, ct_2) = (a(t_1 + t_2), b(t_1 + t_2), c(t_1 + t_2))$$

is in S. Also, for (x, y, z) in S then $k(x, y, z) = (kx, ky, kz) = (a(kt), b(kt), c(kt))$ is also in S.

24. (a) The assumption $c_1 p_1 + c_2 p_2 = 0$ is equivalent to $(c_1 + c_2)x + (c_1 - c_2) = 0$. Thus $c_1 + c_2 = 0$, $c_1 - c_2 = 0$. The only solution of this system is $c_1 = 0$, $c_2 = 0$.

(b) Solving the system $c_1 + c_2 = 5$, $c_1 - c_2 = 2$ gives $c_1 = \frac{7}{2}$, $c_2 = \frac{3}{2}$. Thus $p(x) = \frac{7}{2} p_1(x) + \frac{3}{2} p_2(x)$

27. Linearly independent

30. $(x, \sin x) = \displaystyle\int_0^{2\pi} x \sin x \, dx = (-x \cos x + \sin x) \Big|_0^{2\pi} = -2\pi$

33. We need to show that $\text{Span}\{x_1, x_2, \ldots, x_n\}$ is closed under vector addition and scalar multiplication. Suppose $\mathbf{u}$ and $\mathbf{v}$ are in $\text{Span}\{x_1, x_2, \ldots, x_n\}$. Then $\mathbf{u} = a_1 x_1 + a_2 x_2 + \cdots + a_n x_n$ and $\mathbf{v} = b_1 x_1 + b_2 x_2 + \cdots + b_n x_n$, so that

$$\mathbf{u} + \mathbf{v} = (a_1 + b_1)x_1 + (a_2 + b_2)x_2 + \cdots + (a_n + b_n)x_n,$$

which is in $\text{Span}\{x_1, x_2, \ldots, x_n\}$. Also, for any real number k,

$$k\mathbf{u} = k(a_1 x_1 + a_2 x_2 + \cdots + a_n x_n) = ka_1 x_1 + ka_2 x_2 + \cdots + ka_n x_n,$$

which is in $\text{Span}\{x_1, x_2, \ldots, x_n\}$. Thus, $\text{Span}\{x_1, x_2, \ldots, x_n\}$ is a subspace of $\mathbf{V}$.

Chapter 7 Review Exercises

3. False; since a normal to the plane is $\langle 2, 3, -4 \rangle$ which is not a multiple of the direction vector $\langle 5, -2, 1 \rangle$ of the line.

6. True

9. True

12. orthogonal

15. $\sqrt{(-12)^2 + 4^2 + 6^2} = 14$

18. The coordinates of $(1, -2, -10)$ satisfy the given equation.

21. $x_2 - 2 = 3$, $x_2 = 5$; $y_2 - 1 = 5$, $y_2 = 6$; $z_2 - 7 = -4$, $z_2 = 3$; $P_2 = (5, 6, 3)$

24. $2\mathbf{b} = \langle -2, 4, 2 \rangle$; $4\mathbf{c} = \langle 0, -8, 8 \rangle$; $\mathbf{a} \cdot (2\mathbf{b} + 4\mathbf{c}) = \langle 3, 1, 0 \rangle \cdot \langle -2, -4, 10 \rangle = -10$

27. $A = |5\mathbf{i} - 4\mathbf{j} - 7\mathbf{k}|/2 = 3\sqrt{10}/2$

30. parallel: $-2c = 5$, $c = -5/2$; orthogonal: $1(-2) + 3(-6) + c(5) = 0$, $c = 4$

33. $\text{comp}_{\mathbf{b}}\mathbf{a} = \mathbf{a} \cdot \mathbf{b}/\|\mathbf{b}\| = \langle 1, 2, -2 \rangle \cdot \langle 4, 3, 0 \rangle / 5 = 2$

36. $\text{comp}_{\mathbf{b}}(\mathbf{a} - \mathbf{b}) = (\mathbf{a} - \mathbf{b}) \cdot \mathbf{b}/\|\mathbf{b}\| = \langle -3, -1, -2 \rangle \cdot \langle 4, 3, 0 \rangle / 5 = -3$

$\text{proj}_{\mathbf{b}}(\mathbf{a} - \mathbf{b}) = (\text{comp}_{\mathbf{b}}(\mathbf{a} - \mathbf{b}))\mathbf{b}/\|\mathbf{b}\| = -3\langle 4, 3, 0\rangle/5 = \langle -12/5, -9/5, 0\rangle$

$\text{proj}_{\mathbf{b}\perp}(\mathbf{a} - \mathbf{b}) = (\mathbf{a} - \mathbf{b}) - \text{proj}_{\mathbf{b}}(\mathbf{a} - \mathbf{b}) = \langle -3, -1, -2\rangle - \langle -12/5, -9/5, 0\rangle = \langle -3/5, 4/5, -10/5\rangle$

39. A direction vector of the given line is $\langle 4, -2, 6\rangle$. A parallel line containing $(7, 3, -5)$ is $(x-7)/4 = (y-3)/(-2) = (z+5)/6$.

42. Vectors in the plane are $\langle 2, 3, 1\rangle$ and $\langle 1, 0, 2\rangle$. A normal vector is $\langle 2, 3, 1\rangle \times \langle 1, 0, 2\rangle = \langle 6, -3, -3\rangle = 3\langle 2, -1, -1\rangle$. An equation of the plane is $2x - y - z = 0$

45. $\mathbf{F} = 10\dfrac{\mathbf{a}}{\|\mathbf{a}\|} = \dfrac{10}{\sqrt{2}}(\mathbf{i} + \mathbf{j}) = 5\sqrt{2}\,\mathbf{i} + 5\sqrt{2}\,\mathbf{j}; \quad \mathbf{d} = \langle 7, 4, 0\rangle - \langle 4, 1, 0\rangle = 3\mathbf{i} + 3\mathbf{j}$

$W = \mathbf{F} \cdot \mathbf{d} = 15\sqrt{2} + 15\sqrt{2} = 30\sqrt{2}$ N-m

48. Let $\|\mathbf{F}_1\| = F_1$ and $\|\mathbf{F}_2\| = F_2$. Then $\mathbf{F}_1 = F_1[(\cos 45°)\mathbf{i} + (\sin 45°)\mathbf{j}]$ and $\mathbf{F}_2 = F_2[(\cos 120°)\mathbf{i} + (\sin 120°)\mathbf{j}]$, or $\mathbf{F}_1 = F_1(\frac{1}{\sqrt{2}}\mathbf{i} + \frac{1}{\sqrt{2}}\mathbf{j})$ and $\mathbf{F}_2 = F_2(-\frac{1}{2}\mathbf{i} + \frac{\sqrt{3}}{2}\mathbf{j})$. Since $\mathbf{w} + \mathbf{F}_1 + \mathbf{F}_2 = \mathbf{0}$,

$$F_1(\frac{1}{\sqrt{2}}\mathbf{i} + \frac{1}{\sqrt{2}}\mathbf{j}) + F_2(-\frac{1}{2}\mathbf{i} + \frac{\sqrt{3}}{2}\mathbf{j}) = 50\mathbf{j}, \qquad (\frac{1}{\sqrt{2}}F_1 - \frac{1}{2}F_2)\mathbf{i} + (\frac{1}{\sqrt{2}}F_1 + \frac{\sqrt{3}}{2}F_2)\mathbf{j} = 50\mathbf{j}$$

and

$$\frac{1}{\sqrt{2}}F_1 - \frac{1}{2}F_2 = 0, \qquad \frac{1}{\sqrt{2}}F_1 + \frac{\sqrt{3}}{2}F_2 = 50.$$

Solving, we obtain $F_1 = 25(\sqrt{6} - \sqrt{2}) \approx 25.9$ lb and $F_2 = 50(\sqrt{3} - 1) \approx 36.6$ lb.

51. Let p_1 and p_2 be in P_n such that $\dfrac{d^2 p_1}{dx^2} = 0$ and $\dfrac{d^2 p_2}{dx^2} = 0$. Since

$$0 = \frac{d^2 p_1}{dx^2} + \frac{d^2 p_2}{dx^2} = \frac{d^2}{dx^2}(p_1 + p_2) \quad \text{and} \quad 0 = k\frac{d^2 p_1}{dx^2} = \frac{d^2}{dx^2}(kp_1)$$

we conclude that the set of polynomials with the given property is a subspace of P_n. A basis for the subspace is $1, x$.

8 Matrices

Exercises 8.1

3. 3×3 **6.** 8×1 **9.** Not equal

12. Solving $x^2 = 9$, $y = 4x$ we obtain $x = 3$, $y = 12$ and $x = -3$, $t = -12$.

15. (a) $\mathbf{A} + \mathbf{B} = \begin{bmatrix} 4-2 & 5+6 \\ -6+8 & 9-10 \end{bmatrix} = \begin{bmatrix} 2 & 11 \\ 2 & -1 \end{bmatrix}$

(b) $\mathbf{B} - \mathbf{A} = \begin{bmatrix} -2-4 & 6-5 \\ 8+6 & -10-9 \end{bmatrix} = \begin{bmatrix} -6 & 1 \\ 14 & -19 \end{bmatrix}$

(c) $2\mathbf{A} + 3\mathbf{B} = \begin{bmatrix} 8 & 10 \\ -12 & 18 \end{bmatrix} + \begin{bmatrix} -6 & 18 \\ 24 & -30 \end{bmatrix} = \begin{bmatrix} 2 & 28 \\ 12 & -12 \end{bmatrix}$

18. (a) $\mathbf{AB} = \begin{bmatrix} -4+4 & 6-12 & -3+8 \\ -20+10 & 30-30 & -15+20 \\ -32+12 & 48-36 & -24+24 \end{bmatrix} = \begin{bmatrix} 0 & -6 & 5 \\ -10 & 0 & 5 \\ -20 & 12 & 0 \end{bmatrix}$

(b) $\mathbf{BA} = \begin{bmatrix} -4+30-24 & -16+60-36 \\ 1-15+16 & 4-30+24 \end{bmatrix} = \begin{bmatrix} 2 & 8 \\ 2 & -2 \end{bmatrix}$

21. (a) $\mathbf{A}^T\mathbf{A} = \begin{bmatrix} 4 & 8 & -10 \end{bmatrix} \begin{bmatrix} 4 \\ 8 \\ -10 \end{bmatrix} = (180)$

(b) $\mathbf{B}^T\mathbf{B} = \begin{bmatrix} 2 \\ 4 \\ 5 \end{bmatrix} \begin{bmatrix} 2 & 4 & 5 \end{bmatrix} = \begin{bmatrix} 4 & 8 & 10 \\ 8 & 16 & 20 \\ 10 & 20 & 25 \end{bmatrix}$

(c) $\mathbf{A} + \mathbf{B}^T = \begin{bmatrix} 4 \\ 8 \\ -10 \end{bmatrix} + \begin{bmatrix} 2 \\ 4 \\ 5 \end{bmatrix} = \begin{bmatrix} 6 \\ 12 \\ -5 \end{bmatrix}$

24. (a) $\mathbf{A}^T + \mathbf{B} = \begin{bmatrix} 5 & -4 \\ 9 & 6 \end{bmatrix} + \begin{bmatrix} -3 & 11 \\ -7 & 2 \end{bmatrix} = \begin{bmatrix} 2 & 7 \\ 2 & 8 \end{bmatrix}$

(b) $2\mathbf{A} + \mathbf{B}^T = \begin{bmatrix} 10 & 18 \\ -8 & 12 \end{bmatrix} + \begin{bmatrix} -3 & -7 \\ 11 & 2 \end{bmatrix} = \begin{bmatrix} 7 & 11 \\ 3 & 14 \end{bmatrix}$

27. $\begin{bmatrix} -19 \\ 18 \end{bmatrix} - \begin{bmatrix} 19 \\ 20 \end{bmatrix} = \begin{bmatrix} -38 \\ -2 \end{bmatrix}$

30. 3×2, since the first matrix has 3 columns and the third matrix has 2 rows.

33. $(\mathbf{AB})^T = \begin{bmatrix} 16 & 40 \\ -8 & -20 \end{bmatrix}^T = \begin{bmatrix} 16 & -8 \\ 40 & -20 \end{bmatrix}$; $\quad \mathbf{B}^T\mathbf{A}^T = \begin{bmatrix} 4 & 2 \\ 10 & 5 \end{bmatrix} \begin{bmatrix} 2 & -3 \\ 4 & 2 \end{bmatrix} = \begin{bmatrix} 16 & -8 \\ 40 & -20 \end{bmatrix}$

36. Using Problem 33 we have $(\mathbf{AA}^T)^T = (\mathbf{A}^T)^T\mathbf{A}^T = \mathbf{AA}^T$, so that $\mathbf{AA}^T$ is symmetric.

39. Since $(\mathbf{A}+\mathbf{B})^2 = (\mathbf{A}+\mathbf{B})(\mathbf{A}+\mathbf{B}) \neq \mathbf{A}^2+\mathbf{AB}+\mathbf{BA}+\mathbf{B}^2$, and $\mathbf{AB} \neq \mathbf{BA}$ in general, $(\mathbf{A}+\mathbf{B})^2 \neq \mathbf{A}^2+2\mathbf{AB}+\mathbf{B}^2$.

42. $\begin{bmatrix} 2 & 6 & 1 \\ 1 & 2 & -1 \\ 5 & 7 & -4 \end{bmatrix} \begin{bmatrix} x_1 \\ x_2 \\ x_3 \end{bmatrix} = \begin{bmatrix} 7 \\ -1 \\ 9 \end{bmatrix}$

45. (a) $M_Y \begin{bmatrix} x \\ y \\ z \end{bmatrix} = \begin{bmatrix} \cos\gamma & \sin\gamma & 0 \\ -\sin\gamma & \cos\gamma & 0 \\ 0 & 0 & 1 \end{bmatrix} \begin{bmatrix} x \\ y \\ z \end{bmatrix} = \begin{bmatrix} x\cos\gamma + y\sin\gamma \\ -x\sin\gamma + y\cos\gamma \\ z \end{bmatrix} = \begin{bmatrix} x_Y \\ y_Y \\ z_Y \end{bmatrix}$

(b) $M_R = \begin{bmatrix} \cos\beta & 0 & -\sin\beta \\ 0 & 1 & 0 \\ \sin\beta & 0 & \cos\beta \end{bmatrix}; \qquad M_P \begin{bmatrix} 1 & 0 & 0 \\ 0 & \cos\alpha & \sin\alpha \\ 0 & -\sin\alpha & \cos\alpha \end{bmatrix}$

(c) $M_P \begin{bmatrix} 1 \\ 1 \\ 1 \end{bmatrix} = \begin{bmatrix} 1 & 0 & 0 \\ 0 & \cos 30° & \sin 30° \\ 0 & -\sin 30° & \cos 30° \end{bmatrix} \begin{bmatrix} 1 \\ 1 \\ 1 \end{bmatrix} = \begin{bmatrix} 1 & 0 & 0 \\ 0 & \frac{\sqrt{3}}{2} & \frac{1}{2} \\ 0 & -\frac{1}{2} & \frac{\sqrt{3}}{2} \end{bmatrix} \begin{bmatrix} 1 \\ 1 \\ 1 \end{bmatrix} = \begin{bmatrix} 1 \\ \frac{1}{2}(\sqrt{3}+1) \\ \frac{1}{2}(\sqrt{3}-1) \end{bmatrix}$

$M_R M_P \begin{bmatrix} 1 \\ 1 \\ 1 \end{bmatrix} = \begin{bmatrix} \cos 45° & 0 & -\sin 45° \\ 0 & 1 & 0 \\ \sin 45° & 0 & \cos 45° \end{bmatrix} \begin{bmatrix} 1 \\ \frac{1}{2}(\sqrt{3}+1) \\ \frac{1}{2}(\sqrt{3}-1) \end{bmatrix} = \begin{bmatrix} \frac{\sqrt{2}}{2} & 0 & -\frac{\sqrt{2}}{2} \\ 0 & 1 & 0 \\ \frac{\sqrt{2}}{2} & 0 & \frac{\sqrt{2}}{2} \end{bmatrix} \begin{bmatrix} 1 \\ \frac{1}{2}(\sqrt{3}+1) \\ \frac{1}{2}(\sqrt{3}-1) \end{bmatrix}$

$= \begin{bmatrix} \frac{1}{4}(3\sqrt{2}-\sqrt{6}) \\ \frac{1}{2}(\sqrt{3}+1) \\ \frac{1}{4}(\sqrt{2}+\sqrt{6}) \end{bmatrix}$

$M_Y M_R M_P \begin{bmatrix} 1 \\ 1 \\ 1 \end{bmatrix} = \begin{bmatrix} \cos 60° & \sin 60° & 0 \\ -\sin 60° & \cos 60° & 0 \\ 0 & 0 & 1 \end{bmatrix} \begin{bmatrix} \frac{1}{4}(3\sqrt{2}-\sqrt{6}) \\ \frac{1}{2}(\sqrt{3}+1) \\ \frac{1}{4}(\sqrt{2}+\sqrt{6}) \end{bmatrix}$

$= \begin{bmatrix} \frac{1}{2} & \frac{\sqrt{3}}{2} & 0 \\ -\frac{\sqrt{3}}{2} & \frac{1}{2} & 0 \\ 0 & 0 & 1 \end{bmatrix} \begin{bmatrix} \frac{1}{4}(3\sqrt{2}-\sqrt{6}) \\ \frac{1}{2}(\sqrt{3}+1) \\ \frac{1}{4}(\sqrt{2}+\sqrt{6}) \end{bmatrix} = \begin{bmatrix} \frac{1}{8}(3\sqrt{2}-\sqrt{6}+6+2\sqrt{3}) \\ \frac{1}{8}(-3\sqrt{6}+3\sqrt{2}+2\sqrt{3}+2) \\ \frac{1}{4}(\sqrt{2}+\sqrt{6}) \end{bmatrix}$

Exercises 8.2

3. $\begin{bmatrix} 9 & 3 & | & -5 \\ 2 & -1 & | & -1 \end{bmatrix} \xrightarrow{\frac{1}{9}R_1} \begin{bmatrix} 1 & \frac{1}{3} & | & -\frac{5}{9} \\ 2 & 1 & | & -1 \end{bmatrix} \xrightarrow{-2R_1+R_2} \begin{bmatrix} 1 & \frac{1}{3} & | & -\frac{5}{9} \\ 0 & \frac{1}{3} & | & \frac{1}{9} \end{bmatrix} \xrightarrow{3R_2} \begin{bmatrix} 1 & \frac{1}{3} & | & -\frac{5}{9} \\ 0 & 1 & | & \frac{1}{3} \end{bmatrix}$

$\xrightarrow{-\frac{1}{3}R_2+R_1} \begin{bmatrix} 1 & 0 & | & -\frac{2}{3} \\ 0 & 1 & | & \frac{1}{3} \end{bmatrix}$

The solution is $x_1 = -\frac{2}{3}$, $x_2 = \frac{1}{3}$.

6. $\begin{bmatrix} 1 & 2 & -1 & | & 0 \\ 2 & 1 & 2 & | & 9 \\ 1 & -1 & 1 & | & 3 \end{bmatrix} \xrightarrow[-R_1+R_3]{-2R_1+R_2} \begin{bmatrix} 1 & 2 & -1 & | & 0 \\ 0 & -3 & 4 & | & 9 \\ 0 & -3 & 2 & | & 3 \end{bmatrix} \xrightarrow{-\frac{1}{3}R_2} \begin{bmatrix} 1 & 2 & -1 & | & 0 \\ 0 & 1 & -\frac{4}{3} & | & -3 \\ 0 & -3 & 2 & | & 3 \end{bmatrix}$

$\xrightarrow[3R_2+R_3]{-2R_2+R_1} \begin{bmatrix} 1 & 0 & \frac{5}{3} & | & 6 \\ 0 & 1 & -\frac{4}{3} & | & -3 \\ 0 & 0 & -2 & | & -6 \end{bmatrix} \xrightarrow{-\frac{1}{2}R_3} \begin{bmatrix} 1 & 0 & \frac{5}{3} & | & 6 \\ 0 & 1 & -\frac{4}{3} & | & -3 \\ 0 & 0 & 1 & | & 3 \end{bmatrix} \xrightarrow[\frac{4}{3}R_3+R_2]{-\frac{5}{3}R_3+R_1} \begin{bmatrix} 1 & 0 & 0 & | & 1 \\ 0 & 1 & 0 & | & 1 \\ 0 & 0 & 1 & | & 3 \end{bmatrix}$

The solution is $x_1 = 1$, $x_2 = 1$, $x_3 = 3$.

9. $\begin{bmatrix} 1 & -1 & -1 & | & 8 \\ 1 & -1 & 1 & | & 3 \\ -1 & 1 & 1 & | & 4 \end{bmatrix} \xrightarrow[\text{operations}]{\text{row}} \begin{bmatrix} 1 & -1 & -1 & | & 8 \\ 0 & 0 & 2 & | & -5 \\ 0 & 0 & 0 & | & 12 \end{bmatrix}$

Since the bottom row implies $0 = 12$, the system is inconsistent.

12. $\begin{bmatrix} 1 & -1 & -2 & | & 0 \\ 2 & 4 & 5 & | & 0 \\ 6 & 0 & -3 & | & 0 \end{bmatrix} \xrightarrow[\text{operations}]{\text{row}} \begin{bmatrix} 1 & -1 & -2 & | & 0 \\ 0 & 1 & \frac{3}{2} & | & 0 \\ 0 & 0 & 0 & | & 0 \end{bmatrix}$

The solution is $x_1 = \frac{1}{2}t$, $x_2 = -\frac{3}{2}t$, $x_3 = t$.

15. $\begin{bmatrix} 1 & 1 & 1 & | & 3 \\ 1 & -1 & -1 & | & -1 \\ 3 & 1 & 1 & | & 5 \end{bmatrix} \xrightarrow[\text{operations}]{\text{row}} \begin{bmatrix} 1 & 1 & 1 & | & 3 \\ 0 & 1 & 1 & | & 2 \\ 0 & 0 & 0 & | & 0 \end{bmatrix}$

If $x_3 = t$ the solution is $x_1 = 1$, $x_2 = 2 - t$, $x_3 = t$.

18. $\begin{bmatrix} 2 & 1 & 1 & 0 & | & 3 \\ 3 & 1 & 1 & 1 & | & 4 \\ 1 & 2 & 2 & 3 & | & 3 \\ 4 & 5 & -2 & 1 & | & 16 \end{bmatrix} \xrightarrow[\text{operations}]{\text{row}} \begin{bmatrix} 1 & \frac{1}{2} & \frac{1}{2} & 0 & | & \frac{3}{2} \\ 0 & 1 & 1 & -2 & | & 1 \\ 0 & 0 & 1 & -1 & | & -1 \\ 0 & 0 & 0 & 1 & | & 0 \end{bmatrix}$

The solution is $x_1 = 1$, $x_2 = 2$, $x_3 = -1$, $x_4 = 0$.

21. $\begin{bmatrix} 1 & 1 & 1 & | & 4.280 \\ 0.2 & -0.1 & -0.5 & | & -1.978 \\ 4.1 & 0.3 & 0.12 & | & 1.686 \end{bmatrix} \xrightarrow[\text{operations}]{\text{row}} \begin{bmatrix} 1 & 1 & 1 & | & 4.28 \\ 0 & 1 & 2.333 & | & 9.447 \\ 0 & 0 & 1 & | & 4.1 \end{bmatrix}$

The solution is $x_1 = 0.3$, $x_2 = -0.12$, $x_3 = 4.1$.

24. From $x_1\text{KClO}_3 \to x_2\text{KCl} + x_3\text{O}_2$ we obtain the system $x_1 = x_2$, $x_1 = x_2$, $3x_1 = 2x_3$. Letting $x_3 = t$ we see that a solution of the system is $x_1 = x_2 = \frac{2}{3}t$, $x_3 = t$. Taking $t = 3$ we obtain the balanced equation

$$2\text{KClO}_3 \to 2\text{KCl} + 3\text{O}_2.$$

27. From $x_1\text{Cu} + x_2\text{HNO}_3 \to x_3\text{Cu(NO}_3)_2 + x_4\text{H}_2\text{O} + x_5\text{NO}$ we obtain the system

$$x_1 = 3, \qquad x_2 = 2x_4, \qquad x_2 = 2x_3 + x_5, \qquad 3x_2 = 6x_3 + x_4 + x_5.$$

Letting $x_4 = t$ we see that $x_2 = 2t$ and

$$
\begin{array}{ccc}
2t = 2x_3 + x_5 & & 2x_3 + x_5 = 2t \\
& \text{or} & \\
6t = 6x_3 + t + x_5 & & 6x_3 + x_5 = 5t.
\end{array}
$$

Then $x_3 = \frac{3}{4}t$ and $x_5 = \frac{1}{2}t$. Finally, $x_1 = x_3 = \frac{3}{4}t$. Taking $t = 4$ we obtain the balanced equation

$$3\text{Cu} + 8\text{HNO}_3 \to 3\text{Cu(NO}_3)_2 + 4\text{H}_2\text{O} + 2\text{NO}.$$

30. The system of equations is

$$
\begin{array}{ccc}
i_1 - i_2 - i_3 = 0 & & i_1 - i_2 - i_3 = 0 \\
52 - i_1 - 5i_2 = 0 & \text{or} & i_1 + 5i_2 = 52 \\
-10i_3 + 5i_2 = 0 & & 5i_2 - 10i_3 = 0
\end{array}
$$

Gaussian elimination gives

$$\begin{bmatrix} 1 & -1 & -1 & | & 0 \\ 1 & 5 & 0 & | & 52 \\ 0 & 5 & -10 & | & 0 \end{bmatrix} \xrightarrow[\text{operations}]{\text{row}} \begin{bmatrix} 1 & -1 & -1 & | & 0 \\ 0 & 1 & 1/6 & | & 26/3 \\ 0 & 0 & 1 & | & 4 \end{bmatrix}.$$

The solution is $i_1 = 12$, $i_2 = 8$, $i_3 = 4$.

33. Add c times row 2 to row 3 in $\mathbf{I}_3$.

36. $\mathbf{EA} = \begin{bmatrix} a_{11} & a_{12} & a_{13} \\ a_{21} & a_{22} & a_{23} \\ ca_{31} & ca_{32} & ca_{33} \end{bmatrix}$

39. The system is equivalent to

$$\begin{bmatrix} 1 & 0 \\ \frac{1}{2} & 1 \end{bmatrix} \begin{bmatrix} 2 & -2 \\ 0 & 3 \end{bmatrix} \mathbf{X} = \begin{bmatrix} 2 \\ 6 \end{bmatrix}.$$

Letting

$$\mathbf{Y} = \begin{bmatrix} y_1 \\ y_2 \end{bmatrix} = \begin{bmatrix} 2 & -2 \\ 0 & 3 \end{bmatrix} \mathbf{X}$$

we have

$$\begin{bmatrix} 1 & 0 \\ \frac{1}{2} & 1 \end{bmatrix} \begin{bmatrix} y_1 \\ y_2 \end{bmatrix} = \begin{bmatrix} 2 \\ 6 \end{bmatrix}.$$

This implies $y_1 = 2$ and $\frac{1}{2}y_1 + y_2 = 1 + y_2 = 6$ or $y_2 = 5$. Then

$$\begin{bmatrix} 2 & -2 \\ 0 & 3 \end{bmatrix} \begin{bmatrix} x_1 \\ x_2 \end{bmatrix} = \begin{bmatrix} 2 \\ 5 \end{bmatrix},$$

which implies $3x_2 = 5$ or $x_2 = \frac{5}{3}$ and $2x_1 - 2x_2 = 2x_1 - \frac{10}{3} = 2$ or $x_1 = \frac{8}{3}$. The solution is $\mathbf{X} = \left(\frac{8}{3}, \frac{5}{3}\right)$.

42. The system is equivalent to

$$\begin{bmatrix} 1 & 0 & 0 \\ 3 & 1 & 0 \\ 1 & 1 & 1 \end{bmatrix} \begin{bmatrix} 1 & 1 & 1 \\ 0 & -2 & -1 \\ 0 & 0 & 1 \end{bmatrix} \mathbf{X} = \begin{bmatrix} 0 \\ 1 \\ 4 \end{bmatrix}.$$

Letting

$$\mathbf{Y} = \begin{bmatrix} y_1 \\ y_2 \\ y_3 \end{bmatrix} = \begin{bmatrix} 1 & 1 & 1 \\ 0 & -2 & -1 \\ 0 & 0 & 1 \end{bmatrix} \mathbf{X}$$

we have

$$\begin{bmatrix} 1 & 0 & 0 \\ 3 & 1 & 0 \\ 1 & 1 & 1 \end{bmatrix} \begin{bmatrix} y_1 \\ y_2 \\ y_3 \end{bmatrix} = \begin{bmatrix} 0 \\ 1 \\ 4 \end{bmatrix}.$$

This implies $y_1 = 0$, $3y_1 + y_2 = y_2 = 1$, and $y_1 + y_2 + y_3 = 0 + 1 + y_3 = 4$ or $y_3 = 3$. Then

$$\begin{bmatrix} 1 & 1 & 1 \\ 0 & -2 & -1 \\ 0 & 0 & 1 \end{bmatrix} \begin{bmatrix} x_1 \\ x_2 \\ x_3 \end{bmatrix} = \begin{bmatrix} 0 \\ 1 \\ 3 \end{bmatrix},$$

which implies $x_3 = 3$, $-2x_2 - x_3 = -2x_2 - 3 = 1$ or $x_2 = -2$, and $x_1 + x_2 + x_3 = x_1 - 2 + 3 = 0$ or $x_1 = -1$. The solution is $\mathbf{X} = (-1, -2, 3)$.

Exercises 8.3

3. $\begin{bmatrix} 2 & 1 & 3 \\ 6 & 3 & 9 \\ -1 & -\frac{1}{2} & -\frac{3}{2} \end{bmatrix} \xrightarrow[\text{operations}]{\text{row}} \begin{bmatrix} 1 & \frac{1}{2} & \frac{3}{2} \\ 0 & 0 & 0 \\ 0 & 0 & 0 \end{bmatrix}$; The rank is 1.

6. $\begin{bmatrix} 3 & -1 & 2 & 0 \\ 6 & 2 & 4 & 5 \end{bmatrix} \xrightarrow[\text{operations}]{\text{row}} \begin{bmatrix} 1 & -\frac{1}{3} & \frac{2}{3} & 0 \\ 0 & 1 & 0 & \frac{5}{4} \end{bmatrix}$; The rank is 2.

9. $\begin{bmatrix} 0 & 2 & 4 & 2 & 2 \\ 4 & 1 & 0 & 5 & 1 \\ 2 & 1 & \frac{2}{3} & 3 & \frac{1}{3} \\ 6 & 6 & 6 & 12 & 0 \end{bmatrix} \xrightarrow[\text{operations}]{\text{row}} \begin{bmatrix} 1 & \frac{1}{2} & \frac{1}{3} & \frac{3}{2} & \frac{1}{6} \\ 0 & 1 & \frac{4}{3} & 1 & -\frac{1}{3} \\ 0 & 0 & 1 & 0 & 2 \\ 0 & 0 & 0 & 0 & 0 \end{bmatrix}$; The rank is 3.

12. $\begin{bmatrix} 2 & 6 & 3 \\ 1 & -1 & 4 \\ 3 & 2 & 1 \\ 2 & 5 & 4 \end{bmatrix} \xrightarrow[\text{operations}]{\text{row}} \begin{bmatrix} 1 & -1 & 4 \\ 0 & 1 & -\frac{5}{8} \\ 0 & 0 & 1 \\ 0 & 0 & 0 \end{bmatrix}$

Since the rank of the matrix is 3 and there are 4 vectors, the vectors are linearly dependent.

15. Since the number of unknowns is $n = 8$ and the rank of the coefficient matrix is $r = 3$, the solution of the system has $n - r = 5$ parameters.

18. Since the rank of **A** is $r = 3$ and the number of equations is $n = 6$, the solution of the system has $n - r = 3$ parameters. Thus, the solution of the system is not unique.

Exercises 8.4

3. $C_{13} = (-1)^{1+3} \begin{vmatrix} 1 & -1 \\ -2 & 3 \end{vmatrix} = 1$

6. $M_{41} = \begin{vmatrix} 2 & 4 & 0 \\ 2 & -2 & 3 \\ 1 & 0 & -1 \end{vmatrix} = 24$

9. -7

12. $-1/2$

15. $\begin{vmatrix} 0 & 2 & 0 \\ 3 & 0 & 1 \\ 0 & 5 & 8 \end{vmatrix} = -3 \begin{vmatrix} 2 & 0 \\ 5 & 8 \end{vmatrix} = -48$

18. $\begin{vmatrix} 1 & -1 & -1 \\ 2 & 2 & -2 \\ 1 & 1 & 9 \end{vmatrix} = \begin{vmatrix} 2 & -2 \\ 1 & 9 \end{vmatrix} - 2 \begin{vmatrix} -1 & -1 \\ 1 & 9 \end{vmatrix} + \begin{vmatrix} -1 & -1 \\ 2 & -2 \end{vmatrix} = 20 - 2(-8) + 4 = 40$

21. $\begin{vmatrix} -2 & -1 & 4 \\ -3 & 6 & 1 \\ -3 & 4 & 8 \end{vmatrix} = -2 \begin{vmatrix} 6 & 1 \\ 4 & 8 \end{vmatrix} + 3 \begin{vmatrix} -1 & 4 \\ 4 & 8 \end{vmatrix} - 3 \begin{vmatrix} -1 & 4 \\ 6 & 1 \end{vmatrix} = -2(44) + 3(-24) - 3(-25) = -85$

24. $\begin{vmatrix} 1 & 1 & 1 \\ x & y & z \\ 2+x & 3+y & 4+z \end{vmatrix} = \begin{vmatrix} y & z \\ 3+y & 4+z \end{vmatrix} - \begin{vmatrix} x & z \\ 2+x & 4+z \end{vmatrix} + \begin{vmatrix} x & y \\ 2+x & 3+y \end{vmatrix}$

$$= (4y + yz - 3z - yz) - (4x + xz - 2z - xz) + (3x + xy - 2y - xy) = -x + 2y - z$$

27. Expanding along the first column in the original matrix and each succeeding minor, we obtain $3(1)(2)(4)(2) = 48$.

30. Solving $-\lambda^3 + 3\lambda^2 - 2\lambda = -\lambda(\lambda - 2)(\lambda - 1) = 0$ we obtain $\lambda = 0$, 1, and 2.

Exercises 8.5

3. Theorem 8.14

6. Theorem 8.11 (twice)

9. Theorem 8.8

12. $\det \mathbf{B} = 2(3)(5) = 30$

15. $\det \mathbf{A} = 6(\frac{2}{3})(-4)(-5) = 80$

18. $\det \mathbf{D} = 4(7)(-2) = -56$

21. $\det \mathbf{AB} = \begin{vmatrix} 0 & -2 & 2 \\ 10 & 7 & 23 \\ 8 & 4 & 16 \end{vmatrix} = -80 = 20(-4) = \det \mathbf{A} \det \mathbf{B}$

24. Using Theorems 8.14 and 8.9,

$$\det \mathbf{A} = \begin{vmatrix} 1 & 1 & 1 \\ x & y & z \\ x+y+z & x+y+z & x+y+z \end{vmatrix} = (x+y+z)\begin{vmatrix} 1 & 1 & 1 \\ x & y & z \\ 1 & 1 & 1 \end{vmatrix} = 0.$$

27. $\begin{vmatrix} -1 & 2 & 3 \\ 4 & -5 & -2 \\ 9 & -9 & 6 \end{vmatrix} = \begin{vmatrix} -1 & 2 & 3 \\ 0 & 3 & 10 \\ 0 & 9 & 33 \end{vmatrix} = \begin{vmatrix} -1 & 2 & 3 \\ 0 & 3 & 10 \\ 0 & 0 & 3 \end{vmatrix} = -1(3)(3) = -9$

30. $\begin{vmatrix} 0 & 1 & 4 & 5 \\ 2 & 5 & 0 & 1 \\ 1 & 2 & 2 & 0 \\ 3 & 1 & 3 & 2 \end{vmatrix} = -\begin{vmatrix} 1 & 2 & 2 & 0 \\ 2 & 5 & 0 & 1 \\ 0 & 1 & 4 & 5 \\ 3 & 1 & 3 & 2 \end{vmatrix} = -\begin{vmatrix} 1 & 2 & 2 & 0 \\ 0 & 1 & -4 & 1 \\ 0 & 1 & 4 & 5 \\ 0 & -5 & -3 & 2 \end{vmatrix} = -\begin{vmatrix} 1 & 2 & 2 & 0 \\ 0 & 1 & -4 & 1 \\ 0 & 0 & 8 & 4 \\ 0 & 0 & -23 & 7 \end{vmatrix} = -\begin{vmatrix} 1 & 2 & 2 & 0 \\ 0 & 1 & -4 & 1 \\ 0 & 0 & 8 & 4 \\ 0 & 0 & -23 & \frac{37}{2} \end{vmatrix}$

$$= -(1)(1)(8)(\frac{37}{2}) = -148$$

33. We first use the second row to reduce the third row. Then we use the first row to reduce the second row.

$$\begin{vmatrix} 1 & 1 & 1 \\ a & b & c \\ 0 & b^2 - ab & c^2 - ac \end{vmatrix} = \begin{vmatrix} 1 & 1 & 1 \\ 0 & b-a & c-a \\ 0 & b(b-a) & c(c-a) \end{vmatrix} = (b-a)(c-a)\begin{vmatrix} 1 & 1 & 1 \\ 0 & 1 & 1 \\ 0 & b & c \end{vmatrix}.$$

Expanding along the first row gives $(b - a)(c - a)(c - b)$.

36. Since $C_{11} + -7$, $C_{12} = -8$, and $C_{13} = -10$ we have $a_{21}C_{11} + a_{22}C_{12} + a_{23}C_{13} = -2(-7) + 3(-8) - 1(-10) = 0$. Since $C_{12} = -8$, $C_{22} = -19$, and $C_{32} = -7$ we have $a_{13}C_{12} + a_{23}C_{22} + a_{33}C_{32} = 5(-8) - 1(-19) - 3(-7) = 0$.

39. Factoring -1 out of each row we see that $\det(-\mathbf{A}) = (-1)^5 \det \mathbf{A} = -\det \mathbf{A}$. Then $-\det \mathbf{A} = \det(-\mathbf{A}) = \det \mathbf{A}^T = \det \mathbf{A}$ and $\det \mathbf{A} = 0$.

Exercises 8.6

3. $\det \mathbf{A} = 9$. $\mathbf{A}$ is nonsingular. $\mathbf{A}^{-1} = \dfrac{1}{9}\begin{bmatrix} 1 & 1 \\ -4 & 5 \end{bmatrix} = \begin{bmatrix} \frac{1}{9} & \frac{1}{9} \\ -\frac{4}{9} & \frac{5}{9} \end{bmatrix}$

6. $\det \mathbf{A} = -3\pi^2$. $\mathbf{A}$ is nonsingular. $\mathbf{A}^{-1} = -\dfrac{1}{3\pi^2}\begin{bmatrix} \pi & \pi \\ \pi & -2\pi \end{bmatrix} = \begin{bmatrix} -\frac{1}{3\pi} & -\frac{1}{3\pi} \\ -\frac{1}{3\pi} & \frac{2}{3\pi} \end{bmatrix}$

9. $\det \mathbf{A} = -30$. $\mathbf{A}$ is nonsingular. $\mathbf{A}^{-1} = -\dfrac{1}{30}\begin{bmatrix} -14 & 13 & 16 \\ -2 & 4 & -2 \\ -4 & -7 & -4 \end{bmatrix} = \begin{bmatrix} \frac{7}{15} & -\frac{13}{30} & -\frac{8}{15} \\ \frac{1}{15} & -\frac{2}{15} & \frac{1}{15} \\ \frac{2}{15} & \frac{7}{30} & \frac{2}{15} \end{bmatrix}$

12. $\det \mathbf{A} = 16$. $\mathbf{A}$ is nonsingular. $\mathbf{A}^{-1} = \dfrac{1}{16}\begin{bmatrix} 0 & 0 & 2 \\ 8 & 0 & 0 \\ 0 & 16 & 0 \end{bmatrix} = \begin{bmatrix} 0 & 0 & \frac{1}{8} \\ \frac{1}{2} & 0 & 0 \\ 0 & 1 & 0 \end{bmatrix}$

15. $\begin{bmatrix} 6 & -2 & | & 1 & 0 \\ 0 & 4 & | & 0 & 1 \end{bmatrix} \xrightarrow[\frac{1}{4}R_2]{\frac{1}{6}R_1} \begin{bmatrix} 1 & -\frac{1}{3} & | & \frac{1}{6} & 0 \\ 0 & 1 & | & 0 & \frac{1}{4} \end{bmatrix} \xrightarrow{\frac{1}{3}R_2 + R_1} \begin{bmatrix} 1 & 0 & | & \frac{1}{6} & \frac{1}{12} \\ 0 & 1 & | & 0 & \frac{1}{4} \end{bmatrix}; \quad \mathbf{A}^{-1} = \begin{bmatrix} \frac{1}{6} & \frac{1}{12} \\ 0 & \frac{1}{4} \end{bmatrix}$

18. $\begin{bmatrix} 2 & -3 & | & 1 & 0 \\ -2 & 4 & | & 0 & 1 \end{bmatrix} \xrightarrow{\frac{1}{2}R_1} \begin{bmatrix} 1 & -\frac{3}{2} & | & \frac{1}{2} & 0 \\ -2 & 4 & | & 0 & 1 \end{bmatrix} \xrightarrow{2R_1 + R_2} \begin{bmatrix} 1 & -\frac{3}{2} & | & \frac{1}{2} & 0 \\ 0 & 1 & | & 1 & 1 \end{bmatrix} \xrightarrow{\frac{3}{2}R_2 + R_1} \begin{bmatrix} 1 & 0 & | & 2 & \frac{3}{2} \\ 0 & 1 & | & 1 & 1 \end{bmatrix};$

$\mathbf{A}^{-1} = \begin{bmatrix} 2 & \frac{3}{2} \\ 1 & 1 \end{bmatrix}$

21. $\begin{bmatrix} 4 & 2 & 3 & | & 1 & 0 & 0 \\ 2 & 1 & 0 & | & 0 & 1 & 0 \\ -1 & -2 & 0 & | & 0 & 0 & 1 \end{bmatrix} \xrightarrow{R_{13}} \begin{bmatrix} -1 & -2 & 0 & | & 0 & 0 & 1 \\ 2 & 1 & 0 & | & 0 & 1 & 0 \\ 4 & 2 & 3 & | & 1 & 0 & 0 \end{bmatrix} \xrightarrow[\text{operations}]{\text{row}} \begin{bmatrix} 1 & 0 & 0 & | & 0 & \frac{2}{3} & \frac{1}{3} \\ 0 & 1 & 0 & | & 0 & -\frac{1}{3} & -\frac{2}{3} \\ 0 & 0 & 1 & | & \frac{1}{3} & -\frac{2}{3} & 0 \end{bmatrix};$

$\mathbf{A}^{-1} = \begin{bmatrix} 0 & \frac{2}{3} & \frac{1}{3} \\ 0 & -\frac{1}{3} & -\frac{2}{3} \\ \frac{1}{3} & -\frac{2}{3} & 0 \end{bmatrix}$

24. $\begin{bmatrix} 1 & 2 & 3 & | & 1 & 0 & 0 \\ 0 & 1 & 4 & | & 0 & 1 & 0 \\ 0 & 0 & 8 & | & 0 & 0 & 1 \end{bmatrix} \xrightarrow[\text{operations}]{\text{row}} \begin{bmatrix} 1 & 0 & 0 & | & 1 & -2 & \frac{5}{8} \\ 0 & 1 & 0 & | & 0 & 1 & -\frac{1}{2} \\ 0 & 0 & 1 & | & 0 & 0 & \frac{1}{8} \end{bmatrix}; \quad \mathbf{A}^{-1} = \begin{bmatrix} 1 & -2 & \frac{5}{8} \\ 0 & 1 & -\frac{1}{2} \\ 0 & 0 & \frac{1}{8} \end{bmatrix}$

27. $(\mathbf{AB})^{-1} = \mathbf{B}^{-1}\mathbf{A}^{-1} = \begin{bmatrix} -\frac{1}{3} & \frac{1}{3} \\ -1 & \frac{10}{3} \end{bmatrix}$

30. $\mathbf{A}^T = \begin{bmatrix} 1 & 2 \\ 4 & 10 \end{bmatrix}$; $(\mathbf{A}^T)^{-1} = \begin{bmatrix} 5 & -1 \\ -2 & \frac{1}{2} \end{bmatrix}$; $\mathbf{A}^{-1} = \begin{bmatrix} 5 & -2 \\ -1 & \frac{1}{2} \end{bmatrix}$; $(\mathbf{A}^{-1})^T = \begin{bmatrix} 5 & -1 \\ -2 & \frac{1}{2} \end{bmatrix}$

33. **(a)** $\mathbf{A}^T = \begin{bmatrix} \sin\theta & -\cos\theta \\ \cos\theta & \sin\theta \end{bmatrix} = \mathbf{A}^{-1}$ **(b)** $\mathbf{A}^T = \begin{bmatrix} \frac{1}{\sqrt{3}} & \frac{1}{\sqrt{3}} & \frac{1}{\sqrt{3}} \\ 0 & \frac{1}{\sqrt{2}} & -\frac{1}{\sqrt{2}} \\ -\frac{2}{\sqrt{6}} & \frac{1}{\sqrt{6}} & \frac{1}{\sqrt{6}} \end{bmatrix} = \mathbf{A}^{-1}$

36. Suppose $\mathbf{A}$ is singular. Then $\det \mathbf{A} = 0$, $\det \mathbf{AB} = \det \mathbf{A} \cdot \det \mathbf{B} = 0$, and $\mathbf{AB}$ is singular.

39. If $\mathbf{A}$ is nonsingular, then $\mathbf{A}^{-1}$ exists, and $\mathbf{AB} = \mathbf{0}$ implies $\mathbf{A}^{-1}\mathbf{AB} = \mathbf{A}^{-1}\mathbf{0}$, so $\mathbf{B} = \mathbf{0}$.

42. **A** is nonsingular if $a_{11}a_{22}a_{33} = 0$ or a_{11}, a_{22}, and a_{33} are all nonzero.

$$\mathbf{A}^{-1} = \begin{bmatrix} 1/a_{11} & 0 & 0 \\ 0 & 1/a_{22} & 0 \\ 0 & 0 & 1/a_{33} \end{bmatrix}$$

For any diagonal matrix, the inverse matrix is obtaining by taking the reciprocals of the diagonal entries and leaving all other entries 0.

45. $\mathbf{A}^{-1} = \begin{bmatrix} \frac{1}{16} & \frac{3}{8} \\ -\frac{1}{8} & \frac{1}{4} \end{bmatrix}$; $\mathbf{A}^{-1} \begin{bmatrix} 6 \\ 1 \end{bmatrix} = \begin{bmatrix} \frac{3}{4} \\ -\frac{1}{2} \end{bmatrix}$; $x_1 = \frac{3}{4}$, $x_2 = -\frac{1}{2}$

48. $\mathbf{A}^{-1} = \begin{bmatrix} \frac{5}{12} & -\frac{1}{12} & \frac{1}{4} \\ -\frac{2}{3} & \frac{1}{3} & 0 \\ -\frac{1}{12} & \frac{5}{12} & -\frac{1}{4} \end{bmatrix}$; $\mathbf{A}^{-1} \begin{bmatrix} 1 \\ 2 \\ -3 \end{bmatrix} = \begin{bmatrix} -\frac{1}{2} \\ 0 \\ \frac{3}{2} \end{bmatrix}$; $x_1 = -\frac{1}{2}$, $x_2 = 0$, $x_3 = \frac{3}{2}$

51. $\begin{bmatrix} 7 & -2 \\ 3 & 2 \end{bmatrix} \begin{bmatrix} x_1 \\ x_2 \end{bmatrix} = \begin{bmatrix} b_1 \\ b_2 \end{bmatrix}$; $\mathbf{A}^{-1} = \begin{bmatrix} \frac{1}{10} & \frac{1}{10} \\ -\frac{3}{20} & \frac{7}{20} \end{bmatrix}$; $\mathbf{X} = \mathbf{A}^{-1} \begin{bmatrix} 5 \\ 4 \end{bmatrix} = \begin{bmatrix} \frac{9}{10} \\ \frac{13}{20} \end{bmatrix}$; $\mathbf{X} = \mathbf{A}^{-1} \begin{bmatrix} 10 \\ 50 \end{bmatrix} = \begin{bmatrix} 6 \\ 16 \end{bmatrix}$;

$\mathbf{X} = \mathbf{A}^{-1} \begin{bmatrix} 0 \\ -20 \end{bmatrix} = \begin{bmatrix} -2 \\ -7 \end{bmatrix}$

54. $\det \mathbf{A} = 0$, so the system has a nontrivial solution.

57. **(a)** $\begin{bmatrix} 1 & 1 & 1 \\ -R_1 & R_2 & 0 \\ 0 & -R_2 & R_3 \end{bmatrix} \begin{bmatrix} i_1 \\ i_2 \\ i_3 \end{bmatrix} = \begin{bmatrix} 0 \\ E_2 - E_1 \\ E_3 - E_2 \end{bmatrix}$

(b) $\det \mathbf{A} = R_1 R_2 + R_1 R_3 + R_2 R_3 > 0$, so **A** is nonsingular.

(c) $\mathbf{A}^{-1} = \dfrac{1}{R_1 R_2 + R_1 R_3 + R_2 R_3} \begin{bmatrix} R_2 R_3 & -R_2 - R_3 & -R_2 \\ R_1 R_3 & R_3 & -R_1 \\ R_1 R_2 & R_2 & R_1 + R_2 \end{bmatrix}$;

$\mathbf{A}^{-1} \begin{bmatrix} 0 \\ E_2 - E_1 \\ E_3 - E_2 \end{bmatrix} = \dfrac{1}{R_1 R_2 + R_1 R_3 + R_2 R_3} \begin{bmatrix} R_2 E_1 - R_2 E_3 + R_3 E_1 - R_3 E_2 \\ R_1 E_2 - R_1 E_3 - R_3 E_1 + R_3 E_2 \\ -R_1 E_2 + R_1 E_3 - R_2 E_1 + R_2 E_3 \end{bmatrix}$

Exercises 8.7

3. $\det \mathbf{A} = 0.3$, $\det \mathbf{A}_1 = 0.03$, $\det \mathbf{A}_2 = -0.09$; $x_1 = \frac{0.03}{0.3} = 0.1$, $x_2 = \frac{-0.09}{0.3} = -0.3$

6. $\det \mathbf{A} = -70$, $\det \mathbf{A}_1 = -14$, $\det \mathbf{A}_2 = 35$; $r = \frac{-14}{-70} = \frac{1}{5}$, $s = \frac{35}{-70} = -\frac{1}{2}$

9. $\det \mathbf{A} = -12$, $\det \mathbf{A}_1 = -48$, $\det \mathbf{A}_2 = -18$, $\det \mathbf{A}_3 = -12$; $u = \frac{48}{12} = 4$, $v = \frac{18}{12} = \frac{3}{2}$, $w = 1$

12. **(a)** $\det \mathbf{A} = \epsilon - 1$, $\det \mathbf{A}_1 = \epsilon - 2$, $\det \mathbf{A}_2 = 1$; $x_1 = \dfrac{\epsilon - 2}{\epsilon - 1} = \dfrac{\epsilon - 1 - 1}{\epsilon - 1} = 1 - \dfrac{1}{\epsilon - 1}$, $x_2 = \dfrac{1}{\epsilon - 1}$

(b) When $\epsilon = 1.01$, $x_1 = -99$ and $x_2 = 100$. When $\epsilon = 0.99$, $x_1 = 101$ and $x_2 = -100$.

15. The system is

$$i_1 + i_2 - i_3 = 0$$
$$r_1 i_1 - r_2 i_2 = E_1 - E_2$$
$$r_2 i_2 + R i_3 = E_2$$

$$\det \mathbf{A} = -r_1 R - r_2 R - r_1 r_2, \quad \det A_3 = -r_1 E_2, \ -r_2 E_1; \quad i_3 = \frac{r_1 E_2 + r_2 E_1}{r_1 R + r_2 R + r_1 r_2}$$

Exercises 8.8

3. $\mathbf{K}_3$ since $\begin{bmatrix} 6 & 3 \\ 2 & 1 \end{bmatrix} \begin{bmatrix} -5 \\ 10 \end{bmatrix} = \begin{bmatrix} 0 \\ 0 \end{bmatrix} = 0 \begin{bmatrix} -5 \\ 10 \end{bmatrix}; \quad \lambda = 0$

6. $\mathbf{K}_2$ since $\begin{bmatrix} -1 & 1 & 0 \\ 1 & 2 & 1 \\ 0 & 3 & -1 \end{bmatrix} \begin{bmatrix} 1 \\ 4 \\ 3 \end{bmatrix} = \begin{bmatrix} 3 \\ 12 \\ 9 \end{bmatrix} = 3 \begin{bmatrix} 1 \\ 4 \\ 3 \end{bmatrix}; \quad \lambda = 3$

9. We solve $\det(\mathbf{A} - \lambda \mathbf{I}) = \begin{vmatrix} -8 - \lambda & -1 \\ 16 & -\lambda \end{vmatrix} = (\lambda + 4)^2 = 0.$

For $\lambda_1 = \lambda_2 = -4$ we have $\begin{bmatrix} -4 & -1 & | & 0 \\ 16 & 4 & | & 0 \end{bmatrix} \implies \begin{bmatrix} 1 & 1/4 & | & 0 \\ 0 & 0 & | & 0 \end{bmatrix}$

so that $k_1 = -\frac{1}{4}k_2$. If $k_2 = 4$ then $\mathbf{K}_1 = \begin{bmatrix} -1 \\ 4 \end{bmatrix}$.

12. We solve $\det(\mathbf{A} - \lambda \mathbf{I}) = \begin{vmatrix} 1 - \lambda & -1 \\ 1 & 1 - \lambda \end{vmatrix} = \lambda^2 - 2\lambda + 2 = 0.$

For $\lambda_1 = 1 - i$ we have $\begin{bmatrix} i & -1 & | & 0 \\ 1 & i & | & 0 \end{bmatrix} \implies \begin{bmatrix} i & -1 & | & 0 \\ 0 & 0 & | & 0 \end{bmatrix}$

so that $k_1 = -ik_2$. If $k_2 = 1$ then $\mathbf{K}_1 = \begin{bmatrix} -i \\ 1 \end{bmatrix}$ and $\mathbf{K}_2 = \overline{\mathbf{K}}_1 = \begin{bmatrix} i \\ 1 \end{bmatrix}$.

15. We solve $\det(\mathbf{A} - \lambda \mathbf{I}) = \begin{vmatrix} 5 - \lambda & -1 & 0 \\ 0 & -5 - \lambda & 9 \\ 5 & -1 & -\lambda \end{vmatrix} = \begin{vmatrix} 4 - \lambda & -1 & 0 \\ 4 - \lambda & -5 - \lambda & 9 \\ 4 - \lambda & -1 & -\lambda \end{vmatrix} = \lambda(4 - \lambda)(\lambda + 4) = 0.$

For $\lambda_1 = 0$ we have $\begin{bmatrix} 5 & -1 & 0 & | & 0 \\ 0 & -5 & 9 & | & 0 \\ 5 & -1 & 0 & | & 0 \end{bmatrix} \implies \begin{bmatrix} 1 & 0 & -9/25 & | & 0 \\ 0 & 1 & -9/5 & | & 0 \\ 0 & 0 & 0 & | & 0 \end{bmatrix}$

so that $k_1 = \frac{9}{25}k_3$ and $k_2 = \frac{9}{5}k_3$. If $k_3 = 25$ then $\mathbf{K}_1 = \begin{bmatrix} 9 \\ 45 \\ 25 \end{bmatrix}$. If $\lambda_2 = 4$ then

$$\begin{bmatrix} 1 & -1 & 0 & | & 0 \\ 0 & -9 & 9 & | & 0 \\ 5 & -1 & -4 & | & 0 \end{bmatrix} \implies \begin{bmatrix} 1 & 0 & -1 & | & 0 \\ 0 & 1 & -1 & | & 0 \\ 0 & 0 & 0 & | & 0 \end{bmatrix}$$

so that $k_1 = k_3$ and $k_2 = k_3$. If $k_3 = 1$ then $\mathbf{K}_2 = \begin{bmatrix} 1 \\ 1 \\ 1 \end{bmatrix}$. If $\lambda_3 = -4$ then

$$\begin{bmatrix} 9 & -1 & 0 & | & 0 \\ 0 & -1 & 9 & | & 0 \\ 5 & -1 & 4 & | & 0 \end{bmatrix} \implies \begin{bmatrix} 1 & 0 & -1 & | & 0 \\ 0 & 1 & -9 & | & 0 \\ 0 & 0 & 0 & | & 0 \end{bmatrix}$$

so that $k_1 = k_3$ and $k_2 = 9k_3$. If $k_3 = 1$ then $\mathbf{K}_3 = \begin{bmatrix} 1 \\ 9 \\ 1 \end{bmatrix}$.

18. We solve $\quad \det(\mathbf{A} - \lambda\mathbf{I}) = \begin{vmatrix} 1-\lambda & 6 & 0 \\ 0 & 2-\lambda & 1 \\ 0 & 1 & 2-\lambda \end{vmatrix} = \begin{vmatrix} 1-\lambda & 6 & 0 \\ 0 & 3-\lambda & 3-\lambda \\ 0 & 1 & 2-\lambda \end{vmatrix} = (3-\lambda)(1-\lambda)^2 = 0.$

For $\lambda_1 = 3$ we have $\quad \begin{bmatrix} -2 & 6 & 0 & | & 0 \\ 0 & 0 & 0 & | & 0 \\ 0 & 1 & -1 & | & 0 \end{bmatrix} \Longrightarrow \begin{bmatrix} 1 & 0 & -3 & | & 0 \\ 0 & 1 & -1 & | & 0 \\ 0 & 0 & 0 & | & 0 \end{bmatrix}$

so that $k_1 = 3k_3$ and $k_2 = k_3$. If $k_3 = 1$ then $\mathbf{K}_1 = \begin{bmatrix} 3 \\ 1 \\ 1 \end{bmatrix}$. For $\lambda_2 = \lambda_3 = 1$ we have

$$\begin{bmatrix} 0 & 6 & 0 & | & 0 \\ 0 & 1 & 1 & | & 0 \\ 0 & 1 & 1 & | & 0 \end{bmatrix} \Longrightarrow \begin{bmatrix} 0 & 1 & 0 & | & 0 \\ 0 & 0 & 1 & | & 0 \\ 0 & 0 & 0 & | & 0 \end{bmatrix}$$

so that $k_2 = 0$ and $k_3 = 0$. If $k_1 = 1$ then $\mathbf{K}_2 = \begin{bmatrix} 1 \\ 0 \\ 0 \end{bmatrix}$.

21. We solve $\quad \det(\mathbf{A} - \lambda\mathbf{I}) = \begin{vmatrix} 1-\lambda & 2 & 3 \\ 0 & 5-\lambda & 6 \\ 0 & 0 & -7-\lambda \end{vmatrix} = -(\lambda-1)(\lambda-5)(\lambda+7) = 0.$

For $\lambda_1 = 1$ we have $\quad \begin{bmatrix} 0 & 2 & 3 & | & 0 \\ 0 & 4 & 6 & | & 0 \\ 0 & 0 & -6 & | & 0 \end{bmatrix} \Longrightarrow \begin{bmatrix} 0 & 1 & 0 & | & 0 \\ 0 & 0 & 1 & | & 0 \\ 0 & 0 & 0 & | & 0 \end{bmatrix}$

so that $k_2 = k_3 = 0$. If $k_1 = 1$ then $\mathbf{K}_1 = \begin{bmatrix} 1 \\ 0 \\ 0 \end{bmatrix}$. For $\lambda_2 = 5$ we have

$$\begin{bmatrix} -4 & 2 & 3 & | & 0 \\ 0 & 0 & 6 & | & 0 \\ 0 & 0 & -12 & | & 0 \end{bmatrix} \Longrightarrow \begin{bmatrix} 1 & -\frac{1}{2} & 0 & | & 0 \\ 0 & 0 & 1 & | & 0 \\ 0 & 0 & 0 & | & 0 \end{bmatrix}$$

so that $k_3 = 0$ and $k_2 = 2k_1$. If $k_1 = 1$ then $\mathbf{K}_2 = \begin{bmatrix} 1 \\ 2 \\ 0 \end{bmatrix}$. For $\lambda_3 = -7$ we have

$$\begin{bmatrix} 8 & 2 & 3 & | & 0 \\ 0 & 12 & 6 & | & 0 \\ 0 & 0 & 0 & | & 0 \end{bmatrix} \Longrightarrow \begin{bmatrix} 1 & 0 & \frac{1}{4} & | & 0 \\ 0 & 1 & \frac{1}{2} & | & 0 \\ 0 & 0 & 0 & | & 0 \end{bmatrix}$$

so that $k_1 = -\frac{1}{4}k_3$ and $k_2 = -\frac{1}{2}k_3$. If $k_3 = 4$ then $\mathbf{K}_3 = \begin{bmatrix} -1 \\ -2 \\ 4 \end{bmatrix}$.

24. The eigenvalues and eigenvectors of $\mathbf{A} = \begin{bmatrix} 1 & 2 & -1 \\ 1 & 0 & 1 \\ 4 & -4 & 5 \end{bmatrix}$ are

$$\lambda_1 = 1, \quad \lambda_2 = 2, \quad \lambda_3 = 3, \quad \mathbf{K}_1 = \begin{bmatrix} -1 \\ 1 \\ 2 \end{bmatrix}, \quad \mathbf{K}_2 = \begin{bmatrix} -2 \\ 1 \\ 4 \end{bmatrix}, \quad \mathbf{K}_3 = \begin{bmatrix} -1 \\ 1 \\ 4 \end{bmatrix}.$$

and the eigenvalues and eigenvectors of $\mathbf{A}^{-1} = \dfrac{1}{6}\begin{bmatrix} 4 & -6 & 2 \\ -1 & 9 & -2 \\ -4 & 12 & -2 \end{bmatrix}$ are

$$\lambda_1 = 1, \quad \lambda_2 = \frac{1}{2}, \quad \lambda_3 = \frac{1}{3}, \quad \mathbf{K}_1 = \begin{bmatrix} -1 \\ 1 \\ 2 \end{bmatrix}, \quad \mathbf{K}_2 = \begin{bmatrix} -2 \\ 1 \\ 4 \end{bmatrix}, \quad \mathbf{K}_3 = \begin{bmatrix} -1 \\ 1 \\ 4 \end{bmatrix}.$$

Exercises 8.9

3. The characteristic equation is $\lambda^2 - 3\lambda - 10 = 0$, with eigenvalues -2 and 5. Substituting the eigenvalues into $\lambda^m = c_0 + c_1\lambda$ generates

$$(-2)^m = c_0 - 2c_1$$
$$5^m = c_0 + 5c_1.$$

Solving the system gives

$$c_0 = \frac{1}{7}[5(-2)^m + 2(5)^m], \qquad c_1 = \frac{1}{7}[-(-2)^m + 5^m].$$

Thus

$$\mathbf{A}^m = c_0\mathbf{I} + c_1\mathbf{A} = \begin{bmatrix} \frac{1}{7}[3(-1)^m 2^{m+1} + 5^m] & \frac{3}{7}[-(-2)^m + 5^m] \\ \frac{2}{7}[-(-2)^m + 5^m] & \frac{1}{7}[(-2)^m + 6(5)^m] \end{bmatrix}$$

and

$$\mathbf{A}^3 = \begin{bmatrix} 11 & 57 \\ 38 & 106 \end{bmatrix}.$$

6. The characteristic equation is $\lambda^2 + 4\lambda + 3 = 0$, with eigenvalues -3 and -1. Substituting the eigenvalues into $\lambda^m = c_0 + c_1\lambda$ generates

$$(-3)^m = c_0 - 3c_1$$
$$(-1)^m = c_0 - c_1.$$

Solving the system gives

$$c_0 = \frac{1}{2}[-(-3)^m + 3(-1)^m], \qquad c_1 = \frac{1}{2}[-(-3)^m + (-1)^m].$$

Thus

$$\mathbf{A}^m = c_0\mathbf{I} + c_1\mathbf{A} = \begin{bmatrix} (-1)^m & -(-3)^m + (-1)^m \\ 0 & (-3)^m \end{bmatrix}$$

and

$$\mathbf{A}^6 = \begin{bmatrix} 1 & -728 \\ 0 & 729 \end{bmatrix}.$$

9. The characteristic equation is $-\lambda^3 + 3\lambda^2 + 6\lambda - 8 = 0$, with eigenvalues -2, 1, and 4. Substituting the eigenvalues into $\lambda^m = c_0 + c_1\lambda + c_2\lambda^2$ generates

$$(-2)^m = c_0 - 2c_1 + 4c_2$$
$$1 = c_0 + c_1 + c_2$$
$$4^m = c_0 + 4c_1 + 16c_2.$$

Solving the system gives

$$c_0 = \frac{1}{9}[8 + (-1)^m 2^{m+1} - 4^m],$$

$$c_1 = \frac{1}{18}[4 - 5(-2)^m + 4^m],$$

$$c_2 = \frac{1}{18}[-2 + (-2)^m + 4^m].$$

Thus

$$\mathbf{A}^m = c_0\mathbf{I} + c_1\mathbf{A} + c_2\mathbf{A}^2 = \begin{bmatrix} \frac{1}{9}[(-2)^m + (-1)^m 2^{m+1} + 3 \cdot 2^{2m+1}] & \frac{1}{3}[-(-2)^m + 4^m] & 0 \\ -\frac{2}{3}[(-2)^m - 4^m] & \frac{1}{3}[(-1)^m 2^{m+1} + 4^m] & 0 \\ \frac{1}{3}[-3 + (-2)^m + 2^{2m+1}] & \frac{1}{3}[-(-2)^m + 4^m] & 1 \end{bmatrix}$$

and

$$\mathbf{A}^{10} = \begin{bmatrix} 699392 & 349184 & 0 \\ 698368 & 350208 & 0 \\ 699391 & 349184 & 1 \end{bmatrix}.$$

12. The characteristic equation is $-\lambda^3 - \lambda^2 + 21\lambda + 45 = 0$, with eigenvalues -3, -3, and 5. Substituting the eigenvalues into $\lambda^m = c_0 + c_1\lambda + c_2\lambda^2$ generates

$$(-3)^m = c_0 - 3c_1 + 9c_2$$
$$(-3)^{m-1}m = c_1 - 6c_2$$
$$5^m = c_0 + 5c_1 + 25c_2.$$

Solving the system gives

$$c_0 = \frac{1}{64}[73(-3)^m - 2(-1)^m 3^{m+2} + 9 \cdot 5^m - 40(-3)^m m],$$

$$c_1 = \frac{1}{96}[-(-1)^m 3^{m+2} + 9 \cdot 5^m - 8(-3)^m m],$$

$$c_2 = \frac{1}{64}[-(-3)^m + 5^m - 8(-3)^{m-1}m].$$

Thus

$$\mathbf{A}^m = c_0\mathbf{I} + c_1\mathbf{A} + c_2\mathbf{A}^2$$

$$= \begin{bmatrix} \frac{1}{32}[31(-3)^m - (-1)^m 3^{m+1} + 4 \cdot 5^m] & \frac{1}{16}[-(-3)^m - (-1)^m 3^{m+1} + 4 \cdot 5^m] & \frac{1}{32}[(-3)^m + (-1)^m 3^{m+1} - 4 \cdot 5^m] \\ \frac{1}{16}[-(-3)^m - (-1)^m 3^{m+1} + 4 \cdot 5^m] & \frac{1}{8}[7(-3)^m - (-1)^m 3^{m+1} + 4 \cdot 5^m] & \frac{1}{16}[(-3)^m + (-1)^m 3^{m+1} - 4 \cdot 5^m] \\ \frac{3}{32}[(-3)^m + (-1)^m 3^{m+1} - 4 \cdot 5^m] & \frac{3}{16}[(-3)^m + (-1)^m 3^{m+1} - 4 \cdot 5^m] & \frac{1}{32}[29(-3)^m - (-1)^m 3^{m+2} + 12 \cdot 5^m] \end{bmatrix}$$

and

$$\mathbf{A}^5 = \begin{bmatrix} 178 & 842 & -421 \\ 842 & 1441 & -842 \\ -1263 & -2526 & 1020 \end{bmatrix}.$$

15. The characteristic equation of $\mathbf{A}$ is $\lambda^2 - 5\lambda + 10 = 0$, so $\mathbf{A}^2 - 5\mathbf{A} + 10\mathbf{I} = \mathbf{0}$ and $\mathbf{I} = -\frac{1}{10}\mathbf{A}^2 + \frac{1}{2}\mathbf{A}$. Multiplying by $\mathbf{A}^{-1}$ we find

$$\mathbf{A}^{-1} = -\frac{1}{10}\mathbf{A} + \frac{1}{2}\mathbf{I} = -\frac{1}{10}\begin{bmatrix} 2 & -4 \\ 1 & 3 \end{bmatrix} + \frac{1}{2}\begin{bmatrix} 1 & 0 \\ 0 & 1 \end{bmatrix} = \begin{bmatrix} \frac{3}{10} & \frac{2}{5} \\ -\frac{1}{10} & \frac{1}{5} \end{bmatrix}.$$

18. (a) If $\mathbf{A}^m = \mathbf{0}$ for some m, then $(\det \mathbf{A})^m = \det \mathbf{A}^m = \det \mathbf{0} = 0$, and $\mathbf{A}$ is a singular matrix.

(b) By (1) of Section 8.8 we have $\mathbf{A}\mathbf{K} = \lambda\mathbf{K}$, $\mathbf{A}^2\mathbf{K} = \lambda\mathbf{A}\mathbf{K} = \lambda^2\mathbf{K}$, $\mathbf{A}^3\mathbf{K} = \lambda^2\mathbf{A}\mathbf{K} = \lambda^3\mathbf{K}$, and, in general, $\mathbf{A}^m\mathbf{K} = \lambda^m\mathbf{K}$. If $\mathbf{A}$ is nilpotent with index m, then $\mathbf{A}^m = \mathbf{0}$ and $\lambda^m = 0$.

Exercises 8.10

3. (a)–(b) $\begin{bmatrix} 5 & 13 & 0 \\ 13 & 5 & 0 \\ 0 & 0 & -8 \end{bmatrix} \begin{bmatrix} \frac{\sqrt{2}}{2} \\ \frac{\sqrt{2}}{2} \\ 0 \end{bmatrix} = \begin{bmatrix} 9\sqrt{2} \\ 9\sqrt{2} \\ 0 \end{bmatrix} = 18\begin{bmatrix} \frac{\sqrt{2}}{2} \\ \frac{\sqrt{2}}{2} \\ 0 \end{bmatrix}; \quad \lambda_1 = 18$

$\begin{bmatrix} 5 & 13 & 0 \\ 13 & 5 & 0 \\ 0 & 0 & -8 \end{bmatrix} \begin{bmatrix} \frac{\sqrt{3}}{3} \\ -\frac{\sqrt{3}}{3} \\ \frac{\sqrt{3}}{3} \end{bmatrix} = \begin{bmatrix} -\frac{8\sqrt{2}}{3} \\ \frac{8\sqrt{3}}{3} \\ -\frac{8\sqrt{3}}{3} \end{bmatrix} = (-8)\begin{bmatrix} \frac{\sqrt{3}}{3} \\ -\frac{\sqrt{3}}{3} \\ \frac{\sqrt{3}}{3} \end{bmatrix}; \quad \lambda_2 = -8$

$\begin{bmatrix} 5 & 13 & 0 \\ 13 & 5 & 0 \\ 0 & 0 & -8 \end{bmatrix} \begin{bmatrix} \frac{\sqrt{6}}{6} \\ -\frac{\sqrt{6}}{6} \\ -\frac{\sqrt{6}}{3} \end{bmatrix} = \begin{bmatrix} -\frac{8\sqrt{6}}{6} \\ \frac{8\sqrt{6}}{6} \\ \frac{8\sqrt{6}}{3} \end{bmatrix} = (-8)\begin{bmatrix} \frac{\sqrt{6}}{6} \\ -\frac{\sqrt{6}}{6} \\ -\frac{\sqrt{6}}{3} \end{bmatrix}; \quad \lambda_3 = -8$

(c) $\mathbf{K}_1^T\mathbf{K}_2 = \begin{bmatrix} \frac{\sqrt{2}}{2} & \frac{\sqrt{2}}{2} & 0 \end{bmatrix}\begin{bmatrix} \frac{\sqrt{3}}{3} \\ -\frac{\sqrt{3}}{3} \\ \frac{\sqrt{3}}{3} \end{bmatrix} = \frac{\sqrt{6}}{6} - \frac{\sqrt{6}}{6} = 0; \quad \mathbf{K}_1^T\mathbf{K}_3 = \begin{bmatrix} \frac{\sqrt{2}}{2} & \frac{\sqrt{2}}{2} & 0 \end{bmatrix}\begin{bmatrix} \frac{\sqrt{6}}{6} \\ -\frac{\sqrt{6}}{6} \\ -\frac{\sqrt{6}}{3} \end{bmatrix} = \frac{\sqrt{12}}{12} - \frac{\sqrt{12}}{12} = 0$

$\mathbf{K}_2^T\mathbf{K}_3 = \begin{bmatrix} \frac{\sqrt{3}}{3} & -\frac{\sqrt{3}}{3} & \frac{\sqrt{3}}{3} \end{bmatrix}\begin{bmatrix} \frac{\sqrt{6}}{6} \\ -\frac{\sqrt{6}}{6} \\ -\frac{\sqrt{6}}{3} \end{bmatrix} = \frac{\sqrt{18}}{18} + \frac{\sqrt{18}}{18} - \frac{\sqrt{18}}{9} = 0$

6. Not orthogonal. Columns one and three are not unit vectors.

9. Not orthogonal. Columns are not unit vectors.

12. $\lambda_1 = 7, \quad \lambda_2 = 4, \quad \mathbf{K}_1 = \begin{bmatrix} 1 \\ 0 \end{bmatrix}, \quad \mathbf{K}_2 = \begin{bmatrix} 0 \\ 1 \end{bmatrix}, \quad \mathbf{P} = \begin{bmatrix} 1 & 0 \\ 0 & 1 \end{bmatrix}$

15. $\lambda_1 = 0, \quad \lambda_2 = 2, \quad \lambda_3 = 1, \quad \mathbf{K}_1 = \begin{bmatrix} -1 \\ 0 \\ 1 \end{bmatrix}, \quad \mathbf{K}_2 = \begin{bmatrix} 1 \\ 0 \\ 1 \end{bmatrix}, \quad \mathbf{K}_3 = \begin{bmatrix} 0 \\ 1 \\ 0 \end{bmatrix}, \quad \mathbf{P} = \begin{bmatrix} -\frac{1}{\sqrt{2}} & \frac{1}{\sqrt{2}} & 0 \\ 0 & 0 & 1 \\ \frac{1}{\sqrt{2}} & \frac{1}{\sqrt{2}} & 0 \end{bmatrix}$

18. $\lambda_1 = -18, \; \lambda_2 = 0, \; \lambda_3 = 9, \; \mathbf{K}_1 = \begin{bmatrix} 1 \\ -2 \\ 2 \end{bmatrix}, \; \mathbf{K}_2 = \begin{bmatrix} -2 \\ 1 \\ 2 \end{bmatrix}, \; \mathbf{K}_3 = \begin{bmatrix} 2 \\ 2 \\ 1 \end{bmatrix}, \; \mathbf{P} = \begin{bmatrix} \frac{1}{3} & -\frac{2}{3} & \frac{2}{3} \\ -\frac{2}{3} & \frac{1}{3} & \frac{2}{3} \\ \frac{2}{3} & \frac{2}{3} & \frac{1}{3} \end{bmatrix}$

21. (a) $\mathbf{AK}_1 = \begin{bmatrix} 7 & 4 & -4 \\ 4 & -8 & -1 \\ -4 & -1 & -8 \end{bmatrix} \begin{bmatrix} 4 \\ 1 \\ -1 \end{bmatrix} = \begin{bmatrix} 36 \\ 9 \\ -9 \end{bmatrix} = 9 \begin{bmatrix} 4 \\ 1 \\ -1 \end{bmatrix} = \lambda_1 \mathbf{K}_1$

$\mathbf{AK}_2 = \begin{bmatrix} 7 & 4 & -4 \\ 4 & -8 & -1 \\ -4 & -1 & -8 \end{bmatrix} \begin{bmatrix} 1 \\ 0 \\ 4 \end{bmatrix} = \begin{bmatrix} -9 \\ 0 \\ -36 \end{bmatrix} = -9 \begin{bmatrix} 1 \\ 0 \\ 4 \end{bmatrix} = \lambda_2 \mathbf{K}_2$

$\mathbf{AK}_3 = \begin{bmatrix} 7 & 4 & -4 \\ 4 & -8 & -1 \\ -4 & -1 & -8 \end{bmatrix} \begin{bmatrix} 1 \\ -4 \\ 0 \end{bmatrix} = \begin{bmatrix} -9 \\ 36 \\ 0 \end{bmatrix} = -9 \begin{bmatrix} 1 \\ -4 \\ 0 \end{bmatrix} = \lambda_3 \mathbf{K}_3$

(b) For $\lambda_2 = \lambda_3 = -9$ we have

$$\begin{bmatrix} 16 & 4 & -4 & | & 0 \\ 4 & 1 & -1 & | & 0 \\ -4 & -1 & 1 & | & 0 \end{bmatrix} \implies \begin{bmatrix} 1 & \frac{1}{4} & -\frac{1}{4} & | & 0 \\ 0 & 0 & 0 & | & 0 \\ 0 & 0 & 0 & | & 0 \end{bmatrix}$$

so that $k_1 = -\frac{1}{4}k_2 + \frac{1}{4}k_3$. The choices $k_2 = 0$, $k_3 = 4$ and $k_2 = -4$, $k_3 = 0$ give the eigenvectors in part (a). These eigenvectors are not orthogonal. However, by choosing $k_2 = k_3 = 1$ and $k_2 = -2$, $k_3 = 2$ we obtain, respectively,

$$\mathbf{K}_2 = \begin{bmatrix} 0 \\ 1 \\ 1 \end{bmatrix}, \quad \mathbf{K}_3 = \begin{bmatrix} 1 \\ -2 \\ 2 \end{bmatrix}.$$

Mutually orthogonal eigenvectors are

$$\begin{bmatrix} 4 \\ 1 \\ -1 \end{bmatrix}, \quad \begin{bmatrix} 0 \\ 1 \\ 1 \end{bmatrix}, \quad \begin{bmatrix} 1 \\ -2 \\ 2 \end{bmatrix}.$$

Exercises 8.11

3. Taking $\mathbf{X}_0 = \begin{bmatrix} 1 \\ 1 \end{bmatrix}$ and computing $\mathbf{AX}_0 = \begin{bmatrix} 6 \\ 16 \end{bmatrix}$, we define $\mathbf{X}_1 = \frac{1}{16} \begin{bmatrix} 6 \\ 16 \end{bmatrix} = \begin{bmatrix} 0.375 \\ 1 \end{bmatrix}$. Continuing in this manner we obtain

$$\mathbf{X}_2 = \begin{bmatrix} 0.3363 \\ 1 \end{bmatrix}, \quad \mathbf{X}_3 = \begin{bmatrix} 0.3335 \\ 1 \end{bmatrix}, \quad \mathbf{X}_4 = \begin{bmatrix} 0.3333 \\ 1 \end{bmatrix}.$$

We conclude that a dominant eigenvector is $\mathbf{K} = \begin{bmatrix} 0.3333 \\ 1 \end{bmatrix}$ with corresponding eigenvalue $\lambda = 14$.

6. Taking $\mathbf{X}_0 = \begin{bmatrix} 1 \\ 1 \\ 1 \end{bmatrix}$ and computing $\mathbf{AX}_0 = \begin{bmatrix} 5 \\ 2 \\ 2 \end{bmatrix}$, we define $\mathbf{X}_1 = \dfrac{1}{5} \begin{bmatrix} 5 \\ 2 \\ 2 \end{bmatrix} = \begin{bmatrix} 1 \\ 0.4 \\ 0.4 \end{bmatrix}$. Continuing in this manner we obtain

$$\mathbf{X}_2 = \begin{bmatrix} 1 \\ 0.2105 \\ 0.2105 \end{bmatrix}, \quad \mathbf{X}_3 = \begin{bmatrix} 1 \\ 0.1231 \\ 0.1231 \end{bmatrix}, \quad \mathbf{X}_4 = \begin{bmatrix} 1 \\ 0.0758 \\ 0.0758 \end{bmatrix}, \quad \mathbf{X}_5 = \begin{bmatrix} 1 \\ 0.0481 \\ 0.0481 \end{bmatrix}.$$

At this point if we restart with $\mathbf{X}_0 = \begin{bmatrix} 1 \\ 0 \\ 0 \end{bmatrix}$ we see that $\mathbf{K} = \begin{bmatrix} 1 \\ 0 \\ 0 \end{bmatrix}$ is a dominant eigenvector with corresponding eigenvalue $\lambda = 3$.

9. Taking $\mathbf{X}_0 = \begin{bmatrix} 1 \\ 1 \\ 1 \end{bmatrix}$ and using scaling we obtain

$$\mathbf{X}_1 = \begin{bmatrix} 1 \\ 0 \\ 1 \end{bmatrix}, \quad \mathbf{X}_2 = \begin{bmatrix} 1 \\ -0.6667 \\ 1 \end{bmatrix}, \quad \mathbf{X}_3 = \begin{bmatrix} 1 \\ -0.9091 \\ 1 \end{bmatrix}, \quad \mathbf{X}_4 = \begin{bmatrix} 1 \\ -0.9767 \\ 1 \end{bmatrix}, \quad \mathbf{X}_5 = \begin{bmatrix} 1 \\ -0.9942 \\ 1 \end{bmatrix}.$$

Taking $\mathbf{K} = \begin{bmatrix} 1 \\ -1 \\ 1 \end{bmatrix}$ as the dominant eigenvector we find $\lambda_1 = 4$. Now the normalized eigenvector is

$$\mathbf{K}_1 = \begin{bmatrix} 0.5774 \\ -0.5774 \\ 0.5774 \end{bmatrix} \text{ and } \mathbf{B} = \begin{bmatrix} 1.6667 & 0.3333 & -1.3333 \\ 0.3333 & 0.6667 & 0.3333 \\ -1.3333 & 0.3333 & 1.6667 \end{bmatrix}. \text{ If } \mathbf{X}_0 = \begin{bmatrix} 1 \\ 1 \\ 1 \end{bmatrix} \text{ is now chosen only one more}$$

eigenvalue is found. Thus, try $\mathbf{X}_0 = \begin{bmatrix} 1 \\ 1 \\ 0 \end{bmatrix}$. Using scaling we obtain

$$\mathbf{X}_1 = \begin{bmatrix} 1 \\ 0.5 \\ -0.5 \end{bmatrix}, \quad \mathbf{X}_2 = \begin{bmatrix} 1 \\ 0.2 \\ -0.8 \end{bmatrix}, \quad \mathbf{X}_3 = \begin{bmatrix} 1 \\ 0.0714 \\ -0.9286 \end{bmatrix}, \quad \mathbf{X}_4 = \begin{bmatrix} 1 \\ 0.0244 \\ -0.9756 \end{bmatrix}, \quad \mathbf{X}_5 = \begin{bmatrix} 1 \\ 0.0082 \\ -0.9918 \end{bmatrix}.$$

Taking $\mathbf{K} = \begin{bmatrix} 1 \\ 0 \\ -1 \end{bmatrix}$ as the eigenvector we find $\lambda_2 = 3$. The normalized eigenvector in this case is

$$\mathbf{K}_2 = \begin{bmatrix} 0.7071 \\ 0 \\ -0.7071 \end{bmatrix} \text{ and } \mathbf{C} = \begin{bmatrix} 0.1667 & 0.3333 & 0.1667 \\ 0.3333 & 0.6667 & 0.3333 \\ 0.1667 & 0.3333 & 0.1667 \end{bmatrix}. \text{ If } \mathbf{X}_0 = \begin{bmatrix} 1 \\ 1 \\ 1 \end{bmatrix} \text{ is chosen, and scaling is used we obtain}$$

$$\mathbf{X}_1 = \begin{bmatrix} 0.5 \\ 1 \\ 0.5 \end{bmatrix}, \mathbf{X}_2 = \begin{bmatrix} 0.5 \\ 1 \\ 0.5 \end{bmatrix}. \text{ Taking } \mathbf{K} = \begin{bmatrix} 0.5 \\ 1 \\ 0.5 \end{bmatrix} \text{ we find } \lambda_3 = 1. \text{ The eigenvalues are 4, 3, and 1. The difficulty}$$

in choosing $\mathbf{X}_0 = \begin{bmatrix} 1 \\ 1 \\ 1 \end{bmatrix}$ to find the second eigenvector results from the fact that this vector is a linear combination

of the eigenvectors corresponding to the other two eigenvalues, with 0 contribution from the second eigenvector. When this occurs the development of the power method, shown in the text, breaks down.

12. The inverse matrix is $\begin{bmatrix} 1 & 3 \\ 4 & 2 \end{bmatrix}$. Taking $\mathbf{X}_0 = \begin{bmatrix} 1 \\ 1 \end{bmatrix}$ and using scaling we obtain

$$\mathbf{X}_1 = \begin{bmatrix} 0.6667 \\ 1 \end{bmatrix}, \quad \mathbf{X}_2 = \begin{bmatrix} 0.7857 \\ 1 \end{bmatrix}, \ldots, \mathbf{X}_{10} = \begin{bmatrix} 0.75 \\ 1 \end{bmatrix}.$$

Using $\mathbf{K} = \begin{bmatrix} 0.75 \\ 1 \end{bmatrix}$ we find $\lambda = 5$. The minimum eigenvalue of $\begin{bmatrix} -0.2 & 0.3 \\ 0.4 & -0.1 \end{bmatrix}$ is $1/5 = 0.2$

Exercises 8.12

3. For $\lambda_1 = \lambda_2 = 1$ we obtain the single eigenvector $\mathbf{K}_1 = \begin{bmatrix} 1 \\ 1 \end{bmatrix}$. Hence $\mathbf{A}$ is not diagonalizable.

6. Distinct eigenvalues $\lambda_1 = -4$, $\lambda_2 = 10$ imply $\mathbf{A}$ is diagonalizable.

$$\mathbf{P} = \begin{bmatrix} -3 & 1 \\ 1 & -5 \end{bmatrix}, \quad \mathbf{D} = \begin{bmatrix} -4 & 0 \\ 0 & 10 \end{bmatrix}$$

9. Distinct eigenvalues $\lambda_1 = -i$, $\lambda_2 = i$ imply $\mathbf{A}$ is diagonalizable.

$$\mathbf{P} = \begin{bmatrix} 1 & 1 \\ -i & i \end{bmatrix}, \quad \mathbf{D} = \begin{bmatrix} -i & 0 \\ 0 & i \end{bmatrix}$$

12. Distinct eigenvalues $\lambda_1 = 3$, $\lambda_2 = 4$, $\lambda_3 = 5$ imply $\mathbf{A}$ is diagonalizable.

$$\mathbf{P} = \begin{bmatrix} 1 & 2 & 0 \\ 0 & 2 & 1 \\ 1 & 1 & -1 \end{bmatrix}, \quad \mathbf{D} = \begin{bmatrix} 3 & 0 & 0 \\ 0 & 4 & 0 \\ 0 & 0 & 5 \end{bmatrix}$$

15. The eigenvalues are $\lambda_1 = \lambda_2 = 1$, $\lambda_3 = 2$. For $\lambda_1 = \lambda_2 = 1$ we obtain the single eigenvector $\mathbf{K}_1 = \begin{bmatrix} 1 \\ 0 \\ 0 \end{bmatrix}$.

Hence $\mathbf{A}$ is not diagonalizable.

18. For $\lambda_1 = \lambda_2 = \lambda_3 = 1$ we obtain the single eigenvector $\mathbf{K}_1 = \begin{bmatrix} 1 \\ -2 \\ 1 \end{bmatrix}$. Hence $\mathbf{A}$ is not diagonalizable.

21. $\lambda_1 = 0$, $\lambda_2 = 2$, $\mathbf{K}_1 = \begin{bmatrix} 1 \\ -1 \end{bmatrix}$, $\mathbf{K}_2 = \begin{bmatrix} 1 \\ 1 \end{bmatrix}$, $\mathbf{P} = \begin{bmatrix} \frac{1}{\sqrt{2}} & \frac{1}{\sqrt{2}} \\ -\frac{1}{\sqrt{2}} & \frac{1}{\sqrt{2}} \end{bmatrix}$, $\mathbf{D} = \begin{bmatrix} 0 & 0 \\ 0 & 2 \end{bmatrix}$

24. $\lambda_1 = -1$, $\lambda_2 = 3$, $\mathbf{K}_1 = \begin{bmatrix} 1 \\ 1 \end{bmatrix}$, $\mathbf{K}_2 = \begin{bmatrix} 1 \\ -1 \end{bmatrix}$, $\mathbf{P} = \begin{bmatrix} \frac{1}{\sqrt{2}} & \frac{1}{\sqrt{2}} \\ \frac{1}{\sqrt{2}} & -\frac{1}{\sqrt{2}} \end{bmatrix}$, $\mathbf{D} = \begin{bmatrix} -1 & 0 \\ 0 & 3 \end{bmatrix}$

27. $\lambda_1 = 3$, $\lambda_2 = 6$, $\lambda_3 = 9$, $\mathbf{K}_1 = \begin{bmatrix} 2 \\ 2 \\ 1 \end{bmatrix}$, $\mathbf{K}_2 = \begin{bmatrix} 2 \\ -1 \\ -2 \end{bmatrix}$, $\mathbf{K}_3 = \begin{bmatrix} 1 \\ -2 \\ 2 \end{bmatrix}$, $\mathbf{P} = \begin{bmatrix} \frac{2}{3} & \frac{2}{3} & \frac{1}{3} \\ \frac{2}{3} & -\frac{1}{3} & -\frac{2}{3} \\ \frac{1}{3} & -\frac{2}{3} & \frac{2}{3} \end{bmatrix}$, $\mathbf{D} = \begin{bmatrix} 3 & 0 & 0 \\ 0 & 6 & 0 \\ 0 & 0 & 9 \end{bmatrix}$

30. $\lambda_1 = \lambda_2 = 0$, $\lambda_3 = -2$, $\lambda_4 = 2$, $\mathbf{K}_1 = \begin{bmatrix} -1 \\ 0 \\ 1 \\ 0 \end{bmatrix}$, $\mathbf{K}_2 = \begin{bmatrix} 0 \\ -1 \\ 0 \\ 1 \end{bmatrix}$, $\mathbf{K}_3 = \begin{bmatrix} 1 \\ -1 \\ 1 \\ -1 \end{bmatrix}$, $\mathbf{K}_4 = \begin{bmatrix} 1 \\ 1 \\ 1 \\ 1 \end{bmatrix}$

$$\mathbf{P} = \begin{bmatrix} -\frac{1}{\sqrt{2}} & 0 & \frac{1}{2} & \frac{1}{2} \\ 0 & -\frac{1}{\sqrt{2}} & -\frac{1}{2} & \frac{1}{2} \\ \frac{1}{\sqrt{2}} & 0 & \frac{1}{2} & \frac{1}{2} \\ 0 & \frac{1}{\sqrt{2}} & -\frac{1}{2} & \frac{1}{2} \end{bmatrix}, \quad \mathbf{D} = \begin{bmatrix} 0 & 0 & 0 & 0 \\ 0 & 0 & 0 & 0 \\ 0 & 0 & -2 & 0 \\ 0 & 0 & 0 & 2 \end{bmatrix}$$

33. The given equation can be written as $\mathbf{X}^T\mathbf{A}\mathbf{X} = 20$: $[x \ \ y] \begin{bmatrix} -3 & 4 \\ 4 & 3 \end{bmatrix} \begin{bmatrix} x \\ y \end{bmatrix} = 20$. Using

$\lambda_1 = 5$, $\lambda_2 = -5$, $\mathbf{K}_1 = \begin{bmatrix} 1 \\ 2 \end{bmatrix}$, $\mathbf{K}_2 = \begin{bmatrix} -2 \\ 1 \end{bmatrix}$, $\mathbf{P} = \begin{bmatrix} \frac{1}{\sqrt{5}} & -\frac{2}{\sqrt{5}} \\ \frac{2}{\sqrt{5}} & \frac{1}{\sqrt{5}} \end{bmatrix}$ and $\mathbf{X} = \mathbf{P}\mathbf{X}'$ we find

$$[X \ \ Y] \begin{bmatrix} 5 & 0 \\ 0 & -5 \end{bmatrix} \begin{bmatrix} X \\ Y \end{bmatrix} = 20 \quad \text{or} \quad 5X^2 - 5Y^2 = 20.$$

The conic section is a hyperbola. Now from $\mathbf{X}' = \mathbf{P}^T\mathbf{X}$ we see that the XY-coordinates of $(1, 2)$ and $(-2, 1)$ are $(\sqrt{5}, 0)$ and $(0, \sqrt{5})$, respectively. From this we conclude that the X-axis and Y-axis are as shown in the accompanying figure.

36. Since eigenvectors are mutually orthogonal we use an orthogonal matrix $\mathbf{P}$ and $\mathbf{A} = \mathbf{P}\mathbf{D}\mathbf{P}^T$.

$$\mathbf{A} = \begin{bmatrix} \frac{1}{\sqrt{3}} & \frac{1}{\sqrt{2}} & \frac{1}{\sqrt{6}} \\ -\frac{1}{\sqrt{3}} & 0 & \frac{2}{\sqrt{6}} \\ \frac{1}{\sqrt{3}} & -\frac{1}{\sqrt{2}} & \frac{1}{\sqrt{6}} \end{bmatrix} \begin{bmatrix} 1 & 0 & 0 \\ 0 & 3 & 0 \\ 0 & 0 & 5 \end{bmatrix} \begin{bmatrix} \frac{1}{\sqrt{3}} & -\frac{1}{\sqrt{3}} & \frac{1}{\sqrt{3}} \\ \frac{1}{\sqrt{2}} & 0 & -\frac{1}{\sqrt{2}} \\ \frac{1}{\sqrt{6}} & \frac{2}{\sqrt{6}} & \frac{1}{\sqrt{6}} \end{bmatrix} = \begin{bmatrix} \frac{8}{3} & \frac{4}{3} & -\frac{1}{3} \\ \frac{4}{3} & \frac{11}{3} & \frac{4}{3} \\ -\frac{1}{3} & \frac{4}{3} & \frac{8}{3} \end{bmatrix}$$

39. $\lambda_1 = 2$, $\lambda_2 = -1$, $\mathbf{K}_1 = \begin{bmatrix} 1 \\ 1 \end{bmatrix}$, $\mathbf{K}_2 = \begin{bmatrix} -1 \\ 2 \end{bmatrix}$, $\mathbf{P} = \begin{bmatrix} 1 & -1 \\ 1 & 2 \end{bmatrix}$, $\mathbf{P}^{-1} = \begin{bmatrix} \frac{2}{3} & \frac{1}{3} \\ -\frac{1}{3} & \frac{1}{3} \end{bmatrix}$

$$\mathbf{A}^5 = \begin{bmatrix} 1 & -1 \\ 1 & 2 \end{bmatrix} \begin{bmatrix} 32 & 0 \\ 0 & -1 \end{bmatrix} \begin{bmatrix} \frac{2}{3} & \frac{1}{3} \\ -\frac{1}{3} & \frac{1}{3} \end{bmatrix} = \begin{bmatrix} 21 & 11 \\ 22 & 10 \end{bmatrix}$$

Exercises 8.13

3. (a) The message is $\mathbf{M} = \begin{bmatrix} 16 & 8 & 15 & 14 & 5 \\ 0 & 8 & 15 & 13 & 5 \end{bmatrix}$. The encoded message is

$$\mathbf{B} = \mathbf{A}\mathbf{M} = \begin{bmatrix} 3 & 5 \\ 2 & 3 \end{bmatrix} \begin{bmatrix} 16 & 8 & 15 & 14 & 5 \\ 0 & 8 & 15 & 13 & 5 \end{bmatrix} = \begin{bmatrix} 48 & 64 & 120 & 107 & 40 \\ 32 & 40 & 75 & 67 & 25 \end{bmatrix}.$$

(b) The decoded message is

$$\mathbf{M} = \mathbf{A}^{-1}\mathbf{B} = \begin{bmatrix} -3 & 5 \\ 2 & -3 \end{bmatrix} \begin{bmatrix} 48 & 64 & 120 & 107 & 40 \\ 32 & 40 & 75 & 67 & 25 \end{bmatrix} = \begin{bmatrix} 16 & 8 & 15 & 14 & 5 \\ 0 & 8 & 15 & 13 & 5 \end{bmatrix}.$$

6. (a) The message is $\mathbf{M} = \begin{bmatrix} 4 & 18 & 0 & 10 & 15 & 8 \\ 14 & 0 & 9 & 19 & 0 & 20 \\ 8 & 5 & 0 & 19 & 16 & 25 \end{bmatrix}$. The encoded message is

$$\mathbf{B} = \mathbf{AM} = \begin{bmatrix} 5 & 3 & 0 \\ 4 & 3 & -1 \\ 5 & 2 & 2 \end{bmatrix} \begin{bmatrix} 4 & 18 & 0 & 10 & 15 & 8 \\ 14 & 0 & 9 & 19 & 0 & 20 \\ 8 & 5 & 0 & 19 & 16 & 25 \end{bmatrix} = \begin{bmatrix} 62 & 90 & 27 & 107 & 75 & 100 \\ 50 & 67 & 27 & 78 & 44 & 67 \\ 64 & 100 & 18 & 126 & 107 & 130 \end{bmatrix}.$$

(b) The decoded message is

$$\mathbf{M} = \mathbf{A}^{-1}\mathbf{B} = \begin{bmatrix} 8 & -6 & -3 \\ -13 & 10 & 5 \\ -7 & 5 & 3 \end{bmatrix} \begin{bmatrix} 62 & 90 & 27 & 107 & 75 & 100 \\ 50 & 67 & 27 & 78 & 44 & 67 \\ 64 & 100 & 18 & 126 & 107 & 130 \end{bmatrix} = \begin{bmatrix} 4 & 18 & 0 & 10 & 15 & 8 \\ 14 & 0 & 9 & 19 & 0 & 20 \\ 8 & 5 & 0 & 19 & 16 & 25 \end{bmatrix}.$$

9. The decoded message is

$$\mathbf{M} = \mathbf{A}^{-1}\mathbf{B} = \begin{bmatrix} 0 & 0 & 1 \\ 0 & 1 & 0 \\ 1 & 0 & -1 \end{bmatrix} \begin{bmatrix} 31 & 21 & 21 & 22 & 20 & 9 \\ 19 & 0 & 9 & 13 & 16 & 15 \\ 13 & 1 & 20 & 8 & 0 & 9 \end{bmatrix} = \begin{bmatrix} 13 & 1 & 20 & 8 & 0 & 9 \\ 19 & 0 & 9 & 13 & 16 & 15 \\ 18 & 20 & 1 & 14 & 20 & 0 \end{bmatrix}.$$

From correspondence (1) we obtain: MATH_IS_IMPORTANT.

12. (a) $\mathbf{M}^T = \begin{bmatrix} 22 & 8 & 19 & 27 & 21 & 3 & 3 & 27 & 21 & 18 & 21 \\ 13 & 3 & 21 & 22 & 3 & 25 & 27 & 6 & 7 & 14 & 23 \\ 2 & 27 & 21 & 7 & 27 & 5 & 21 & 17 & 2 & 25 & 7 \end{bmatrix}$

(b) $\mathbf{B}^T = \mathbf{M} = \begin{bmatrix} 1 & 1 & 0 \\ 1 & 0 & 1 \\ 1 & 1 & -1 \end{bmatrix} = \begin{bmatrix} 37 & 38 & 61 & 56 & 51 & 33 & 51 & 50 & 30 & 57 & 51 \\ 24 & 35 & 40 & 34 & 48 & 8 & 24 & 44 & 23 & 43 & 28 \\ 11 & -24 & 0 & 15 & -24 & 20 & 6 & -11 & 5 & -11 & 16 \end{bmatrix}$

(c) $\mathbf{BA}^{-1} = \mathbf{B} \begin{bmatrix} -1 & 1 & 1 \\ 2 & -1 & -1 \\ 1 & 0 & -1 \end{bmatrix} = \mathbf{M}$

Exercises 8.14

3. $[0 \ \ 0 \ \ 0 \ \ 1 \ \ 1]$

6. $[0 \ \ 1 \ \ 1 \ \ 0 \ \ 1 \ \ 0 \ \ 1 \ \ 0]$

9. Parity error

12. Parity error

In Problems 15 and 18, $\mathbf{D} = [c_1 \quad c_2 \quad c_3]$ and $\mathbf{P} = \begin{bmatrix} 1 & 1 & 0 & 1 \\ 1 & 0 & 1 & 1 \\ 0 & 1 & 1 & 1 \end{bmatrix}$.

15. $\mathbf{D}^T = \mathbf{P}[0 \ \ 1 \ \ 0 \ \ 1]^T = [0 \ \ 1 \ \ 0]^T$; $\quad \mathbf{C} = [0 \ \ 1 \ \ 0 \ \ 0 \ \ 1 \ \ 0 \ \ 1]$

18. $\mathbf{D}^T = \mathbf{P}[1 \ \ 1 \ \ 0 \ \ 0]^T = [0 \ \ 1 \ \ 1]^T$; $\quad \mathbf{C} = [0 \ \ 1 \ \ 1 \ \ 1 \ \ 1 \ \ 0 \ \ 0]$

In Problems 21-27, **W** represents the correctly decoded message.

21. $\mathbf{S} = \mathbf{H}\mathbf{R}^T = \mathbf{H}\begin{bmatrix} 1 & 1 & 0 & 1 & 1 & 0 & 1 \end{bmatrix} = \begin{bmatrix} 1 & 0 & 1 \end{bmatrix}^T$; not a code word. The error is in the fifth bit.
   $\mathbf{W} = \begin{bmatrix} 0 & 0 & 0 & 1 \end{bmatrix}$

24. $\mathbf{S} = \mathbf{H}\mathbf{R}^T = \mathbf{H}\begin{bmatrix} 1 & 1 & 0 & 0 & 1 & 1 & 0 \end{bmatrix} = \begin{bmatrix} 0 & 0 & 0 \end{bmatrix}^T$; a code word. $\mathbf{W} = \begin{bmatrix} 0 & 1 & 1 & 0 \end{bmatrix}$

27. $\mathbf{S} = \mathbf{H}\mathbf{R}^T = \mathbf{H}\begin{bmatrix} 1 & 0 & 1 & 1 & 0 & 1 & 1 \end{bmatrix} = \begin{bmatrix} 1 & 1 & 1 \end{bmatrix}^T$; not a code word. The error is in the seventh bit.
   $\mathbf{W} = \begin{bmatrix} 1 & 0 & 1 & 0 \end{bmatrix}$

30. (a) $c_4 = 0, c_3 = 1, c_2 = 1, c_1 = 0$; $\begin{bmatrix} 0 & 1 & 1 & 0 & 0 & 1 & 1 & 0 \end{bmatrix}$

   (b) $\mathbf{H} = \begin{bmatrix} 0 & 0 & 0 & 0 & 1 & 1 & 1 & 1 \\ 0 & 0 & 1 & 1 & 0 & 0 & 1 & 1 \\ 0 & 1 & 0 & 1 & 0 & 1 & 0 & 1 \\ 1 & 1 & 1 & 1 & 1 & 1 & 1 & 1 \end{bmatrix}$

   (c) $\mathbf{S} = \mathbf{H}\mathbf{R}^T = \mathbf{H}\begin{bmatrix} 0 & 0 & 1 & 1 & 1 & 1 & 0 & 0 \end{bmatrix}^T = \begin{bmatrix} 0 & 0 & 0 & 0 \end{bmatrix}^T$

Exercises 8.15

3. We have $\mathbf{Y}^T = \begin{bmatrix} 1 & 1.5 & 3 & 4.5 & 5 \end{bmatrix}$ and $\mathbf{A}^T = \begin{bmatrix} 1 & 2 & 3 & 4 & 5 \\ 1 & 1 & 1 & 1 & 1 \end{bmatrix}$.

Now $\mathbf{A}^T\mathbf{A} = \begin{bmatrix} 55 & 15 \\ 15 & 5 \end{bmatrix}$ and $(\mathbf{A}^T\mathbf{A})^{-1} = \dfrac{1}{50}\begin{bmatrix} 5 & -15 \\ -15 & 55 \end{bmatrix}$

so $\mathbf{X} = (\mathbf{A}^T\mathbf{A})^{-1}\mathbf{A}^T\mathbf{Y} = \begin{bmatrix} 1.1 \\ -0.3 \end{bmatrix}$ and the least squares line is $y = 1.1x - 0.3$.

6. We have $\mathbf{Y}^T = \begin{bmatrix} 2 & 2.5 & 1 & 1.5 & 2 & 3.2 & 5 \end{bmatrix}$ and $\mathbf{A}^T = \begin{bmatrix} 1 & 2 & 3 & 4 & 5 & 6 & 7 \\ 1 & 1 & 1 & 1 & 1 & 1 & 1 \end{bmatrix}$.

Now $\mathbf{A}^T\mathbf{A} = \begin{bmatrix} 140 & 28 \\ 28 & 7 \end{bmatrix}$ and $(\mathbf{A}^T\mathbf{A})^{-1} = \dfrac{1}{196}\begin{bmatrix} 7 & -28 \\ -28 & 140 \end{bmatrix}$

so $\mathbf{X} = (\mathbf{A}^T\mathbf{A})^{-1}\mathbf{A}^T\mathbf{Y} = \begin{bmatrix} 0.407143 \\ 0.828571 \end{bmatrix}$ and the least squares line is $y = 0.407143x + 0.828571$.

Exercises 8.16

3. (a) We use the fact that the element τ_{ij} in the transfer matrix **T** is the rate of transfer from compartment j to compartment i, and the fact that the sum of each column in **T** is 1. The initial state and the transfer matrix are

$$\mathbf{X}_0 = \begin{bmatrix} 100 \\ 0 \\ 0 \end{bmatrix} \quad \text{and} \quad \mathbf{T} = \begin{bmatrix} 0.2 & 0.5 & 0 \\ 0.3 & 0.1 & 0 \\ 0.5 & 0.4 & 1 \end{bmatrix}.$$

   (b) We have

$$\mathbf{X}_1 = \mathbf{T}\mathbf{X}_0 = \begin{bmatrix} 20 \\ 30 \\ 50 \end{bmatrix} \quad \text{and} \quad \mathbf{X}_2 = \mathbf{T}\mathbf{X}_1 = \begin{bmatrix} 19 \\ 9 \\ 72 \end{bmatrix}.$$

(c) From $\mathbf{T\hat{X}} - \mathbf{\hat{X}} = (\mathbf{T} - \mathbf{I})\mathbf{\hat{X}} = \mathbf{0}$ and the fact that the system is closed we obtain

$$-0.8x_1 + 0.5x_2 \qquad = 0$$
$$0.3x_1 - 0.9x_2 \qquad = 0$$
$$x_1 + \quad x_2 + x_3 = 100.$$

The solution is $x_1 = x_2 = 0$, $x_3 = 100$, so the equilibrium state is $\mathbf{\hat{X}} = \begin{bmatrix} 0 \\ 0 \\ 100 \end{bmatrix}$.

6. From $\mathbf{T\hat{X}} = 1\mathbf{\hat{X}}$ we see that the equilibrium state vector $\mathbf{\hat{X}}$ is the eigenvector of the transfer matrix $\mathbf{T}$ corresponding to the eigenvalue 1. It has the properties that its components add up to the sum of the components of the initial state vector.

Chapter 8 Review Exercises

3. $\mathbf{AB} = \begin{bmatrix} 3 & 4 \\ 6 & 8 \end{bmatrix}$; $\mathbf{BA} = [11]$

6. True

9. 0

12. True

15. False; if the characteristic equation of an $n \times n$ matrix has repeated roots, there may not be n linearly independent eigenvectors.

18. True

21. $\mathbf{A} = \frac{1}{2}(\mathbf{A} + \mathbf{A}^T) + \frac{1}{2}(\mathbf{A} - \mathbf{A}^T)$ where $\frac{1}{2}(\mathbf{A} + \mathbf{A}^T)$ is symmetric and $\frac{1}{2}(\mathbf{A} - \mathbf{A}^T)$ is skew-symmetric.

24. (a) $\sigma_x\sigma_y = \begin{bmatrix} i & 0 \\ 0 & -i \end{bmatrix} = -\sigma_y\sigma_x$; $\sigma_x\sigma_z = \begin{bmatrix} 0 & -1 \\ 1 & 0 \end{bmatrix} = -\sigma_z\sigma_x$; $\sigma_y\sigma_z = \begin{bmatrix} 0 & i \\ i & 0 \end{bmatrix} = -\sigma_z\sigma_y$

(b) We first note that for anticommuting matrices $\mathbf{AB} = -\mathbf{BA}$, so $\mathbf{C} = 2\mathbf{AB}$. Then $\mathbf{C}_{xy} = \begin{bmatrix} 2i & 0 \\ 0 & -2i \end{bmatrix}$, $\mathbf{C}_{yz} = \begin{bmatrix} 0 & 2i \\ 2i & 0 \end{bmatrix}$, and $\mathbf{C}_{zx} = \begin{bmatrix} 0 & 2 \\ -2 & 0 \end{bmatrix}$.

27. Multiplying the second row by abc we obtain the third row. Thus the determinant is 0.

30. $(-3)(6)(9)(1) = -162$

33. From $x_1\mathrm{I}_2 + x_2\mathrm{HNO}_3 \to x_3\mathrm{HIO}_3 + x_4\mathrm{NO}_2 + x_5\mathrm{H}_2\mathrm{O}$ we obtain the system $2x_1 = x_3$, $x_2 = x_3 + 2x_5$, $x_2 = x_4$, $3x_2 = 3x_3 + 2x_4 + x_5$. Letting $x_4 = x_2$ in the fourth equation we obtain $x_2 = 3x_3 + x_5$. Taking $x_1 = t$ we see that $x_3 = 2t$, $x_2 = 2t + 2x_5$, and $x_2 = 6t + x_5$. From the latter two equations we get $x_5 = 4t$. Taking $t = 1$ we have $x_1 = 1$, $x_2 = 10$, $x_3 = 2$, $x_4 = 10$, and $x_5 = 4$. The balanced equation is $\mathrm{I}_2 + 10\mathrm{HNO}_3 \to 2\mathrm{HIO}_3 + 10\mathrm{NO}_2 + 4\mathrm{H}_2\mathrm{O}$.

36. $\det = 4$, $\det \mathbf{A}_1 = 16$, $\det \mathbf{A}_2 = -4$, $\det \mathbf{A}_3 = 0$; $x_1 = \frac{16}{4} = 4$, $x_2 = \frac{-4}{4} = -1$, $x_3 = \frac{0}{4} = 0$

39. $\mathbf{AX} = \mathbf{B}$ is $\begin{bmatrix} 2 & 3 & -1 \\ 1 & -2 & 0 \\ -2 & 0 & 1 \end{bmatrix} \begin{bmatrix} x_1 \\ x_2 \\ x_3 \end{bmatrix} = \begin{bmatrix} 6 \\ -3 \\ 9 \end{bmatrix}$. Since $\mathbf{A}^{-1} = -\frac{1}{3} \begin{bmatrix} -2 & -3 & -2 \\ -1 & 0 & -1 \\ -4 & -6 & -7 \end{bmatrix}$, we have

$\mathbf{X} = \mathbf{A}^{-1}\mathbf{B} = \begin{bmatrix} 7 \\ 5 \\ 23 \end{bmatrix}$.

42. From the characteristic equation $\lambda^2 = 0$ we see that the eigenvalues are $\lambda_1 = \lambda_2 = 0$. For $\lambda_1 = \lambda_2 = 0$ we have $4k_1 = 0$ and $\mathbf{K}_1 = \begin{bmatrix} 0 \\ 1 \end{bmatrix}$ is a single eigenvector.

45. From the characteristic equation $-\lambda^3 - \lambda^2 + 21\lambda + 45 = -(\lambda + 3)^2(\lambda - 5) = 0$ we see that the eigenvalues are $\lambda_1 = \lambda_2 = -3$ and $\lambda_3 = 5$. For $\lambda_1 = \lambda_2 = -3$ we have

$$\begin{bmatrix} 1 & 2 & -3 & | & 0 \\ 2 & 4 & -6 & | & 0 \\ -1 & -2 & 3 & | & 0 \end{bmatrix} \xrightarrow[\text{operations}]{\text{row}} \begin{bmatrix} 1 & 2 & -3 & | & 0 \\ 0 & 0 & 0 & | & 0 \\ 0 & 0 & 0 & | & 0 \end{bmatrix}.$$

Thus $\mathbf{K}_1 = \begin{bmatrix} -2 & 1 & 0 \end{bmatrix}^T$ and $\mathbf{K}_2 = \begin{bmatrix} 3 & 0 & 1 \end{bmatrix}^T$. For $\lambda_3 = 5$ we have

$$\begin{bmatrix} -7 & 2 & -3 & | & 0 \\ 2 & -4 & -6 & | & 0 \\ -1 & -2 & -5 & | & 0 \end{bmatrix} \xrightarrow[\text{operations}]{\text{row}} \begin{bmatrix} 1 & -\frac{2}{7} & \frac{3}{7} & | & 0 \\ 0 & 1 & 2 & | & 0 \\ 0 & 0 & 0 & | & 0 \end{bmatrix}.$$

Thus $\mathbf{K}_3 = \begin{bmatrix} -1 & -2 & 1 \end{bmatrix}^T$.

48. (a) Eigenvalues are $\lambda_1 = \lambda_2 = 0$ and $\lambda_3 = 5$ with corresponding eigenvectors $\mathbf{K}_1 = \begin{bmatrix} 0 & 1 & 0 \end{bmatrix}^T$, $\mathbf{K}_2 = \begin{bmatrix} 2 & 0 & 1 \end{bmatrix}^T$, and $\mathbf{K}_3 = \begin{bmatrix} -1 & 0 & 2 \end{bmatrix}^T$. Since $\|\mathbf{K}_1\| = 1$, $\|\mathbf{K}_2\| = \sqrt{5}$, and $\|\mathbf{K}_3\| = \sqrt{5}$, we have

$$\mathbf{P} = \begin{bmatrix} 0 & \frac{2}{\sqrt{5}} & -\frac{1}{\sqrt{5}} \\ 1 & 0 & 0 \\ 0 & \frac{1}{\sqrt{5}} & \frac{2}{\sqrt{5}} \end{bmatrix} \quad \text{and} \quad \mathbf{P}^{-1} = \mathbf{P}^T = \begin{bmatrix} 0 & 1 & 0 \\ \frac{2}{\sqrt{5}} & 0 & \frac{1}{\sqrt{5}} \\ -\frac{1}{\sqrt{5}} & 0 & \frac{2}{\sqrt{5}} \end{bmatrix}.$$

(b) $\mathbf{P}^{-1}\mathbf{AP} = \begin{bmatrix} 0 & 0 & 0 \\ 0 & 0 & 0 \\ 0 & 0 & 5 \end{bmatrix}$

51. The encoded message is

$$\mathbf{B} = \mathbf{AM} = \begin{bmatrix} 10 & 1 \\ 9 & 1 \end{bmatrix} \begin{bmatrix} 19 & 1 & 20 & 5 & 12 & 12 & 9 & 20 & 5 & 0 & 12 & 1 & 21 \\ 14 & 3 & 8 & 5 & 4 & 0 & 15 & 14 & 0 & 6 & 18 & 9 & 0 \end{bmatrix}$$

$$= \begin{bmatrix} 204 & 13 & 208 & 55 & 124 & 120 & 105 & 214 & 50 & 6 & 138 & 19 & 210 \\ 185 & 12 & 188 & 50 & 112 & 108 & 96 & 194 & 45 & 6 & 126 & 18 & 189 \end{bmatrix}.$$

54. The decoded message is

$$\mathbf{M} = \mathbf{A}^{-1}\mathbf{B} = \begin{bmatrix} -3 & 2 & -1 \\ 1 & 0 & 0 \\ 2 & -1 & 1 \end{bmatrix} \begin{bmatrix} 5 & 2 & 21 \\ 27 & 17 & 40 \\ 21 & 13 & -2 \end{bmatrix} = \begin{bmatrix} 18 & 15 & 19 \\ 5 & 2 & 21 \\ 4 & 0 & 0 \end{bmatrix}.$$

From correspondence (1) we obtain: ROSEBUD__.

9 Vector Calculus

3.

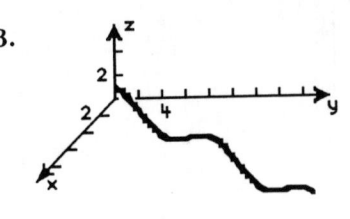

6.

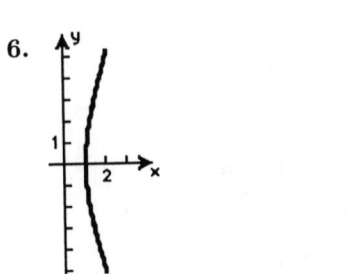

9.

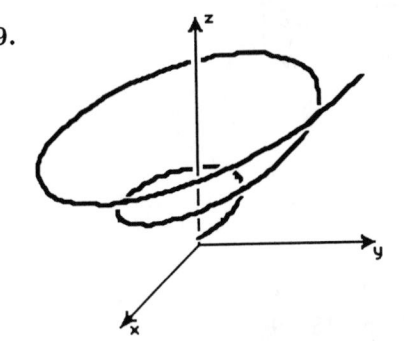

Note: the scale is distorted in this graph. For $t = 0$, the graph starts at $(1, 0, 1)$. The upper loop shown intersects the xz-plane at about $(286751, 0, 286751)$.

12. $x = t$, $y = 2t$, $z = \pm\sqrt{t^2 + 4t^2 + 1} = \pm\sqrt{5t^2 - 1}$; $\mathbf{r}(t) = t\mathbf{i} + 2t\mathbf{j} \pm \sqrt{5t^2 - 1}\,\mathbf{k}$

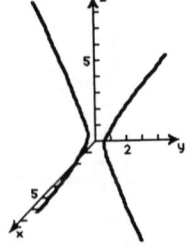

15. $\mathbf{r}(t) = \dfrac{\sin 2t}{t}\,\mathbf{i} + (t-2)^5\mathbf{j} + \dfrac{\ln t}{1/t}\,\mathbf{k}$. Using L'Hôpital's Rule,

$$\lim_{t \to 0^+} \mathbf{r}(t) = \left[\frac{2\cos 2t}{1}\,\mathbf{i} + (t-2)^5\mathbf{j} + \frac{1/t}{-1/t^2}\,\mathbf{k}\right] = 2\mathbf{i} - 32\mathbf{j}.$$

18. $\mathbf{r}'(t) = \langle -t\sin t, 1 - \sin t \rangle$; $\mathbf{r}''(t) = \langle -t\cos t - \sin t, -\cos t \rangle$

21. $\mathbf{r}'(t) = -2\sin t\mathbf{i} + 6\cos t\mathbf{j}$

$\mathbf{r}'(\pi/6) = -\mathbf{i} + 3\sqrt{3}\,\mathbf{j}$

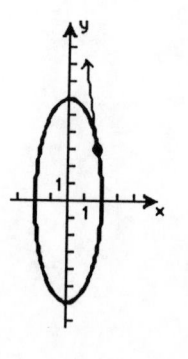

24. $\mathbf{r}'(t) = -3\sin t\mathbf{i} + 3\cos t\mathbf{j} + 2\mathbf{k}$

$\mathbf{r}'(\pi/4) = \dfrac{-3\sqrt{2}}{2}\mathbf{i} + \dfrac{3\sqrt{2}}{2}\mathbf{j} + 2\mathbf{k}$

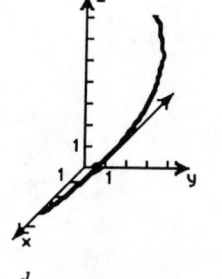

27. $\dfrac{d}{dt}[\mathbf{r}(t) \times \mathbf{r}'(t)] = \mathbf{r}(t) \times \mathbf{r}''(t) + \mathbf{r}'(t) \times \mathbf{r}'(t) = \mathbf{r}(t) \times \mathbf{r}''(t)$

30. $\dfrac{d}{dt}[\mathbf{r}_1(t) \times (\mathbf{r}_2(t) \times \mathbf{r}_3(t))] = \mathbf{r}_1(t) \times \dfrac{d}{dt}(\mathbf{r}_2(t) \times \mathbf{r}_3(t)) + \mathbf{r}_1'(t) \times (\mathbf{r}_2(t) \times \mathbf{r}_3(t))$

$$= \mathbf{r}_1(t) \times (\mathbf{r}_2(t) \times \mathbf{r}_3'(t) + \mathbf{r}_2'(t) \times \mathbf{r}_3(t)) + \mathbf{r}_1'(t) \times (\mathbf{r}_2(t) \times \mathbf{r}_3(t))$$

$$= \mathbf{r}_1(t) \times (\mathbf{r}_2(t) \times \mathbf{r}_3'(t)) + \mathbf{r}_1(t) \times (\mathbf{r}_2'(t) \times \mathbf{r}_3(t)) + \mathbf{r}_1(t) \times (\mathbf{r}_2(t) \times \mathbf{r}_3(t))$$

33. $\displaystyle\int_{-1}^{2} \mathbf{r}(t)\,dt = \left[\int_{-1}^{2} t\,dt\right]\mathbf{i} + \left[\int_{-1}^{2} 3t^2\,dt\right]\mathbf{j} + \left[\int_{-1}^{2} 4t^3\,dt\right]\mathbf{k} = \frac{1}{2}t^2\,\Big|_{-1}^{2}\,\mathbf{i} + t^3\,\Big|_{-1}^{2}\,\mathbf{j} + t^4\,\Big|_{-1}^{2}\,\mathbf{k} = \frac{3}{2}\mathbf{i} + 9\mathbf{j} + 15\mathbf{k}$

36. $\displaystyle\int \mathbf{r}(t)\,dt = \left[\int \frac{1}{1+t^2}\,dt\right]\mathbf{i} + \left[\int \frac{t}{1+t^2}\,dt\right]\mathbf{j} + \left[\int \frac{t^2}{1+t^2}\,dt\right]\mathbf{k}$

$$= [\tan^{-1}t + c_1]\mathbf{i} + \left[\frac{1}{2}\ln(1+t^2) + c_2\right]\mathbf{j} + \left[\int\left(1 - \frac{1}{1+t^2}\right)dt\right]\mathbf{k}$$

$$= [\tan^{-1}t + c_1]\mathbf{i} + \left[\frac{1}{2}\ln(1+t^2) + c_2\right]\mathbf{j} + [t - \tan^{-1}t + c_3]\mathbf{k}$$

$$= \tan^{-1}t\,\mathbf{i} + \frac{1}{2}\ln(1+t^2)\mathbf{j} + (t - \tan^{-1}t)\mathbf{k} + \mathbf{c},$$

where $\mathbf{c} = c_1\mathbf{i} + c_2\mathbf{j} + c_3\mathbf{k}$.

39. $\mathbf{r}'(t) = \displaystyle\int \mathbf{r}''(t)\,dt = \left[\int 12t\,dt\right]\mathbf{i} + \left[\int -3t^{-1/2}\,dt\right]\mathbf{j} + \left[\int 2\,dt\right]\mathbf{k} = [6t^2 + c_1]\mathbf{i} + [-6t^{1/2} + c_2]\mathbf{j} + [2t + c_3]\mathbf{k}$

Since $\mathbf{r}'(1) = \mathbf{j} = (6 + c_1)\mathbf{i} + (-6 + c_2)\mathbf{j} + (2 + c_3)\mathbf{k}$, $c_1 = -6$, $c_2 = 7$, and $c_3 = -2$. Thus,

$$\mathbf{r}'(t) = (6t^2 - 6)\mathbf{i} + (-6t^{1/2} + 7)\mathbf{j} + (2t - 2)\mathbf{k}.$$

$$\mathbf{r}(t) = \int \mathbf{r}'(t)\,dt = \left[\int (6t^2 - 6)\,dt\right]\mathbf{i} + \left[\int (-6t^{1/2} + 7)\,dt\right]\mathbf{j} + \left[\int (2t - 2)\,dt\right]\mathbf{k}$$

$$= [2t^3 - 6t + c_4]\mathbf{i} + [-4t^{3/2} + 7t + c_5]\mathbf{j} + [t^2 - 2t + c_6]\mathbf{k}.$$

Since

$$\mathbf{r}(1) = 2\mathbf{i} - \mathbf{k} = (-4 + c_4)\mathbf{i} + (3 + c_5)\mathbf{j} + (-1 + c_6)\mathbf{k},$$

$c_4 = 6$, $c_5 = -3$, and $c_6 = 0$. Thus,

$$\mathbf{r}(t) = (2t^3 - 6t + 6)\mathbf{i} + (-4t^{3/2} + 7t - 3)\mathbf{j} + (t^2 - 2t)\mathbf{k}.$$

42. $\mathbf{r}'(t) = \mathbf{i} + (\cos t - t\sin t)\mathbf{j} + (\sin t + t\cos t)\mathbf{k}$

$\|\mathbf{r}'(t)\| = \sqrt{1^2 + (\cos t - t\sin t)^2 + (\sin t + t\cos t)^2} = \sqrt{2 + t^2}$

$s = \displaystyle\int_0^\pi \sqrt{2 + t^2}\,dt = \left(\frac{t}{2}\sqrt{2 + t^2} + \ln\left|t + \sqrt{2 + t^2}\right|\right)\Big|_0^\pi = \frac{\pi}{2}\sqrt{2 + \pi^2} + \ln(\pi + \sqrt{2 + \pi^2}) - \ln\sqrt{2}$

45. $\mathbf{r}'(t) = -a\sin t\mathbf{i} + a\cos t\mathbf{j}$; $\|\mathbf{r}'(t)\| = \sqrt{a^2\sin^2 t + a^2\cos^2 t} = a,\ a > 0$; $s = \displaystyle\int_0^t a\,du = at$

$\mathbf{r}(s) = a\cos(s/a)\mathbf{i} + a\sin(s/a)\mathbf{j}$; $\mathbf{r}'(s) = -\sin(s/a)\mathbf{i} + \cos(s/a)\mathbf{j}$

$\|\mathbf{r}'(s)\| = \sqrt{\sin^2(s/a) + \cos^2(s/a)} = 1$

48. Since $\|\mathbf{r}(t)\|$ is the length of $\mathbf{r}(t)$, $\|\mathbf{r}(t)\| = c$ represents a curve lying on a sphere of radius c centered at the origin.

51. $\dfrac{d}{dt}[\mathbf{r}_1(t) \times \mathbf{r}_2(t)] = \lim\limits_{h\to 0} \dfrac{\mathbf{r}_1(t+h) \times \mathbf{r}_2(t+h) - \mathbf{r}_1(t) \times \mathbf{r}_2(t)}{h}$

$= \lim\limits_{h\to 0} \dfrac{\mathbf{r}_1(t+h) \times \mathbf{r}_2(t+h) - \mathbf{r}_1(t+h) \times \mathbf{r}_2(t) + \mathbf{r}_1(t+h) \times \mathbf{r}_2(t) - \mathbf{r}_1(t) \times \mathbf{r}_2(t)}{h}$

$= \lim\limits_{h\to 0} \dfrac{\mathbf{r}_1(t+h) \times [\mathbf{r}_2(t+h) - \mathbf{r}_2(t)]}{h} + \lim\limits_{h\to 0} \dfrac{[\mathbf{r}_1(t+h) - \mathbf{r}_1(t)] \times \mathbf{r}_2(t)}{h}$

$= \mathbf{r}_1(t) \times \left(\lim\limits_{h\to 0} \dfrac{\mathbf{r}_2(t+h) - \mathbf{r}_2(t)}{h}\right) + \left(\lim\limits_{h\to 0} \dfrac{\mathbf{r}_1(t+h) - \mathbf{r}_1(t)}{h}\right) \times \mathbf{r}_2(t)$

$= \mathbf{r}_1(t) \times \mathbf{r}_2'(t) + \mathbf{r}_1'(t) \times \mathbf{r}_2(t)$

Exercises 9.2

3. $\mathbf{v}(t) = -2\sinh 2t\mathbf{i} + 2\cosh 2t\mathbf{j}$; $\mathbf{v}(0) = 2\mathbf{j}$; $\|\mathbf{v}(0)\| = 2$;
$\mathbf{a}(t) = -4\cosh 2t\mathbf{i} + 4\sinh 2t\mathbf{j}$; $\mathbf{a}(0) = -4\mathbf{i}$

6. $\mathbf{v}(t) = \mathbf{i}+\mathbf{j}+3t^2\mathbf{k}$; $\mathbf{v}(2) = \mathbf{i}+\mathbf{j}+12\mathbf{k}$; $\|\mathbf{v}(2)\| = \sqrt{1+1+144} = \sqrt{146}$; $\mathbf{a}(t) = 6t\mathbf{k}$;
$\mathbf{a}(2) = 12\mathbf{k}$

9. The particle passes through the xy-plane when $z(t) = t^2 - 5t = 0$ or $t = 0,\ 5$ which gives us the points $(0,0,0)$
and $(25, 115, 0)$. $\mathbf{v}(t) = 2t\mathbf{i}+(3t^2 - 2)\mathbf{j}+(2t-5)\mathbf{k}$; $\mathbf{v}(0) = -2\mathbf{j}-5\mathbf{k}$, $\mathbf{v}(5) = 10\mathbf{i}+73\mathbf{j}+5\mathbf{k}$; $\mathbf{a}(t) = 2\mathbf{i}+6t\mathbf{j}+2\mathbf{k}$;
$\mathbf{a}(0) = 2\mathbf{i}+2\mathbf{k}$, $\mathbf{a}(5) = 2\mathbf{i}+30\mathbf{j}+2\mathbf{k}$

12. Initially we are given $\mathbf{s}_0 = 1600\mathbf{j}$ and $\mathbf{v}_0 = (480\cos 30°)\mathbf{i} + (480\sin 30°)\mathbf{j} = 240\sqrt{3}\,\mathbf{i} + 240\mathbf{j}$. Using $\mathbf{a}(t) = -32\mathbf{j}$
we find

$$\mathbf{v}(t) = \int \mathbf{a}(t)\,dt = -32t\mathbf{j} + \mathbf{c}$$

$$240\sqrt{3}\,\mathbf{i} + 240\mathbf{j} = \mathbf{v}(0) = \mathbf{c}$$

$$\mathbf{v}(t) = -32t\mathbf{j} + 240\sqrt{3}\,\mathbf{i} + 240\mathbf{j} = 240\sqrt{3}\,\mathbf{i} + (240 - 32t)\mathbf{j}$$

$$\mathbf{r}(t) = \int \mathbf{v}(t)\,dt = 240\sqrt{3}\,t\mathbf{i} + (240t - 16t^2)\mathbf{j} + \mathbf{b}$$

$$1600\mathbf{j} = \mathbf{r}(0) = \mathbf{b}.$$

(a) The shell's trajectory is given by $\mathbf{r}(t) = 240\sqrt{3}\,t\mathbf{i} + (240t - 16t^2 + 1600)\mathbf{j}$ or $x = 240\sqrt{3}\,t$,
$y = 240t - 16t^2 + 1600$.

(b) Solving $dy/dt = 240 - 32t = 0$, we see that y is maximum when $t = 15/2$. The maximum
altitude is $y(15/2) = 2500$ ft.

(c) Solving $y(t) = -16t^2 + 240t + 1600 = -16(t - 20)(t + 5) = 0$, we see that the shell hits the ground
when $t = 20$. The range of the shell is $x(20) = 4800\sqrt{3} \approx 8314$ ft.

(d) From **(c)**, impact is when $t = 20$. The speed at impact is

$$\|\mathbf{v}(20)\| = |240\sqrt{3}\,\mathbf{i} + (240 - 32\cdot 20)\mathbf{j}| = \sqrt{240^2\cdot 3 + (-400)^2} = 160\sqrt{13} \approx 577 \text{ ft/s}.$$

15. Let s be the initial speed. Then $\mathbf{v}(0) = s\cos 45°\mathbf{i} + s\sin 45°\mathbf{j} = \dfrac{s\sqrt{2}}{2}\mathbf{i} + \dfrac{s\sqrt{2}}{2}\mathbf{j}$. Using $\mathbf{a}(t) = -32\mathbf{j}$, we have

$$\mathbf{v}(t) = \int \mathbf{a}(t)\,dt = -32t\mathbf{j} + \mathbf{c}$$

$$\frac{s\sqrt{2}}{2}\mathbf{i} + \frac{s\sqrt{2}}{2}\mathbf{j} = \mathbf{v}(0) = \mathbf{c}$$

$$\mathbf{v}(t) = \frac{s\sqrt{2}}{2}\mathbf{i} + \left(\frac{s\sqrt{2}}{2} - 32t\right)\mathbf{j}$$

$$\mathbf{r}(t) = \frac{s\sqrt{2}}{2}t\mathbf{i} + \left(\frac{s\sqrt{2}}{2}t - 16t^2\right)\mathbf{j} + \mathbf{b}.$$

Since $\mathbf{r}(0) = \mathbf{0}$, $\mathbf{b} = \mathbf{0}$ and

$$\mathbf{r}(t) = \frac{s\sqrt{2}}{2}t\mathbf{i} + \left(\frac{s\sqrt{2}}{2}t - 16t^2\right)\mathbf{j}.$$

Setting $y(t) = s\sqrt{2}\,t/2 - 16t^2 = t(s\sqrt{2}/2 - 16t) = 0$ we see that the ball hits the ground when $t = \sqrt{2}\,s/32$. Thus, using $x(t) = s\sqrt{2}\,t/2$ and the fact that 100 yd = 300 ft, $300 = x(t) = \dfrac{s\sqrt{2}}{2}(\sqrt{2}\,s/32) = \dfrac{s^2}{32}$ and $s = \sqrt{9600} \approx 97.98$ ft/s.

18. The initial angle is $\theta = 0$, the initial height is 1024 ft, and the initial speed is $s = 180(5280)/3600 = 264$ ft/s. Then $x(t) = 264t$ and $y(t) = -16t^2 + 1024$. Solving $y(t) = 0$ we see that the pack hits the ground at $t = 8$ seconds The horizontal distance travelled is $x(8) = 2112$ feet. From the figure in the text, $\tan\alpha = 1024/2112 = 16/33$ and $\alpha \approx 0.45$ radian or $25.87°$.

21. By Problem 19, $a = v^2/r_0 = 1530^2/(4000 \cdot 5280) \approx 0.1108$. We are given $mg = 192$, so $m = 192/32$ and $w_e = 192 - (192/32)(0.1108) \approx 191.33$ lb.

24. Since the projectile is launched from ground level, $s_0 = 0$. To find the maximum height we maximize $y(t) = -\frac{1}{2}gt^2 + (v_0\sin\theta)t$. Solving $y'(t) = -gt + v_0\sin\theta = 0$, we see that $t = (v_0/g)\sin\theta$ is a critical point. Since $y''(t) = -g < 0$,

$$H = y\left(\frac{v_0\sin\theta}{g}\right) = -\frac{1}{2}g\frac{v_0^2\sin^2\theta}{g^2} + v_0\sin\theta\frac{v_0\sin\theta}{g} = \frac{v_0^2\sin^2\theta}{2g}$$

is the maximum height. To find the range we solve $y(t) = -\frac{1}{2}gt^2 + (v_0\sin\theta)t = t(v_0\sin\theta - \frac{1}{2}gt) = 0$. The positive solution of this equation is $t = (2v_0\sin\theta)/g$. The range is thus

$$x(t) = (v_0\cos\theta)\frac{2v_0\sin\theta}{g} = \frac{v_0^2\sin 2\theta}{g}.$$

27. **(a)** Since $\mathbf{F}$ is directed along $\mathbf{r}$ we have $\mathbf{F} = c\mathbf{r}$ for some constant c. Then

$$\boldsymbol{\tau} = \mathbf{r} \times \mathbf{F} = \mathbf{r} \times (c\mathbf{r}) = c(\mathbf{r} \times \mathbf{r}) = \mathbf{0}.$$

(b) If $\boldsymbol{\tau} = \mathbf{0}$ then $d\mathbf{L}/dt = \mathbf{0}$ and $\mathbf{L}$ is constant.

Exercises 9.3

3. We assume $a > 0$. $\mathbf{r}'(t) = -a\sin t\,\mathbf{i} + a\cos t\,\mathbf{j} + c\mathbf{k}$; $|\mathbf{r}'(t)| = \sqrt{a^2\sin^2 t + a^2\cos^2 t + c^2} = \sqrt{a^2 + c^2}$;

$$\mathbf{T}(t) = -\frac{a\sin t}{\sqrt{a^2 + c^2}}\mathbf{i} + \frac{a\cos t}{\sqrt{a^2 + c^2}}\mathbf{j} + \frac{c}{\sqrt{a^2 + c^2}}\mathbf{k}; \quad \frac{d\mathbf{T}}{dt} = -\frac{a\cos t}{\sqrt{a^2 + c^2}}\mathbf{i} - \frac{a\sin t}{\sqrt{a^2 + c^2}}\mathbf{j},$$

$$\left|\frac{d\mathbf{T}}{dt}\right| = \sqrt{\frac{a^2\cos^2 t}{a^2 + c^2} + \frac{a^2\sin^2 t}{a^2 + c^2}} = \frac{a}{\sqrt{a^2 + c^2}}; \quad \mathbf{N} = -\cos t\,\mathbf{i} - \sin t\,\mathbf{j};$$

$$\mathbf{B} = \mathbf{T} \times \mathbf{N} = \begin{vmatrix} \mathbf{i} & \mathbf{j} & \mathbf{k} \\ -\dfrac{a\sin t}{\sqrt{a^2 + c^2}} & \dfrac{a\cos t}{\sqrt{a^2 + c^2}} & \dfrac{c}{\sqrt{a^2 + c^2}} \\ -\cos t & -\sin t & 0 \end{vmatrix} = \frac{c\sin t}{\sqrt{a^2 + c^2}}\mathbf{i} - \frac{c\cos t}{\sqrt{a^2 + c^2}}\mathbf{j} + \frac{a}{\sqrt{a^2 + c^2}}\mathbf{k};$$

$$\kappa = \frac{|d\mathbf{T}/dt|}{|\mathbf{r}'(t)|} = \frac{a/\sqrt{a^2 + c^2}}{\sqrt{a^2 + c^2}} = \frac{a}{a^2 + c^2}$$

6. From Problem 4, a normal to the osculating plane is $\mathbf{B}(1) = \frac{1}{\sqrt{6}}(\mathbf{i} - 2\mathbf{j} + \mathbf{k})$. The point on the curve when $t = 1$ is $(1, 1/2, 1/3)$. An equation of the plane is $(x - 1) - 2(y - 1/2) + (z - 1/3) = 0$ or $x - 2y + z = 1/3$.

9. $\mathbf{v}(t) = 2t\mathbf{i} + 2t\mathbf{j} + 4t\mathbf{k}$, $|\mathbf{v}(t)| = 2\sqrt{6}\,t$, $t > 0$; $\mathbf{a}(t) = 2\mathbf{i} + 2\mathbf{j} + 4\mathbf{k}$; $\mathbf{v} \cdot \mathbf{a} = 24t$, $\mathbf{v} \times \mathbf{a} = \mathbf{0}$;

$$a_T = \frac{24t}{2\sqrt{6}\,t} = 2\sqrt{6}, \quad a_N = 0, \, t > 0$$

12. $\mathbf{v}(t) = \dfrac{1}{1+t^2}\mathbf{i} + \dfrac{t}{1+t^2}\mathbf{j}$, $|\mathbf{v}(t)| = \dfrac{\sqrt{1+t^2}}{1+t^2}$; $\mathbf{a}(t) = -\dfrac{2t}{(1+t^2)^2}\mathbf{i} + \dfrac{1-t^2}{(1+t^2)^2}\mathbf{j}$;

$\mathbf{v}\cdot\mathbf{a} = -\dfrac{2t}{(1+t^2)^3} + \dfrac{t-t^3}{(1+t^2)^3} = -\dfrac{t}{(1+t^2)^2}$; $\mathbf{v}\times\mathbf{a} = \dfrac{1}{(1+t^2)^2}\mathbf{k}$, $|\mathbf{v}\times\mathbf{a}| = \dfrac{1}{(1+t^2)^2}$;

$a_T = -\dfrac{t/(1+t^2)^2}{\sqrt{1+t^2}/(1+t^2)} = -\dfrac{t}{(1+t^2)^{3/2}}$, $a_N = \dfrac{1/(1+t^2)^2}{\sqrt{1+t^2}/(1+t^2)} = \dfrac{1}{(1+t^2)^{3/2}}$

15. $\mathbf{v}(t) = -e^{-t}(\mathbf{i}+\mathbf{j}+\mathbf{k})$, $|\mathbf{v}(t)| = \sqrt{3}\,e^{-t}$; $\mathbf{a}(t) = e^{-t}(\mathbf{i}+\mathbf{j}+\mathbf{k})$; $\mathbf{v}\cdot\mathbf{a} = -3e^{-2t}$; $\mathbf{v}\times\mathbf{a} = \mathbf{0}$, $|\mathbf{v}\times\mathbf{a}| = 0$; $a_T = -\sqrt{3}\,e^{-t}$, $a_N = 0$

18. (a) $\mathbf{v}(t) = -a\sin t\,\mathbf{i} + b\cos t\,\mathbf{j}$, $|\mathbf{v}(t)| = \sqrt{a^2\sin^2 t + b^2\cos^2 t}$; $\mathbf{a}(t) = -a\cos t\,\mathbf{i} - b\sin t\,\mathbf{j}$;

$\mathbf{v}\times\mathbf{a} = ab\mathbf{k}$; $|\mathbf{v}\times\mathbf{a}| = ab$; $\kappa = \dfrac{ab}{(a^2\sin^2 t + b^2\cos^2 t)^{3/2}}$

(b) When $a = b$, $|\mathbf{v}(t)| = a$, $|\mathbf{v}\times\mathbf{a}| = a^2$, and $\kappa = a^2/a^3 = 1/a$.

21. $\mathbf{v}(t) = f'(t)\mathbf{i} + g'(t)\mathbf{j}$, $|\mathbf{v}(t)| = \sqrt{[f'(t)]^2 + [g'(t)]^2}$; $\mathbf{a}(t) = f''(t)\mathbf{i} + g''(t)\mathbf{j}$;

$\mathbf{v}\times\mathbf{a} = [f'(t)g''(t) - g'(t)f''(t)]\mathbf{k}$, $|\mathbf{v}\times\mathbf{a}| = |f'(t)g''(t) - g'(t)f''(t)|$;

$\kappa = \dfrac{|\mathbf{v}\times\mathbf{a}|}{|\mathbf{v}|^3} = \dfrac{|f'(t)g''(t) - g'(t)f''(t)|}{([f'(t)]^2 + [g'(t)]^2)^{3/2}}$

24. $F(x) = x^3$, $F(-1) = -1$, $F(1/2) = 1/8$; $F'(x) = 3x^2$, $F'(-1) = 3$, $F'(1/2) = 3/4$; $F''(x) = 6x$,

$F''(-1) = -6$, $F''(1/2) = 3$; $\kappa(-1) = \dfrac{|-6|}{(1+3^2)^{3/2}} = \dfrac{6}{10\sqrt{10}} = \dfrac{3}{5\sqrt{10}} \approx 0.19$;

$\rho(-1) = \dfrac{5\sqrt{10}}{3} \approx 5.27$; $\kappa\left(\dfrac{1}{2}\right) = \dfrac{3}{[1+(3/4)^2]^{3/2}} = \dfrac{3}{125/64} = \dfrac{192}{125} \approx 1.54$; $\rho\left(\dfrac{1}{2}\right) = \dfrac{125}{192} \approx 0.65$

Since $1.54 > 0.19$, the curve is "sharper" at $(1/2, 1/8)$.

Exercises 9.4

3. $x^2 - y^2 = 1 + c^2$

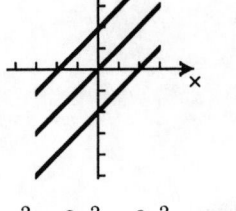

6. $y = x + \tan c$, $-\pi/x < c < \pi/2$

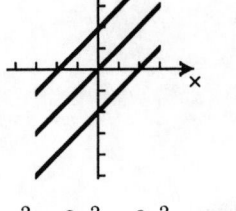

9. $x^2 + 3y^2 + 6z^2 = c$; ellipsoid

12. Setting $x = -4$, $y = 2$, and $z = -3$ in $x^2/16 + y^2/4 + z^2/9 = c$ we obtain $c = 3$. The equation of the surface is $x^2/16 + y^2/4 + z^2/9 = 3$. Setting $y = z = 0$ we find the x-intercepts are $\pm 4\sqrt{3}$. Similarly, the y-intercepts are $\pm 2\sqrt{3}$ and the z-intercepts are $\pm 3\sqrt{3}$.

15. $z_x = 20x^3y^3 - 2xy^6 + 30x^4$; $\quad z_y = 15x^4y^2 - 6x^2y^5 - 4$

18. $z_x = 12x^2 - 10x + 8$; $\quad z_y = 0$

21. $z_x = 2(\cos 5x)(-\sin 5x)(5) = -10\sin 5x \cos 5x$; $\quad z_y = 2(\sin 5y)(\cos 5y)(5) = 10 \sin 5y \cos 5y$

24. $f_\theta = \phi^2 \left(\cos \dfrac{\theta}{\phi} \right) \left(\dfrac{1}{\phi} \right) = \phi \cos \dfrac{\theta}{\phi}$; $\quad f_\phi = \phi^2 \left(\cos \dfrac{\theta}{\phi} \right) \left(-\dfrac{\theta}{\phi^2} \right) + 2\phi \sin \dfrac{\theta}{\phi} = -\theta \cos \dfrac{\theta}{\phi} + 2\phi \sin \dfrac{\theta}{\phi}$

27. $g_u = \dfrac{8u}{4u^2 + 5v^3}$; $\quad g_v = \dfrac{15v^2}{4u^2 + 5v^3}$

30. $w_x = xy \left(\dfrac{1}{x} \right) + (\ln xz)y = y + y \ln xz$; $\quad w_y = x \ln xz$; $\quad w_z = \dfrac{xy}{z}$

33. $\dfrac{\partial z}{\partial x} = \dfrac{2x}{x^2 + y^2}$, $\quad \dfrac{\partial^2 z}{\partial x^2} = \dfrac{(x^2 + y^2)2 - 2x(2x)}{(x^2 + y^2)^2} = \dfrac{2y^2 - 2x^2}{(x^2 + y^2)^2}$; $\quad \dfrac{\partial z}{\partial y} = \dfrac{2y}{x^2 + y^2}$,

$\dfrac{\partial^2 z}{\partial y^2} = \dfrac{(x^2 + y^2)2 - 2y(2y)}{(x^2 + y^2)^2} = \dfrac{2x^2 - 2y^2}{(x^2 + y^2)^2}$; $\quad \dfrac{\partial^2 z}{\partial x^2} + \dfrac{\partial^2 z}{\partial y^2} = \dfrac{2y^2 - 2x^2 + 2x^2 - 2y^2}{(x^2 + y^2)^2} = 0$

36. $\dfrac{\partial u}{\partial x} = -\sin(x + at) + \cos(x - at)$, $\quad \dfrac{\partial^2 u}{\partial x^2} = -\cos(x + at) - \sin(x - at)$;

$\dfrac{\partial u}{\partial t} = -a\sin(x + at) - a\cos(x - at)$, $\quad \dfrac{\partial^2 u}{\partial t^2} = -a^2 \cos(x + at) - a^2 \sin(x - at)$;

$a^2 \dfrac{\partial^2 u}{\partial x^2} = -a^2 \cos(x + at) - a^2 \sin(x - at) = \dfrac{\partial^2 u}{\partial t^2}$

39. $z_x = v^2 e^{uv^2}(3x^2) + 2uve^{uv^2}(1) = 3x^2 v^2 e^{uv^2} + 2uve^{uv^2}$; $\quad z_y = v^2 e^{uv^2}(0) + 2uve^{uv^2}(-2y) = -4yuve^{uv^2}$

42. $z_u = \dfrac{2y}{(x + y)^2} \left(\dfrac{1}{v} \right) + \dfrac{-2x}{(x + y)^2} \left(-\dfrac{v^2}{u^2} \right) = \dfrac{2y}{v(x + y)^2} + \dfrac{2xv^2}{u^2(x + y)^2}$

$z_v = \dfrac{2y}{(x + y)^2} \left(-\dfrac{u}{v^2} \right) + \dfrac{-2x}{(x + y)^2} \left(\dfrac{2v}{u} \right) = -\dfrac{2yu}{v^2(x + y)^2} - \dfrac{4xv}{u(x + y)^2}$

45. $R_u = s^2 t^4 (e^{v^2}) + 2rst^4(-2uve^{-u^2}) + 4rs^2t^3(2uv^2 e^{u^2 v^2}) = s^2 t^4 e^{v^2} - 4uvrst^4 e^{-u^2} + 8uv^2 rs^2 t^3 e^{u^2 v^2}$

$R_v = s^2 t^4 (2uve^{v^2}) + 2rst^4(e^{-u^2}) + 4rs^2 t^3(2u^2 ve^{u^2 v^2}) = 2s^2 t^4 uve^{v^2} + 2rst^4 e^{-u^2} + 8rs^2 t^3 u^2 ve^{u^2 v^2}$

48. $s_\phi = 2pe^{3\theta} + 2q[-\sin(\phi + \theta)] - 2r\theta^2 + 4(2) = 2pe^{3\theta} - 2q\sin(\phi + \theta) - 2r\theta^2 + 8$

$s_\theta = 2p(3\phi e^{3\theta}) + 2q[-\sin(\phi + \theta)] - 2r(2\phi\theta) + 4(8) = 6p\phi e^{3\theta} - 2q\sin(\phi + \theta) - 4r\phi\theta + 32$

51. $\dfrac{dw}{dt} = -3\sin(3u + 4v)(2) - 4\sin(3u + 4v)(-1)$; $\quad u(\pi) = 5\pi/2$, $\quad v(\pi) = -5\pi/4$

$\dfrac{dw}{dt} \bigg|_\pi = -6\sin\left(\dfrac{15\pi}{2} - 5\pi \right) + 4\sin\left(\dfrac{15\pi}{2} - 5\pi \right) = -2\sin\dfrac{5\pi}{2} = -2$

54. $\dfrac{dP}{dt} = \dfrac{(V - 0.0427)(0.08)dT/dt}{(V - 0.0427)^2} - \dfrac{0.08T(dV/dt)}{(V - 0.0427)^2} + \dfrac{3.6}{V^3} \dfrac{dV}{dt}$

$= \dfrac{0.08}{V - 0.0427} \dfrac{dT}{dt} + \left(\dfrac{3.6}{V^3} - \dfrac{0.08T}{(V - 0.0427)^2} \right) \dfrac{dV}{dt}$

57. Since the height of the triangle is $x \sin \theta$, the area is given by $A = \frac{1}{2}xy \sin \theta$. Then

$$\dfrac{dA}{dt} \dfrac{\partial A}{\partial x} \dfrac{dx}{dt} + \dfrac{\partial A}{\partial y} \dfrac{dy}{dt} + \dfrac{\partial A}{\partial \theta} \dfrac{d\theta}{dt} = \dfrac{1}{2}y \sin \theta \dfrac{dx}{dt} + \dfrac{1}{2}x \sin \theta \dfrac{dy}{dt} + \dfrac{1}{2}xy \cos \theta \dfrac{d\theta}{dt}.$$

When $x = 10$, $y = 8$, $\theta = \pi/6$, $dx/dt = 0.3$, $dy/dt = 0.5$, and $d\theta/dt = 0.1$,

$$\frac{dA}{dt} = \frac{1}{2}(8)\left(\frac{1}{2}\right)(0.3) + \frac{1}{2}(10)\left(\frac{1}{2}\right)(0.5) + \frac{1}{2}(10)(8)\left(\frac{\sqrt{3}}{2}\right)(0.1)$$

$$= 0.6 + 1.25 + 2\sqrt{3} = 1.85 + 2\sqrt{3} \approx 5.31 \text{ cm}^2/\text{s}.$$

Exercises 9.5

3. $\nabla F = \dfrac{y^2}{z^3}\mathbf{i} + \dfrac{2xy}{z^3}\mathbf{j} - \dfrac{3xy^2}{z^4}\mathbf{k}$

6. $\nabla f = \dfrac{3x^2}{2\sqrt{x^3y - y^4}}\mathbf{i} + \dfrac{x^3 - 4y^3}{2\sqrt{x^3y - y^4}}\mathbf{j};\quad \nabla f(3,2) = \dfrac{27}{\sqrt{38}}\mathbf{i} - \dfrac{5}{2\sqrt{38}}\mathbf{j}$

9. $D_\mathbf{u}f(x,y) = \lim\limits_{h\to 0} \dfrac{f(x + h\sqrt{3}/2, y + h/2) - f(x,y)}{h} = \lim\limits_{h\to 0}\dfrac{(x + h\sqrt{3}/2)^2 + (y + h/2)^2 - x^2 - y^2}{h}$

$= \lim\limits_{h\to 0}\dfrac{h\sqrt{3}\,x + 3h^2/4 + hy + h^2/4}{h} = \lim\limits_{h\to 0}(\sqrt{3}\,x + 3h/4 + y + h/4) = \sqrt{3}\,x + y$

12. $\mathbf{u} = \dfrac{\sqrt{2}}{2}\mathbf{i} + \dfrac{\sqrt{2}}{2}\mathbf{j};\quad \nabla f = (4 + y^2)\mathbf{i} + (2xy - 5)\mathbf{j};\quad \nabla f(3,-1) = 5\mathbf{i} - 11\mathbf{j};$

$D_\mathbf{u}f(3,-1) = \dfrac{5\sqrt{2}}{2} - \dfrac{11\sqrt{2}}{2} = -3\sqrt{2}$

15. $\mathbf{u} = (2\mathbf{i} + \mathbf{j})/\sqrt{5};\quad \nabla f = 2y(xy + 1)\mathbf{i} + 2x(xy + 1)\mathbf{j};\quad \nabla f(3,2) = 28\mathbf{i} + 42\mathbf{j}$

$D_\mathbf{u}f(3,2) = \dfrac{2(28)}{\sqrt{5}} + \dfrac{42}{\sqrt{5}} = \dfrac{98}{\sqrt{5}}$

18. $\mathbf{u} = \dfrac{1}{\sqrt{6}}\mathbf{i} - \dfrac{2}{\sqrt{6}}\mathbf{j} + \dfrac{1}{\sqrt{6}}\mathbf{k};\quad \nabla F = \dfrac{2x}{z^2}\mathbf{i} - \dfrac{2y}{z^2}\mathbf{j} + \dfrac{2y^2 - 2x^2}{z^3}\mathbf{k};\quad \nabla F(2,4,-1) = 4\mathbf{i} - 8\mathbf{j} - 24\mathbf{k}$

$D_\mathbf{u}F(2,4,-1) = \dfrac{4}{\sqrt{6}} - \dfrac{16}{\sqrt{6}} - \dfrac{24}{\sqrt{6}} = -6\sqrt{6}$

21. $\mathbf{u} = (-4\mathbf{i} - \mathbf{j})/\sqrt{17};\quad \nabla f = 2(x - y)\mathbf{i} - 2(x - y)\mathbf{j};\quad \nabla f(4,2) = 4\mathbf{i} - 4\mathbf{j};\quad D_\mathbf{u}F(4,2) = -\dfrac{16}{\sqrt{17}}\dfrac{4}{\sqrt{17}} = -\dfrac{12}{\sqrt{17}}$

24. $\nabla f = (xye^{x-y} + ye^{x-y})\mathbf{i} + (-xye^{x-y} + xe^{x-y})\mathbf{j};\quad \nabla f(5,5) = 30\mathbf{i} - 20\mathbf{j}$

The maximum $D_\mathbf{u}$ is $[30^2 + (-20)^2]^{1/2} = 10\sqrt{13}$ in the direction $30\mathbf{i} - 20\mathbf{j}$.

27. $\nabla f = 2x\sec^2(x^2 + y^2)\mathbf{i} + 2y\sec^2(x^2 + y^2)\mathbf{j};$

$\nabla f(\sqrt{\pi/6}, \sqrt{\pi/6}) = 2\sqrt{\pi/6}\sec^2(\pi/3)(\mathbf{i} + \mathbf{j}) = 8\sqrt{\pi/6}\,(\mathbf{i} + \mathbf{j})$

The minimum $D_\mathbf{u}$ is $-8\sqrt{\pi/6}\,(1^2 + 1^2)^{1/2} = -8\sqrt{\pi/3}$ in the direction $-(\mathbf{i} + \mathbf{j})$.

30. $\nabla F = \dfrac{1}{x}\mathbf{i} + \dfrac{1}{y}\mathbf{j} - \dfrac{1}{z}\mathbf{k};\quad \nabla F(1/2, 1/6, 1/3) = 2\mathbf{i} + 6\mathbf{j} - 3\mathbf{k}$

The minimum $D_\mathbf{u}$ is $-[2^2 + 6^2 + (-3)^2]^{1/2} = -7$ in the direction $-2\mathbf{i} - 6\mathbf{j} + 3\mathbf{k}$.

33. **(a)** Vectors perpendicular to $4\mathbf{i} + 3\mathbf{j}$ are $\pm(3\mathbf{i} - 4\mathbf{j})$. Take $\mathbf{u} = \pm\left(\dfrac{3}{5}\mathbf{i} - \dfrac{4}{5}\mathbf{j}\right)$.

(b) $\mathbf{u} = (4\mathbf{i} + 3\mathbf{j})/\sqrt{16 + 9} = \dfrac{4}{5}\mathbf{i} + \dfrac{3}{5}\mathbf{j}$

(c) $\mathbf{u} = -\dfrac{4}{5}\mathbf{i} - \dfrac{3}{5}\mathbf{j}$

36. $\nabla U = \dfrac{Gmx}{(x^2 + y^2)^{3/2}}\,\mathbf{i} + \dfrac{Gmy}{(x^2 + y^2)^{3/2}}\,\mathbf{j} = \dfrac{Gm}{(x^2 + y^2)^{3/2}}(x\mathbf{i} + y\mathbf{j})$

The maximum and minimum values of $D_{\mathbf{u}}U(x, y)$ are obtained when $\mathbf{u}$ is in the directions ∇U and $-\nabla U$, respectively. Thus, at a point (x, y), not $(0, 0)$, the directions of maximum and minimum increase in U are $x\mathbf{i} + y\mathbf{j}$ and $-x\mathbf{i} - y\mathbf{j}$, respectively. A vector at (x, y) in the direction $\pm(x\mathbf{i} + y\mathbf{j})$ lies on a line through the origin.

39. $\nabla T = 4x\mathbf{i} + 2y\mathbf{j}$; $\nabla T(4, 2) = 16\mathbf{i} + 4\mathbf{j}$. The minimum change in temperature (that is, the maximum decrease in temperature) is in the direction $-\nabla T(4, 3) = -16\mathbf{i} - 4\mathbf{j}$.

42. Substituting $x = 0$, $y = 0$, $z = 1$, and $T = 500$ into $T = k/(x^2 + y^2 + z^2)$ we see that $k = 500$ and $T(x, y, z) = 500/(x^2 + y^2 + z^2)$.

(a) $\mathbf{u} = \dfrac{1}{3}\langle 1, -2, -2 \rangle = \dfrac{1}{3}\mathbf{i} - \dfrac{2}{3}\mathbf{j} - \dfrac{2}{3}\mathbf{k}$

$$\nabla T = -\dfrac{1000x}{(x^2 + y^2 + z^2)^2}\,\mathbf{i} - \dfrac{1000y}{(x^2 + y^2 + z^2)^2}\,\mathbf{j} - \dfrac{1000z}{(x^2 + y^2 + z^2)^2}\,\mathbf{k}$$

$$\nabla T(2, 3, 3) = -\dfrac{500}{121}\mathbf{i} - \dfrac{750}{121}\mathbf{j} - \dfrac{750}{121}\mathbf{k}$$

$$D_{\mathbf{u}}T(2, 3, 3) = \dfrac{1}{3}\left(-\dfrac{500}{121}\right) - \dfrac{2}{3}\left(-\dfrac{750}{121}\right) - \dfrac{2}{3}\left(-\dfrac{750}{121}\right) = \dfrac{2500}{363}$$

(b) The direction of maximum increase is $\nabla T(2, 3, 3) = -\dfrac{500}{121}\mathbf{i} - \dfrac{750}{121}\mathbf{j} - \dfrac{750}{121}\mathbf{k} = \dfrac{250}{121}(-2\mathbf{i} - 3\mathbf{j} - 3\mathbf{k})$.

(c) The maximum rate of change of T is $|\nabla T(2, 3, 3)| = \dfrac{250}{121}\sqrt{4 + 9 + 9} = \dfrac{250\sqrt{22}}{121}$.

45. $\nabla(cf) = \dfrac{\partial}{\partial x}(cf)\mathbf{i} + \dfrac{\partial}{\partial y}(cf)\mathbf{j} = cf_x\mathbf{i} + cf_y\mathbf{j} = c(f_x\mathbf{i} + f_y\mathbf{j}) = c\nabla f$

48. $\nabla(f/g) = [(gf_x - fg_x)/g^2]\mathbf{i} + [(gf_y - fg_y)/g^2]\mathbf{j} = g(f_x\mathbf{i} + f_y\mathbf{j})/g^2 - f(g_x\mathbf{i} + g_y\mathbf{j})/g^2$

$\qquad\quad = g\nabla f/g^2 - f\nabla g/g^2 = (g\nabla f - f\nabla g)/g^2$

Exercises 9.6

3. Since $f(2, 5) = 1$, the level curve is $y = x^2 + 1$.

$\nabla f = -2x\mathbf{i} + \mathbf{j}$; $\nabla f(2, 5) = -10\mathbf{i} + \mathbf{j}$

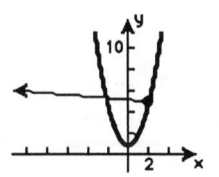

6. Since $f(2, 2) = 2$, the level curve is $y^2 = 2x$, $x \neq 0$. $\nabla f = -\dfrac{y^2}{x^2}\mathbf{i} + \dfrac{2y}{x}\mathbf{j}$;

$\nabla f(2, 2) = -\mathbf{i} + 2\mathbf{j}$

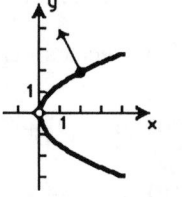

9. Since $F(3,1,1) = 2$, the level surface is $y + z = 2$. $\nabla F = \mathbf{j} + \mathbf{k}$;
$\nabla F(3,1,1) = \mathbf{j} + \mathbf{k}$

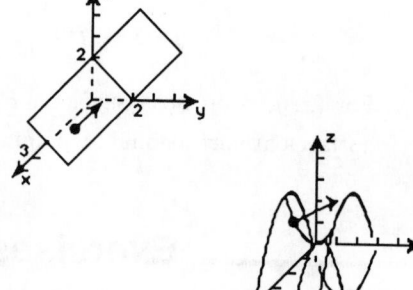

12. Since $F(0,-1,1) = 0$, the level surface is $x^2 - y^2 + z = 0$ or $z = y^2 - x^2$.
$\nabla F = 2x\mathbf{i} - 2y\mathbf{j} + \mathbf{k}$; $\nabla F(0,-1,1) = 2\mathbf{j} + \mathbf{k}$

15. $F(x,y,z) = x^2 + y^2 + z^2$; $\nabla F = 2x\mathbf{i} + 2y\mathbf{j} + 2z\mathbf{k}$. $\nabla F(-2,2,1) = -4\mathbf{i} + 4\mathbf{j} + 2\mathbf{k}$. The equation of the tangent plane is $-4(x+2) + 4(y-2) + 2(z-1) = 0$ or $-2x + 2y + z = 9$.

18. $F(x,y,z) = xy + yz + zx$; $\nabla F = (y+z)\mathbf{i} + (x+z)\mathbf{j} + (y+x)\mathbf{k}$; $\nabla F(1,-3,-5) = -8\mathbf{i} - 4\mathbf{j} - 2\mathbf{k}$. The equation of the tangent plane is $-8(x-1) - 4(y+3) - 2(z+5) = 0$ or $4x + 2y + z = -7$.

21. $F(x,y,z) = \cos(2x+y) - z$; $\nabla F = -2\sin(2x+y)\mathbf{i} - \sin(2x+y)\mathbf{j} - \mathbf{k}$; $\nabla F(\pi/2, \pi/4, -1/\sqrt{2}) = \sqrt{2}\mathbf{i} + \dfrac{\sqrt{2}}{2}\mathbf{j} - \mathbf{k}$.
The equation of the tangent plane is $\sqrt{2}\left(x - \dfrac{\pi}{2}\right) + \dfrac{\sqrt{2}}{2}\left(y - \dfrac{\pi}{4}\right) - \left(z + \dfrac{1}{\sqrt{2}}\right) = 0$,

$$2\left(x - \frac{\pi}{2}\right) + \left(y - \frac{\pi}{4}\right) - \sqrt{2}\left(z + \frac{1}{\sqrt{2}}\right) = 0, \text{ or } 2x + y - \sqrt{2}z = \frac{5\pi}{4} + 1.$$

24. $F(x,y,z) = 8e^{-2y}\sin 4x - z$; $\nabla F = 32e^{-2y}\cos 4x\mathbf{i} - 16e^{-2y}\sin 4x\mathbf{j} - \mathbf{k}$; $\nabla F(\pi/24, 0, 4) = 16\sqrt{3}\,\mathbf{i} - 8\mathbf{j} - \mathbf{k}$. The equation of the tangent plane is

$$16\sqrt{3}(x - \pi/24) - 8(y-0) - (z-4) = 0 \quad \text{or} \quad 16\sqrt{3}\,x - 8y - z = \frac{2\sqrt{3}\,\pi}{3} - 4.$$

27. The gradient of $F(x,y,z) = x^2 + 4x + y^2 + z^2 - 2z$ is $\nabla F = (2x+4)\mathbf{i} + 2y\mathbf{j} + (2z-2)\mathbf{k}$, so a normal to the surface at (x_0, y_0, z_0) is $(2x_0+4)\mathbf{i} + 2y_0\mathbf{j} + (2z_0-2)\mathbf{k}$. A horizontal plane has normal $c\mathbf{k}$ for $c \neq 0$. Thus, we want $2x_0 + 4 = 0$, $2y_0 = 0$, $2z_0 - 2 = c$ or $x_0 = -2$, $y_0 = 0$, $z_0 = c + 1$. since (x_0, y_0, z_0) is on the surface, $(-2)^2 + 4(-2) + (c+1)^2 - 2(c+1) = c^2 - 5 = 11$ and $c = \pm 4$. The points on the surface are $(-2, 0, 5)$ and $(-2, 0, -3)$.

30. If (x_0, y_0, z_0) is on $x^2/a^2 - y^2/b^2 + z^2/c^2 = 1$, then $x_0^2/a^2 - y_0^2/b^2 + z_0^2/c^2 = 1$ and (x_0, y_0, z_0) is on the plane $xx_0/a^2 - yy_0/b^2 + zz_0/c^2 = 1$. A normal to the surface at (x_0, y_0, z_0) is

$$\nabla F(x_0, y_0, z_0) = (2x_0/a^2)\mathbf{i} - (2y_0/b^2)\mathbf{j} + (2z_0/c^2)\mathbf{k}.$$

A normal to the plane is $(x_0/a^2)\mathbf{i} - (y_0/b^2)\mathbf{j} + (z_0/c^2)\mathbf{k}$. Since the normal to the surface is a multiple of the normal to the plane, the normal vectors are parallel, and the plane is tangent to the surface.

33. $F(x,y,z) = x^2 + 2y^2 + z^2$; $\nabla F = 2x\mathbf{i} + 4y\mathbf{j} + 2z\mathbf{k}$; $\nabla F(1,-1,1) = 2\mathbf{i} - 4\mathbf{j} + 2\mathbf{k}$. Parametric equations of the line are $x = 1 + 2t$, $y = -1 - 4t$, $z = 1 + 2t$.

36. $F(x,y,z) = x^2 + y^2 - z^2$; $\nabla F = 2x\mathbf{i} + 2y\mathbf{j} - 2z\mathbf{k}$; $\nabla F(3,4,5) = 6\mathbf{i} + 8\mathbf{j} - 10\mathbf{k}$. Symmetric equations of the line are $\dfrac{x-3}{6} = \dfrac{y-4}{8} = \dfrac{z-5}{-10}$.

39. Let $F(x, y, z) = x^2 + y^2 + z^2 - 25$ and $G(x, y, z) = -x^2 + y^2 + z^2$. Then

$$F_x G_x + F_y G_y + F_z g_z = (2x)(-2x) + (2y)(2y) + (2z)(2z) = 4(-x^2 + y^2 + z^2).$$

For (x, y, z) on both surfaces, $F(x, y, z) = G(x, y, z) = 0$. Thus, $F_x G_x + F_y G_y + F_z G_z = 4(0) = 0$ and the surfaces are orthogonal at points of intersection.

Exercises 9.7

3.

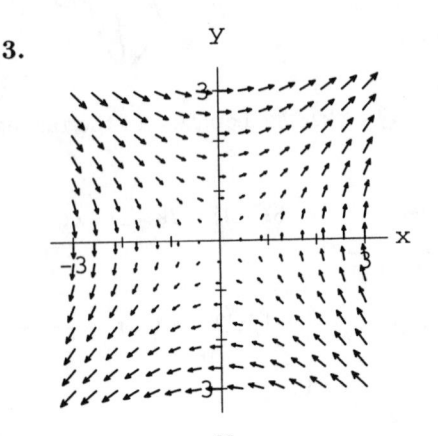

6.

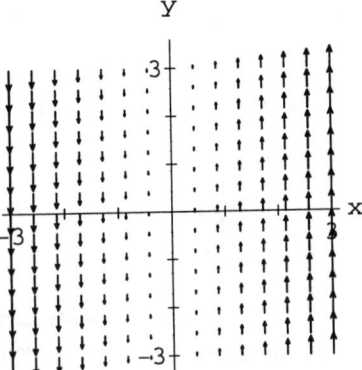

9. curl $\mathbf{F} = 0$; div $\mathbf{F} = 4y + 8z$

12. curl $\mathbf{F} = -x^3 z \mathbf{i} + (3x^2 yz - z)\mathbf{j} + \left(\frac{3}{2}x^2 y^2 - y - 15y^2\right)\mathbf{k}$; div $\mathbf{F} = (x^3 y - x) - (x^3 y - x) = 0$

15. curl $\mathbf{F} = (xy^2 e^y + 2xy e^y + x^3 y e^z + x^3 yz e^z)\mathbf{i} - y^2 e^y \mathbf{j} + (-3x^2 yz e^z - x e^x)\mathbf{k}$; div $\mathbf{F} = xy e^x + y e^x - x^3 z e^z$

18. curl $\mathbf{r} = \begin{vmatrix} \mathbf{i} & \mathbf{j} & \mathbf{k} \\ \partial/\partial x & \partial/\partial y & \partial/\partial z \\ x & y & z \end{vmatrix} = 0\mathbf{i} - 0\mathbf{j} + 0\mathbf{k} = \mathbf{0}$

21. $\nabla \cdot (\mathbf{a} \times \mathbf{r}) = \begin{vmatrix} \partial/\partial x & \partial/\partial y & \partial/\partial z \\ a_1 & a_2 & a_3 \\ x & y & z \end{vmatrix} = \dfrac{\partial}{\partial x}(a_2 z - a_3 y) - \dfrac{\partial}{\partial y}(a_1 z - a_3 x) + \dfrac{\partial}{\partial z}(a_1 y - a_2 x) = 0$

24. $\mathbf{r} \cdot \mathbf{a} = a_1 x + a_2 y + a_3 z$; $\mathbf{r} \cdot \mathbf{r} = x^2 + y^2 + z^2$; $\nabla \cdot [(\mathbf{r} \cdot \mathbf{r})\mathbf{a}] = 2xa_1 + 2ya_2 + 2za_3 = 2(\mathbf{r} \cdot \mathbf{a})$

27. $\nabla \cdot (f\mathbf{F}) = \nabla \cdot (fP\mathbf{i} + fQ\mathbf{j} + fR\mathbf{k}) = fP_x + Pf_x + fQ_y + Qf_y + fR_z + Rf_z$
$$= f(P_x + Q_y + R_z) + (Pf_x + Qf_y + Rf_z) = f(\nabla \cdot \mathbf{F}) + \mathbf{F} \cdot (\nabla f)$$

30. Assuming continuous second partial derivatives,

$$\text{div (curl } \mathbf{F}) = \nabla \cdot [(R_y - Q_z)\mathbf{i} - (R_x - P_z)\mathbf{j} + (Q_x - P_y)\mathbf{k}]$$
$$= (R_{yx} - Q_{zx} - (R_{xy} - P_{zy}) + (Q_{xz} - P_{yz}) = 0.$$

33. $\nabla \cdot \nabla f = \nabla \cdot (f_x \mathbf{i} + f_y \mathbf{j} + f_z \mathbf{k}) = f_{xx} + f_{yy} + f_{zz}$

36. (a) For $\mathbf{F} = P\mathbf{i} + Q\mathbf{j} + R\mathbf{k}$,

$$\text{curl (curl } \mathbf{F}) = (Q_{xy} - P_{yy} - P_{zz} + R_{xz})\mathbf{i} + (R_{yz} - Q_{zz} - Q_{xx} + P_{yx})\mathbf{j}$$
$$+ (P_{zx} - R_{xx} - R_{yy} + Q_{zy})\mathbf{k}$$

and

$$-\nabla^2 \mathbf{F} + \text{grad (div } \mathbf{F}) = -(P_{xx} + P_{yy} + P_{zz})\mathbf{i} - (Q_{xx} + Q_{yy} + Q_{zz})\mathbf{j} - (R_{xx} + R_{yy} + R_{zz})\mathbf{k}$$
$$+ \text{grad } (P_x + Q_y + R_z)$$
$$= -P_{xx}\mathbf{i} - Q_{yy}\mathbf{j} - R_{zz}\mathbf{k} + (-P_{yy} - P_{zz})\mathbf{i} + (-Q_{xx} - Q_{zz})\mathbf{j}$$
$$+ (-R_{xx} - R_{yy})\mathbf{k} + (P_{xx} + Q_{yx} + R_{zx})\mathbf{i} + (P_{xy} + Q_{yy} + R_{zy})\mathbf{j}$$
$$+ (P_{xz} + Q_{yz} + R_{zz})\mathbf{k}$$
$$= (-P_{yy} - P_{zz} + Q_{yx} + R_{zx})\mathbf{i} + (-Q_{xx} - Q_{zz} + P_{xy} + R_{zy})\mathbf{j}$$
$$+ (-R_{xx} - R_{yy} + P_{xz} + Q_{yz})\mathbf{k}.$$

Thus, curl (curl $\mathbf{F}$) $= -\nabla^2 \mathbf{F} + \text{grad (div } \mathbf{F})$.

(b) For $\mathbf{F} = xy\mathbf{i} + 4yz^2\mathbf{j} + 2xz\mathbf{k}$, $\nabla^2 \mathbf{F} = 0\mathbf{i} + 8y\mathbf{j} + 0\mathbf{k}$, div $\mathbf{F} = y + 4z^2 + 2x$, and grad (div $\mathbf{F}$) $= 2\mathbf{i} + \mathbf{j} + 8z\mathbf{k}$. Then curl (curl $\mathbf{F}$) $= -8y\mathbf{j} + 2\mathbf{i} + \mathbf{j} + 8z\mathbf{k} = 2\mathbf{i} + (1 - 8y)\mathbf{j} + 8z\mathbf{k}$.

39. $\text{curl } \mathbf{F} = -Gm_1m_2 \begin{vmatrix} \mathbf{i} & \mathbf{j} & \mathbf{k} \\ \partial/\partial x & \partial/\partial y & \partial/\partial z \\ x/|\mathbf{r}|^3 & y/|\mathbf{r}|^3 & z/|\mathbf{r}|^3 \end{vmatrix}$

$$= -Gm_1m_2[(-3yz/|\mathbf{r}|^5 + 3yz/|\mathbf{r}|^5)\mathbf{i} - (-3xz/|\mathbf{r}|^5 + 3xz/|\mathbf{r}|^5)\mathbf{j} + (-3xy/|\mathbf{r}|^5 + 3xy/|\mathbf{r}|^5)\mathbf{k}]$$
$$= \mathbf{0}$$

$$\text{div } \mathbf{F} = -Gm_1m_2 \left[\frac{-2x^2 + y^2 + z^2}{|\mathbf{r}|^{5/2}} + \frac{x^2 - 2y^2 + z^2}{|\mathbf{r}|^{5/2}} + \frac{x^2 + y^2 - 2z^2}{|\mathbf{r}|^{5/2}} \right] = 0$$

42. Recall that $\mathbf{a} \cdot (\mathbf{a} \times \mathbf{b}) = 0$. Then, using Problems 31, 29, and 28,

$$\nabla \cdot \mathbf{F} = \text{div } (\nabla f \times f \nabla g) = f \nabla g \cdot (\text{curl } \nabla f) - \nabla f \cdot (\text{curl } f \nabla g) = f \nabla g \cdot \mathbf{0} - \nabla f \cdot (\nabla \times f \nabla g)$$
$$= -\nabla f \cdot [f(\nabla \times \nabla g) + (\nabla f \times \nabla g)] = -\nabla f \cdot [f\text{curl } \nabla g + (\nabla f \times \nabla g)]$$
$$= -\nabla f \cdot [f\mathbf{0} + (\nabla f \times \nabla g)] = -\nabla f \cdot (\nabla f \times \nabla g) = 0.$$

45. We note that div $\mathbf{F} = 2xyz - 2xyz + 1 = 1 \neq 0$. If $\mathbf{F} = \text{curl } \mathbf{G}$, then div (curl $\mathbf{G}$) $=$ div $\mathbf{F} = 1$. But, by Problem 30, for any vector field $\mathbf{G}$, div (curl $\mathbf{G}$) $= 0$. Thus, $\mathbf{F}$ cannot be the curl of $\mathbf{G}$.

Exercises 9.8

3. $\displaystyle\int_C (3x^2 + 6y^2)\,dx = \int_{-1}^0 [3x^2 + 6(2x+1)^2]\,dx = \int_{-1}^0 (27x^2 + 24x + 6)\,dx = (9x^3 + 12x^2 + 6x)\,\Big|_{-1}^0$

$$= -(-9 + 12 - 6) = 3$$

$\displaystyle\int_C (3x^2 + 6y^2)\,dy = \int_{-1}^0 [3x^2 + 6(2x+1)^2]2\,dx = 6$

$\displaystyle\int_C (3x^2 + 6y^2)\,ds = \int_{-1}^0 [3x^2 + 6(2x+1)^2]\sqrt{1+4}\,dx = 3\sqrt{5}$

6. $\displaystyle\int_C 4xyz\,dx = \int_0^1 4\left(\frac{1}{3}t^3\right)(t^2)(2t)t^2\,dt = \frac{8}{3}\int_0^1 t^8\,dt = \frac{8}{27}t^9\,\Big|_0^1 = \frac{8}{27}$

$\displaystyle\int_C 4xyz\,dy = \int_0^1 4\left(\frac{1}{3}t^3\right)(t^2)(2t)2t\,dt = \frac{16}{3}\int_0^1 t^7\,dt = \frac{2}{3}t^8\,\Big|_0^1 = \frac{2}{3}$

$\displaystyle\int_C 4xyz\,dz = \int_0^1 4\left(\frac{1}{3}t^3\right)(t^2)(2t)2\,dt = \frac{16}{3}\int_0^1 t^6\,dt = \frac{16}{21}t^7\,\Big|_0^1 = \frac{16}{21}$

$\displaystyle\int_C 4xyz\,ds = \int_0^1 4\left(\frac{1}{3}t^3\right)(t^2)(2t)\sqrt{t^4 + 4t^2 + 4}\,dt = \frac{8}{3}\int_0^1 t^6(t^2 + 2)\,dt = \frac{8}{3}\left(\frac{1}{9}t^9 + \frac{2}{7}t^7\right)\Big|_0^1 = \frac{200}{189}$

9. From $(-1, 2)$ to $(2, 2)$ we use x as a parameter with $y = 2$ and $dy = 0$. From $(2, 2)$ to $(2, 5)$ we use y as a parameter with $x = 2$ and $dx = 0$.

$$\int_C (2x + y)\,dx + xy\,dy = \int_{-1}^2 (2x + 2)\,dx + \int_2^5 2y\,dy = (x^2 + 2x)\,\Big|_{-1}^2 + y^2\,\Big|_2^5 = 9 + 21 = 30$$

12. Using x as the parameter, $dy = dx$.

$$\int_C y\,dx + x\,dy = \int_0^1 x\,dx + \int_0^1 x\,dx = \int_0^1 2x\,dx = x^2\,\Big|_0^1 = 1$$

15. $\displaystyle\int_C (6x^2 + 2y^2)\,dx + 4xy\,dy = \int_4^9 (6t + 2t^2)\frac{1}{2}t^{-1/2}\,dt + \int_4^9 4\sqrt{t}\,t\,dt = \int_4^9 (3t^{1/2} + 5t^{3/2})\,dt$

$$= (2t^{3/2} + 2t^{5/2})\,\Big|_4^9 = 460$$

18. $\displaystyle\int_C 4x\,dx + 2y\,dy = \int_{-1}^2 4(y^3 + 1)3y^2\,dy + \int_{-1}^2 2y\,dy = \int_{-1}^2 (12y^5 + 12y^2 + 2y)\,dy$

$$= (2y^6 + 4y^3 + y^2)\,\Big|_{-1}^2 = 165$$

21. From $(1, 1)$ to $(-1, 1)$ and $(-1, -1)$ to $(1, -1)$ we use x as a parameter with $y = 1$ and $y = -1$, respectively, and $dy = 0$. From $(-1, 1)$ to $(-1, -1)$ and $(1, -1)$ to $(1, 1)$ we use y as a parameter with $x = -1$ and $x = 1$, respectively, and $dx = 0$.

$$\oint_C x^2 y^3\,dx - xy^2\,dy = \int_1^{-1} x^2(1)\,dx + \int_1^{-1} -(-1)y^2\,dy + \int_{-1}^1 x^2(-1)^3\,dx + \int_{-1}^1 -(1)y^2\,dy$$

$$= \frac{1}{3}x^3\,\Big|_1^{-1} + \frac{1}{3}y^3\,\Big|_1^{-1} - \frac{1}{3}x^3\,\Big|_{-1}^1 - \frac{1}{3}y^3\,\Big|_{-1}^1 = -\frac{8}{3}$$

24. $\displaystyle\oint_C y\,dx - x\,dy = \int_0^\pi 3\sin t(-2\sin t)\,dt - \int_0^\pi 2\cos t(3\cos t)\,dt = -6\int_0^\pi (\sin^2 t + \cos^2 t)\,dt = -6\int_0^\pi dt = -6\pi$

Thus, $\displaystyle\int_{-C} y\,dx - x\,dy = 6\pi.$

27. From $(0,0,0)$ to $(6,0,0)$ we use x as a parameter with $y = dy = 0$ and $z = dz = 0$. From $(6,0,0)$ to $(6,0,5)$ we use z as a parameter with $x = 6$ and $dx = 0$ and $y = dy = 0$. From $(6,0,5)$ to $(6,8,5)$ we use y as a parameter with $x = 6$ and $dx = 0$ and $z = 5$ and $dz = 0$.

$$\int_C y\,dx + z\,dy + x\,dz = \int_0^6 0 + \int_0^5 6\,dz + \int_0^8 5\,dy = 70$$

30. $\mathbf{F} = e^t\mathbf{i} + te^{t^3}\mathbf{j} + t^3 e^{t^6}\mathbf{k}; \quad d\mathbf{r} = (\mathbf{i} + 2t\mathbf{j} + 3t^2\mathbf{k})\,dt;$

$\displaystyle\int_C \mathbf{F}\cdot d\mathbf{r} = \int_0^1 (e^t + 2t^2 e^{t^3} + 3t^5 e^{t^6})\,dt = \left(e^t + \frac{2}{3}e^{t^3} + \frac{1}{2}e^{t^6}\right)\Big|_0^1 = \frac{13}{6}(e - 1)$

33. Let $\mathbf{r}_1 = (1 + 2t)\mathbf{i} + \mathbf{j}$, $\mathbf{r}_2 = 3\mathbf{i} + (1 + t)\mathbf{j}$, and $\mathbf{r}_3 = (3 - 2t)\mathbf{i} + (2 - t)\mathbf{j}$ for $0 \le t \le 1$. Then

$$d\mathbf{r}_1 = 2\mathbf{i}, \qquad d\mathbf{r}_2 = \mathbf{j}, \qquad d\mathbf{r}_3 = -2\mathbf{i} - \mathbf{j},$$
$$\mathbf{F}_1 = (1 + 2t + 2)\mathbf{i} + (6 - 2 - 4t)\mathbf{j} = (3 + 2t)\mathbf{i} + (4 - 4t)\mathbf{j},$$
$$\mathbf{F}_2 = (3 + 2 + 2t)\mathbf{i} + (6 + 6t - 6)\mathbf{j} = (5 + 2t)\mathbf{i} + 6t\mathbf{j},$$
$$\mathbf{F}_3 = (3 - 2t + 4 - 2t)\mathbf{i} + (12 - 6t - 6 + 4t)\mathbf{j} = (7 - 4t)\mathbf{i} + (6 - 2t)\mathbf{j},$$

and

$$W = \int_{C_1} \mathbf{F}_1 \cdot d\mathbf{r}_1 + \int_{C_2} \mathbf{F}_2 \cdot d\mathbf{r}_2 + \int_{C_3} \mathbf{F}_3 \cdot d\mathbf{r}_3$$
$$= \int_0^1 (6 + 4t)\,dt + \int_0^1 6t\,dt + \int_0^1 (-14 + 8t - 6 + 2t)\,dt$$
$$= \int_0^1 (-14 + 20t)\,dt = (-14t + 10t^2)\Big|_0^1 = -4.$$

36. Let $\mathbf{r} = t\mathbf{i} + t\mathbf{j} + t\mathbf{k}$ for $1 \le t \le 3$. Then $d\mathbf{r} = \mathbf{i} + \mathbf{j} + \mathbf{k}$, and

$$\mathbf{F} = \frac{c}{|\mathbf{r}|^3}(t\mathbf{i} + t\mathbf{j} + t\mathbf{k}) = \frac{ct}{(\sqrt{3t^2}\,)^3}(\mathbf{i} + \mathbf{j} + \mathbf{k}) = \frac{c}{3\sqrt{3}\,t^2}(\mathbf{i} + \mathbf{j} + \mathbf{k}),$$

$$W = \int_C \mathbf{F}\cdot d\mathbf{r} = \int_1^3 \frac{c}{3\sqrt{3}\,t^2}(1 + 1 + 1)\,dt = \frac{c}{\sqrt{3}}\int_1^3 \frac{1}{t^2}\,dt = \frac{c}{\sqrt{3}}\left(-\frac{1}{t}\right)\Big|_1^3 = \frac{c}{\sqrt{3}}\left(-\frac{1}{3} + 1\right) = \frac{2c}{3\sqrt{3}}.$$

39. Since $\mathbf{v}\cdot\mathbf{v} = v^2$, $\dfrac{d}{dt}v^2 = \dfrac{d}{dt}(\mathbf{v}\cdot\mathbf{v}) = \mathbf{v}\cdot\dfrac{d\mathbf{v}}{dt} + \dfrac{d\mathbf{v}}{dt}\cdot\mathbf{v} = 2\dfrac{d\mathbf{v}}{dt}\cdot\mathbf{v}$. Then

$$W = \int_C \mathbf{F}\cdot d\mathbf{r} = \int_a^b m\mathbf{a}\cdot\left(\frac{d\mathbf{r}}{dt}\,dt\right) = m\int_a^b \frac{d\mathbf{v}}{dt}\cdot\mathbf{v}\,dt = m\int_a^b \frac{1}{2}\left(\frac{d}{dt}v^2\right)dt$$
$$= \frac{1}{2}m(v^2)\Big|_a^b = \frac{1}{2}m[v(b)]^2 - \frac{1}{2}m[v(a)]^2.$$

42. On C_1, $\mathbf{T} = \mathbf{i}$ and $\mathbf{F}\cdot\mathbf{T} = \text{comp}_{\mathbf{T}}\mathbf{F} \approx 1$. On C_2, $\mathbf{T} = -\mathbf{j}$ and $\mathbf{F}\cdot\mathbf{T} = \text{comp}_{\mathbf{T}}\mathbf{F} \approx 2$. On C_3, $\mathbf{T} = -\mathbf{i}$ and $\mathbf{F}\cdot\mathbf{T} = \text{comp}_{\mathbf{T}}\mathbf{F} \approx 1.5$. Using the fact that the lengths of C_1, C_2, and C_3 are 4, 5, and 5, respectively, we have

$$W = \int_C \mathbf{F}\cdot\mathbf{T}\,ds = \int_{C_1} \mathbf{F}\cdot\mathbf{T}\,ds + \int_{C_2} \mathbf{F}\cdot\mathbf{T}\,ds + \int_{C_3} \mathbf{F}\cdot\mathbf{T}\,ds \approx 1(4) + 2(5) + 1.5(5) = 21.5 \text{ ft-lb}.$$

3. (a) $P_y = 2 = Q_x$ and the integral is independent of path. $\phi_x = x + 2y$, $\phi = \dfrac{1}{2}x^2 + 2xy + g(y)$,

$\phi_y = 2x + g'(y) = 2x - y$, $g(y) = -\dfrac{1}{2}y^2$, $\phi = \dfrac{1}{2}x^2 + 2xy - \dfrac{1}{2}y^2$,

$$\int_{(1,0)}^{(3,2)} (x+2y)\,dx + (2x-y)\,dy = \left(\dfrac{1}{2}x^2 + 2xy - \dfrac{1}{2}y^2\right)\Big|_{(1,0)}^{(3,2)} = 14$$

(b) Use $y = x - 1$ for $1 \le x \le 3$.

$$\int_{(1,0)}^{(3,2)} (x+2y)\,dx + (2x-y)\,dy = \int_1^3 [x + 2(x-1) + 2x - (x-1)\,dx$$

$$= \int_1^3 (4x - 1)\,dx = (2x^2 - x)\Big|_1^3 = 14$$

6. (a) $P_y = -xy(x^2+y^2)^{-3/2} = Q_x$ and the integral is independent of path. $\phi_x = \dfrac{x}{\sqrt{x^2+y^2}}$,

$\phi = \sqrt{x^2+y^2} + g(y)$, $\phi_y = \dfrac{y}{\sqrt{x^2+y^2}} + g'(y) = \dfrac{y}{\sqrt{x^2+y^2}}$, $g(y) = 0$, $\phi = \sqrt{x^2+y^2}$,

$$\int_{(1,0)}^{(3,4)} \dfrac{x\,dx + y\,dy}{\sqrt{x^2+y^2}} = \sqrt{x^2+y^2}\,\Big|_{(1,0)}^{(3,4)} = 4$$

(b) Use $y = 2x - 2$ for $1 \le x \le 3$.

$$\int_{(1,0)}^{(3,4)} \dfrac{x\,dx + y\,dy}{\sqrt{x^2+y^2}} = \int_1^3 \dfrac{x + (2x-2)2}{\sqrt{x^2 + (2x-2)^2}}\,dx = \int_1^3 \dfrac{5x-4}{\sqrt{5x^2-8x+4}} = \sqrt{5x^2 - 8x + 4}\,\Big|_1^3 = 4$$

9. (a) $P_y = 3y^2 + 3x^2 = Q_x$ and the integral is independent of path. $\phi_x = y^3 + 3x^2y$,

$\phi = xy^3 + x^3y + g(y)$, $\phi_y = 3xy^2 + x^3 + g'(y) = x^3 + 3y^2x + 1$, $g(y) = y$, $\phi = xy^3 + x^3y + y$,

$$\int_{(0,0)}^{(2,8)} (y^3 + 3x^2y)\,dx + (x^3 + 3y^2x + 1)\,dy = (xy^3 + x^3y + y)\Big|_{(0,0)}^{(2,8)} = 1096$$

(b) Use $y = 4x$ for $0 \le x \le 2$.

$$\int_{(0,0)}^{(2,8)} (y^3 + 3x^2y)\,dx + (x^3 + 3y^2x + 1)\,dy = \int_0^2 [(64x^3 + 12x^3) + (x^3 + 48x^3 + 1)(4)]\,dx$$

$$= \int_0^2 (272x^3 + 4)\,dx = (68x^4 + 4x)\Big|_0^2 = 1096$$

12. $P_y = 6xy^2 = Q_x$ and the vector field is a gradient field. $\phi_x = 2xy^3$, $\phi = x^2y^3 + g(y)$,

$\phi_y = 3x^2y^2 + g'(y) = 3x^2y^2 + 3y^2$, $g(y) = y^3$, $\phi = x^2y^3 + y^3$

15. $P_y = 1 = Q_x$ and the vector field is a gradient field. $\phi_x = x^3 + y$, $\phi = \dfrac{1}{4}x^4 + xy + g(y)$, $\phi_y = x + g'(y) = x + y^3$,

$g(y) - \dfrac{1}{4}y^4$, $\phi = \dfrac{1}{4}x^4 + xy + \dfrac{1}{4}y^4$

18. Since $P_y = -e^{-y} = Q_x$, $\mathbf{F}$ is conservative and $\displaystyle\int_C \mathbf{F} \cdot d\mathbf{r}$ is independent of the path. Thus, instead of the given

curve we may use the simpler curve C_1: $y = 0$, $-2 \le -x \le 2$. Then $dy = 0$ and

$$W = \int_{C_1} (2x + e^{-y})\,dx + (4y - xe^{-y})\,dy = \int_2^{-2} (2x + 1)\,dx = (x^2 + x)\Big|_2^{-2} = (4 - 2) - (4 + 2) = -4.$$

21. $P_y = 2x \cos y = Q_x$, $Q_z = 0 = R_y$, $R_x = 3e^{3z} = P_z$, and the integral is independent of path. Integrating $\phi_x = 2x \sin y + e^{3z}$ we find $\phi = x^2 \sin y + xe^{3z} + g(y, z)$. Then $\phi_y = x^2 \cos y + g_y = Q = x^2 \cos y$, so $g_y = 0$, $g(y, z) = h(z)$, and $\phi = x^2 \sin y + xe^{3z} + h(z)$. Now $\phi_z = 3xe^{3z} + h'(z) = R = 3xe^{3z} + 5$, so $h'(z) = 5$ and $h(z) = 5z$. Thus $\phi = x^2 \sin y + xe^{3z} + 5z$ and

$$\int_{(1,0,0)}^{(2,\pi/2,1)} (2x \sin y + e^{3z}) \, dx + x^2 \cos y \, dy + (3xe^{3z} + 5) \, dz$$

$$= (x^2 \sin y + xe^{3z} + 5z) \Big|_{(1,0,0)}^{(2,\pi/2,1)} = [4(1) + 2e^3 + 5] - [0 + 1 + 0] = 8 + 2e^3.$$

24. $P_y = 0 = Q_x$, $Q_z = 2y = R_y$, $R_x = 2x = P_z$ and the integral is independent of path. Parameterize the line segment between the points by $x = -2(1 - t)$, $y = 3(1 - t)$, $z = 1 - t$, for $0 \le t \le 1$. Then $dx = 2 \, dt$, $dy = -3 \, dt$, $dz = -dt$, and

$$\int_{(-2,3,1)}^{(0,0,0)} 2xz \, dx + 2yz \, dy + (x^2 + y^2) \, dz$$

$$= \int_0^1 [-4(1 - t)^2(2) + 6(1 - t)^2(-3) + 4(1 - t)^2(-1) + 9(1 - t)^2(-1)] \, dt$$

$$= \int_0^1 -39(1 - t)^2 \, dt = 13(1 - t)^3 \Big|_0^1 = -13.$$

27. Since $P_y = Gm_1m_2(2xy/|\mathbf{r}|^5) = Q_x$, $Q_z = Gm_1m_2(2yz/|\mathbf{r}|^5) = R_y$, and $R_x = Gm_1m_2(2xz/|\mathbf{r}|^5) = P_z$, the force field is conservative.

$$\phi_x = -Gm_1m_2 \frac{x}{(x^2 + y^2 + z^2)^{3/2}}, \quad \phi = Gm_1m_2(x^2 + y^2 + z^2)^{-1/2} + g(y, z),$$

$$\phi_y = -Gm_1m_2 \frac{y}{(x^2 + y^2 + z^2)^{3/2}} + g_y(y, z) = -Gm_1m_2 \frac{y}{(x^2 + y^2 + z^2)^{3/2}}, \quad g(y, z) = h(z),$$

$$\phi = Gm_1m_2(x^2 + y^2 + z^2)^{-1/2} + h(z),$$

$$\phi_z = -Gm_1m_2 \frac{z}{(x^2 + y^2 + z^2)^{3/2}} + h'(z) = -Gm_1m_2 \frac{z}{(x^2 + y^2 + z^2)^{3/2}},$$

$$h(z) = 0, \quad \phi = \frac{Gm_1m_2}{\sqrt{x^2 + y^2 + z^2}} = \frac{Gm_1m_2}{|\mathbf{r}|}$$

30. From $\mathbf{F} = (x^2 + y^2)^{n/2}(x\mathbf{i} + y\mathbf{j})$ we obtain $P_y = nxy(x^2 + y^2)^{n/2-1} = Q_x$, so that $\mathbf{F}$ is conservative. From $\phi_x = x(x^2 + y^2)^{n/2}$ we obtain the potential function $\phi = (x^2 + y^2)^{(n+2)/2}/(n + 2)$. Then

$$W = \int_{(x_1,y_1)}^{(x_2,y_2)} \mathbf{F} \cdot d\mathbf{r} = \left(\frac{(x^2 + y^2)^{(n+2)/2}}{n + 2} \right) \Big|_{(x_1,y_1)}^{(x_2,y_2)} = \frac{1}{n + 2} \left[(x_2^2 + y_2^2)^{(n+2)/2} - (x_1^2 + y_1^2)^{(n+2)/2} \right].$$

Exercises 9.10

3. $\displaystyle\int_1^{3x} x^3 e^{xy}\, dy = x^2 e^{xy}\,\Big|_1^{3x} = x^2(e^{3x^2} - e^x)$

6. $\displaystyle\int_{x^3}^{x} e^{2y/x}\, dy = \frac{x}{2}\,e^{2y/x}\,\Big|_{x^3}^{x} = \frac{x}{2}(e^{2x/x} - e^{2x^3/x}) = \frac{x}{2}\left(e^2 - e^{2x^2}\right)$

9.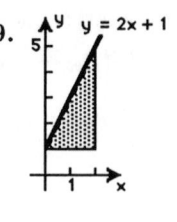

12.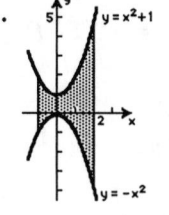

15. $\displaystyle\iint_R (2x + 4y + 1)\, dA = \int_0^1 \int_{x^3}^{x^2} (2x + 4y + 1)\, dy\, dx = \int_0^1 (2xy + 2y^2 + y)\,\Big|_{x^3}^{x^2}\, dx$

$\displaystyle\qquad\qquad = \int_0^1 [(2x^3 + 2x^4 + x^2) - (2x^4 + 2x^6 + x^3)]\, dx$

$\displaystyle\qquad\qquad = \int_0^1 (x^3 + x^2 - 2x^6)\, dx = \left(\frac{1}{4}x^4 + \frac{1}{3}x^3 - \frac{2}{7}x^7\right)\Big|_0^1$

$\displaystyle\qquad\qquad = \frac{1}{4} + \frac{1}{3} - \frac{2}{7} = \frac{25}{84}$

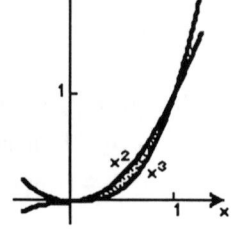

18. $\displaystyle\iint_R \frac{x}{\sqrt{y}}\, dA = \int_{-1}^{1} \int_{x^2+1}^{3-x^2} xy^{-1/2}\, dy\, dx = \int_{-1}^{1} 2x\sqrt{y}\,\Big|_{x^2+1}^{3-x^2}\, dx$

$\displaystyle\qquad\qquad = 2\int_{-1}^{1} \left(x\sqrt{3 - x^2} - x\sqrt{x^2 + 1}\,\right) dx$

$\displaystyle\qquad\qquad = 2\left[-\frac{1}{3}(3 - x^2)^{3/2} - \frac{1}{3}(x^2 + 1)^{3/2}\right]\Big|_{-1}^{1}$

$\displaystyle\qquad\qquad = -\frac{2}{3}[(2^{3/2} + 2^{3/2}) - (2^{3/2} + 2^{3/2})] = 0$

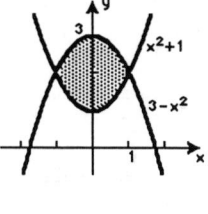

21. $\displaystyle\iint_R \sqrt{x^2 + 1}\, dA = \int_0^{\sqrt{3}} \int_{-x}^{x} \sqrt{x^2 + 1}\, dy\, dx = \int_0^{\sqrt{3}} y\sqrt{x^2 + 1}\,\Big|_{-x}^{x}\, dx$

$\displaystyle\qquad\qquad = \int_0^{\sqrt{3}} (x\sqrt{x^2 + 1} + x\sqrt{x^2 + 1}\,)\, dx = \int_0^{\sqrt{3}} 2x\sqrt{x^2 + 1}\, dx$

$\displaystyle\qquad\qquad = \frac{2}{3}(x^2 + 1)^{3/2}\,\Big|_0^{\sqrt{3}} = \frac{2}{3}(4^{3/2} - 1^{3/2}) = \frac{14}{3}$

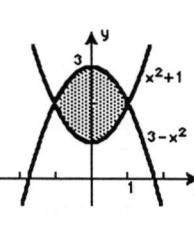

24. The correct integral is **(b)**.

$\displaystyle A = 8\int_0^2 \int_0^{\sqrt{4-y^2}} (4 - y^2)^{1/2} dx\, dy = 8\int_0^2 (4 - y^2)^{1/2} x\,\Big|_0^{\sqrt{4-y^2}}\, dy = 8\int_0^2 (4 - y^2)\, dy$

$\displaystyle\qquad = 8\left(4y - \frac{1}{3}y^3\right)\Big|_0^2 = \frac{128}{3}$

27. Solving for z, we have $x = 2 - \frac{1}{2}x + \frac{1}{2}y$. Setting $z = 0$, we see that this surface (plane) intersects the xy-plane in the line $y = x - 4$. since $z(0,0) = 2 > 0$, the surface lies above the xy-plane over the quarter-circular region.

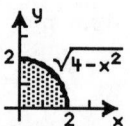
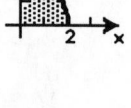

$$V = \int_0^2 \int_0^{\sqrt{4-x^2}} \left(2 - \frac{1}{2}x + \frac{1}{2}y\right) dy\,dx = \int_0^2 \left(2y - \frac{1}{2}xy + \frac{1}{4}y^2\right)\Big|_0^{\sqrt{4-x^2}} dx$$

$$= \int_0^2 \left(2\sqrt{4-x^2} - \frac{1}{2}x\sqrt{4-x^2} + 1 - \frac{1}{4}x^2\right) dx = \left[x\sqrt{4-x^2} + 4\sin^{-1}\frac{x}{2} + \frac{1}{6}(4-x^2)^{3/2} + x - \frac{1}{12}x^3\right]\Big|_0^2$$

$$= \left(2\pi + 2 - \frac{2}{3}\right) - \frac{4}{3} = 2\pi$$

30. In the first octant, $z = x + y$ is nonnegative. Then

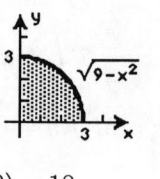

$$V = \int_0^3 \int_0^{\sqrt{9-x^2}} (x+y)\,dy\,dx = \int_0^3 \left(xy + \frac{1}{2}y^2\right)\Big|_0^{\sqrt{9-x^2}} dx$$

$$= \int_0^3 \left(x\sqrt{9-x^2} + \frac{9}{2} - \frac{1}{2}x^2\right) dx = \left[-\frac{1}{3}(9-x^2)^{3/2} + \frac{9}{2}x - \frac{1}{6}x^3\right]\Big|_0^3 = \left(\frac{27}{2} - \frac{9}{2}\right) - (-9) = 18.$$

33. Note that $z = 4 - y^2$ is positive for $|y| \le 1$. Using symmetry,

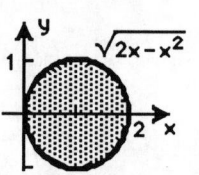

$$V = 2\int_0^2 \int_0^{\sqrt{2x-x^2}} (4-y^2)\,dy\,dx = 2\int_0^2 \left(4y - \frac{1}{3}y^3\right)\Big|_0^{\sqrt{2x-x^2}} dx$$

$$= 2\int_0^2 \left[4\sqrt{2x-x^2} - \frac{1}{3}(2x-x^2)\sqrt{2x-x^2}\right] dx$$

$$= 2\int_0^2 \left(4\sqrt{1-(x-1)^2} - \frac{1}{3}[1-(x-1)^2]\sqrt{1-(x-1)^2}\right) dx \qquad \boxed{u = x-1,\ du = dx}$$

$$= 2\int_{-1}^1 \left[4\sqrt{1-u^2} - \frac{1}{3}(1-u^2)\sqrt{1-u^2}\right] du = 2\int_{-1}^1 \left(\frac{11}{3}\sqrt{1-u^2} + \frac{1}{3}u^2\sqrt{1-u^2}\right) du$$

$$\boxed{\text{Trig substitution}}$$

$$= 2\left[\frac{11}{6}u\sqrt{1-u^2} + \frac{11}{6}\sin u + \frac{1}{24}x(2x^2-1)\sqrt{1-u^2} + \frac{1}{24}\sin^{-1}u\right]\Big|_{-1}^1$$

$$= 2\left[\left(\frac{11}{6}\frac{\pi}{2} + \frac{1}{24}\frac{\pi}{2}\right) - \left(-\frac{11}{6}\frac{\pi}{2} - \frac{1}{24}\frac{\pi}{2}\right)\right] = \frac{15\pi}{4}.$$

36. $\displaystyle\int_0^1 \int_{2y}^2 e^{-y/x}\,dx\,dy = \int_0^2 \int_0^{x/2} e^{-y/x}\,dy\,dx = \int_0^2 -xe^{-y/x}\Big|_0^{x/2} = \int_0^2 (-xe^{-1/2} + x)\,dx$

$$= \int_0^2 (1-e^{-1/2})x\,dx = \frac{1}{2}(1-e^{-1/2})x^2\Big|_0^2 = 2(1-e^{-1/2})$$

39. $\displaystyle\int_0^1 \int_x^1 \frac{1}{1+y^4}\,dy\,dx = \int_0^1 \int_0^y \frac{1}{1+y^4}\,dx\,dy = \int_0^1 \frac{x}{1+y^4}\Big|_0^y dy = \int_0^1 \frac{y}{1+y^4}\,dy$

$$= \frac{1}{2}\tan^{-1}y^2\Big|_0^1 = \frac{\pi}{8}$$

42. $m = \int_0^2 \int_0^{4-2x} x^2 \, dy \, dx = \int_0^2 x^2 y \Big|_0^{4-2x} dx = \int_0^2 x^2(4-2x) \, dx$

$= \int_0^2 (4x^2 - 2x^3) \, dx = \left(\frac{4}{3}x^3 - \frac{1}{2}x^4 \right) \Big|_0^2 = \frac{32}{3} - 8 = \frac{8}{3}$

$M_y = \int_0^2 \int_0^{4-2x} x^3 \, dy \, dx = \int_0^2 x^3 y \Big|_0^{4-2x} dx = \int_0^2 x^3(4-2x) \, dx = \int_0^2 (4x^3 - 2x^4) \, dx$

$= \left(x^4 - \frac{2}{5}x^5 \right) \Big|_0^2 = 16 - \frac{64}{5} = \frac{16}{5}$

$M_x = \int_0^2 \int_0^{4-2x} x^2 y \, dy \, dx = \int_0^2 \frac{1}{2} x^2 y^2 \Big| \int_0^{4-2x} dx = \frac{1}{2} \int_0^2 x^2(4-2x)^2 \, dx = \frac{1}{2} \int_0^2 (16x^2 - 16x^3 + 4x^4) \, dx$

$= 2 \int_0^2 (4x^2 - 4x^3 + x^4) \, dx = 2 \left(\frac{4}{3}x^3 - x^4 + \frac{1}{5}x^5 \right) \Big|_0^2 = 2 \left(\frac{32}{3} - 16 + \frac{32}{5} \right) = \frac{32}{15}$

$\bar{x} = M_y/m = \dfrac{16/5}{8/3} = 6/5; \quad \bar{y} = M_x/m = \dfrac{32/15}{8/3} = 4/5.$ The center of mass is $(6/5, 4/5)$.

45. $m = \int_0^1 \int_0^{x^2} (x+y) \, dy \, dx = \int_0^1 \left(xy + \frac{1}{2}y^2 \right) \Big|_0^{x^2} dx = \int_0^1 \left(x^3 + \frac{1}{2}x^4 \right) dx$

$= \left(\frac{1}{4}x^4 + \frac{1}{10}x^5 \right) \Big|_0^1 = \frac{7}{20}$

$M_y = \int_0^1 \int_0^{x^2} (x^2 + xy) \, dy \, dx = \int_0^1 \left(x^2 y + \frac{1}{2}xy^2 \right) \Big|_0^{x^2} dx = \int_0^1 \left(x^4 + \frac{1}{2}x^5 \right) dx$

$= \left(\frac{1}{5}x^5 + \frac{1}{12}x^6 \right) \Big|_0^1 = \frac{17}{60}$

$M_x = \int_0^1 \int_0^{x^2} (xy + y^2) \, dy \, dx = \int_0^1 \left(\frac{1}{2}xy^2 + \frac{1}{3}y^3 \right) \Big|_0^{x^2} = \int_0^1 \left(\frac{1}{2}x^5 + \frac{1}{3}x^6 \right) dx = \left(\frac{1}{12}x^6 + \frac{1}{21}x^7 \right) \Big|_0^1 = \frac{11}{84}$

$\bar{x} = M_y/m = \dfrac{17/60}{7/20} = 17/21; \quad \bar{y} = M_x/m = \dfrac{11/84}{7/20} = 55/147.$ The center of mass is $(17/21, 55/147)$.

48. The density is $\rho = kx$.

$m = \int_0^\pi \int_0^{\sin x} kx \, dy \, dx = \int_0^\pi kxy \Big|_0^{\sin x} dx = \int_0^\pi kx \sin x \, dx$

$\boxed{\text{Integration by parts}}$

$= k(\sin x - x \cos x) \Big|_0^\pi = k\pi$

$M_y = \int_0^\pi \int_0^{\sin x} kx^2 \, dy \, dx = \int_0^\pi kx^2 y \Big|_0^{\sin x} dx = \int_0^\pi kx^2 \sin x \, dx \qquad \boxed{\text{Integration by parts}}$

$= k(-x^2 \cos x + 2 \cos x + 2x \sin x) \Big|_0^\pi = k[(\pi^2 - 2) - 2] = k(\pi^2 - 4)$

$$M_x = \int_0^\pi \int_0^{\sin x} kxy\, dy\, dx = \int_0^\pi \frac{1}{2}kxy^2 \Big|_0^{\sin x} dx = \int_0^\pi \frac{1}{2}kx\sin^2 x\, dx = \int_0^\pi \frac{1}{4}kx(1-\cos 2x)\, dx$$

$$= \frac{1}{4}k\left[\int_0^\pi x\, dx - \int_0^\pi x\cos 2x\, dx\right] \qquad \boxed{\text{Integration by parts}}$$

$$= \frac{1}{4}k\left[\frac{1}{2}x^2 \Big|_0^\pi - \frac{1}{4}(\cos 2x + 2x\sin 2x)\Big|_0^\pi\right] = \frac{1}{4}k\left(\frac{1}{2}\pi^2\right) = \frac{1}{8}k\pi^2$$

$\bar{x} = M_y/m = \dfrac{k(\pi^2-4)}{k\pi} = \pi - 4/\pi; \quad \bar{y} = M_x/m = \dfrac{k\pi^2/8}{k\pi} = \pi/8.$ The center of mass is $(\pi - 4/\pi, \pi/8)$.

51. $I_x = \displaystyle\int_0^1 \int_0^{y-y^2} 2xy^2\, dx\, dy = \int_0^1 x^2 y^2 \Big|_0^{y-y^2} dy = \int_0^1 (y-y^2)^2\, y^2\, dy$

$$= \int_0^1 (y^4 - 2y^5 + y^6)\, dy = \left(\frac{1}{5}y^5 - \frac{1}{3}y^6 + \frac{1}{7}y^7\right)\Big|_0^1 = \frac{1}{105}$$

54. $I_x = \displaystyle\int_0^2 \int_0^{\sqrt{4-x^2}} y^3\, dy\, dx = \int_0^2 \frac{1}{4}y^4 \Big|_0^{\sqrt{4-x^2}} dx = \frac{1}{4}\int_0^2 (4-x^2)^2\, dx$

$$= \frac{1}{4}\int_0^2 (16 - 8x^2 + x^4)\, dx = \frac{1}{4}\left(16x - \frac{8}{3}x^3 + \frac{1}{5}x^5\right)\Big|_0^2 = \frac{1}{4}\left(32 - \frac{64}{3} + \frac{32}{5}\right)$$

$$= 8\left(1 - \frac{2}{3} + \frac{1}{5}\right) = \frac{64}{15}$$

57. $I_y = \displaystyle\int_0^1 \int_y^3 (4x^3 + 3x^2 y)\, dx\, dy = \int_0^1 (x^4 + x^3 y)\Big|_y^3 dy = \int_0^1 (81 + 27y - 2y^4)\, dy$

$$= \left(81y + \frac{27}{2}y^2 - \frac{2}{5}y^5\right)\Big|_0^1 = \frac{941}{10}$$

60. $m = \displaystyle\int_0^a \int_0^{a-x} k\, dy\, dx = \int_0^a ky\Big|_0^{a-x} dx = k\int_0^a (a-x)\, dx = k\left(ax - \frac{1}{2}x^2\right)\Big|_0^a = \frac{1}{2}ka^2$

$I_x = \displaystyle\int_0^a \int_0^{a-x} ky^2\, dy\, dx = \int_0^a \frac{1}{3}ky^3 \Big|_0^{a-x} dx = \frac{1}{3}k\int_0^a (a-x)^3\, dx$

$$= \frac{1}{3}k\int_0^a (a^3 - 3a^2 x + 3ax^2 - x^3)\, dx = \frac{1}{3}k\left(a^3 x - \frac{3}{2}a^2 x^2 + ax^3 - \frac{1}{4}x^4\right)\Big|_0^a = \frac{1}{12}ka^4$

$$R_g = \sqrt{\frac{I_x}{m}} = \sqrt{\frac{ka^4/12}{ka^2/2}} = \sqrt{\frac{1}{6}}\, a$$

63. From Problem 60, $m = \frac{1}{2}ka^2$ and $I_x = \frac{1}{12}ka^4$.

$I_y = \displaystyle\int_0^a \int_0^{a-x} kx^2\, dy\, dx = \int_0^a kx^2 y\Big|_0^{a-x} dx = k\int_0^a x^2(a-x)\, dx$

$$= k\left(\frac{1}{3}ax^3 - \frac{1}{4}x^4\right)\Big|_0^a = \frac{1}{12}ka^4$$

$$I_0 = I_x + I_y = \frac{1}{12}ka^4 + \frac{1}{12}ka^4 = \frac{1}{6}ka^4$$

66. $I_0 = \displaystyle\int_0^3 \int_y^4 k(x^2 + y^2)\, dx\, dy = k\int_0^3 \left(\frac{1}{3}x^3 + xy^2\right)\Big|_y^4 dy$

$$= k\int_0^3 \left(\frac{64}{3} + 4y^2 - \frac{1}{3}y^3 - y^3\right) dy = k\left(\frac{64}{3}y + \frac{4}{3}y^3 - \frac{1}{3}y^4\right)\Big|_0^3 = 73k$$

Exercises 9.11

3. Solving $r = 2\sin\theta$ and $r = 1$, we obtain $\sin\theta = 1/2$ or $\theta = \pi/6$. Using symmetry,

$$A = 2\int_0^{\pi/6}\int_0^{2\sin\theta} r\,dr\,d\theta + 2\int_{\pi/6}^{\pi/2}\int_0^1 r\,dr\,d\theta$$

$$= 2\int_0^{\pi/6}\frac{1}{2}r^2\Big|_0^{2\sin\theta} d\theta + 2\int_{\pi/6}^{\pi/2}\frac{1}{2}r^2\Big|_0^1 d\theta = \int_0^{\pi/6} 4\sin^2\theta\,d\theta + \int_{\pi/6}^{\pi/2} d\theta$$

$$= (2\theta - \sin 2\theta)\Big|_0^{\pi/6} + \left(\frac{\pi}{2} - \frac{\pi}{6}\right) = \frac{\pi}{3} - \frac{\sqrt{3}}{2} + \frac{\pi}{3} = \frac{4\pi - 3\sqrt{3}}{6}$$

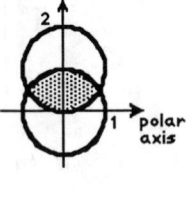

6. $V = \displaystyle\int_0^{2\pi}\int_0^2 \sqrt{9-r^2}\,r\,dr\,d\theta = \int_0^{2\pi} -\frac{1}{3}(9-r^2)^{3/2}\Big|_0^2 d\theta$

$$= -\frac{1}{3}\int_0^{2\pi}(5^{3/2} - 27)\,d\theta = \frac{1}{3}(27 - 5^{3/2})2\pi = \frac{2\pi(27 - 5\sqrt{5})}{3}$$

9. $V = \displaystyle\int_0^{\pi/2}\int_0^{1+\cos\theta}(r\sin\theta)r\,dr\,d\theta = \int_0^{\pi/2}\frac{1}{3}r^3\sin\theta\Big|_0^{1+\cos\theta} d\theta$

$$= \frac{1}{3}\int_0^{\pi/2}(1+\cos\theta)^3\sin\theta\,d\theta = \frac{1}{3}\left[-\frac{1}{4}(1+\cos\theta)^4\right]\Big|_0^{\pi/2} = -\frac{1}{12}(1-2^4) = \frac{5}{4}$$

12. The interior of the upper-half circle is traced from $\theta = 0$ to $\pi/2$. The density is kr. Since both the region and the density are symmetric about the polar axis, $\bar{y} = 0$.

$$m = \int_0^{\pi/2}\int_0^{\cos\theta} kr^2\,dr\,d\theta = k\int_0^{\pi/2}\frac{1}{3}r^3\Big|_0^{\cos\theta} d\theta = \frac{k}{3}\int_0^{\pi/2}\cos^3\theta\,d\theta$$

$$= \frac{k}{3}\left(\frac{2}{3} + \frac{1}{3}\cos^2\theta\right)\sin\theta\Big|_0^{\pi/2} = \frac{2k}{9}$$

$$M_y = k\int_0^{\pi/2}\int_0^{\cos\theta}(r\cos\theta)(r)(r\,dr\,d\theta) = k\int_0^{\pi/2}\int_0^{\cos\theta} r^3\cos\theta\,dr\,d\theta = k\int_0^{\pi/2}\frac{1}{4}r^4\cos\theta\Big|_0^{\cos\theta} d\theta$$

$$= \frac{k}{4}\int_0^{\pi/2}\cos^5\theta\,d\theta = \frac{k}{4}\left(\sin\theta - \frac{2}{3}\sin^3\theta + \frac{1}{5}\sin^5\theta\right)\Big|_0^{\pi/2} = \frac{2k}{15}$$

Thus, $\bar{x} = \dfrac{2k/15}{2k/9} = 3/5$ and the center of mass is $(3/5, 0)$.

15. The density is $\rho = k/r$.

$$m = \int_0^{\pi/2}\int_2^{2+2\cos\theta}\frac{k}{r}\,r\,dr\,d\theta = k\int_0^{\pi/2}\int_2^{2+2\cos\theta} dr\,d\theta$$

$$= k\int_0^{\pi/2} 2\cos\theta\,d\theta = 2k(\sin\theta)\Big|_0^{\pi/2} = 2k$$

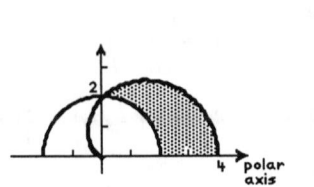

$$M_y = \int_0^{\pi/2} \int_2^{2+2\cos\theta} x\frac{k}{r}r\,dr\,d\theta = k\int_0^{\pi/2}\int_2^{2+2\cos\theta} r\cos\theta\,dr\,d\theta = k\int_0^{\pi/2}\frac{1}{2}r^2\,\bigg|_2^{2+2\cos\theta}\cos\theta\,d\theta$$

$$= \frac{1}{2}k\int_0^{\pi/2}(8\cos\theta + 4\cos^2\theta)\cos\theta\,d\theta = 2k\int_0^{\pi/2}(2\cos^2\theta + \cos\theta - \sin^2\theta\cos\theta)\,d\theta$$

$$= 2k\left(\theta + \frac{1}{2}\sin 2\theta + \sin\theta - \frac{1}{3}\sin^3\theta\right)\bigg|_0^{\pi/2} = 2k\left(\frac{\pi}{2} + \frac{2}{3}\right) = \frac{3\pi+4}{3}k$$

$$M_x = \int_0^{\pi/2}\int_2^{2+2\cos\theta} y\frac{k}{r}r\,dr\,d\theta = k\int_0^{\pi/2}\int_2^{2+2\cos\theta} r\sin\theta\,dr\,d\theta = k\int_0^{\pi/2}\frac{1}{2}r^2\,\bigg|_2^{2+2\cos\theta}\sin\theta\,d\theta$$

$$= \frac{1}{2}k\int_0^{\pi/2}(8\cos\theta + 4\cos^2\theta)\sin\theta\,d\theta = \frac{1}{2}k\left(-4\cos^2\theta - \frac{4}{3}\cos^3\theta\right)\bigg|_0^{\pi/2} = \frac{1}{2}k\left[-\left(-4 - \frac{4}{3}\right)\right] = \frac{8}{3}k$$

$$\bar{x} = M_y/m = \frac{(3\pi+4)k/3}{2k} = \frac{3\pi+4}{6}; \quad \bar{y} = M_x/m = \frac{8k/3}{2k} = \frac{4}{3}.$$ The center of mass is $((3\pi+4)/6, 4/3)$.

18. $I_x = \displaystyle\int_0^{2\pi}\int_0^a y^2\frac{1}{1+r^4}r\,dr\,d\theta = \int_0^{2\pi}\int_0^a \frac{r^3}{1+r^4}\sin^2\theta\,dr\,d\theta$

$$= \int_0^{2\pi}\frac{1}{4}\ln(1+r^4)\,\bigg|_0^a\sin^2\theta\,d\theta = \frac{1}{4}\ln(1+a^4)\left(\frac{1}{2}\theta - \frac{1}{4}\sin 2\theta\right)\bigg|_0^{2\pi} = \frac{\pi}{4}\ln(1+a^4)$$

21. From Problem 17, $I_x = k\pi a^4/4$. By symmetry, $I_y = I_x$. Thus $I_0 = k\pi a^4/2$.

24. $I_0 = \displaystyle\int_0^\pi\int_0^{2a\cos\theta} r^2 kr\,dr\,d\theta = k\int_0^\pi\frac{1}{4}r^4\,\bigg|_0^{2a\cos\theta}d\theta = 4ka^4\int_0^\pi\cos^4\theta\,d\theta$

$$= 4ka^4\left(\frac{3}{8}\theta + \frac{1}{4}\sin 2\theta + \frac{1}{32}\sin 4\theta\right)\bigg|_0^\pi = 4ka^4\left(\frac{3\pi}{8}\right) = \frac{3k\pi a^4}{2}$$

27. $\displaystyle\int_0^1\int_0^{\sqrt{1-y^2}} e^{x^2+y^2}\,dx\,dy = \int_0^{\pi/2}\int_0^1 e^{r^2}r\,dr\,d\theta = \int_0^{\pi/2}\frac{1}{2}e^{r^2}\,\bigg|_0^1 d\theta$

$$= \frac{1}{2}\int_0^{\pi/2}(e-1)\,d\theta = \frac{\pi(e-1)}{4}$$

30. $\displaystyle\int_0^1\int_0^{\sqrt{2y-y^2}}(1-x^2-y^2)\,dx\,dy$

$$= \int_0^{\pi/4}\int_0^{2\sin\theta}(1-r^2)r\,dr\,d\theta + \int_{\pi/4}^{\pi/2}\int_0^{\csc\theta}(1-r^2)r\,dr\,d\theta$$

$$= \int_0^{\pi/4}\left(\frac{1}{2}r^2 - \frac{1}{4}r^4\right)\bigg|_0^{2\sin\theta}d\theta + \int_{\pi/4}^{\pi/2}\left(\frac{1}{2}r^2 - \frac{1}{4}r^4\right)\bigg|_0^{\csc\theta}d\theta$$

$$= \int_0^{\pi/4}(2\sin^2\theta - 4\sin^4\theta)\,d\theta + \int_{\pi/4}^{\pi/2}\left(\frac{1}{2}\csc^2\theta - \frac{1}{4}\csc^4\theta\right)d\theta$$

$$= \left[\theta - \frac{1}{2}\sin 2\theta - \left(\frac{3}{2}\theta - \sin 2\theta + \frac{1}{8}\sin 4\theta\right)\right] + \left[-\frac{1}{2}\cot\theta - \frac{1}{4}\left(-\cot\theta - \frac{1}{3}\cot^3\theta\right)\right]\bigg|_{\pi/4}^{\pi/2}$$

$$= \left(-\frac{\pi}{8} + \frac{1}{2}\right) + \left[0 - \left(-\frac{1}{4} + \frac{1}{12}\right)\right] = \frac{16-3\pi}{24}$$

141

33. The volume of the cylindrical portion of the tank is $V_c = \pi(4.2)^2 19.3 \approx 1069.56$ m^3. We take the equation of the ellipsoid to be

$$\frac{x^2}{(4.2)^2} + \frac{z^2}{(5.15)^2} = 1 \quad \text{or} \quad z = \pm\frac{5.15}{4.2}\sqrt{(4.2)^2 - x^2 - y^2}.$$

The volume of the ellipsoid is

$$V_e = 2\left(\frac{5.15}{4.2}\right) \iint_R \sqrt{(4.2)^2 - x^2 - y^2}\, dx\, dy = \frac{10.3}{4.2} \int_0^{2\pi} \int_0^{4.2} [(4.2)^2 - r^2]^{1/2} r\, dr\, d\theta$$

$$= \frac{10.3}{4.2} \int_0^{2\pi} \left[\left(-\frac{1}{2}\right)\frac{2}{3}[(4.2)^2 - r^2]^{3/2}\,\Big|_0^{4.2}\right] d\theta = \frac{10.3}{4.2}\frac{1}{3} \int_0^{2\pi} (4.2)^3\, d\theta$$

$$= \frac{2\pi}{3}\frac{10.3}{4.2}(4.2)^3 \approx 380.53.$$

The volume of the tank is approximately $1069.56 + 380.53 = 1450.09$ m^3.

Exercises 9.12

3. $\oint_C -y^2\, dx + x^2\, dy = \int_0^{2\pi} (-9\sin^2 t)(-3\sin t)\, dt + \int_0^{2\pi} 9\cos^2 t(3\cos t)\, dt$

$$= 27 \int_0^{2\pi} [(1 - \cos^2 t)\sin t + (1 - \sin^2 t)\cos t]\, dt$$

$$= 27\left(-\cos t + \frac{1}{3}\cos^3 t + \sin t - \frac{1}{3}\sin^3 t\right)\Big|_0^{2\pi} = 27(0) = 0$$

$\iint_R (2x + 2y)\, dA = 2 \int_0^{2\pi} \int_0^3 (r\cos\theta + r\sin\theta)r\, dr\, d\theta = 2 \int_0^{2\pi} \int_0^3 r^2(\cos\theta + \sin\theta)\, dr\, d\theta$

$$= 2 \int_0^{2\pi} \left[\frac{1}{3}r^3(\cos\theta + \sin\theta)\right]\Big|_0^3 d\theta = 18 \int_0^{2\pi} (\cos\theta + \sin\theta)\, d\theta$$

$$= 18(\sin\theta - \cos\theta)\Big|_0^{2\pi} = 18(0) = 0$$

6. $P = x + y^2$, $P_y = 2y$, $Q = 2x^2 - y$, $Q_x = 4x$

$\oint_C (x + y^2)\, dx + (2x^2 - y)\, dy = \iint_R (4x - 2y)\, dA = \int_{-2}^2 \int_{x^2}^4 (4x - 2y)\, dy\, dx$

$$= \int_{-2}^2 (4xy - y^2)\Big|_{x^2}^4 dx = \int_{-2}^2 (16x - 16 - 4x^3 + x^4)\, dx$$

$$= \left(8x^2 - 16x - x^4 + \frac{1}{5}x^5\right)\Big|_{-2}^2 = -\frac{96}{5}$$

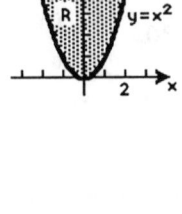

9. $P = 2xy$, $P_y = 2x$, $Q = 3xy^2$, $Q_x = 3y^2$

$\oint_C 2xy\, dx + 3xy^2\, dy = \iint_R (3y^2 - 2x)\, dA = \int_1^2 \int_2^{2x} (3y^2 - 2x)\, dy\, dx$

$$= \int_1^2 (y^3 - 2xy)\Big|_2^{2x} dx = \int_1^2 (8x^3 - 4x^2 - 8 + 4x)\, dx$$

$$= \left(2x^4 - \frac{4}{3}x^3 - 8x + 2x^2\right)\Big|_1^2 = \frac{40}{3} - \left(-\frac{16}{3}\right) = \frac{56}{3}$$

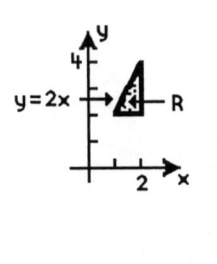

12. $P = e^{x^2}$, $P_y = 0$, $Q = 2\tan^{-1} x$, $Q_x = \dfrac{2}{1+x^2}$

$$\oint_C e^{x^2}\,dx + 2\tan^{-1} x\,dy = \iint_R \frac{2}{1+x^2}\,dA = \int_{-1}^0 \int_{-x}^1 \frac{2}{1+x^2}\,dy\,dx$$

$$= \int_{-1}^0 \left(\frac{2y}{1+x^2}\right)\Big|_{-x}^1 \,dx = \int_{-1}^0 \left(\frac{2}{1+x^2} + \frac{2x}{1+x^2}\right) dx$$

$$= [2\tan^{-1} x + \ln(1+x^2)]\Big|_{-1}^0 = 0 - \left(-\frac{\pi}{2} + \ln 2\right) = \frac{\pi}{2} - \ln 2$$

15. $P = ay$, $P_y = a$, $Q = bx$, $Q_x = b$. $\oint_C ay\,dx + bx\,dy = \iint_R (b-a)\,dA = (b-a) \times$ (area bounded by C)

18. $P = -y$, $P_y = -1$, $Q = x$, $Q_x = 1$. $\dfrac{1}{2}\oint_C -y\,dx + x\,dy = \dfrac{1}{2}\iint_R 2\,dA = \iint_R dA =$ area of R

21. (a) Parameterize C by $x = x_1 + (x_2 - x_1)t$ and $y = y_1 + (y_2 - y_1)t$ for $0 \le t \le 1$. Then

$$\int_C -y\,dx + x\,dy = \int_0^1 -[y_1 + (y_2 - y_1)t](x_2 - x_1)\,dt + \int_0^1 [x_1 + (x_2 - x_1)t](y_2 - y_1)\,dt$$

$$= -(x_2 - x_1)\left[y_1 t + \frac{1}{2}(y_2 - y_1)t^2\right]\Big|_0^1 + (y_2 - y_1)\left[x_1 t + \frac{1}{2}(x_2 - x_1)t^2\right]\Big|_0^1$$

$$= -(x_2 - x_1)\left[y_1 + \frac{1}{2}(y_2 - y_1)\right] + (y_2 - y_1)\left[x_1 + \frac{1}{2}(x_2 - x_1)\right] = x_1 y_2 - x_2 y_1.$$

(b) Let C_i be the line segment from (x_i, y_i) to (x_{i+1}, y_{i+1}) for $i = 1, 2, \ldots, n-1$, and C_2 the line segment from (x_n, y_n) to (x_1, y_1). Then

$$A = \frac{1}{2}\oint_C -y\,dx + x\,dy \qquad \boxed{\text{Using Problem 18}}$$

$$= \frac{1}{2}\left[\int_{C_1} -y\,dx + x\,dy + \int_{C_2} -y\,dx + x\,dy + \cdots + \int_{C_{n-1}} -y\,dx + x\,dy + \int_{C_n} -y\,dx + x\,dy\right]$$

$$= \frac{1}{2}(x_1 y_2 - x_2 y_1) + \frac{1}{2}(x_2 y_3 - x_3 y_2) + \frac{1}{2}(x_{n-1} y_n - x_n y_{n-1}) + \frac{1}{2}(x_n y_1 - x_1 y_n).$$

24. $P = \cos x^2 - y$, $P_y = -1$; $Q = \sqrt{y^3 + 1}$, $Q_x = 0$

$$\oint_C (\cos x^2 - y)\,dx + \sqrt{y^3 + 1}\,dy = \iint_R (0 + 1)\,dA \iint_R dA = (6\sqrt{2})^2 - \pi(2)(4) = 72 - 8\pi$$

27. Writing $\iint_R x^2\,dA = \iint_R (Q_x - P_y)\,dA$ we identify $Q = 0$ and $P = -x^2 y$. Then, with C: $x = 3\cos t$, $y = 2\sin t$, $0 \le t \le 2\pi$, we have

$$\iint_R x^2\,dA = \oint_C P\,dx + Q\,dy = \oint_C -x^2 y\,dx = -\int_0^{2\pi} 9\cos^2 t(2\sin t)(-3\sin t)\,dt$$

$$= \frac{54}{4}\int_0^{2\pi} 4\sin^2 t\cos^2 t\,dt = \frac{27}{2}\int_0^{2\pi} \sin^2 2t\,dt = \frac{27}{4}\int_0^{2\pi} (1 - \cos 4t)\,dt$$

$$= \frac{27}{4}\left(t - \frac{1}{4}\sin 4t\right)\Big|_0^{2\pi} = \frac{27\pi}{2}.$$

30. $P = -xy^2$, $P_y = -2xy$, $Q = x^2 y$, $Q_x = 2xy$. Using polar coordinates,

$$W = \oint_C \mathbf{F} \cdot d\mathbf{r} = \iint_R 4xy \, dA = \int_0^{\pi/2} \int_1^2 4(r\cos\theta)(r\sin\theta) r \, dr \, d\theta = \int_0^{\pi/2} (r^4 \cos\theta \sin\theta) \Big|_1^2 \, d\theta$$

$$= 15 \int_0^{\pi/2} \sin\theta \cos\theta \, d\theta = \frac{15}{2} \sin^2\theta \Big|_0^{\pi/2} = \frac{15}{2}.$$

33. Using Green's Theorem,

$$W = \oint_C \mathbf{F} \cdot d\mathbf{r} = \oint_C -y \, dx + x \, dy = \iint_R 2 \, dA = 2 \int_0^{2\pi} \int_0^{1+\cos\theta} r \, dr \, d\theta$$

$$= 2 \int_0^{2\pi} \left(\frac{1}{2} r^2 \right) \Big|_0^{1+\cos\theta} d\theta = \int_0^{2\pi} (1 + 2\cos\theta + \cos^2\theta) \, d\theta$$

$$= \left(\theta + 2\sin\theta + \frac{1}{2}\theta + \frac{1}{4}\sin 2\theta \right) \Big|_0^{2\pi} = 3\pi.$$

Exercises 9.13

3. Using $f(x,y) = z = \sqrt{16 - x^2}$ we see that for $0 \le x \le 2$ and $0 \le y \le 5$, $z > 0$.
Thus, the surface is entirely above the region. Now $f_x = -\dfrac{x}{\sqrt{16-x^2}}$, $f_y = 0$,

$$1 + f_x^2 + f_y^2 = 1 + \frac{x^2}{16 - x^2} = \frac{16}{16 - x^2} \quad \text{and}$$

$$A = \int_0^5 \int_0^2 \frac{4}{\sqrt{16 - x^2}} \, dx \, dy = 4 \int_0^5 \sin^{-1}\frac{x}{4} \Big|_0^2 \, dy = 4 \int_0^5 \frac{\pi}{6} \, dy = \frac{10\pi}{3}.$$

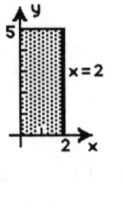

6. The surfaces $x^2 + y^2 + z^2 = 2$ and $z^2 = x^2 + y^2$ intersect on the cylinder $2x^2 + 2y^2 = 2$ or $x^2 + y^2 = 1$. There are portions of the sphere within the cone both above and below the xy-plane. Using $f(x,y) = \sqrt{2 - x^2 - y^2}$ we have $f_x = -\dfrac{x}{\sqrt{2 - x^2 - y^2}}$,

$f_y = -\dfrac{y}{\sqrt{2 - x^2 - y^2}}$, $1 + f_x^2 + f_y^2 = \dfrac{2}{2 - x^2 - y^2}$. Then

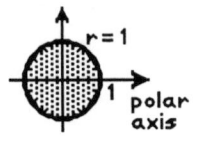

$$A = 2 \left[\int_0^{2\pi} \int_0^1 \frac{\sqrt{2}}{\sqrt{2 - r^2}} r \, dr \, d\theta \right] = 2\sqrt{2} \int_0^{2\pi} -\sqrt{2 - r^2} \Big|_0^1 \, d\theta$$

$$= 2\sqrt{2} \int_0^{2\pi} (\sqrt{2} - 1) \, d\theta = 4\pi\sqrt{2}(\sqrt{2} - 1).$$

9. There are portions of the sphere within the cylinder both above and below the xy-plane. Using $f(x,y) = z = \sqrt{a^2 - x^2 - y^2}$ we have $f_x = -\dfrac{x}{\sqrt{1^2 - x^2 - y^2}}$, $f_y = -\dfrac{y}{\sqrt{a^2 - x^2 - y^2}}$,

$1 + f_x^2 + f_y^2 = \dfrac{a^2}{a^2 - x^2 - y^2}$. Then, using symmetry,

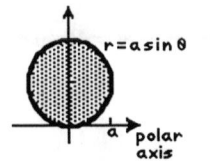

$$A = 2 \left[2 \int_0^{\pi/2} \int_0^{a\sin\theta} \frac{a}{\sqrt{a^2 - r^2}} r \, dr \, d\theta \right] = 4a \int_0^{\pi/2} -\sqrt{a^2 - r^2} \Big|_0^{a\sin\theta} \, d\theta$$

$$= 4a \int_0^{\pi/2} (a - a\sqrt{1 - \sin^2\theta}) \, d\theta = 4a^2 \int_0^{\pi/2} (1 - \cos\theta) \, d\theta$$

$$= 4a^2(\theta - \sin\theta)\Big|_0^{\pi/2} = 4a^2\left(\frac{\pi}{2} - 1\right) = 2a^2(\pi - 2).$$

12. From Example 1, the area of the portion of the hemisphere within $x^2 + y^2 = b^2$ is $2\pi a(a - \sqrt{a^2 - b^2})$. Thus, the area of the sphere is $A = 2\lim_{b\to a} 2\pi a(a - \sqrt{a^2 - b^2}) = 2(2\pi a^2) = 4\pi a^2$.

15. $z_x = -2x$, $z_y = 0$; $dS = \sqrt{1 + 4x^2}\, dA$

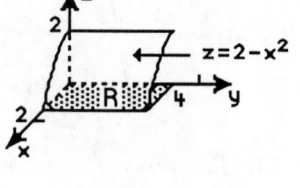

$$\iint_S x\, dS = \int_0^4 \int_0^{\sqrt{2}} x\sqrt{1 + 4x^2}\, dx\, dy = \int_0^4 \frac{1}{12}(1 + 4x^2)^{3/2}\Big|_0^{\sqrt{2}}\, dy$$

$$= \int_0^4 \frac{13}{6}\, dy = \frac{26}{3}$$

18. $z_x = \dfrac{x}{\sqrt{x^2 + y^2}}$, $z_y = \dfrac{y}{\sqrt{x^2 + y^2}}$; $dS = \sqrt{2}\, dA$.

Using polar coordinates,

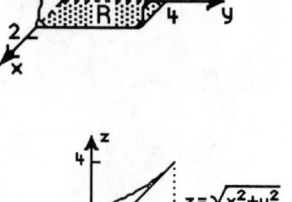

$$\iint_S (x + y + z)\, dS = \iint_R (x + y + \sqrt{x^2 + y^2})\sqrt{2}\, dA$$

$$= \sqrt{2}\int_0^{2\pi}\int_1^4 (r\cos\theta + r\sin\theta + r)r\, dr\, d\theta$$

$$= \sqrt{2}\int_0^{2\pi}\int_1^4 r^2(1 + \cos\theta + \sin\theta)\, dr\, d\theta = \sqrt{2}\int_0^{2\pi} \frac{1}{3}r^3(1 + \cos\theta + \sin\theta)\Big|_1^4\, d\theta$$

$$= \frac{63\sqrt{2}}{3}\int_0^{2\pi}(1 + \cos\theta + \sin\theta)\, d\theta = 21\sqrt{2}(\theta + \sin\theta - \cos\theta)\Big|_0^{2\pi} = 42\sqrt{2}\,\pi.$$

21. $z_x = -x$, $z_y = -y$; $dS = \sqrt{1 + x^2 + y^2}\, dA$

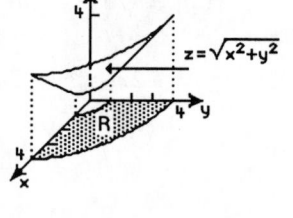

$$\iint_S xy\, dS = \int_0^1\int_0^1 xy\sqrt{1 + x^2 + y^2}\, dx\, dy = \int_0^1 \frac{1}{3}y(1 + x^2 + y^2)^{3/2}\Big|_0^1\, dy$$

$$= \int_0^1 \left[\frac{1}{3}y(2 + y^2)^{3/2} - \frac{1}{3}y(1 + y^2)^{3/2}\right] dy$$

$$= \left[\frac{1}{15}(2 + y^2)^{5/2} - \frac{1}{15}(1 + y^2)^{5/2}\right]\Big|_0^1 = \frac{1}{15}(3^{5/2} - 2^{7/2} + 1)$$

24. $x_y = -2y$, $x_z = -2z$; $dS = \sqrt{1 + 4y^2 + 4z^2}\, dA$.

Using polar coordinates,

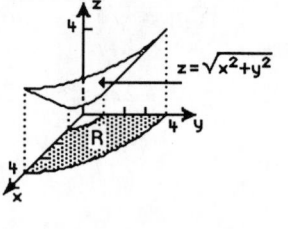

$$\iint_S (1 + 4y^2 + 4z^2)^{1/2}\, dS = \int_0^{\pi/2}\int_1^2 (1 + 4r^2)r\, dr\, d\theta$$

$$= \int_0^{\pi/2} \frac{1}{16}(1 + 4r^2)^2\Big|_1^2\, d\theta = \frac{1}{16}\int_0^{\pi/2} 12\, d\theta = \frac{3\pi}{8}.$$

27. The density is $\rho = kx^2$. The surface is $z = 1 - x - y$. Then $z_x = -1$, $z_y = -1$; $dS = \sqrt{3}\, dA$.

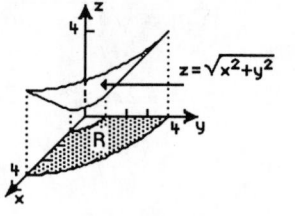

$$m = \iint_S kx^2\, dS = k\int_0^1\int_0^{1-x} x^2\sqrt{3}\, dy\, dx = \sqrt{3}\,k\int_0^1 \frac{1}{3}x^3\Big|_0^{1-x}\, dx$$

$$= \frac{\sqrt{3}}{3}k\int_0^1 (1 - x)^3\, dx = \frac{\sqrt{3}}{3}k\left[-\frac{1}{4}(1 - x)^4\right]\Big|_0^1 = \frac{\sqrt{3}}{12}k$$

30. The surface is $g(x, y, z) = x^2 + y^2 + z - 5 = 0$. $\nabla g = 2x\mathbf{i} + 2y\mathbf{j} + \mathbf{k}$,

$$|\nabla g| = \sqrt{1 + 4x^2 + 4y^2}\,; \quad \mathbf{n} = \frac{2x\mathbf{i} + 2y\mathbf{j} + \mathbf{k}}{\sqrt{1 + 4x^2 + 4y^2}}\,; \quad \mathbf{F} \cdot \mathbf{n} = \frac{z}{\sqrt{1 + 4x^2 + 4y^2}}\,;$$

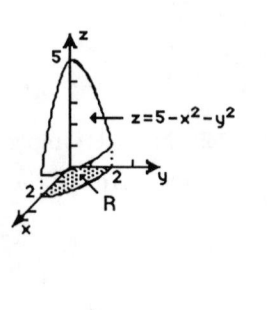

$z_x = -2x$, $z_y = -2y$, $dS = \sqrt{1 + 4x^2 + 4y^2}\, dA$. Using polar coordinates,

$$\text{Flux} = \iint_S \mathbf{F} \cdot \mathbf{n}\, dS = \iint_R \frac{z}{\sqrt{1 + 4x^2 + 4y^2}}\, \sqrt{1 + 4x^2 + 4y^2}\, dA = \iint_R (5 - x^2 - y^2)\, dA$$

$$= \int_0^{2\pi} \int_0^2 (5 - r^2) r\, dr\, d\theta = \int_0^{2\pi} \left(\frac{5}{2}r^2 - \frac{1}{4}r^4 \right) \Big|_0^2\, d\theta = \int_0^{2\pi} 6\, d\theta = 12\pi.$$

33. The surface is $g(x, y, z) = x^2 + y^2 + z - 4$. $\nabla g = 2x\mathbf{i} + 2y\mathbf{j} + \mathbf{k}$,

$$|\nabla g| = \sqrt{4x^2 + 4y^2 + 1}\,; \quad \mathbf{n} = \frac{2x\mathbf{i} + 2y\mathbf{j} + \mathbf{k}}{\sqrt{4x^2 + 4y^2 + 1}}\,; \quad \mathbf{F} \cdot \mathbf{n} = \frac{x^3 + y^3 + z}{\sqrt{4x^2 + 4y^2 + 1}}\,;$$

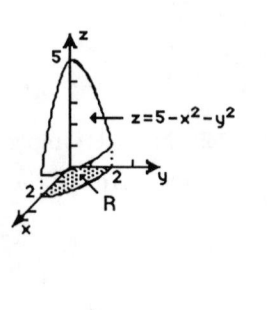

$z_x = -2x$, $z_y = -2y$, $dS = \sqrt{1 + 4x^2 + 4y^2}\, dA$. Using polar coordinates,

$$\text{Flux} = \iint_S \mathbf{F} \cdot \mathbf{n}\, dS = \iint_R (x^3 + y^3 + z)\, dA = \iint_R (4 - x^2 - y^2 + x^3 + y^3)\, dA$$

$$= \int_0^{2\pi} \int_0^2 (4 - r^2 + r^3 \cos^3 \theta + r^3 \sin^3 \theta) r\, dr\, d\theta$$

$$= \int_0^{2\pi} \left(2r^2 - \frac{1}{4}r^4 + \frac{1}{5}r^5 \cos^3 \theta + \frac{1}{5}r^5 \sin^3 \theta \right) \Big|_0^2\, d\theta$$

$$= \int_0^{2\pi} \left(4 + \frac{32}{5} \cos^3 \theta + \frac{32}{5} \sin^3 \theta \right) d\theta = 4\theta \Big|_0^{2\pi} + 0 + 0 = 8\pi.$$

36. For S_1: $g(x, y, z) = x^2 + y^2 + z - 4$, $\nabla g = 2x\mathbf{i} + 2y\mathbf{j} + \mathbf{k}$, $|\nabla g| = \sqrt{4x^2 + 4y^2 + 1}\,$;

$$\mathbf{n}_1 = \frac{2x\mathbf{i} + 2y\mathbf{j} + \mathbf{k}}{\sqrt{4x^2 + 4y^2 + 1}}\,; \quad \mathbf{F} \cdot \mathbf{n}_1 = 6z^2 / \sqrt{4x^2 + 4y^2 + 1}\,; \ z_x = -2x, \ z_y = -2y,$$

$dS_1 = \sqrt{1 + 4x^2 + 4y^2}\, dA$. For S_2: $g(x, y, z) = x^2 + y^2 - z$, $\nabla g = 2x\mathbf{i} + 2y\mathbf{j} - \mathbf{k}$,

$$|\nabla g| = \sqrt{4x^2 + 4y^2 + 1}\,; \quad \mathbf{n}_2 = \frac{2x\mathbf{i} + 2y\mathbf{j} - \mathbf{k}}{\sqrt{4x^2 + y^2 + 1}}\,; \ \mathbf{F} \cdot \mathbf{n}_2 = -6z^2 / \sqrt{4x^2 + 4y^2 + 1}\,; \ z_x = 2x, \ z_y = 2y,$$

$dS_2 = \sqrt{1 + 4x^2 + 4y^2}\, dA$. Using polar coordinates and R: $x^2 + y^2 \le 2$ we have

$$\text{Flux} = \iint_{S_1} \mathbf{F} \cdot \mathbf{n}_1\, dS_1 + \iint_{S_1} \mathbf{F} \cdot \mathbf{n}_2\, dS_2 = \iint_R 6z^2\, dA + \iint -6z^2\, dA$$

$$= \iint_R [6(4 - x^2 - y^2)^2 - 6(x^2 + y^2)^2]\, dA = 6 \int_0^{2\pi} \int_0^{\sqrt{2}} [(4 - r^2)^2 - r^4] r\, dr\, d\theta$$

$$= 6 \int_0^{2\pi} \left[-\frac{1}{6}(4 - r^2)^3 - \frac{1}{6}r^6 \right] \Big|_0^{\sqrt{2}}\, d\theta = -\int_0^{2\pi} [(2^3 - 4^3) + (\sqrt{2})^6]\, d\theta = \int_0^{2\pi} 48\, d\theta = 96\pi.$$

39. Using $g(x, y, z) = x^2 + y^2 + z^2 - a^2 = 0$, we have $\mathbf{n} = \nabla g / |\nabla g| = \dfrac{x\mathbf{i} + y\mathbf{j} + z\mathbf{k}}{\sqrt{x^2 + y^2 + z^2}}$ and $dS = \dfrac{a}{\sqrt{a^2 - x^2 - y^2}}\, dA$.

Then

$$\mathbf{F} \cdot \mathbf{n} = kq\, \frac{\mathbf{r}}{|\mathbf{r}|^3} \cdot \frac{\mathbf{r}}{|\mathbf{r}|} = \frac{kq}{|\mathbf{r}|^4} |\mathbf{r}|^2 = \frac{kq}{|\mathbf{r}|^2} = \frac{kq}{x^2 + y^2 + z^2} = \frac{kq}{a^2}$$

and

$$\text{Flux} = \iint_S \mathbf{F} \cdot \mathbf{n} \, dS = \iint_S \frac{kq}{a^2} \, dS = \frac{kq}{a^2} \times \text{area} = \frac{kq}{a^2}(4\pi a^2) = 4\pi kq.$$

42. The area of the hemisphere is $A(s) = 2\pi a^2$. By symmetry, $\bar{x} = \bar{y} = 0$.

$$z_x = -\frac{x}{\sqrt{a^2 - x^2 - y^2}}, \; z_y = -\frac{y}{\sqrt{a^2 - x^2 - y^2}};$$

$$dS = \sqrt{1 + \frac{x^2}{a^2 - x^2 - y^2} + \frac{y^2}{a^2 - x^2 - y^2}} \, dA = \frac{a}{\sqrt{a^2 - x^2 - y^2}} \, dA$$

Using polar coordinates,

$$z = \iint_S \frac{z \, dS}{2\pi a^2} = \frac{1}{2\pi a^2} \iint_R \sqrt{a^2 - x^2 - y^2} \, \frac{a}{\sqrt{a^2 - x^2 - y^2}} \, dA = \frac{1}{2\pi a} \int_0^{2\pi} \int_0^a r \, dr \, d\theta$$

$$= \frac{1}{2\pi a} \int_0^{2\pi} \frac{1}{2} r^2 \Big|_0^a \, d\theta = \frac{1}{2\pi a} \int_0^{2\pi} \frac{1}{2} s^2 \, d\theta = \frac{a}{2}.$$

The centroid is $(0, 0, a/2)$.

Exercises 9.14

3. Surface Integral: curl $\mathbf{F} = \mathbf{i} + \mathbf{j} + \mathbf{k}$. Letting $g(x, y, z) = 2x + y + 2z - 6$, we have
$\nabla g = 2\mathbf{i} + \mathbf{j} + 2\mathbf{k}$ and $\mathbf{n} = (2\mathbf{i} + \mathbf{j} + 2\mathbf{k})/3$. Then $\iint_S (\text{curl } \mathbf{F}) \cdot \mathbf{n} \, dS = \iint_S \frac{5}{3} \, dS$. Letting
the surface be $z = 3 - \frac{1}{2}y - x$ we have $z_x = -1$, $z_y = -\frac{1}{2}$, and
$dS = \sqrt{1 + (-1)^2 + (-\frac{1}{2})^2} \, dA = \frac{3}{2} \, dA$. Then

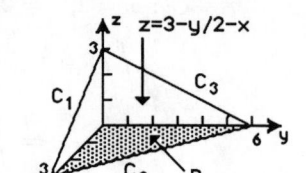

$$\iint_S (\text{curl } \mathbf{F}) \cdot \mathbf{n} \, dS = \iint_R \frac{5}{3}\left(\frac{3}{2}\right) dA = \frac{5}{2} \times (\text{area of } R) = \frac{5}{2}(9) = \frac{45}{2}.$$

Line Integral: C_1: $z = 3 - x$, $0 \leq x \leq 3$, $y = 0$; C_2: $y = 6 - 2x$, $3 \geq x \geq 0$, $z = 0$; C_3: $z = 3 - y/2$, $6 \geq y \geq 0$, $x = 0$.

$$\oint_C z \, dx + x \, dy + y \, dz = \iint_{C_1} z \, dx + \int_{C_2} x \, dy + \int_{C_3} y \, dz$$

$$= \int_0^3 (3 - x) \, dx + \int_3^0 x(-2 \, dx) + \int_6^0 y(-dy/2)$$

$$= \left(3x - \frac{1}{2}x^2\right) \Big|_0^3 - x^2 \Big|_3^0 - \frac{1}{4}y^2 \Big|_6^0 = \frac{9}{2} - (0 - 9) - \frac{1}{4}(0 - 36) = \frac{45}{2}$$

6. curl $\mathbf{F} = -2xz\mathbf{i} + z^2\mathbf{k}$. A unit vector normal to the plane is $\mathbf{n} = (\mathbf{j} + \mathbf{k})/\sqrt{2}$. From $z = 1 - y$, we have $z_x = 0$
and $z_y = -1$. Thus, $dS = \sqrt{1 + 1} \, dA = \sqrt{2} \, dA$ and

$$\oint_C \mathbf{F} \cdot d\mathbf{r} = \iint_S (\text{curl } \mathbf{F}) \cdot \mathbf{n} \, dS = \iint_R \frac{1}{\sqrt{2}} z^2 \sqrt{2} \, dA = \iint_R (1 - y)^2 \, dA$$

$$= \int_0^2 \int_0^1 (1 - y)^2 \, dy \, dx = \int_0^2 -\frac{1}{3}(1 - y)^3 \Big|_0^1 \, dx = \int_0^2 \frac{1}{3} \, dx = \frac{2}{3}.$$

9. curl $\mathbf{F} = (-3x^2 - 3y^2)\mathbf{k}$. A unit vector normal to the plane is $\mathbf{n} = (\mathbf{i}+\mathbf{j}+\mathbf{k})/\sqrt{3}$. From $z = 1 - x - y$, we have $z_x = z_y = -1$ and $dS = \sqrt{3}\, dA$. Then, using polar coordinates,

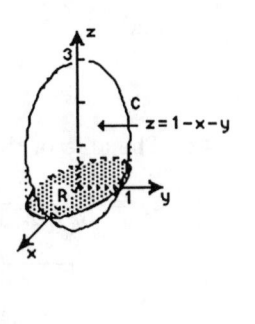

$$\oint_C \mathbf{F} \cdot d\mathbf{r} = \iint_S (\text{curl } \mathbf{F}) \cdot \mathbf{n}\, dS = \iint_R (-\sqrt{3}\,x^2 - \sqrt{3}\,y^2)\sqrt{3}\, dA$$

$$= 3 \iint_R (-x^2 - y^2)\, dA = 3 \int_0^{2\pi} \int_0^1 (-r^2) r\, dr\, d\theta$$

$$= 3 \int_0^{2\pi} -\frac{1}{4}r^4 \bigg|_0^1 d\theta = 3 \int_0^{2\pi} -\frac{1}{4}\, d\theta = -\frac{3\pi}{2}.$$

12. curl $\mathbf{F} = \mathbf{i} + \mathbf{j} + \mathbf{k}$. Taking the surface S bounded by C to be the portion of the plane $x + y + z = 0$ inside C, we have $\mathbf{n} = (\mathbf{i}+\mathbf{j}+\mathbf{k})/\sqrt{3}$ and $dS = \sqrt{3}\, dA$.

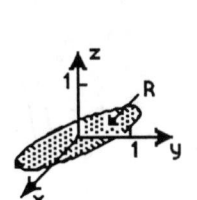

$$\oint_C \mathbf{F} \cdot d\mathbf{r} = \iint_S (\text{curl } \mathbf{F}) \cdot \mathbf{n}\, dS = \iint_S \sqrt{3}\, dS = \sqrt{3} \iint_R \sqrt{3}\, dA = 3 \times (\text{area of } R)$$

The region R is obtained by eliminating z from the equations of the plane and the sphere. This gives $x^2 + xy + y^2 = \frac{1}{2}$. Rotating axes, we see that R is enclosed by the ellipse $X^2/(1/3) + Y^2/1 = 1$ in a rotated coordinate system. Thus,

$$\oint_C \mathbf{F} \cdot d\mathbf{r} = 3 \times (\text{area of } R) = 3\left(\pi \frac{1}{\sqrt{3}} 1\right) = \sqrt{3}\,\pi.$$

15. Parameterize C by C_1: $x = 0$, $z = 0$, $2 \ge y \ge 0$; C_2: $z = x$, $y = 0$, $0 \le x \le 2$; C_3: $x = 2$, $z = 2$, $0 \le y \le 2$; C_4: $z = x$, $y = 2$, $2 \ge x \ge 0$. Then

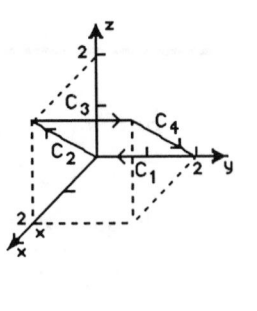

$$\iint_S (\text{curl } \mathbf{F}) \cdot \mathbf{n}\, dS = \oint_C \mathbf{F} \cdot \mathbf{r} = \oint_C 3x^2\, dx + 8x^3 y\, dy + 3x^2 y\, dz$$

$$= \int_{C_1} 0\, dx + 0\, dy + 0\, dz + \int_{C_2} 3x^2\, dx + \int_{C_3} 64\, dy + \int_{C_4} 3x^2\, dx + 6x^2\, dx$$

$$= \int_0^2 3x^2\, dx + \int_0^2 64\, dy + \int_2^0 9x^2\, dx = x^3 \bigg|_0^2 + 64y \bigg|_0^2 + 3x^3 \bigg|_2^0 = 112.$$

18. (a) curl $\mathbf{F} = xz\mathbf{i} - yz\mathbf{j}$. A unit vector normal to the surface is $\mathbf{n} = \dfrac{2x\mathbf{i} + 2y\mathbf{j} + \mathbf{k}}{\sqrt{4x^2 + 4y^2 + 1}}$ and

$dS = \sqrt{1 + 4x^2 + 4y^2}\, dA$. Then, using $x = \cos t$, $y = \sin t$, $0 \le t \le 2\pi$, we have

$$\iint_S (\text{curl } \mathbf{F}) \cdot \mathbf{n}\, dS = \iint_R (2x^2 z - 2y^2 z)\, dA = \iint_R (2x^2 - 2y^2)(1 - x^2 - y^2)\, dA$$

$$= \iint_R (2x^2 - 2y^2 - 2x^4 + 2y^4)\, dA$$

$$= \int_0^{2\pi} \int_0^1 (2r^2 \cos^2\theta - 2r^2 \sin^2\theta - 2r^4 \cos^4\theta + 2r^4 \cos^4\theta) r\, dr\, d\theta$$

$$= 2 \int_0^{2\pi} \int_0^1 [r^3 \cos 2\theta - r^5 (\cos^2\theta - \sin^2\theta)(\cos^2\theta + \sin^2\theta)]\, dr\, d\theta$$

$$= 2 \int_0^{2\pi} \int_0^1 (r^3 \cos 2\theta - r^5 \cos 2\theta)\, dr\, d\theta = 2 \int_0^{2\pi} \cos 2\theta \left(\frac{1}{4}r^4 - \frac{1}{6}r^6\right) \bigg|_0^1 d\theta$$

$$= \frac{1}{6} \int_0^{2\pi} \cos 2\theta\, d\theta = 0.$$

(b) We take the surface to be $z = 0$. Then $\mathbf{n} = \mathbf{k}$, curl $\mathbf{F} \cdot \mathbf{n} = $ curl $\mathbf{F} \cdot \mathbf{k} = 0$ and $\displaystyle\iint_S (\text{curl } \mathbf{F}) \cdot \mathbf{n}\, dS = 0$.

(c) By Stoke's Theorem, using $z = 0$, we have

$$\iint_S (\text{curl } \mathbf{F}) \cdot \mathbf{n}\, dS = \oint_C \mathbf{F} \cdot d\mathbf{r} = \oint_C xyz\, dz = \oint_C xy(0)\, dz = 0.$$

Exercises 9.15

3. $\displaystyle\int_0^6 \int_0^{6-x} \int_0^{6-x-z} dy\, dz\, dx = \int_0^6 \int_0^{6-x} (6 - x - z)dz\, dx = \int_0^6 \left(6z - xz - \frac{1}{2}z^2 \right) \Big|_0^{6-x} dx$

$\displaystyle = \int_0^6 \left[6(6 - x) - x(6 - x) - \frac{1}{2}(6 - x)^2 \right] dx = \int_0^6 \left(18 - 6x + \frac{1}{2}x^2 \right) dx$

$\displaystyle = \left(18x - 3x^2 + \frac{1}{6}x^3 \right) \Big|_0^6 = 36$

6. $\displaystyle\int_0^{\sqrt{2}} \int_{\sqrt{y}}^2 \int_0^{e^{x^2}} x\, dz\, dx\, dy = \int_0^{\sqrt{2}} \int_{\sqrt{y}}^2 xe^{x^2}\, dx\, dy = \int_0^{\sqrt{2}} \frac{1}{2}e^{x^2} \Big|_{\sqrt{y}}^2 dy = \frac{1}{2}\int_0^{\sqrt{2}} (e^4 - e^y)dy$

$\displaystyle = \frac{1}{2}(ye^4 - e^y) \Big|_0^{\sqrt{2}} = \frac{1}{2}[(e^4\sqrt{2} - e^{\sqrt{2}}) - (-1)] = \frac{1}{2}(1 + e^4\sqrt{2} - e^{\sqrt{2}})$

9. $\displaystyle\iiint_D z\, dV = \int_0^5 \int_1^3 \int_y^{y+2} z\, dx\, dy\, dz = \int_0^5 \int_1^3 xz \Big|_y^{y+2} dy\, dz = \int_0^5 \int_1^3 2z\, dy\, dz$

$\displaystyle = \int_0^5 2yz \Big|_1^3 dz = \int_0^5 4z\, dz = 2z^2 \Big|_0^5 = 50$

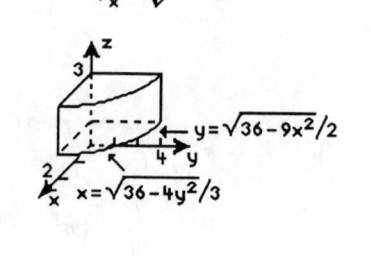

12. The other five integrals are $\displaystyle\int_0^3 \int_0^{\sqrt{36-4y^2}/3} \int_1^3 F(x, y, z)\, dz\, dx\, dy,$

$\displaystyle\int_1^3 \int_0^2 \int_0^{\sqrt{36-9x^2}/2} F(x, y, z)\, dy\, dx\, dz, \quad \int_1^3 \int_0^3 \int_0^{\sqrt{36-4y^2}/3} F(x, y, z)\, dx\, dy\, dz,$

$\displaystyle\int_0^3 \int_1^3 \int_0^{\sqrt{36-4y^2}/3} F(x, y, z)\, dx\, dz\, dy, \quad \int_0^2 \int_1^3 \int_0^{\sqrt{36-9x^2}/2} F(x, y, z)\, dy\, dz\, dx.$

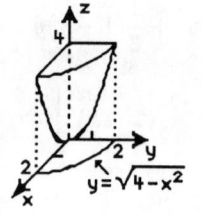

15.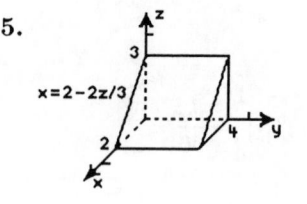

18.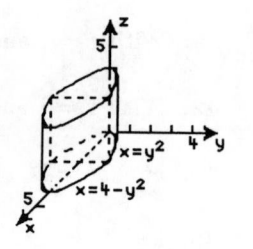

21. Solving $x = y^2$ and $4 - x = y^2$, we obtain $x = 2$, $y = \pm\sqrt{2}$. Using symmetry,

$$V = 2\int_0^3 \int_0^{\sqrt{2}} \int_{y^2}^{4-y^2} dx\, dy\, dz = 2\int_0^3 \int_0^{\sqrt{2}} (4 - 2y^2)dy\, dz$$

$$= 2\int_0^3 \left(4y - \frac{2}{3}y^3 \right) \Big|_0^{\sqrt{2}} dz = 2\int_0^3 \frac{8\sqrt{2}}{3}\, dz = 16\sqrt{2}.$$

24. Solving $x = 2$, $y = x$, and $z = x^2 + y^2$, we obtain the point $(2, 2, 8)$.

$$V = \int_0^2 \int_0^x \int_0^{x^2+y^2} dz\, dy\, dx = \int_0^2 \int_0^x (x^2 + y^2)\, dy\, dx = \int_0^2 \left(x^2 y + \frac{1}{3} y^3 \right) \Big|_0^x dx$$

$$= \int_0^2 \frac{4}{3} x^3\, dx = \frac{1}{3} x^4 \Big|_0^2 = \frac{16}{3}.$$

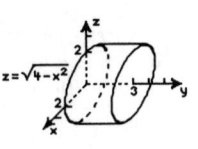

27. The density is $\rho(x, y, z) = ky$. Since both the region and the density function are symmetric with respect to the xy-and yz-planes, $\bar{x} = \bar{z} = 0$. Using symmetry,

$$m = 4 \int_0^3 \int_0^2 \int_0^{\sqrt{4-x^2}} ky\, dz\, dx\, dy = 4k \int_0^3 \int_0^2 yz \Big|_0^{\sqrt{4-x^2}} dx\, dy = 4k \int_0^3 \int_0^2 y\sqrt{4 - x^2}\, dx\, dy$$

$$= 4k \int_0^3 y \left(\frac{x}{2} \sqrt{4 - x^2} + 2\sin^{-1} \frac{x}{2} \right) \Big|_0^2 dy = 4k \int_0^3 \pi y\, dy = 4\pi k \left(\frac{1}{2} y^2 \right) \Big|_0^3 = 18\pi k$$

$$M_{xz} = 4 \int_0^3 \int_0^2 \int_0^{\sqrt{4-x^2}} ky^2\, dz\, dx\, dy = 4k \int_0^3 \int_0^2 y^2 z \Big|_0^{\sqrt{4-x^2}} dx\, dy = 4k \int_0^3 \int_0^2 y^2 \sqrt{4 - x^2}\, dx\, dy$$

$$= 4k \int_0^3 y^2 \left(\frac{x}{2} \sqrt{4 - x^2} + 2\sin^{-1} \frac{x}{2} \right) \Big|_0^2 dy = 4k \int_0^3 \pi y^2\, dy = 4\pi k \left(\frac{1}{3} y^3 \right) \Big|_0^3 = 36\pi k.$$

$\bar{y} = M_{xz}/m = \dfrac{36\pi k}{18\pi k} = 2$. The center of mass is $(0, 2, 0)$.

30. Both the region and the density function are symmetric with respect to the xz- and yz-planes. Thus, $m = 4 \displaystyle\int_{-1}^2 \int_0^{\sqrt{1+z^2}} \int_0^{\sqrt{1+z^2-y^2}} z^2\, dx\, dy\, dz$.

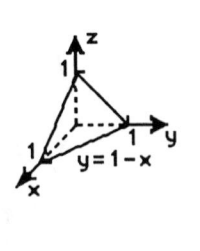

33. $I_z = k \displaystyle\int_0^1 \int_0^{1-x} \int_0^{1-x-y} (x^2 + y^2)\, dz\, dy\, dx = k \int_0^1 \int_0^{1-x} (x^2 + y^2)(1 - x - y)\, dy\, dx$

$$= k \int_0^1 \int_0^{1-x} (x^2 - x^3 - x^2 y + y^2 - xy^2 - y^3)\, dy\, dx$$

$$= k \int_0^1 \left[(x^2 - x^3)y - \frac{1}{2} x^2 y^2 + \frac{1}{3}(1 - x)y^3 - \frac{1}{4} y^4 \right] \Big|_0^{1-x} dx$$

$$= k \int_0^1 \left[\frac{1}{2} x^2 - x^3 + \frac{1}{2} x^4 + \frac{1}{12}(1 - x)^4 \right] dx = k \left[\frac{1}{6} x^6 - \frac{1}{4} x^4 + \frac{1}{10} x^5 - \frac{1}{60}(1 - x)^5 \right] \Big|_0^1 = \frac{k}{30}$$

36. $x = 2\cos 5\pi/6 = -\sqrt{3}$; $y = 2\sin 5\pi/6 = 1$; $(-\sqrt{3}, 1, -3)$

39. With $x = 1$ and $y = -1$ we have $r^2 = 2$ and $\tan\theta = -1$. The point is $(\sqrt{2}, -\pi/4, -9)$.

42. With $x = 1$ and $y = 2$ we have $r^2 = 5$ and $\tan\theta = 2$. The point is $(\sqrt{5}, \tan^{-1} 2, 7)$.

45. $r^2 - z^2 = 1$

48. $z = 2y$

51. The equations are $r^2 = 4$, $r^2 + z^2 = 16$, and $z = 0$.

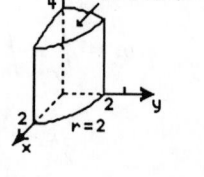

$$V = \int_0^{2\pi} \int_0^2 \int_0^{\sqrt{16-r^2}} r \, dz \, dr \, d\theta = \int_0^{2\pi} \int_0^2 r\sqrt{16-r^2} \, dr \, d\theta$$

$$= \int_0^{2\pi} -\frac{1}{3}(16-r^2)^{3/2} \Big|_0^2 \, d\theta = \int_0^{2\pi} \frac{1}{3}(64 - 24\sqrt{3}) \, d\theta = \frac{2\pi}{3}(64 - 24\sqrt{3})$$

54. Substituting the first equation into the second, we see that the surfaces intersect in the plane $y = 4$. Using polar coordinates in the xz-plane, the equations of the surfaces become $y = r^2$ and $y = \frac{1}{2}r^2 + 2$.

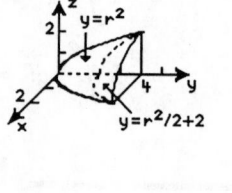

$$V = \int_0^{2\pi} \int_0^2 \int_{r^2}^{r^2/2+2} r \, dy \, dr \, d\theta = \int_0^{2\pi} \int_0^2 r\left(\frac{r^2}{2} + 2 - r^2\right) dr \, d\theta$$

$$= \int_0^{2\pi} \int_0^2 \left(2r - \frac{1}{2}r^3\right) dr \, d\theta = \int_0^{2\pi} \left(r^2 - \frac{1}{8}r^4\right)\Big|_0^2 \, d\theta = \int_0^{2\pi} 2 \, d\theta = 4\pi$$

57. The equation is $z = \sqrt{9-r^2}$ and the density is $\rho = k/r^2$. When $z = 2$, $r = \sqrt{5}$.

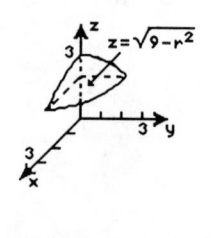

$$I_z = \int_0^{2\pi} \int_0^{\sqrt{5}} \int_2^{\sqrt{9-r^2}} r^2(k/r^2) r \, dz \, dr \, d\theta = k \int_0^{2\pi} \int_0^{\sqrt{5}} rz \Big|_2^{\sqrt{9-r^2}} \, dr \, d\theta$$

$$= k \int_0^{2\pi} \int_0^{\sqrt{5}} (r\sqrt{9-r^2} - 2r) \, dr \, d\theta = k \int_0^{2\pi} \left[-\frac{1}{3}(9-r^2)^{3/2} - r^2\right]\Big|_0^{\sqrt{5}} \, d\theta$$

$$= k \int_0^{2\pi} \frac{4}{3} \, d\theta = \frac{8}{3}\pi k$$

60. (a) $x = 5\sin(5\pi/4)\cos(2\pi/3) = 5\sqrt{2}/4$; $\quad y = 5\sin(5\pi/4)\sin(2\pi/3) = -5\sqrt{6}/4$;
   $z = 5\cos(5\pi/4) = -5\sqrt{2}/2$; $\quad (5\sqrt{2}/4, -5\sqrt{6}/4, -5\sqrt{2}/2)$

(b) With $x = 5\sqrt{2}/4$ and $y = -5\sqrt{6}/4$ we have $r^2 = 25/2$ and $\tan\theta = -\sqrt{3}$.
   The point is $(5/\sqrt{2}, 2\pi/3, -5\sqrt{2}/2)$.

63. With $x = -5$, $y = -5$, and $z = 0$, we have $\rho^2 = 50$, $\tan\theta = 1$, and $\cos\phi = 0$. The point is $(5\sqrt{2}, \pi/2, 5\pi/4)$.

66. With $x = -\sqrt{3}/2$, $y = 0$, and $z = -1/2$, we have $\rho^2 = 1$, $\tan\theta = 0$, and $\cos\phi = -1/2$.
   The point is $(1, 2\pi/3, 0)$.

69. $4z^2 = 3x^2 + 3y^2 + 3z^2$; $\quad 4\rho^2\cos^2\phi = 3\rho^2$; $\quad \cos\phi = \pm\sqrt{3}/2$; $\quad \phi = \pi/6, 5\pi/6$

72. $\cos\phi = 1/2$; $\quad \rho^2\cos^2\phi = \rho^2/4$; $\quad 4z^2 = x^2 + y^2 + z^2$; $\quad x^2 + y^2 = 3z^2$

75. The equations are $\phi = \pi/4$ and $\rho = 3$.

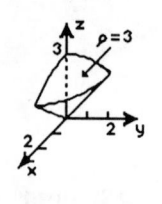

$$V = \int_0^{2\pi} \int_0^{\pi/4} \int_0^3 \rho^2 \sin\phi \, d\rho \, d\phi \, d\theta = \int_0^{2\pi} \int_0^{\pi/4} \frac{1}{3}\rho^3 \sin\phi \Big|_0^3 \, d\phi \, d\theta = \int_0^{2\pi} \int_0^{\pi/4} 9\sin\phi \, d\phi \, d\theta$$

$$= \int_0^{2\pi} -9\cos\phi \Big|_0^{\pi/4} \, d\theta = -9 \int_0^{2\pi} \left(\frac{\sqrt{2}}{2} - 1\right) d\theta = 9\pi(2 - \sqrt{2})$$

78. The equations are $\rho = 1$ and $\phi = \pi/4$. We find the volume above the xy-plane and double.

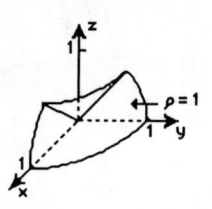

$$V = 2 \int_0^{2\pi} \int_{\pi/4}^{\pi/2} \int_0^1 \rho^2 \sin\phi \, d\rho \, d\phi \, d\theta = 2 \int_0^{2\pi} \int_{\pi/4}^{\pi/2} \frac{1}{3}\rho^3 \sin\phi \Big|_0^1 \, d\phi \, d\theta$$

$$= \frac{2}{3} \int_0^{2\pi} \int_{\pi/4}^{\pi/2} \sin\phi \, d\phi \, d\theta = \frac{2}{3} \int_0^{2\pi} -\cos\phi \Big|_{\pi/4}^{\pi/2} \, d\theta = \frac{2}{3} \int_0^{2\pi} \frac{\sqrt{2}}{2} \, d\theta = \frac{2\pi\sqrt{2}}{3}$$

81. We are given density $= k/\rho$.

$$m = \int_0^{2\pi} \int_0^{\cos^{-1} 4/5} \int_{4\sec\phi}^5 \frac{k}{\rho}\rho^2 \sin\phi \, d\rho \, d\phi \, d\theta = k \int_0^{2\pi} \int_0^{\cos^{-1} 4/5} \frac{1}{2}\rho^2 \sin\phi \, \Big|_{4\sec\phi}^5 \, d\phi \, d\theta$$

$$= \frac{1}{2}k \int_0^{2\pi} \int_0^{\cos^{-1} 4/5} (25\sin\phi - 16\tan\phi\sec\phi) \, d\phi \, d\theta$$

$$= \frac{1}{2}k \int_0^{2\pi} (-25\cos\phi - 16\sec\phi) \, \Big|_0^{\cos^{-1} 4/5} \, d\theta = \frac{1}{2}k \int_0^{2\pi} [-25(4/5) - 16(5/4) - (-25 - 16)] \, d\theta$$

$$= \frac{1}{2}k \int_0^{2\pi} d\theta = k\pi$$

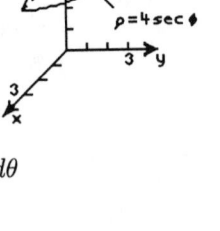

Exercises 9.16

3. div $\mathbf{F} = 3x^2 + 3y^2 + 3z^2$. Using spherical coordinates,

$$\iint_S \mathbf{F} \cdot \mathbf{n} \, dS = \iiint_D 3(x^2 + y^2 + z^2) \, dV = \int_0^{2\pi} \int_0^\pi \int_0^a 3\rho^2 \rho^2 \sin\phi \, d\rho \, d\phi \, d\theta$$

$$= \int_0^{2\pi} \int_0^\pi \frac{3}{5}\rho^5 \sin\phi \, \Big|_0^a \, d\phi \, d\theta = \frac{3a^5}{5} \int_0^{2\pi} \int_0^\pi \sin\phi \, d\phi \, d\theta$$

$$= \frac{3a^5}{5} \int_0^{2\pi} -\cos\phi \, \Big|_0^\pi \, d\theta = \frac{6a^5}{5} \int_0^{2\pi} d\theta = \frac{12\pi a^5}{5}.$$

6. div $\mathbf{F} = 2x + 2z + 12z^2$.

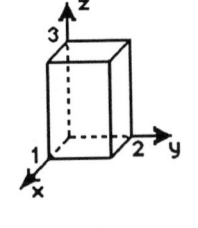

$$\iint_S \mathbf{F} \cdot \mathbf{n} \, dS = \iiint_D \text{div } \mathbf{F} \, dV = \int_0^3 \int_0^2 \int_0^1 (2x + 2z + 12z^2) \, dx \, dy \, dz$$

$$= \int_0^3 \int_0^2 (x^2 + 2xz + 12xz^2) \, \Big|_0^1 \, dy \, dz = \int_0^3 \int_0^2 (1 + 2z + 12z^2) \, dy \, dz$$

$$= \int_0^3 2(1 + 2z + 12z^2) \, dz = (2z + 2z^2 + 8z^3) \, \Big|_0^3 = 240$$

9. div $\mathbf{F} = \dfrac{1}{x^2 + y^2 + z^2}$. Using spherical coordinates,

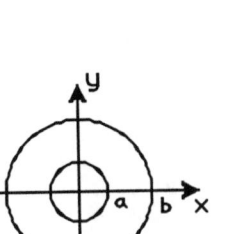

$$\iint_S \mathbf{F} \cdot \mathbf{n} \, dS = \iiint_D \text{div } \mathbf{F} \, dV = \int_0^{2\pi} \int_0^\pi \int_a^b \frac{1}{\rho^2}\rho^2 \sin\phi \, d\rho \, d\phi \, d\theta$$

$$= \int_0^{2\pi} \int_0^\pi (b - a) \sin\phi \, d\phi \, d\theta = (b - a) \int_0^{2\pi} -\cos\phi \, \Big|_0^\pi \, d\theta$$

$$= (b - a) \int_0^{2\pi} 2 \, d\theta = 4\pi(b - a).$$

12. div $\mathbf{F} = 30xy$.

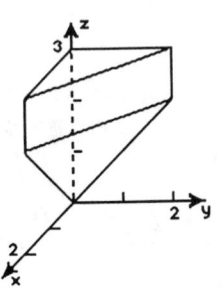

$$\iint_S \mathbf{F} \cdot \mathbf{n} \, dS = \iiint_D 30xy \, dV = \int_0^2 \int_0^{2-x} \int_{x+y}^3 30xy \, dz \, dy \, dx$$

$$= \int_0^2 \int_0^{2-x} 30xyz \, \Big|_{x+y}^3 \, dy \, dx$$

$$= \int_0^2 \int_0^{2-x} (90xy - 30x^2y - 30xy^2) \, dy \, dx$$

$$= \int_0^2 (45xy^2 - 15x^2y^2 - 10xy^3) \Big|_0^{2-x} dx$$

$$= \int_0^2 (-5x^4 + 45x^3 - 120x^2 + 100x) \, dx = \left(-x^5 + \frac{45}{4}x^4 - 40x^3 + 50x^2\right)\Big|_0^2 = 28$$

15. (a) $\operatorname{div} \mathbf{E} = q\left[\dfrac{-2x^2 + y^2 + z^2}{(x^2+y^2+z^2)^{5/2}} + \dfrac{x^2 - 2y^2 + z^2}{(x^2+y^2+z^2)^{5/2}} + \dfrac{x^2 + y^2 - 2z^2}{(x^2+y^2+z^2)^{5/2}}\right] = 0$

$$\iint_{S \cup S_a} (\mathbf{E} \cdot \mathbf{n}) \, dS = \iiint_D \operatorname{div} \mathbf{E} \, dV = \iiint_D 0 \, dV = 0$$

(b) From (a), $\displaystyle\iint_S (\mathbf{E} \cdot \mathbf{n}) \, dS + \iint_{S_a} (\mathbf{E} \cdot \mathbf{n}) \, dS = 0$ and $\displaystyle\iint_S (\mathbf{E} \cdot \mathbf{n}) \, dS = -\iint_{S_a} (\mathbf{E} \cdot \mathbf{n}) \, dS.$ On S_a,

$|\mathbf{r}| = a$, $\mathbf{n} = -(x\mathbf{i} + y\mathbf{j} + z\mathbf{k})/a = -\mathbf{r}/a$ and $\mathbf{E} \cdot \mathbf{n} = (q\mathbf{r}/a^3) \cdot (-\mathbf{r}/a) = -qa^2/a^4 = -q/a^2$. Thus

$$\iint_S (\mathbf{E} \cdot \mathbf{n}) \, dS = -\iint_{S_a} \left(-\frac{q}{a^2}\right) dS = \frac{q}{a^2} \iint_{S_a} dS = \frac{q}{a^2} \times (\text{area of } S_a) = \frac{q}{a^2}(4\pi a^2) = 4\pi q.$$

18. By the Divergence Theorem and Problem 30 in Section 9.7,

$$\iint_S (\operatorname{curl} \mathbf{F} \cdot \mathbf{n}) \, dS = \iiint_D \operatorname{div} (\operatorname{curl} \mathbf{F}) \, dV = \iiint_D 0 \, dV = 0.$$

21. If $G(x, y, z)$ is a vector valued function then we define surface integrals and triple integrals of $\mathbf{G}$ component-wise. In this case, if $\mathbf{a}$ is a constant vector it is easily shown that

$$\iint_S \mathbf{a} \cdot \mathbf{G} \, dS = \mathbf{a} \cdot \iint_S \mathbf{G} \, dS \quad \text{and} \quad \iiint_D \mathbf{a} \cdot \mathbf{G} \, dV = \mathbf{a} \cdot \iiint_D \mathbf{G} \, dV.$$

Now let $\mathbf{F} = f\mathbf{a}$. Then

$$\iint_S \mathbf{F} \cdot \mathbf{n} \, dS = \iint_S (f\mathbf{a}) \cdot \mathbf{n} \, dS = \iint_S \mathbf{a} \cdot (f\mathbf{n}) \, dS$$

and, using Problem 27 in Section 8.7 and the fact that $\nabla \cdot \mathbf{a} = 0$, we have

$$\iiint_D \operatorname{div} \mathbf{F} \, dV = \iiint_D \nabla \cdot (f\mathbf{a}) \, dV = \iiint_D [f(\nabla \cdot \mathbf{a}) + \mathbf{a} \cdot \nabla f] \, dV = \iiint_D \mathbf{a} \cdot \nabla f \, dV.$$

By the Divergence Theorem,

$$\iint_S \mathbf{a} \cdot (f\mathbf{n}) \, dS = \iint_S \mathbf{F} \cdot \mathbf{n} \, dS = \iiint_D \operatorname{div} \mathbf{F} \, dV = \iiint_D \mathbf{a} \cdot \nabla f \, dV$$

and

$$\mathbf{a} \cdot \left(\iint_S f\mathbf{n} \, dS\right) = \mathbf{a} \cdot \left(\iiint_D \nabla f \, dV\right) \quad \text{or} \quad \mathbf{a} \cdot \left(\iint_S f\mathbf{n} \, dS - \iiint_D \nabla f \, dV\right) = 0.$$

Since $\mathbf{a}$ is arbitrary,

$$\iint_S f\mathbf{n} \, dS - \iiint_D \nabla f \, dV = 0 \quad \text{and} \quad \iint_S f\mathbf{n} \, dS = \iiint_D \nabla f \, dV.$$

Exercises 9.17

3. The uv-corner points $(0,0)$, $(2,0)$, $(2,2)$ correspond to xy-points $(0,0)$, $(4,2)$, $(6,-4)$.

$v = 0$: $x = 2u$, $y = u \implies y = x/2$

$u = 2$: $x = 4 + v$, $y = 2 - 3v \implies$
$$y = 2 - 3(x - 4) = -3x + 14$$

$v = u$: $x = 3u$, $y = -2u \implies y = -2x/3$

6. The uv-corner points $(1,1)$, $(2,1)$, $(2,2)$, $(1,2)$ correspond to the xy-points $(1,1)$, $(2,1)$, $(4,4)$, $(2,4)$.

$v = 1$: $x = u$, $y = 1 \implies y = 1$, $1 \le x \le 2$

$u = 2$: $x = 2v$, $y = v^2 \implies y = x^2/4$

$v = 2$: $x = 2u$, $y = 4 \implies y = 4$, $2 \le x \le 4$

$u = 1$: $x = v$, $y = v^2 \implies y = x^2$

9. $\dfrac{\partial(u, v)}{\partial(x, y)} = \begin{vmatrix} -2y/x^3 & 1/x^2 \\ -y^2/x^2 & 2y/x \end{vmatrix} = -\dfrac{3y^2}{x^4} = -3\left(\dfrac{y}{x^2}\right)^2 = -3u^2$; $\dfrac{\partial(x, y)}{\partial(u, v)} = \dfrac{1}{-3u^2} = -\dfrac{1}{3u^2}$

12. $\dfrac{\partial(x, y)}{\partial(u, v)} = \begin{vmatrix} 1 - v & v \\ -u & u \end{vmatrix} = u$. The transformation is 0 when u is 0, for $0 \le v \le 1$.

15. $R1$: $y = x^2 \implies u = 1$
$R2$: $x = y^2 \implies v = 1$
$R3$: $y = \frac{1}{2}x^2 \implies u = 2$
$R4$: $x = \frac{1}{2}y^2 \implies v = 2$

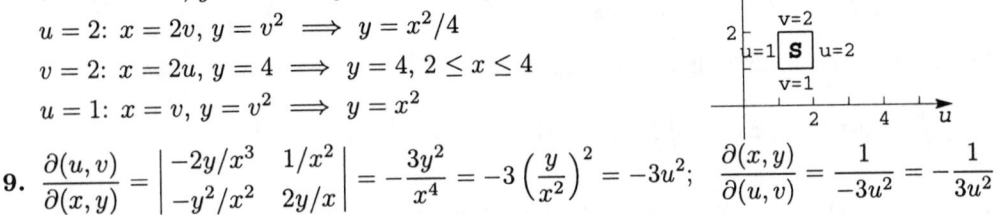

$\dfrac{\partial(u, v)}{\partial(x, y)} = \begin{vmatrix} 2x/y & -x^2/y^2 \\ -y^2/x^2 & 2y/x \end{vmatrix} = 3 \implies \dfrac{\partial(x, y)}{\partial(u, v)} = \dfrac{1}{3}$

$$\iint_R \dfrac{y^2}{x}\, dA = \iint_S v\left(\dfrac{1}{3}\right) dA' = \dfrac{1}{3}\int_1^2 \int_1^2 v\, du\, dv = \dfrac{1}{3}\int_1^2 v\, dv = \dfrac{1}{6}v^2 \Big|_1^2 = \dfrac{1}{2}$$

18. $R1$: $xy = -2 \implies v = -2$
$R2$: $x^2 - y^2 = 9 \implies u = 9$
$R3$: $xy = 2 \implies v = 2$
$R4$: $x^2 - y^2 = 1 \implies u = 1$

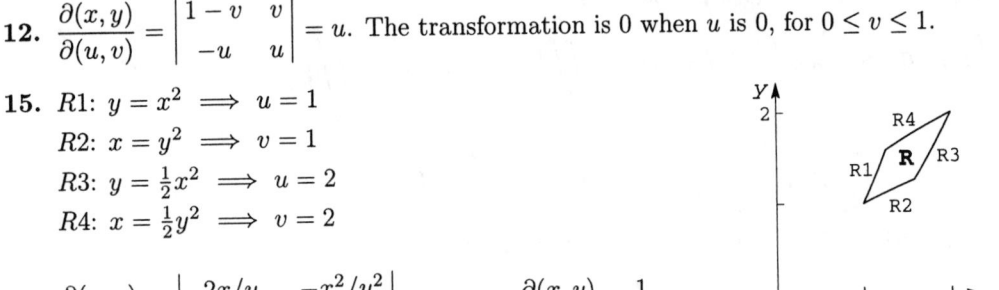

$\dfrac{\partial(u, v)}{\partial(x, y)} = \begin{vmatrix} 2x & -2y \\ y & x \end{vmatrix} = 2(x^2 + y^2)$

$\implies \dfrac{\partial(x, y)}{\partial(u, v)} = \dfrac{1}{2(x^2 + y^2)}$

$$\iint_R (x^2 + y^2)\sin xy\, dA = \iint_S (x^2 + y^2)\sin v\left(\dfrac{1}{2(x^2 + y^2)}\right) dA' = \dfrac{1}{2}\int_{-2}^2 \int_1^9 \sin v\, du\, dv = \dfrac{1}{2}\int_{-2}^2 8\sin v\, dv = 0$$

21. $R1: y = 1/x \implies u = 1$

$\quad R2: y = x \implies v = 1$

$\quad R3: y = 4/x \implies u = 4$

$\quad R4: y = 4x \implies v = 4$

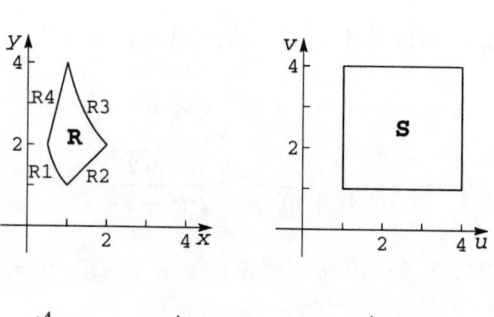

$$\frac{\partial(u,v)}{\partial(x,y)} = \begin{vmatrix} y & x \\ -y/x^2 & 1/x \end{vmatrix} = \frac{2y}{x} \implies \frac{\partial(x,y)}{\partial(u,v)} = \frac{x}{2y}$$

$$8\iint_R y^4\,dA = \iint_S u^2 v^2 \left(\frac{1}{2v}\right) du\,dv = \frac{1}{2}\int_1^4 u^2 v\,du\,dv = \frac{1}{2}\int_1^4 \frac{1}{3}u^3 v\Big|_1^4\,dv = \frac{1}{6}\int_1^4 63v\,dv = \frac{21}{4}v^2\Big|_1^4 = \frac{315}{4}$$

24. We let $u = y - x$ and $v = y$.

$\quad R1: y = 0 \implies v = 0,\ u = -x \implies v = 0,\ 0 \le u \le 2$

$\quad R2: x = 0 \implies v = u$

$\quad R3: y = x + 2 \implies u = 2$

$$\frac{\partial(u,v)}{\partial(x,y)} = \begin{vmatrix} -1 & 1 \\ 0 & 1 \end{vmatrix} = -1 \implies \frac{\partial(x,y)}{\partial(u,v)} = -1$$

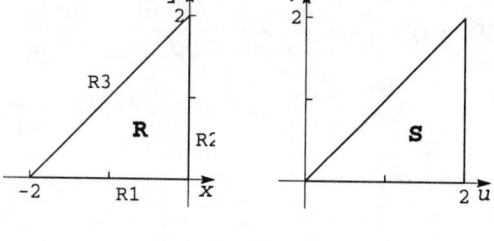

$$\iint_R e^{y^2 - 2xy + x^2}\,dA = \iint_S e^{u^2}|-1|\,dA' = \int_0^2 \int_0^u e^{u^2}\,dv\,du = \int_0^2 ue^{u^2}\,du = \frac{1}{2}e^{u^2}\Big|_0^2 = \frac{1}{2}(e^4 - 1)$$

27. Let $u = xy$ and $v = xy^{1.4}$. Then $xy^{1.4} = c \implies v = c;\ xy = b \implies u = b;\ xy^{1.4} = d \implies v = d;$

$\quad xy = a \implies u = a.$

$$\frac{\partial(u,v)}{\partial(x,y)} = \begin{vmatrix} y & x \\ y^{1.4} & 1.4xy^{0.4} \end{vmatrix} = 0.4xy^{1.4} = 0.4v \implies \frac{\partial(x,y)}{\partial(u,v)} = \frac{5}{2v}$$

$$\iint_R dA = \iint_S \frac{5}{2v}\,dA' = \int_c^d \int_a^b \frac{5}{2v}\,du\,dv = \frac{5}{2}(b-a)\int_c^d \frac{dv}{v} = \frac{5}{2}(b-a)(\ln d - \ln c)$$

30. $\dfrac{\partial(x,y,z)}{\partial(\rho,\phi,\theta)} = \begin{vmatrix} \sin\phi\cos\theta & \rho\cos\phi\cos\theta & -\rho\sin\phi\sin\theta \\ \sin\phi\sin\theta & \rho\cos\phi\sin\theta & \rho\sin\phi\cos\theta \\ \cos\phi & -\rho\sin\phi & 0 \end{vmatrix}$

$\quad = \cos\phi(\rho^2\sin\phi\cos\phi\cos^2\theta + \rho^2\sin\phi\cos\phi\sin^2\theta) + \rho\sin\phi(\rho\sin^2\phi\cos^2\theta + \rho\sin^2\phi\sin^2\theta)$

$\quad = \rho^2\sin\phi\cos^2\phi(\cos^2\theta + \sin^2\theta) + \rho^2\sin^3\phi(\cos^2\theta + \sin^2\theta) = \rho^2\sin\phi(\cos^2\phi + \sin^2\phi) = \rho^2\sin\phi$

Chapter 9 Review Exercises

3. True

6. False; consider $f(x,y) = xy$ at $(0,0)$.

9. False; $\displaystyle\int_C x\,dx + x^2\,dy = 0$ from $(-1,0)$ to $(1,0)$ along the x-axis and along the semicircle $y = \sqrt{1-x^2}$, but since $x\,dx + x^2\,dy$ is not exact, the integral is not independent of path.

12. True $\qquad\qquad\qquad$ **15.** True $\qquad\qquad\qquad$ **18.** True

21. $\mathbf{v}(t) = 6\mathbf{i} + \mathbf{j} + 2t\mathbf{k};\ \ \mathbf{a}(t) = 2\mathbf{k}.$ To find when the particle passes through the plane, we solve $-6t + t + t^2 = -4$ or $t^2 - 5t + 4 = 0$. This gives $t = 1$ and $t = 4$. $\mathbf{v}(1) = 6\mathbf{i} + \mathbf{j} + 2\mathbf{k},\ \ \mathbf{a}(1) = 2\mathbf{k};\ \ \mathbf{v}(4) = 6\mathbf{i} + \mathbf{j} + 8\mathbf{k},\ \ \mathbf{a}(4) = 2\mathbf{k}$

24. $\mathbf{v}(t) = t\mathbf{i} + t^2\mathbf{j} - t\mathbf{k}$; $\quad |\mathbf{v}| = t\sqrt{t^2 + 2}$, $t > 0$; $\quad \mathbf{a}(t) = \mathbf{i} + 2t\mathbf{j} - \mathbf{k}$; $\quad \mathbf{v} \cdot \mathbf{a} = t + 2t^3 + t = 2t + 2t^3$;

$$\mathbf{v} \times \mathbf{a} = t^2\mathbf{i} + t^2\mathbf{k}, \quad |\mathbf{v} \times \mathbf{a}| = t^2\sqrt{2}; \quad a_T = \frac{2t + 2t^3}{t\sqrt{t^2 + 2}} = \frac{2 + 2t^2}{\sqrt{t^2 + 2}}, \quad a_N = \frac{t^2\sqrt{2}}{t\sqrt{t^2 + 2}} = \frac{\sqrt{2}\,t}{\sqrt{t^2 + 2}};$$

$$\kappa = \frac{t^2\sqrt{2}}{t^3(t^2 + 2)^{3/2}} = \frac{\sqrt{2}}{t(t^2 + 2)^{3/2}}$$

27. $\nabla f = (2xy - y^2)\mathbf{i} + (x^2 - 2xy)\mathbf{j}$; $\quad \mathbf{u} = \dfrac{2}{\sqrt{40}}\mathbf{i} + \dfrac{6}{\sqrt{40}}\mathbf{j} = \dfrac{1}{\sqrt{10}}(\mathbf{i} + 3\mathbf{j})$;

$$D_{\mathbf{u}}f = \frac{1}{\sqrt{10}}(2xy - y^2 + 3x^2 - 6xy) = \frac{1}{\sqrt{10}}(3x^2 - 4xy - y^2)$$

30. (a) $\dfrac{dw}{dt} = \dfrac{\partial w}{\partial x}\dfrac{dx}{dt} + \dfrac{\partial w}{\partial y}\dfrac{dy}{dt} + \dfrac{\partial w}{\partial z}\dfrac{dz}{dt}$

$$= \frac{x}{\sqrt{x^2 + y^2 + z^2}}\,6\cos 2t + \frac{y}{\sqrt{x^2 + y^2 + z^2}}(-8\sin 2t) + \frac{z}{\sqrt{x^2 + y^2 + z^2}}\,15t^2$$

$$= \frac{(6x\cos 2t - 8y\sin 2t + 15zt^2)}{\sqrt{x^2 + y^2 + z^2}}$$

(b) $\dfrac{\partial w}{\partial t} = \dfrac{\partial w}{\partial x}\dfrac{\partial x}{\partial t} + \dfrac{\partial w}{\partial y}\dfrac{\partial y}{\partial t} + \dfrac{\partial w}{\partial z}\dfrac{\partial z}{\partial t}$

$$= \frac{x}{\sqrt{x^2 + y^2 + z^2}}\frac{6}{r}\cos\frac{2t}{r} + \frac{y}{\sqrt{x^2 + y^2 + z^2}}\left(\frac{8r}{t^2}\sin\frac{2r}{t}\right) + \frac{z}{\sqrt{x^2 + y^2 + z^2}}\,15t^2 r^3$$

$$= \frac{\left(\dfrac{6x}{r}\cos\dfrac{2t}{r} + \dfrac{8yr}{t^2}\sin\dfrac{2r}{t} + 15zt^2 r^3\right)}{\sqrt{x^2 + y^2 + z^2}}$$

33. (a) $V = \displaystyle\int_0^1 \int_x^{2x} \sqrt{1 - x^2}\,dy\,dx = \int_0^1 y\sqrt{1 - x^2}\,\Big|_x^{2x}\,dx = \int_0^1 x\sqrt{1 - x^2}\,dx = -\frac{1}{3}(1 - x^2)^{3/2}\,\Big|_0^1 = \frac{1}{3}$

(b) $V = \displaystyle\int_0^1 \int_{y/2}^y \sqrt{1 - x^2}\,dx\,dy + \int_1^2 \int_{y/2}^1 \sqrt{1 - x^2}\,dx\,dy$

36. (a) Using symmetry,

$$V = 8\int_0^a \int_0^{\sqrt{a^2 - x^2}} \int_0^{\sqrt{a^2 - x^2 - y^2}} dz\,dy\,dx = 8\int_0^a \int_0^{\sqrt{a^2 - x^2}} \sqrt{a^2 - x^2 - y^2}\,dy\,dx$$

$$\boxed{\text{Trig substitution}}$$

$$= 8\int_0^a \left(\frac{y}{2}\sqrt{a^2 - x^2 - y^2} + \frac{a^2 - x^2}{2}\sin^{-1}\frac{y}{\sqrt{a^2 - x^2}}\right)\Bigg|_0^{\sqrt{a^2 - x^2}} dx = 8\int_0^a \frac{\pi}{2}\frac{a^2 - x^2}{2}\,dx$$

$$= 2\pi\left(a^2 x - \frac{1}{3}x^3\right)\Bigg|_0^a = \frac{4}{3}\pi a^3$$

(b) Using symmetry,

$$V = 2\int_0^{2\pi} \int_0^a \int_0^{\sqrt{a^2 - r^2}} r\,dz\,dr\,d\theta = 2\int_0^{2\pi} \int_0^a r\sqrt{a^2 - r^2}\,dr\,d\theta$$

$$= 2\int_0^{2\pi} -\frac{1}{3}(a^2 - r^2)^{3/2}\,\Big|_0^a\,d\theta = \frac{2}{3}\int_0^{2\pi} a^3\,d\theta = \frac{4}{3}\pi a^3$$

(c) $V = \int_0^{2\pi} \int_0^\pi \int_0^a \rho^2 \sin\phi \, d\rho \, d\phi \, d\theta = \int_0^{2\pi} \int_0^\pi \frac{1}{3}\rho^3 \sin\phi \, \Big|_0^a \, d\phi \, d\theta$

$= \frac{1}{3} \int_0^{2\pi} \int_0^\pi a^3 \sin\phi \, d\phi \, d\theta = \frac{1}{3} \int_0^{2\pi} -a^3 \cos\phi \, \Big|_0^\pi \, d\theta = \frac{1}{3} \int_0^{2\pi} 2a^3 \, d\theta = \frac{4}{3}\pi a^3$

39. $2xy + 2xy + 2xy = 6xy$

42. $\nabla(6xy) = 6y\mathbf{i} + 6x\mathbf{j}$

45. Since $P_y = 6x^2 y = Q_x$, the integral is independent of path.

$\phi_x = 3x^2 y^2$, $\phi = x^3 y^2 + g(y)$, $\phi_y = 2x^3 y + g'(y) = 2x^3 y - 3y^2$; $g(y) = -y^3$; $\phi = x^3 y^2 - y^3$;

$\int_{(0,0)}^{(1,-2)} 3x^2 y^2 \, dx + (2x^3 y - 3y^2) \, dy = (x^3 y^2 - y^3) \, \Big|_{(0,0)}^{(1,-2)} = 12$

48. Parameterize C by $x = \cos t$, $y = \sin t$; $0 \le t \le 2\pi$. Then

$$\oint_C \mathbf{F} \cdot d\mathbf{r} = \int_0^{2\pi} [4\sin t(-\sin t \, dt) + 6\cos t(\cos t) \, dt] = \int_0^{2\pi} (6\cos^2 t - 4\sin^2 t) \, dt$$

$$= \int_0^{2\pi} (10\cos^2 t - 4) \, dt = \left(5t + \frac{5}{2}\sin 2t - 4t\right) \Big|_0^{2\pi} = 2\pi.$$

Using Green's Theorem, $Q_x - P_y = 6 - 4 = 2$ and $\oint_C \mathbf{F} \cdot d\mathbf{r} = \iint_R 2 \, dA = 2(\pi \cdot 1^2) = 2\pi$.

51. $z_x = 2x$, $z_y = 0$; $dS = \sqrt{1 + 4x^2} \, dA$

$\iint_S \frac{z}{xy} \, dS = \int_1^3 \int_1^2 \frac{x^2}{xy}\sqrt{1 + 4x^2} \, dx \, dy = \int_1^3 \frac{1}{y}\left[\frac{1}{12}(1 + 4x^2)^{3/2}\right] \Big|_1^2 \, dy$

$= \frac{1}{12}\int_1^3 \frac{17^{3/2} - 5^{3/2}}{y} \, dy = \frac{17\sqrt{17} - 5\sqrt{5}}{12}\ln y \, \Big|_1^3$

$= \frac{17\sqrt{17} - 5\sqrt{5}}{12}\ln 3$

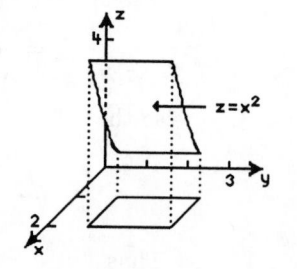

54. In Problem 53, $\mathbf{F}$ is not continuous at $(0,0,0)$ which is in any acceptable region containing the sphere.

57. Identify $\mathbf{F} = -2y\mathbf{i} + 3x\mathbf{j} + 10z\mathbf{k}$. Then curl $\mathbf{F} = 5\mathbf{k}$. The curve C lies in the plane $z = 3$, so $\mathbf{n} = \mathbf{k}$ and $dS = dA$. Thus,

$$\oint_C \mathbf{F} \cdot d\mathbf{r} = \iint_S (\text{curl } \mathbf{F}) \cdot \mathbf{n} \, dS = \iint_R 5 \, dA = 5 \times (\text{area of } R) = 5(25\pi) = 125\pi.$$

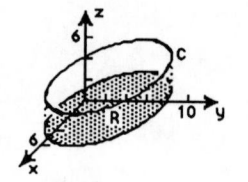

60. div $\mathbf{F} = x^2 + y^2 + z^2$. Using cylindrical coordinates,

$$\iint_S \mathbf{F} \cdot \mathbf{n} \, dS = \iiint_D \text{div } \mathbf{F} \, dV = \iiint_D (x^2 + y^2 + z^2) \, dV = \int_0^{2\pi} \int_0^1 \int_0^1 (r^2 + z^2) r \, dz \, dr \, d\theta$$

$$= \int_0^{2\pi} \int_0^1 \left(r^3 z + \frac{1}{3}rz^3\right) \Big|_0^1 \, dr \, d\theta = \int_0^{2\pi} \int_0^1 \left(r^3 + \frac{1}{3}r\right) \, dr \, d\theta$$

$$= \int_0^{2\pi} \left(\frac{1}{4}r^4 + \frac{1}{6}r^2\right) \Big|_0^1 \, d\theta = \int_0^{2\pi} \frac{5}{12} \, d\theta = \frac{5\pi}{6}.$$

63. $x = 0 \implies u = 0,\ v = -y^2 \implies u = 0,\ -1 \le v \le 0$

$x = 1 \implies u = 2y,\ v = 1 - y^2 = 1 - u^2/4$

$y = 0 \implies u = 0,\ v = x^2 \implies u = 0,\ 0 \le v \le 1$

$y = 1 \implies u = 2x,\ v = x^2 - 1 = u^2/4 - 1$

$$\frac{\partial(u,v)}{\partial(x,y)} = \begin{vmatrix} 2y & 2x \\ 2x & -2y \end{vmatrix} = -4(x^2 + y^2) \implies \frac{\partial(x,y)}{\partial(u,v)} = -\frac{1}{4(x^2+y^2)}$$

$$\iint_R (x^2+y^2)\sqrt[3]{x^2-y^2}\, dA = \iint_S (x^2+y^2)\sqrt[3]{v}\left| -\frac{1}{4(x^2+y^2)} \right| dA' = \frac{1}{4}\int_0^2 \int_{u^2/4-1}^{1-u^2/4} v^{1/3}\, dv\, du$$

$$= \frac{1}{4}\int_0^2 \frac{3}{4} v^{4/3}\Big|_{u^2/4-1}^{1-u^2/4}\, du = \frac{3}{16}\int_0^2 \left[(1-u^2/4)^{4/3} - (u^2/4-1)^{4/3} \right] du$$

$$= \frac{3}{16}\int_0^2 \left[(1-u^2/4)^{4/3} - (1-u^2/4)^{4/3} \right] du = 0$$

66. (a) Both states span 7 degrees of longitude and 4 degrees of latitude, but Colorado is larger because it lies to the south of Wyoming. Lines of longitude converge as they go north, so the east-west dimensions of Wyoming are shorter than those of Colorado.

(b) We use the function $f(x,y) = \sqrt{R^2 - x^2 - y^2}$ to describe the northern hemisphere, where $R \approx 3960$ miles is the radius of the Earth. We need to compute the surface area over a polar rectangle P of the form $\theta_1 \le \theta \le \theta_2$, $R\cos\phi_2 \le r \le R\cos\phi_1$. We have

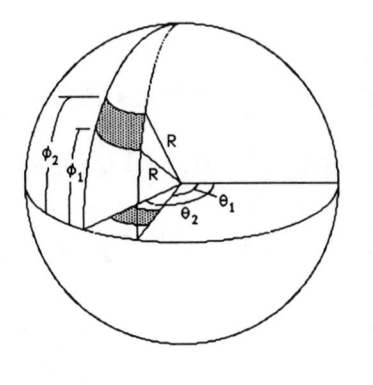

$$f_x = \frac{-x}{\sqrt{R^2-x^2-y^2}} \quad \text{and} \quad f_y = \frac{-y}{\sqrt{R^2-x^2-y^2}}$$

so that

$$\sqrt{1+f_x^2+f_y^2} = \sqrt{1 + \frac{x^2+y^2}{R^2-x^2-y^2}} = \frac{R}{\sqrt{R^2-r^2}}.$$

Thus

$$A = \iint_P \sqrt{1+f_x^2+f_y^2}\, dA = \int_{\theta_1}^{\theta_2} \int_{R\cos\phi_2}^{R\cos\phi_1} \frac{R}{\sqrt{R^2-r^2}}\, r\, dr\, d\theta$$

$$= (\theta_2-\theta_1)R\sqrt{R^2-r^2}\,\Big|_{R\cos\phi_1}^{R\cos\phi_2} = (\theta_2-\theta_1)R^2(\sin\phi_2 - \sin\phi_1).$$

The ratio of Wyoming to Colorado is then $\dfrac{\sin 45° - \sin 41°}{\sin 41° - \sin 37°} \approx 0.941$. Thus Wyoming is about 6% smaller than Colorado.

(c) $97{,}914/104{,}247 \approx 0.939$, which is close to the theoretical value of 0.941. (Our formula for the area says that the area of Colorado is approximately 103,924 square miles, while the area of Wyoming is approximately 97,801 square miles.)

10 Systems of Linear Differential Equations

_____ **Exercises 10.1** _____

3. Let $\mathbf{X} = \begin{pmatrix} x \\ y \\ z \end{pmatrix}$. Then

$$\mathbf{X}' = \begin{pmatrix} -3 & 4 & -9 \\ 6 & -1 & 0 \\ 10 & 4 & 3 \end{pmatrix} \mathbf{X}.$$

6. Let $\mathbf{X} = \begin{pmatrix} x \\ y \\ z \end{pmatrix}$. Then

$$\mathbf{X}' = \begin{pmatrix} -3 & 4 & 0 \\ 5 & 0 & 9 \\ 0 & 1 & 6 \end{pmatrix} \mathbf{X} + \begin{pmatrix} e^{-t}\sin 2t \\ 4e^{-t}\cos 2t \\ -e^{-t} \end{pmatrix}$$

9. $\dfrac{dx}{dt} = x - y + 2z + e^{-t} - 3t; \quad \dfrac{dy}{dt} = 3x - 4y + z + 2e^{-t} + t; \quad \dfrac{dz}{dt} = -2x + 5y + 6z + 2e^{-t} - t$

12. Since

$$\mathbf{X}' = \begin{pmatrix} 5\cos t - 5\sin t \\ 2\cos t - 4\sin t \end{pmatrix} e^t \quad \text{and} \quad \begin{pmatrix} -2 & 5 \\ -2 & 4 \end{pmatrix} \mathbf{X} = \begin{pmatrix} 5\cos t - 5\sin t \\ 2\cos t - 4\sin t \end{pmatrix} e^t$$

we see that

$$\mathbf{X}' = \begin{pmatrix} -2 & 5 \\ -2 & 4 \end{pmatrix} \mathbf{X}.$$

15. Since

$$\mathbf{X}' = \begin{pmatrix} 0 \\ 0 \\ 0 \end{pmatrix} \quad \text{and} \quad \begin{pmatrix} 1 & 2 & 1 \\ 6 & -1 & 0 \\ -1 & -2 & -1 \end{pmatrix} \mathbf{X} = \begin{pmatrix} 0 \\ 0 \\ 0 \end{pmatrix}$$

we see that

$$\mathbf{X}' = \begin{pmatrix} 1 & 2 & 1 \\ 6 & -1 & 0 \\ -1 & -2 & -1 \end{pmatrix} \mathbf{X}.$$

18. Yes, since $W(\mathbf{X}_1, \mathbf{X}_2) = 8e^{2t} \neq 0$ and $\mathbf{X}_1$ and $\mathbf{X}_2$ are linearly independent on $-\infty < t < \infty$.

21. Since

$$\mathbf{X}_p' = \begin{pmatrix} 2 \\ -1 \end{pmatrix} \quad \text{and} \quad \begin{pmatrix} 1 & 4 \\ 3 & 2 \end{pmatrix} \mathbf{X}_p + \begin{pmatrix} 2 \\ -4 \end{pmatrix} t + \begin{pmatrix} -7 \\ -18 \end{pmatrix} = \begin{pmatrix} 2 \\ -1 \end{pmatrix}$$

we see that

$$\mathbf{X}_p' = \begin{pmatrix} 1 & 4 \\ 3 & 2 \end{pmatrix} \mathbf{X}_p + \begin{pmatrix} 2 \\ -4 \end{pmatrix} t + \begin{pmatrix} -7 \\ -18 \end{pmatrix}.$$

24. Since

$$\mathbf{X}_p' = \begin{pmatrix} 3\cos 3t \\ 0 \\ -3\sin 3t \end{pmatrix} \quad \text{and} \quad \begin{pmatrix} 1 & 2 & 3 \\ -4 & 2 & 0 \\ -6 & 1 & 0 \end{pmatrix} \mathbf{X}_p + \begin{pmatrix} -1 \\ 4 \\ 3 \end{pmatrix} \sin 3t = \begin{pmatrix} 3\cos 3t \\ 0 \\ -3\sin 3t \end{pmatrix}$$

159

we see that

$$\mathbf{X}'_p = \begin{pmatrix} 1 & 2 & 3 \\ -4 & 2 & 0 \\ -6 & 1 & 0 \end{pmatrix} \mathbf{X}_p + \begin{pmatrix} -1 \\ 4 \\ 3 \end{pmatrix} \sin 3t.$$

Exercises 10.2

3. The system is

$$\mathbf{X}' = \begin{pmatrix} -4 & 2 \\ -5/2 & 2 \end{pmatrix} \mathbf{X}$$

and $\det(\mathbf{A} - \lambda\mathbf{I}) = (\lambda - 1)(\lambda + 3) = 0$. For $\lambda_1 = 1$ we obtain

$$\begin{pmatrix} -5 & 2 & | & 0 \\ -5/2 & 1 & | & 0 \end{pmatrix} \implies \begin{pmatrix} -5 & 2 & | & 0 \\ 0 & 0 & | & 0 \end{pmatrix} \quad \text{so that} \quad \mathbf{K}_1 = \begin{pmatrix} 2 \\ 5 \end{pmatrix}.$$

For $\lambda_2 = -3$ we obtain

$$\begin{pmatrix} -1 & 2 & | & 0 \\ -5/2 & 5 & | & 0 \end{pmatrix} \implies \begin{pmatrix} -1 & 2 & | & 0 \\ 0 & 0 & | & 0 \end{pmatrix} \quad \text{so that} \quad \mathbf{K}_2 = \begin{pmatrix} 2 \\ 1 \end{pmatrix}.$$

Then

$$\mathbf{X} = c_1 \begin{pmatrix} 2 \\ 5 \end{pmatrix} e^t + c_2 \begin{pmatrix} 2 \\ 1 \end{pmatrix} e^{-3t}.$$

6. The system is

$$\mathbf{X}' = \begin{pmatrix} -6 & 2 \\ -3 & 1 \end{pmatrix} \mathbf{X}$$

and $\det(\mathbf{A} - \lambda\mathbf{I}) = \lambda(\lambda + 5) = 0$. For $\lambda_1 = 0$ we obtain

$$\begin{pmatrix} -6 & 2 & | & 0 \\ -3 & 1 & | & 0 \end{pmatrix} \implies \begin{pmatrix} 1 & -1/3 & | & 0 \\ 0 & 0 & | & 0 \end{pmatrix} \quad \text{so that} \quad \mathbf{K}_1 = \begin{pmatrix} 1 \\ 3 \end{pmatrix}.$$

For $\lambda_2 = -5$ we obtain

$$\begin{pmatrix} -1 & 2 & | & 0 \\ -3 & 6 & | & 0 \end{pmatrix} \implies \begin{pmatrix} 1 & -2 & | & 0 \\ 0 & 0 & | & 0 \end{pmatrix} \quad \text{so that} \quad \mathbf{K}_2 = \begin{pmatrix} 2 \\ 1 \end{pmatrix}.$$

Then

$$\mathbf{X} = c_1 \begin{pmatrix} 1 \\ 3 \end{pmatrix} + c_2 \begin{pmatrix} 2 \\ 1 \end{pmatrix} e^{-5t}.$$

9. We have $\det(\mathbf{A} - \lambda\mathbf{I}) = -(\lambda + 1)(\lambda - 3)(\lambda + 2) = 0$. For $\lambda_1 = -1$, $\lambda_2 = 3$, and $\lambda_3 = -2$ we obtain

$$\mathbf{K}_1 = \begin{pmatrix} -1 \\ 0 \\ 1 \end{pmatrix}, \quad \mathbf{K}_2 = \begin{pmatrix} 1 \\ 4 \\ 3 \end{pmatrix}, \quad \text{and} \quad \mathbf{K}_3 = \begin{pmatrix} 1 \\ -1 \\ 3 \end{pmatrix},$$

so that

$$\mathbf{X} = c_1 \begin{pmatrix} -1 \\ 0 \\ 1 \end{pmatrix} e^{-t} + c_2 \begin{pmatrix} 1 \\ 4 \\ 3 \end{pmatrix} e^{3t} + c_3 \begin{pmatrix} 1 \\ -1 \\ 3 \end{pmatrix} e^{-2t}.$$

12. We have $\det(\mathbf{A} - \lambda\mathbf{I}) = (\lambda - 3)(\lambda + 5)(6 - \lambda) = 0$. For $\lambda_1 = 3$, $\lambda_2 = -5$, and $\lambda_3 = 6$ we obtain

$$\mathbf{K}_1 = \begin{pmatrix} 1 \\ 1 \\ 0 \end{pmatrix}, \quad \mathbf{K}_2 = \begin{pmatrix} 1 \\ -1 \\ 0 \end{pmatrix}, \quad \text{and} \quad \mathbf{K}_3 = \begin{pmatrix} 2 \\ -2 \\ 11 \end{pmatrix},$$

so that

$$\mathbf{X} = c_1 \begin{pmatrix} 1 \\ 1 \\ 0 \end{pmatrix} e^{3t} + c_2 \begin{pmatrix} 1 \\ -1 \\ 0 \end{pmatrix} e^{-5t} + c_3 \begin{pmatrix} 2 \\ -2 \\ 11 \end{pmatrix} e^{6t}.$$

21. We have $\det(\mathbf{A} - \lambda\mathbf{I}) = (\lambda - 2)^2 = 0$. For $\lambda_1 = 2$ we obtain

$$\mathbf{K} = \begin{pmatrix} 1 \\ 1 \end{pmatrix}.$$

A solution of $(\mathbf{A} - \lambda_1\mathbf{I})\mathbf{P} = \mathbf{K}$ is

$$\mathbf{P} = \begin{pmatrix} -1/3 \\ 0 \end{pmatrix}$$

so that

$$\mathbf{X} = c_1 \begin{pmatrix} 1 \\ 1 \end{pmatrix} e^{2t} + c_2 \left[\begin{pmatrix} 1 \\ 1 \end{pmatrix} te^{2t} + \begin{pmatrix} -1/3 \\ 0 \end{pmatrix} e^{2t} \right].$$

24. We have $\det(\mathbf{A} - \lambda\mathbf{I}) = (\lambda - 8)(\lambda + 1)^2 = 0$. For $\lambda_1 = 8$ we obtain

$$\mathbf{K}_1 = \begin{pmatrix} 2 \\ 1 \\ 2 \end{pmatrix}.$$

For $\lambda_2 = -1$ we obtain

$$\mathbf{K}_2 = \begin{pmatrix} 0 \\ -2 \\ 1 \end{pmatrix} \quad \text{and} \quad \mathbf{K}_3 = \begin{pmatrix} 1 \\ -2 \\ 0 \end{pmatrix}.$$

Then

$$\mathbf{X} = c_1 \begin{pmatrix} 2 \\ 1 \\ 2 \end{pmatrix} e^{8t} + c_2 \begin{pmatrix} 0 \\ -2 \\ 1 \end{pmatrix} e^{-t} + c_3 \begin{pmatrix} 1 \\ -2 \\ 0 \end{pmatrix} e^{-t}.$$

27. We have $\det(\mathbf{A} - \lambda\mathbf{I}) = -(\lambda - 1)^3 = 0$. For $\lambda_1 = 1$ we obtain

$$\mathbf{K} = \begin{pmatrix} 0 \\ 1 \\ 1 \end{pmatrix}.$$

Solutions of $(\mathbf{A} - \lambda_1\mathbf{I})\mathbf{P} = \mathbf{K}$ and $(\mathbf{A} - \lambda_1\mathbf{I})\mathbf{Q} = \mathbf{P}$ are

$$\mathbf{P} = \begin{pmatrix} 0 \\ 1 \\ 0 \end{pmatrix} \quad \text{and} \quad \mathbf{Q} = \begin{pmatrix} 1/2 \\ 0 \\ 0 \end{pmatrix}$$

so that

$$\mathbf{X} = c_1 \begin{pmatrix} 0 \\ 1 \\ 1 \end{pmatrix} + c_2 \left[\begin{pmatrix} 0 \\ 1 \\ 1 \end{pmatrix} te^t + \begin{pmatrix} 0 \\ 1 \\ 0 \end{pmatrix} e^t \right] + c_3 \left[\begin{pmatrix} 0 \\ 1 \\ 1 \end{pmatrix} \frac{t^2}{2} e^t + \begin{pmatrix} 0 \\ 1 \\ 0 \end{pmatrix} te^t + \begin{pmatrix} 1/2 \\ 0 \\ 0 \end{pmatrix} e^t \right].$$

30. We have $\det(\mathbf{A} - \lambda\mathbf{I}) = -(\lambda + 1)(\lambda - 1)^2 = 0$. For $\lambda_1 = -1$ we obtain

$$\mathbf{K}_1 = \begin{pmatrix} -1 \\ 0 \\ 1 \end{pmatrix}.$$

For $\lambda_2 = 1$ we obtain

$$\mathbf{K}_2 = \begin{pmatrix} 1 \\ 0 \\ 1 \end{pmatrix} \quad \text{and} \quad \mathbf{K}_3 = \begin{pmatrix} 0 \\ 1 \\ 0 \end{pmatrix}$$

so that

$$\mathbf{X} = c_1 \begin{pmatrix} -1 \\ 0 \\ 1 \end{pmatrix} e^{-t} + c_2 \begin{pmatrix} 1 \\ 0 \\ 1 \end{pmatrix} e^{t} + c_3 \begin{pmatrix} 0 \\ 1 \\ 0 \end{pmatrix} e^{t}.$$

If

$$\mathbf{X}(0) = \begin{pmatrix} 1 \\ 2 \\ 5 \end{pmatrix}$$

then $c_1 = 2$, $c_2 = 3$, and $c_3 = 2$.

In Problems 33-45 the form of the answer will vary according to the choice of eigenvector. For example, in Problem 33, if $\mathbf{K}_1$ is chosen to be $\begin{pmatrix} 1 \\ 2 - i \end{pmatrix}$ the solution has the form

$$\mathbf{X} = c_1 \begin{pmatrix} \cos t \\ 2\cos t + \sin t \end{pmatrix} e^{4t} + c_2 \begin{pmatrix} \sin t \\ 2\sin t - \cos t \end{pmatrix} e^{4t}.$$

33. We have $\det(\mathbf{A} - \lambda\mathbf{I}) = \lambda^2 - 8\lambda + 17 = 0$. For $\lambda_1 = 4 + i$ we obtain

$$\mathbf{K}_1 = \begin{pmatrix} 2 + i \\ 5 \end{pmatrix}$$

so that

$$\mathbf{X}_1 = \begin{pmatrix} 2 + i \\ 5 \end{pmatrix} e^{(4+i)t} = \begin{pmatrix} 2\cos t - \sin t \\ 5\cos t \end{pmatrix} e^{4t} + i \begin{pmatrix} \cos t + 2\sin t \\ 5\sin t \end{pmatrix} e^{4t}.$$

Then

$$\mathbf{X} = c_1 \begin{pmatrix} 2\cos t - \sin t \\ 5\cos t \end{pmatrix} e^{4t} + c_2 \begin{pmatrix} 2\sin t + \cos t \\ 5\sin t \end{pmatrix} e^{4t}.$$

36. We have $\det(\mathbf{A} - \lambda\mathbf{I}) = \lambda^2 - 10\lambda + 34 = 0$. For $\lambda_1 = 5 + 3i$ we obtain

$$\mathbf{K}_1 = \begin{pmatrix} 1 - 3i \\ 2 \end{pmatrix}$$

so that

$$\mathbf{X}_1 = \begin{pmatrix} 1 - 3i \\ 2 \end{pmatrix} e^{(5+3i)t} = \begin{pmatrix} \cos 3t + 3\sin 3t \\ 2\cos 3t \end{pmatrix} e^{5t} + i \begin{pmatrix} \sin 3t - 3\cos 3t \\ 2\cos 3t \end{pmatrix} e^{5t}.$$

Then

$$\mathbf{X} = c_1 \begin{pmatrix} \cos 3t + 3\sin 3t \\ 2\cos 3t \end{pmatrix} e^{5t} + c_2 \begin{pmatrix} \sin 3t - 3\cos 3t \\ 2\cos 3t \end{pmatrix} e^{5t}.$$

39. We have $\det(\mathbf{A} - \lambda\mathbf{I}) = -\lambda\left(\lambda^2 + 1\right) = 0$. For $\lambda_1 = 0$ we obtain

$$\mathbf{K}_1 = \begin{pmatrix} 1 \\ 0 \\ 0 \end{pmatrix}.$$

For $\lambda_2 = i$ we obtain

$$\mathbf{K}_2 = \begin{pmatrix} -i \\ i \\ 1 \end{pmatrix}$$

so that

$$\mathbf{X}_2 = \begin{pmatrix} -i \\ i \\ 1 \end{pmatrix} e^{it} = \begin{pmatrix} \sin t \\ -\sin t \\ \cos t \end{pmatrix} + i \begin{pmatrix} -\cos t \\ \cos t \\ \sin t \end{pmatrix}.$$

Then

$$\mathbf{X} = c_1 \begin{pmatrix} 1 \\ 0 \\ 0 \end{pmatrix} + c_2 \begin{pmatrix} \sin t \\ -\sin t \\ \cos t \end{pmatrix} + c_3 \begin{pmatrix} -\cos t \\ \cos t \\ \sin t \end{pmatrix}.$$

42. We have $\det(\mathbf{A} - \lambda\mathbf{I}) = -(\lambda - 6)(\lambda^2 - 8\lambda + 20) = 0$. For $\lambda_1 = 6$ we obtain

$$\mathbf{K}_1 = \begin{pmatrix} 0 \\ 1 \\ 0 \end{pmatrix}.$$

For $\lambda_2 = 4 + 2i$ we obtain

$$\mathbf{K}_2 = \begin{pmatrix} -i \\ 0 \\ 2 \end{pmatrix}$$

so that

$$\mathbf{X}_2 = \begin{pmatrix} -i \\ 0 \\ 2 \end{pmatrix} e^{(4+2i)t} = \begin{pmatrix} \sin 2t \\ 0 \\ 2\cos 2t \end{pmatrix} e^{4t} + i \begin{pmatrix} -\cos 2t \\ 0 \\ 2\sin 2t \end{pmatrix} e^{4t}.$$

Then

$$\mathbf{X} = c_1 \begin{pmatrix} 0 \\ 1 \\ 0 \end{pmatrix} e^{6t} + c_2 \begin{pmatrix} \sin 2t \\ 0 \\ 2\cos 2t \end{pmatrix} e^{4t} + c_3 \begin{pmatrix} -\cos 2t \\ 0 \\ 2\sin 2t \end{pmatrix} e^{4t}.$$

45. We have $\det(\mathbf{A} - \lambda\mathbf{I}) = (1 - \lambda)(\lambda^2 + 25) = 0$. For $\lambda_1 = 1$ we obtain

$$\mathbf{K}_1 = \begin{pmatrix} 25 \\ -7 \\ 6 \end{pmatrix}.$$

For $\lambda_2 = 5i$ we obtain

$$\mathbf{K}_2 = \begin{pmatrix} 1 + 5i \\ 1 \\ 1 \end{pmatrix}$$

so that

$$\mathbf{X}_2 = \begin{pmatrix} 1+5i \\ 1 \\ 1 \end{pmatrix} e^{5it} = \begin{pmatrix} \cos 5t - 5\sin 5t \\ \cos 5t \\ \cos 5t \end{pmatrix} + i \begin{pmatrix} \sin 5t + 5\cos 5t \\ \sin 5t \\ \sin 5t \end{pmatrix}.$$

Then

$$\mathbf{X} = c_1 \begin{pmatrix} 25 \\ -7 \\ 6 \end{pmatrix} e^t + c_2 \begin{pmatrix} \cos 5t - 5\sin 5t \\ \cos 5t \\ \cos 5t \end{pmatrix} + c_3 \begin{pmatrix} \sin 5t + 5\cos 5t \\ \sin 5t \\ \sin 5t \end{pmatrix}.$$

If

$$\mathbf{X}(0) = \begin{pmatrix} 4 \\ 6 \\ -7 \end{pmatrix}$$

then $c_1 = c_2 = -1$ and $c_3 = 6$.

51.

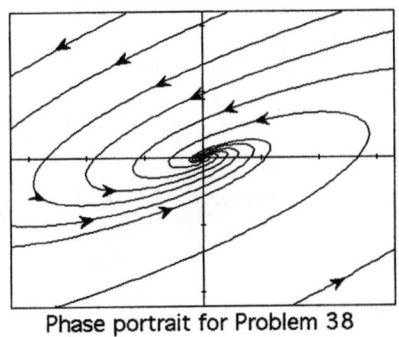

Phase portrait for Problem 36   Phase portrait for Problem 37   Phase portrait for Problem 38

Suppose the eigenvalues are $\alpha \pm i\beta$, $\beta > 0$. In Problem 36 the eigenvalues are $5 \pm 3i$, in Problem 37 they are $\pm 3i$, and in Problem 38 they are $-1 \pm 2i$. From the above pictures we deduce that the phase portrait will consist of a family of closed curves when $\alpha = 0$ and spirals when $\alpha \neq 0$. The origin will be a repellor when $\alpha > 0$, and an attractor when $\alpha < 0$.

Exercises 10.3

3. $\lambda_1 = \frac{1}{2}$, $\lambda_2 = \frac{3}{2}$, $\mathbf{K}_1 = \begin{bmatrix} 1 \\ -2 \end{bmatrix}$, $\mathbf{K}_2 = \begin{bmatrix} 1 \\ 2 \end{bmatrix}$, $\mathbf{P} = \begin{bmatrix} 1 & 1 \\ -2 & 2 \end{bmatrix}$;

$$\mathbf{X} = \mathbf{PY} = \begin{bmatrix} 1 & 1 \\ -2 & 2 \end{bmatrix} \begin{bmatrix} c_1 e^{t/2} \\ c_2 e^{3t/2} \end{bmatrix} = \begin{bmatrix} c_1 e^{t/2} + c_2 e^{3t/2} \\ -2c_1 e^{t/2} + 2c_2 e^{3t/2} \end{bmatrix} = c_1 \begin{bmatrix} 1 \\ -2 \end{bmatrix} e^{t/2} + c_2 \begin{bmatrix} 1 \\ 2 \end{bmatrix} e^{3t/2}$$

6. $\lambda_1 = -1$, $\lambda_2 = 1$, $\lambda_3 = 4$, $\mathbf{K}_1 = \begin{bmatrix} -1 \\ 0 \\ 1 \end{bmatrix}$, $\mathbf{K}_2 = \begin{bmatrix} 1 \\ -2 \\ 1 \end{bmatrix}$, $\mathbf{K}_3 = \begin{bmatrix} 1 \\ 1 \\ 1 \end{bmatrix}$, $\mathbf{P} = \begin{bmatrix} -1 & 1 & 1 \\ 0 & -2 & 1 \\ 1 & 1 & 1 \end{bmatrix}$;

$$\mathbf{X} = \mathbf{PY} = \begin{bmatrix} -1 & 1 & 1 \\ 0 & -2 & 1 \\ 1 & 1 & 1 \end{bmatrix} \begin{bmatrix} c_1 e^{-t} \\ c_2 e^t \\ c_3 e^{4t} \end{bmatrix} = \begin{bmatrix} -c_1 + c_2 e^t + c_3 e^{4t} \\ -2c_2 e^t + c_3 e^{4t} \\ c_1 e^{-t} + c_2 e^t + c_3 e^{4t} \end{bmatrix} = c_1 \begin{bmatrix} -1 \\ 0 \\ 1 \end{bmatrix} e^{-t} + c_2 \begin{bmatrix} 1 \\ -2 \\ 1 \end{bmatrix} e^t + c_3 \begin{bmatrix} 1 \\ 1 \\ 1 \end{bmatrix} e^{4t}$$

9. $\lambda_1 = 1$, $\lambda_2 = 2$, $\lambda_3 = 3$, $\mathbf{K}_1 = \begin{bmatrix} 1 \\ 1 \\ 1 \end{bmatrix}$, $\mathbf{K}_2 = \begin{bmatrix} 2 \\ 2 \\ 3 \end{bmatrix}$, $\mathbf{K}_3 = \begin{bmatrix} 3 \\ 4 \\ 5 \end{bmatrix}$, $\mathbf{P} = \begin{bmatrix} 1 & 2 & 3 \\ 1 & 2 & 4 \\ 1 & 3 & 5 \end{bmatrix}$;

$$\mathbf{X} = \mathbf{PY} = \begin{bmatrix} 1 & 2 & 3 \\ 1 & 2 & 4 \\ 1 & 3 & 5 \end{bmatrix} \begin{bmatrix} c_1 e^t \\ c_2 e^{2t} \\ c_3 e^{3t} \end{bmatrix} = \begin{bmatrix} c_1 e^t + 2c_2 e^{2t} + 3c_3 e^{3t} \\ c_1 e^t + 2c_2 e^{2t} + 4c_3 e^{3t} \\ c_1 e^t + 3c_2 e^{2t} + 5c_3 e^{3t} \end{bmatrix} = c_1 \begin{bmatrix} 1 \\ 1 \\ 1 \end{bmatrix} e^t + c_2 \begin{bmatrix} 2 \\ 2 \\ 3 \end{bmatrix} e^{2t} + c_3 \begin{bmatrix} 3 \\ 4 \\ 5 \end{bmatrix} e^{3t}$$

Exercises 10.4

3. From

$$\mathbf{X}' = \begin{pmatrix} 3 & -5 \\ 3/4 & -1 \end{pmatrix} \mathbf{X} + \begin{pmatrix} 1 \\ -1 \end{pmatrix} e^{t/2}$$

we obtain

$$\mathbf{X}_c = c_1 \begin{pmatrix} 10 \\ 3 \end{pmatrix} e^{3t/2} + c_2 \begin{pmatrix} 2 \\ 1 \end{pmatrix} e^{t/2}.$$

Then

$$\mathbf{\Phi} = \begin{pmatrix} 10e^{3t/2} & 2e^{t/2} \\ 3e^{3t/2} & e^{t/2} \end{pmatrix} \quad \text{and} \quad \mathbf{\Phi}^{-1} = \begin{pmatrix} \frac{1}{4}e^{-3t/2} & -\frac{1}{2}e^{-3t/2} \\ -\frac{3}{4}e^{-t/2} & \frac{5}{2}e^{-t/2} \end{pmatrix}$$

so that

$$\mathbf{U} = \int \mathbf{\Phi}^{-1}\mathbf{F}\,dt = \int \begin{pmatrix} \frac{3}{4}e^{-t} \\ -\frac{13}{4} \end{pmatrix} dt = \begin{pmatrix} -\frac{3}{4}e^{-t} \\ -\frac{13}{4}t \end{pmatrix}$$

and

$$\mathbf{X}_p = \mathbf{\Phi}\mathbf{U} = \begin{pmatrix} -13/2 \\ -13/4 \end{pmatrix} te^{t/2} + \begin{pmatrix} -15/2 \\ -9/4 \end{pmatrix} e^{t/2}.$$

6. From

$$\mathbf{X}' = \begin{pmatrix} 0 & 2 \\ -1 & 3 \end{pmatrix} \mathbf{X} + \begin{pmatrix} 2 \\ e^{-3t} \end{pmatrix}$$

we obtain

$$\mathbf{X}_c = c_1 \begin{pmatrix} 2 \\ 1 \end{pmatrix} e^t + c_2 \begin{pmatrix} 1 \\ 1 \end{pmatrix} e^{2t}.$$

Then

$$\mathbf{\Phi} = \begin{pmatrix} 2e^t & e^{2t} \\ e^t & e^{2t} \end{pmatrix} \quad \text{and} \quad \mathbf{\Phi}^{-1} = \begin{pmatrix} e^{-t} & -e^{-t} \\ -e^{-2t} & 2e^{-2t} \end{pmatrix}$$

so that

$$\mathbf{U} = \int \mathbf{\Phi}^{-1}\mathbf{F}\,dt = \int \begin{pmatrix} 2e^{-t} - e^{-4t} \\ -2e^{-2t} + 2e^{-5t} \end{pmatrix} dt = \begin{pmatrix} -2e^{-t} + \frac{1}{4}e^{-4t} \\ e^{-2t} - \frac{2}{5}e^{-5t} \end{pmatrix}$$

and

$$\mathbf{X}_p = \mathbf{\Phi}\mathbf{U} = \begin{pmatrix} \frac{1}{10}e^{-3t} - 3 \\ -\frac{3}{20}e^{-3t} - 1 \end{pmatrix}.$$

9. From

$$\mathbf{X}' = \begin{pmatrix} 3 & 2 \\ -2 & -1 \end{pmatrix} \mathbf{X} + \begin{pmatrix} 2 \\ 1 \end{pmatrix} e^{-t}$$

we obtain

$$\mathbf{X}_c = c_1 \begin{pmatrix} 1 \\ -1 \end{pmatrix} e^t + c_2 \left[\begin{pmatrix} 1 \\ -1 \end{pmatrix} te^t + \begin{pmatrix} 0 \\ 1/2 \end{pmatrix} e^t \right].$$

Then

$$\mathbf{\Phi} = \begin{pmatrix} e^t & te^t \\ -e^t & \frac{1}{2}e^t - te^t \end{pmatrix} \quad \text{and} \quad \mathbf{\Phi}^{-1} = \begin{pmatrix} e^{-t} - 2te^{-t} & -2te^{-t} \\ 2e^{-t} & 2e^{-t} \end{pmatrix}$$

so that

$$\mathbf{U} = \int \mathbf{\Phi}^{-1}\mathbf{F}\,dt = \int \begin{pmatrix} 2e^{-2t} - 6te^{-2t} \\ 6e^{-2t} \end{pmatrix} dt = \begin{pmatrix} \frac{1}{2}e^{-2t} + 3te^{-2t} \\ -3e^{-2t} \end{pmatrix}$$

and
$$\mathbf{X}_p = \mathbf{\Phi U} = \begin{pmatrix} 1/2 \\ -2 \end{pmatrix} e^{-t}.$$

12. From
$$\mathbf{X}' = \begin{pmatrix} 1 & -1 \\ 1 & 1 \end{pmatrix} \mathbf{X} + \begin{pmatrix} 3 \\ 3 \end{pmatrix} e^t$$

we obtain
$$\mathbf{X}_c = c_1 \begin{pmatrix} -\sin t \\ \cos t \end{pmatrix} e^t + c_2 \begin{pmatrix} \cos t \\ \sin t \end{pmatrix} e^t.$$

Then
$$\mathbf{\Phi} = \begin{pmatrix} -\sin t & \cos t \\ \cos t & \sin t \end{pmatrix} e^t \quad \text{and} \quad \mathbf{\Phi}^{-1} = \begin{pmatrix} -\sin t & \cos t \\ \cos t & \sin t \end{pmatrix} e^{-t}$$

so that
$$\mathbf{U} = \int \mathbf{\Phi}^{-1}\mathbf{F}\, dt = \int \begin{pmatrix} -3\sin t + 3\cos t \\ 3\cos t + 3\sin t \end{pmatrix} dt = \begin{pmatrix} 3\cos t + 3\sin t \\ 3\sin t - 3\cos t \end{pmatrix}$$

and
$$\mathbf{X}_p = \mathbf{\Phi U} = \begin{pmatrix} -3 \\ 3 \end{pmatrix} e^t.$$

15. From
$$\mathbf{X}' = \begin{pmatrix} 0 & 1 \\ -1 & 0 \end{pmatrix} \mathbf{X} + \begin{pmatrix} 0 \\ \sec t \tan t \end{pmatrix}$$

we obtain
$$\mathbf{X}_c = c_1 \begin{pmatrix} \cos t \\ -\sin t \end{pmatrix} + c_2 \begin{pmatrix} \sin t \\ \cos t \end{pmatrix}.$$

Then
$$\mathbf{\Phi} = \begin{pmatrix} \cos t & \sin t \\ -\sin t & \cos t \end{pmatrix} t \quad \text{and} \quad \mathbf{\Phi}^{-1} = \begin{pmatrix} \cos t & -\sin t \\ \sin t & \cos t \end{pmatrix}$$

so that
$$\mathbf{U} = \int \mathbf{\Phi}^{-1}\mathbf{F}\, dt = \int \begin{pmatrix} -\tan^2 t \\ \tan t \end{pmatrix} dt = \begin{pmatrix} t - \tan t \\ \ln|\sec t| \end{pmatrix}$$

and
$$\mathbf{X}_p = \mathbf{\Phi U} = \begin{pmatrix} \cos t \\ -\sin t \end{pmatrix} t + \begin{pmatrix} -\sin t \\ \sin t \tan t \end{pmatrix} + \begin{pmatrix} \sin t \\ \cos t \end{pmatrix} \ln|\sec t|.$$

18. From
$$\mathbf{X}' = \begin{pmatrix} 1 & -2 \\ 1 & -1 \end{pmatrix} \mathbf{X} + \begin{pmatrix} \tan t \\ 1 \end{pmatrix}$$

we obtain
$$\mathbf{X}_c = c_1 \begin{pmatrix} \cos t - \sin t \\ \cos t \end{pmatrix} + c_2 \begin{pmatrix} \cos t + \sin t \\ \sin t \end{pmatrix}.$$

Then
$$\mathbf{\Phi} = \begin{pmatrix} \cos t - \sin t & \cos t + \sin t \\ \cos t & \sin t \end{pmatrix} \quad \text{and} \quad \mathbf{\Phi}^{-1} = \begin{pmatrix} -\sin t & \cos t + \sin t \\ \cos t & \sin t - \cos t \end{pmatrix}$$

so that
$$\mathbf{U} = \int \mathbf{\Phi}^{-1}\mathbf{F}\, dt = \int \begin{pmatrix} 2\cos t + \sin t - \sec t \\ 2\sin t - \cos t \end{pmatrix} dt = \begin{pmatrix} 2\sin t - \cos t - \ln|\sec t + \tan t| \\ -2\cos t - \sin t \end{pmatrix}$$

and
$$\mathbf{X}_p = \mathbf{\Phi U} = \begin{pmatrix} 3\sin t \cos t - \cos^2 t - 2\sin^2 t + (\sin t - \cos t)\ln|\sec t + \tan t| \\ \sin^2 t - \cos^2 t - \cos t(\ln|\sec t + \tan t|) \end{pmatrix}.$$

21. From

$$\mathbf{X}' = \begin{pmatrix} 3 & -1 \\ -1 & 3 \end{pmatrix} \mathbf{X} + \begin{pmatrix} 4e^{2t} \\ 4e^{4t} \end{pmatrix}$$

we obtain

$$\boldsymbol{\Phi} = \begin{pmatrix} -e^{4t} & e^{2t} \\ e^{4t} & e^{2t} \end{pmatrix}, \quad \boldsymbol{\Phi}^{-1} = \begin{pmatrix} -\frac{1}{2}e^{-4t} & \frac{1}{2}e^{-4t} \\ \frac{1}{2}e^{-2t} & \frac{1}{2}e^{-2t} \end{pmatrix},$$

and

$$\mathbf{X} = \boldsymbol{\Phi}\boldsymbol{\Phi}^{-1}(0)\mathbf{X}(0) + \boldsymbol{\Phi}\int_0^t \boldsymbol{\Phi}^{-1}\mathbf{F}\, ds = \boldsymbol{\Phi}\cdot\begin{pmatrix} 0 \\ 1 \end{pmatrix} + \boldsymbol{\Phi}\cdot\begin{pmatrix} e^{-2t} + 2t - 1 \\ e^{2t} + 2t - 1 \end{pmatrix}$$

$$= \begin{pmatrix} 2 \\ 2 \end{pmatrix} te^{2t} + \begin{pmatrix} -1 \\ 1 \end{pmatrix} e^{2t} + \begin{pmatrix} -2 \\ 2 \end{pmatrix} te^{4t} + \begin{pmatrix} 2 \\ 0 \end{pmatrix} e^{4t}.$$

24. $\lambda_1 = -1, \lambda_2 = 4,\ \mathbf{K}_1 = \begin{bmatrix} 3 \\ -2 \end{bmatrix},\ \mathbf{K}_2 = \begin{bmatrix} 1 \\ 1 \end{bmatrix},\ \mathbf{P} = \begin{bmatrix} 3 & 1 \\ -2 & 1 \end{bmatrix},\ \mathbf{P}^{-1} = \frac{1}{5}\begin{bmatrix} 1 & -1 \\ 2 & 3 \end{bmatrix},\ \mathbf{P}^{-1}\mathbf{F} = \begin{bmatrix} 0 \\ e^t \end{bmatrix};$

$$\mathbf{Y}' = \begin{bmatrix} -1 & 0 \\ 0 & 4 \end{bmatrix}\mathbf{Y} + \begin{bmatrix} 0 \\ e^t \end{bmatrix}$$

$$y_1 = c_1 e^{-t}, \qquad y_2 = -\frac{1}{3}e^t + c_2 e^{4t}$$

$$\mathbf{X} = \mathbf{PY} = \begin{bmatrix} 3 & 1 \\ -2 & 1 \end{bmatrix}\begin{bmatrix} c_1 e^{-t} \\ -\frac{1}{3}e^t + c_2 e^{4t} \end{bmatrix} = \begin{bmatrix} -\frac{1}{3}e^t + 3c_1 e^{-t} + c_2 e^{4t} \\ -\frac{1}{3}e^t - 2c_1 e^{-t} + c_2 e^{4t} \end{bmatrix} = c_1\begin{bmatrix} 3 \\ -2 \end{bmatrix}e^{-t} + c_2\begin{bmatrix} 1 \\ 1 \end{bmatrix}e^{4t} - \frac{1}{3}\begin{bmatrix} 1 \\ 1 \end{bmatrix}e^t$$

27. Let $\mathbf{I} = \begin{pmatrix} i_1 \\ i_2 \end{pmatrix}$ so that

$$\mathbf{I}' = \begin{pmatrix} -11 & 3 \\ 3 & -3 \end{pmatrix}\mathbf{I} + \begin{pmatrix} 100\sin t \\ 0 \end{pmatrix}$$

and

$$\mathbf{X}_c = c_1\begin{pmatrix} 1 \\ 3 \end{pmatrix}e^{-2t} + c_2\begin{pmatrix} 3 \\ -1 \end{pmatrix}e^{-12t}.$$

Then

$$\boldsymbol{\Phi} = \begin{pmatrix} e^{-2t} & 3e^{-12t} \\ 3e^{-2t} & -e^{-12t} \end{pmatrix}, \quad \boldsymbol{\Phi}^{-1} = \begin{pmatrix} \frac{1}{10}e^{2t} & \frac{3}{10}e^{2t} \\ \frac{3}{10}e^{12t} & -\frac{1}{10}e^{12t} \end{pmatrix},$$

$$\mathbf{U} = \int \boldsymbol{\Phi}^{-1}\mathbf{F}\, dt = \int \begin{pmatrix} 10e^{2t}\sin t \\ 30e^{12t}\sin t \end{pmatrix} dt = \begin{pmatrix} 2e^{2t}(2\sin t - \cos t) \\ \frac{6}{29}e^{12t}(12\sin t - \cos t) \end{pmatrix},$$

and

$$\mathbf{I}_p = \boldsymbol{\Phi}\mathbf{U} = \begin{pmatrix} \frac{332}{29}\sin t - \frac{76}{29}\cos t \\ \frac{276}{29}\sin t - \frac{168}{29}\cos t \end{pmatrix}$$

so that

$$\mathbf{I} = c_1\begin{pmatrix} 1 \\ 3 \end{pmatrix}e^{-2t} + c_2\begin{pmatrix} 3 \\ -1 \end{pmatrix}e^{-12t} + \mathbf{I}_p.$$

If $\mathbf{I}(0) = \begin{pmatrix} 0 \\ 0 \end{pmatrix}$ then $c_1 = 2$ and $c_2 = \frac{6}{29}$.

Exercises 10.5

3. For

$$\mathbf{A} = \begin{pmatrix} 1 & 1 & 1 \\ 1 & 1 & 1 \\ -2 & -2 & -2 \end{pmatrix}$$

we have

$$\mathbf{A}^2 = \begin{pmatrix} 1 & 1 & 1 \\ 1 & 1 & 1 \\ -2 & -2 & -2 \end{pmatrix}\begin{pmatrix} 1 & 1 & 1 \\ 1 & 1 & 1 \\ -2 & -2 & -2 \end{pmatrix} = \begin{pmatrix} 0 & 0 & 0 \\ 0 & 0 & 0 \\ 0 & 0 & 0 \end{pmatrix}.$$

Thus, $\mathbf{A}^3 = \mathbf{A}^4 = \mathbf{A}^5 = \cdots = \mathbf{0}$ and

$$e^{\mathbf{A}t} = \mathbf{I} + \mathbf{A}t = \begin{pmatrix} 1 & 0 & 0 \\ 0 & 1 & 0 \\ 0 & 0 & 1 \end{pmatrix} + \begin{pmatrix} t & t & t \\ t & t & t \\ -2t & -2t & -2t \end{pmatrix} = \begin{pmatrix} t+1 & t & t \\ t & t+1 & t \\ -2t & -2t & -2t+1 \end{pmatrix}.$$

6. Using the result of Problem 2

$$\mathbf{X} = \begin{pmatrix} \cosh t & \sinh t \\ \sinh t & \cosh t \end{pmatrix}\begin{pmatrix} c_1 \\ c_2 \end{pmatrix} = c_1 \begin{pmatrix} \cosh t \\ \sinh t \end{pmatrix} + c_2 \begin{pmatrix} \sinh t \\ \cosh t \end{pmatrix}.$$

9. To solve

$$\mathbf{X}' = \begin{pmatrix} 1 & 0 \\ 0 & 2 \end{pmatrix}\mathbf{X} + \begin{pmatrix} 3 \\ -1 \end{pmatrix}$$

we identify $t_0 = 0$, $\mathbf{F}(s) = \begin{pmatrix} 3 \\ -1 \end{pmatrix}$, and use the results of Problem 1 and equation (6) in the text.

$$\mathbf{X}(t) = e^{\mathbf{A}t}\mathbf{C} + e^{\mathbf{A}t}\int_{t_0}^{t} e^{-\mathbf{A}s}\mathbf{F}(s)\,ds$$

$$= \begin{pmatrix} e^t & 0 \\ 0 & e^{2t} \end{pmatrix}\begin{pmatrix} c_1 \\ c_2 \end{pmatrix} + \begin{pmatrix} e^t & 0 \\ 0 & e^{2t} \end{pmatrix}\int_0^t \begin{pmatrix} e^{-s} & 0 \\ 0 & e^{-2s} \end{pmatrix}\begin{pmatrix} 3 \\ -1 \end{pmatrix}\,ds$$

$$= \begin{pmatrix} c_1 e^t \\ c_2 e^{2t} \end{pmatrix} + \begin{pmatrix} e^t & 0 \\ 0 & e^{2t} \end{pmatrix}\int_0^t \begin{pmatrix} 3e^{-s} \\ -e^{-2s} \end{pmatrix}\,ds$$

$$= \begin{pmatrix} c_1 e^t \\ c_2 e^{2t} \end{pmatrix} + \begin{pmatrix} e^t & 0 \\ 0 & e^{2t} \end{pmatrix}\begin{pmatrix} -3e^{-s} \\ \frac{1}{2}e^{-2s} \end{pmatrix}\Big|_0^t$$

$$= \begin{pmatrix} c_1 e^t \\ c_2 e^{2t} \end{pmatrix} + \begin{pmatrix} e^t & 0 \\ 0 & e^{2t} \end{pmatrix}\begin{pmatrix} -3e^{-t} - 3 \\ \frac{1}{2}e^{-2t} - \frac{1}{2} \end{pmatrix}$$

$$= \begin{pmatrix} c_1 e^t \\ c_2 e^{2t} \end{pmatrix} + \begin{pmatrix} -3 - 3e^t \\ \frac{1}{2} - \frac{1}{2}e^{2t} \end{pmatrix} = c_3 \begin{pmatrix} 1 \\ 0 \end{pmatrix}e^t + c_4 \begin{pmatrix} 0 \\ 1 \end{pmatrix}e^{2t} + \begin{pmatrix} -3 \\ \frac{1}{2} \end{pmatrix}.$$

12. To solve

$$\mathbf{X}' = \begin{pmatrix} 0 & 1 \\ 1 & 0 \end{pmatrix}\mathbf{X} + \begin{pmatrix} \cosh t \\ \sinh t \end{pmatrix}$$

we identify $t_0 = 0$, $\mathbf{F}(s) = \begin{pmatrix} \cosh t \\ \sinh t \end{pmatrix}$, and use the results of Problem 2 and equation (6) in the text.

$$\mathbf{X}(t) = e^{\mathbf{A}t}\mathbf{C} + e^{\mathbf{A}t} \int_{t_0}^{t} e^{-\mathbf{A}s}\mathbf{F}(s)\,ds$$

$$= \begin{pmatrix} \cosh t & \sinh t \\ \sinh t & \cosh t \end{pmatrix}\begin{pmatrix} c_1 \\ c_2 \end{pmatrix} + \begin{pmatrix} \cosh t & \sinh t \\ \sinh t & \cosh t \end{pmatrix} \int_0^t \begin{pmatrix} \cosh s & -\sinh s \\ -\sinh s & \cosh s \end{pmatrix}\begin{pmatrix} \cosh s \\ \sinh s \end{pmatrix}\,ds$$

$$= \begin{pmatrix} c_1\cosh t + c_2\sinh t \\ c_1\sinh t + c_2\cosh t \end{pmatrix} + \begin{pmatrix} \cosh t & \sinh t \\ \sinh t & \cosh t \end{pmatrix} \int_0^t \begin{pmatrix} 1 \\ 0 \end{pmatrix}\,ds$$

$$= \begin{pmatrix} c_1\cosh t + c_2\sinh t \\ c_1\sinh t + c_2\cosh t \end{pmatrix} + \begin{pmatrix} \cosh t & \sinh t \\ \sinh t & \cosh t \end{pmatrix} \begin{pmatrix} s \\ 0 \end{pmatrix}\Big|_0^t$$

$$= \begin{pmatrix} c_1\cosh t + c_2\sinh t \\ c_1\sinh t + c_2\cosh t \end{pmatrix} + \begin{pmatrix} \cosh t & \sinh t \\ \sinh t & \cosh t \end{pmatrix} \begin{pmatrix} t \\ 0 \end{pmatrix}$$

$$= \begin{pmatrix} c_1\cosh t + c_2\sinh t \\ c_1\sinh t + c_2\cosh t \end{pmatrix} + \begin{pmatrix} t\cosh t \\ t\sinh t \end{pmatrix} = c_1\begin{pmatrix} \cosh t \\ \sinh t \end{pmatrix} + c_2\begin{pmatrix} \sinh t \\ \cosh t \end{pmatrix} + t\begin{pmatrix} \cosh t \\ \sinh t \end{pmatrix}.$$

15. From $s\mathbf{I} - \mathbf{A} = \begin{pmatrix} s-4 & -3 \\ 4 & s+4 \end{pmatrix}$ we find

$$(s\mathbf{I} - \mathbf{A})^{-1} = \begin{pmatrix} \dfrac{3/2}{s-2} - \dfrac{1/2}{s+2} & \dfrac{3/4}{s-2} - \dfrac{3/4}{s+2} \\[3mm] \dfrac{-1}{s-2} + \dfrac{1}{s+2} & \dfrac{-1/2}{s-2} + \dfrac{3/2}{s+2} \end{pmatrix}$$

and

$$e^{\mathbf{A}t} = \begin{pmatrix} \frac{3}{2}e^{2t} - \frac{1}{2}e^{-2t} & \frac{3}{4}e^{2t} - \frac{3}{4}e^{-2t} \\[2mm] -e^{2t} + e^{-2t} & -\frac{1}{2}e^{2t} + \frac{3}{2}e^{-2t} \end{pmatrix}.$$

The general solution of the system is then

$$\mathbf{X} = e^{\mathbf{A}t}\mathbf{C} = \begin{pmatrix} \frac{3}{2}e^{2t} - \frac{1}{2}e^{-2t} & \frac{3}{4}e^{2t} - \frac{3}{4}e^{-2t} \\[2mm] -e^{2t} + e^{-2t} & -\frac{1}{2}e^{2t} + \frac{3}{2}e^{-2t} \end{pmatrix}\begin{pmatrix} c_1 \\ c_2 \end{pmatrix}$$

$$= c_1\begin{pmatrix} 3/2 \\ -1 \end{pmatrix}e^{2t} + c_1\begin{pmatrix} -1/2 \\ 1 \end{pmatrix}e^{-2t} + c_2\begin{pmatrix} 3/4 \\ -1/2 \end{pmatrix}e^{2t} + c_2\begin{pmatrix} -3/4 \\ 3/2 \end{pmatrix}e^{-2t}$$

$$= \left(\frac{1}{2}c_1 + \frac{1}{4}c_2\right)\begin{pmatrix} 3 \\ -2 \end{pmatrix}e^{2t} + \left(-\frac{1}{2}c_1 - \frac{3}{4}c_2\right)\begin{pmatrix} 1 \\ -2 \end{pmatrix}e^{-2t}$$

$$= c_3\begin{pmatrix} 3 \\ -2 \end{pmatrix}e^{2t} + c_4\begin{pmatrix} 1 \\ -2 \end{pmatrix}e^{-2t}.$$

18. From $s\mathbf{I} - \mathbf{A} = \begin{pmatrix} s & -1 \\ 2 & s+2 \end{pmatrix}$ we find

$$(s\mathbf{I} - \mathbf{A})^{-1} = \begin{pmatrix} \dfrac{s+1+1}{(s+1)^2+1} & \dfrac{1}{(s+1)^2+1} \\[3mm] \dfrac{-2}{(s+1)^2+1} & \dfrac{s+1+1}{(s+1)^2+1} \end{pmatrix}$$

and

$$e^{\mathbf{A}t} = \begin{pmatrix} e^{-t}\cos t + e^{-t}\sin t & e^{-t}\sin t \\[2mm] -2e^{-t}\sin t & e^{-t}\cos t - e^{-t}\sin t \end{pmatrix}.$$

The general solution of the system is then

$$\mathbf{X} = e^{\mathbf{A}t}\mathbf{C} = \begin{pmatrix} e^{-t}\cos t + e^{-t}\sin t & e^{-t}\sin t \\ -2e^{-t}\sin t & e^{-t}\cos t - e^{-t}\sin t \end{pmatrix}\begin{pmatrix} c_1 \\ c_2 \end{pmatrix}$$

$$= c_1\begin{pmatrix} 1 \\ 0 \end{pmatrix}e^{-t}\cos t + c_1\begin{pmatrix} 1 \\ -2 \end{pmatrix}e^{-t}\sin t + c_2\begin{pmatrix} 0 \\ 1 \end{pmatrix}e^{-t}\cos t + c_2\begin{pmatrix} 1 \\ -1 \end{pmatrix}e^{-t}\sin t$$

$$= c_1\begin{pmatrix} \cos t + \sin t \\ -2\sin t \end{pmatrix}e^{-t} + c_2\begin{pmatrix} \sin t \\ \cos t - \sin t \end{pmatrix}e^{-t}.$$

21. The eigenvalues are $\lambda_1 = -1$ and $\lambda_2 = 3$. This leads to the system

$$e^{-t} = b_0 - b_1$$

$$e^{3t} = b_0 + 3b_1,$$

which has the solution $b_0 = \frac{3}{4}e^{-t} + \frac{1}{4}e^{3t}$ and $b_1 = -\frac{1}{4}e^{-t} + \frac{1}{4}e^{3t}$. Then

$$e^{\mathbf{A}t} = b_0\mathbf{I} + b_1\mathbf{A} = \begin{pmatrix} e^{3t} & -2e^{-t} + 2e^{3t} \\ 0 & e^{-t} \end{pmatrix}.$$

The general solution of the system is then

$$\mathbf{X} = e^{\mathbf{A}t}\mathbf{C} = \begin{pmatrix} e^{3t} & -2e^{-t} + 2e^{3t} \\ 0 & e^{-t} \end{pmatrix}\begin{pmatrix} c_1 \\ c_2 \end{pmatrix} = c_1\begin{pmatrix} 1 \\ 0 \end{pmatrix}e^{3t} + c_2\begin{pmatrix} -2 \\ 1 \end{pmatrix}e^{-t} + c_2\begin{pmatrix} 2 \\ 0 \end{pmatrix}e^{3t}$$

$$= c_2\begin{pmatrix} -2 \\ 1 \end{pmatrix}e^{-t} + (c_1 + 2c_2)\begin{pmatrix} 1 \\ 0 \end{pmatrix}e^{3t} = c_3\begin{pmatrix} -2 \\ 1 \end{pmatrix}e^{-t} + c_4\begin{pmatrix} 1 \\ 0 \end{pmatrix}e^{3t}.$$

24. From equation (2) in the text

$$e^{\mathbf{D}t} = \begin{bmatrix} 1 & 0 & \cdots & 0 \\ 0 & 1 & \cdots & 0 \\ \vdots & \vdots & \ddots & \vdots \\ 0 & 0 & \cdots & 1 \end{bmatrix} + \begin{bmatrix} \lambda_1 & 0 & \cdots & 0 \\ 0 & \lambda_2 & \cdots & 0 \\ \vdots & \vdots & \ddots & \vdots \\ 0 & 0 & \cdots & \lambda_n \end{bmatrix} + \frac{1}{2!}t^2\begin{bmatrix} \lambda_1^2 & 0 & \cdots & 0 \\ 0 & \lambda_2^2 & \cdots & 0 \\ \vdots & \vdots & \ddots & \vdots \\ 0 & 0 & \cdots & \lambda_n^2 \end{bmatrix}$$

$$+ \frac{1}{3!}t^3\begin{bmatrix} \lambda_1^3 & 0 & \cdots & 0 \\ 0 & \lambda_2^3 & \cdots & 0 \\ \vdots & \vdots & \ddots & \vdots \\ 0 & 0 & \cdots & \lambda_n^3 \end{bmatrix} + \cdots$$

$$= \begin{bmatrix} 1 + \lambda_1 t + \frac{1}{2!}(\lambda_1 t)^2 + \cdots & 0 & \cdots & 0 \\ 0 & 1 + \lambda_2 t + \frac{1}{2!}(\lambda_2 t)^2 + \cdots & \cdots & 0 \\ \vdots & \vdots & \ddots & \vdots \\ 0 & 0 & \cdots & 1 + \lambda_n t + \frac{1}{2!}(\lambda_n t)^2 + \cdots \end{bmatrix}$$

$$= \begin{bmatrix} e^{\lambda_1 t} & 0 & \cdots & 0 \\ 0 & e^{\lambda_2 t} & \cdots & 0 \\ \vdots & \vdots & \ddots & \vdots \\ 0 & 0 & \cdots & e^{\lambda_n t} \end{bmatrix}$$

Chapter 10 Review Exercises

3. Since

$$\begin{pmatrix} 4 & 6 & 6 \\ 1 & 3 & 2 \\ -1 & -4 & -3 \end{pmatrix} \begin{pmatrix} 3 \\ 1 \\ -1 \end{pmatrix} = \begin{pmatrix} 12 \\ 4 \\ -4 \end{pmatrix} = 4 \begin{pmatrix} 3 \\ 1 \\ -1 \end{pmatrix},$$

we see that $\lambda = 4$ is an eigenvalue with eigenvector $\mathbf{K}_3$. The corresponding solution is $\mathbf{X} = \mathbf{K}_3 e^{4t}$.

6. We have $\det(\mathbf{A} - \lambda\mathbf{I}) = (\lambda + 6)(\lambda + 2) = 0$ so that

$$\mathbf{X} = c_1 \begin{pmatrix} 1 \\ -1 \end{pmatrix} e^{-6t} + c_2 \begin{pmatrix} 1 \\ 1 \end{pmatrix} e^{-2t}.$$

9. We have $\det(\mathbf{A} - \lambda\mathbf{I}) = -(\lambda - 2)(\lambda - 4)(\lambda + 3) = 0$ so that

$$\mathbf{X} = c_1 \begin{pmatrix} -2 \\ 3 \\ 1 \end{pmatrix} e^{2t} + c_2 \begin{pmatrix} 0 \\ 1 \\ 1 \end{pmatrix} e^{4t} + c_3 \begin{pmatrix} 7 \\ 12 \\ -16 \end{pmatrix} e^{-3t}.$$

12. We have

$$\mathbf{X}_c = c_1 \begin{pmatrix} 2\cos t \\ -\sin t \end{pmatrix} e^t + c_2 \begin{pmatrix} 2\sin t \\ \cos t \end{pmatrix} e^t.$$

Then

$$\mathbf{\Phi} = \begin{pmatrix} 2\cos t & 2\sin t \\ -\sin t & \cos t \end{pmatrix} e^t, \quad \mathbf{\Phi}^{-1} = \begin{pmatrix} \frac{1}{2}\cos t & -\sin t \\ \frac{1}{2}\sin t & \cos t \end{pmatrix} e^{-t},$$

and

$$\mathbf{U} = \int \mathbf{\Phi}^{-1}\mathbf{F}\,dt = \int \begin{pmatrix} \cos t - \sec t \\ \sin t \end{pmatrix} dt = \begin{pmatrix} \sin t - \ln|\sec t + \tan t| \\ -\cos t \end{pmatrix},$$

so that

$$\mathbf{X}_p = \mathbf{\Phi}\mathbf{U} = \begin{pmatrix} -2\cos t \ln|\sec t + \tan t| \\ -1 + \sin t \ln|\sec t + \tan t| \end{pmatrix}$$

15. (a) Letting

$$\mathbf{K} = \begin{pmatrix} k_1 \\ k_2 \\ k_3 \end{pmatrix}$$

we note that $(\mathbf{A} - 2\mathbf{I})\mathbf{K} = \mathbf{0}$ implies that $3k_1 + 3k_2 + 3k_3 = 0$, so $k_1 = -(k_2 + k_3)$. Choosing $k_2 = 0$, $k_3 = 1$ and then $k_2 = 1$, $k_3 = 0$ we get

$$\mathbf{K}_1 = \begin{pmatrix} -1 \\ 0 \\ 1 \end{pmatrix} \quad \text{and} \quad \mathbf{K}_2 = \begin{pmatrix} -1 \\ 1 \\ 0 \end{pmatrix},$$

respectively. Thus,

$$\mathbf{X}_1 = \begin{pmatrix} -1 \\ 0 \\ 1 \end{pmatrix} e^{2t} \quad \text{and} \quad \mathbf{X}_2 = \begin{pmatrix} -1 \\ 1 \\ 0 \end{pmatrix} e^{2t}$$

are two solutions.

(b) From $\det(\mathbf{A} - \lambda\mathbf{I}) = \lambda^2(3 - \lambda) = 0$ we see that $\lambda_1 = 3$, and 0 is an eigenvalue of multiplicity two. Letting

$$\mathbf{K} = \begin{pmatrix} k_1 \\ k_2 \\ k_3 \end{pmatrix},$$

as in part **(a)**, we note that $(\mathbf{A} - 0\mathbf{I})\mathbf{K} = \mathbf{A}\mathbf{K} = \mathbf{0}$ implies that $k_1 + k_2 + k_3 = 0$, so $k_1 = -(k_2 + k_3)$. Choosing $k_2 = 0$, $k_3 = 1$, and then $k_2 = 1$, $k_3 = 0$ we get

$$\mathbf{K}_2 = \begin{pmatrix} -1 \\ 0 \\ 1 \end{pmatrix} \quad \text{and} \quad \mathbf{K}_3 = \begin{pmatrix} -1 \\ 1 \\ 0 \end{pmatrix},$$

respectively. Since the eigenvector corresponding to $\lambda_1 = 3$ is

$$\mathbf{K}_1 \begin{pmatrix} 1 \\ 1 \\ 1 \end{pmatrix},$$

the general solution of the system is

$$\mathbf{X} = c_1 \begin{pmatrix} 1 \\ 1 \\ 1 \end{pmatrix} e^{3t} + c_2 \begin{pmatrix} -1 \\ 0 \\ 1 \end{pmatrix} + c_3 \begin{pmatrix} -1 \\ 1 \\ 0 \end{pmatrix}.$$

11 Systems of Nonlinear Differential Equations

Exercises 11.1

3. The corresponding plane autonomous system is

$$x' = y, \quad y' = x^2 - y(1 - x^3).$$

If (x, y) is a critical point, $y = 0$ and so $x^2 - y(1 - x^3) = x^2 = 0$. Therefore $(0, 0)$ is the sole critical point.

6. The corresponding plane autonomous system is

$$x' = y, \quad y' = -x + \epsilon x |x|.$$

If (x, y) is a critical point, $y = 0$ and $-x + \epsilon x |x| = x(-1 + \epsilon |x|) = 0$. Hence $x = 0, 1/\epsilon, -1/\epsilon$. The critical points are $(0, 0)$, $(1/\epsilon, 0)$ and $(-1/\epsilon, 0)$.

9. From $x - y = 0$ we have $y = x$. Substituting into $3x^2 - 4y = 0$ we obtain $3x^2 - 4x = x(3x - 4) = 0$. It follows that $(0, 0)$ and $(4/3, 4/3)$ are the critical points of the system.

12. Adding the two equations we obtain $10 - 15 \dfrac{y}{y + 5} = 0$. It follows that $y = 10$, and from $-2x + y + 10 = 0$ we may conclude that $x = 10$. Therefore $(10, 10)$ is the sole critical point of the system.

15. From $x(1 - x^2 - 3y^2) = 0$ we have $x = 0$ or $x^2 + 3y^2 = 1$. If $x = 0$, then substituting into $y(3 - x^2 - 3y^2)$ gives $y(3 - 3y^2) = 0$. Therefore $y = 0, 1, -1$. Likewise $x^2 = 1 - 3y^2$ yields $2y = 0$ so that $y = 0$ and $x^2 = 1 - 3(0)^2 = 1$. The critical points of the system are therefore $(0, 0)$, $(0, 1)$, $(0, -1)$, $(1, 0)$, and $(-1, 0)$.

18. (a) From Exercises 10.2, Problem 6, $x = c_1 + 2c_2 e^{-5t}$ and $y = 3c_1 + c_2 e^{-5t}$.

(b) From $\mathbf{X}(0) = (3, 4)$ it follows that $c_1 = c_2 = 1$. Therefore $x = 1 + 2e^{-5t}$ and $y = 3 + e^{-5t}$.

(c)

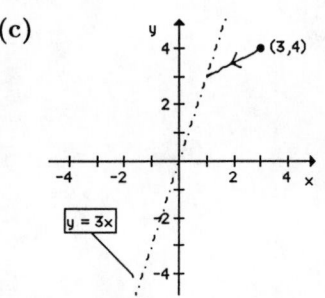

21. (a) From Exercises 10.2, Problem 35, $x = c_1(\sin t - \cos t)e^{4t} + c_2(-\sin t - \cos t)e^{4t}$ and $y = 2c_1(\cos t) e^{4t} + 2c_2(\sin t) e^{4t}$. Because of the presence of e^{4t}, there are no periodic solutions.

(b) From $\mathbf{X}(0) = (-1, 2)$ it follows that $c_1 = 1$ and $c_2 = 0$. Therefore $x = (\sin t - \cos t)e^{4t}$ and $y = 2(\cos t) e^{4t}$.

(c)

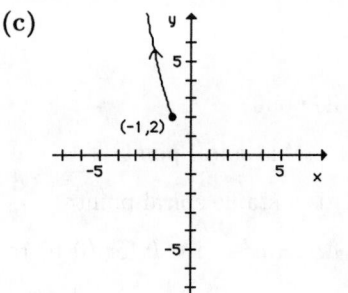

24. Switching to polar coordinates,

$$\frac{dr}{dt} = \frac{1}{r}\left(x\frac{dx}{dt} + y\frac{dy}{dt}\right) = \frac{1}{r}(xy - x^2r^2 - xy + y^2r^2) = r^3$$

$$\frac{d\theta}{dt} = \frac{1}{r^2}\left(-y\frac{dx}{dt} + x\frac{dy}{dt}\right) = \frac{1}{r^2}(-y^2 - xyr^2 - x^2 + xyr^2) = -1.$$

If we use separation of variables, it follows that

$$r = \frac{1}{\sqrt{-2t + c_1}} \quad \text{and} \quad \theta = -t + c_2.$$

Since $\mathbf{X}(0) = (4, 0)$, $r = 4$ and $\theta = 0$ when $t = 0$. It follows that $c_2 = 0$ and $c_1 = \frac{1}{16}$. The final solution may be written as

$$r = \frac{4}{\sqrt{1 - 32t}}, \quad \theta = -t.$$

Note that $r \to \infty$ as $t \to \left(\frac{1}{32}\right)^-$. Because $0 \le t \le \frac{1}{32}$, the curve is not a spiral.

27. The system has no critical points, so there are no periodic solutions.

30. The system has no critical points, so there are no periodic solutions.

Exercises 11.2

3. (a) All solutions are unstable spirals which become unbounded as t increases.

(b)

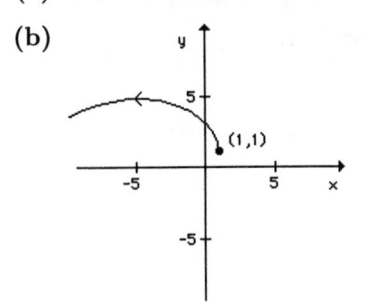

6. (a) All solutions become unbounded and $y = x/2$ serves as the asymptote.

(b)

9. Since $\Delta = -41 < 0$, we may conclude from Figure 11.18 that $(0, 0)$ is a saddle point.

12. Since $\Delta = 1$ and $\tau = -1$, $\tau^2 - 4\Delta = -3$ and so from Figure 11.18, $(0, 0)$ is a stable spiral point.

15. Since $\Delta = 0.01$ and $\tau = -0.03$, $\tau^2 - 4\Delta < 0$ and so from Figure 11.18, $(0, 0)$ is a stable spiral point.

18. Note that $\Delta = 1$ and $\tau = \mu$. Therefore we need both $\tau = \mu < 0$ and $\tau^2 - 4\Delta = \mu^2 - 4 < 0$ for $(0, 0)$ to be a stable spiral point. These two conditions may be written as $-2 < \mu < 0$.

21. $\mathbf{AX}_1 + \mathbf{F} = \mathbf{0}$ implies that $\mathbf{AX}_1 = -\mathbf{F}$ or $\mathbf{X}_1 = -\mathbf{A}^{-1}\mathbf{F}$. Since $\mathbf{X}_p(t) = -\mathbf{A}^{-1}\mathbf{F}$ is a particular solution, it follows from Theorem 10.6 that $\mathbf{X}(t) = \mathbf{X}_c(t) + \mathbf{X}_1$ is the general solution to $\mathbf{X}' = \mathbf{AX} + \mathbf{F}$. If $\tau < 0$ and $\Delta > 0$ then $\mathbf{X}_c(t)$ approaches $(0,0)$ by Theorem 11.1(a). It follows that $\mathbf{X}(t)$ approaches $\mathbf{X}_1$ as $t \to \infty$.

24. **(a)** The critical point is $\mathbf{X}_1 = (-1, -2)$.

(b) From the graph, $\mathbf{X}_1$ appears to be a stable node or a degenerate stable node.

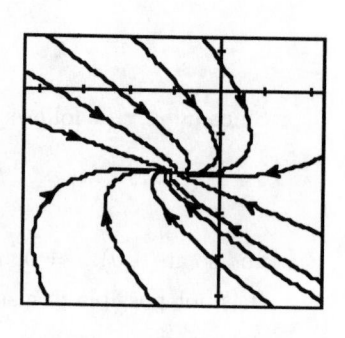

(c) Since $\tau = -16$, $\Delta = 64$, and $\tau^2 - 4\Delta = 0$, $(0,0)$ is a degenerate stable node.

Exercises 11.3

3. The critical points are $x = 0$ and $x = n + 1$. Since $g'(x) = k(n+1) - 2kx$, $g'(0) = k(n+1) > 0$ and $g'(n+1) = -k(n+1) < 0$. Therefore $x = 0$ is unstable while $x = n+1$ is asymptotically stable. See Theorem 11.2.

6. The only critical point is $v = mg/k$. Now $g(v) = g - (k/m)v$ and so $g'(v) = -k/m < 0$. Therefore $v = mg/k$ is an asymptotically stable critical point by Theorem 11.2.

9. Critical points occur at $P = a/b$, c but not at $P = 0$. Since $g'(P) = (a - bP) + (P - c)(-b)$,

$$g'(a/b) = (a/b - c)(-b) = -a + bc \quad \text{and} \quad g'(c) = a - bc.$$

Since $a < bc$, $-a + bc > 0$ and $a - bc < 0$. Therefore $P = a/b$ is unstable while $P = c$ is asymptotically stable.

12. Critical points are $(1, 0)$ and $(-1, 0)$, and

$$\mathbf{g}'(\mathbf{X}) = \begin{pmatrix} 2x & -2y \\ 0 & 2 \end{pmatrix}.$$

At $\mathbf{X} = (1, 0)$, $\tau = 4$, $\Delta = 4$, and so $\tau^2 - 4\Delta = 0$. We may conclude that $(1, 0)$ is unstable but we are unable to classify this critical point any further. At $\mathbf{X} = (-1, 0)$, $\Delta = -4 < 0$ and so $(-1, 0)$ is a saddle point.

15. Since $x^2 - y^2 = 0$, $y^2 = x^2$ and so $x^2 - 3x + 2 = (x-1)(x-2) = 0$. It follows that the critical points are $(1, 1)$, $(1, -1)$, $(2, 2)$, and $(2, -2)$. We next use the Jacobian

$$\mathbf{g}'(\mathbf{X}) = \begin{pmatrix} -3 & 2y \\ 2x & -2y \end{pmatrix}$$

to classify these four critical points. For $\mathbf{X} = (1, 1)$, $\tau = -5$, $\Delta = 2$, and so $\tau^2 - 4\Delta = 17 > 0$. Therefore $(1, 1)$ is a stable node. For $\mathbf{X} = (1, -1)$, $\Delta = -2 < 0$ and so $(1, -1)$ is a saddle point. For $\mathbf{X} = (2, 2)$, $\Delta = -4 < 0$ and so we have another saddle point. Finally, if $\mathbf{X} = (2, -2)$, $\tau = 1$, $\Delta = 4$, and so $\tau^2 - 4\Delta = -15 < 0$. Therefore $(2, -2)$ is an unstable spiral point.

18. We found that $(0, 0)$, $(0, 1)$, $(0, -1)$, $(1, 0)$ and $(-1, 0)$ were the critical points in Exercise 15, Section 11.1. The Jacobian is

$$\mathbf{g}'(\mathbf{X}) = \begin{pmatrix} 1 - 3x^2 - 3y^2 & -6xy \\ -2xy & 3 - x^2 - 9y^2 \end{pmatrix}.$$

For $\mathbf{X} = (0, 0)$, $\tau = 4$, $\Delta = 3$ and so $\tau^2 - 4\Delta = 4 > 0$. Therefore $(0, 0)$ is an unstable node. Both $(0, 1)$ and $(0, -1)$ give $\tau = -8$, $\Delta = 12$, and $\tau^2 - 4\Delta = 16 > 0$. These two critical points are therefore stable nodes. For $\mathbf{X} = (1, 0)$ or $(-1, 0)$, $\Delta = -4 < 0$ and so saddle points occur.

21. The corresponding plane autonomous system is

$$\theta' = y, \quad y' = \left(\cos\theta - \frac{1}{2}\right)\sin\theta.$$

Since $|\theta| < \pi$, it follows that critical points are $(0, 0)$, $(\pi/3, 0)$ and $(-\pi/3, 0)$. The Jacobian matrix is

$$\mathbf{g}'(\mathbf{X}) = \begin{pmatrix} 0 & 1 \\ \cos 2\theta - \frac{1}{2}\cos\theta & 0 \end{pmatrix}$$

and so at $(0, 0)$, $\tau = 0$ and $\Delta = -1/2$. Therefore $(0, 0)$ is a saddle point. For $\mathbf{X} = (\pm\pi/3, 0)$, $\tau = 0$ and $\Delta = 3/4$. It is not possible to classify either critical point in this borderline case.

24. The corresponding plane autonomous system is

$$x' = y, \quad y' = -\frac{4x}{1 + x^2} - 2y$$

and the only critical point is $(0, 0)$. Since the Jacobian matrix is

$$\mathbf{g}'(\mathbf{X}) = \begin{pmatrix} 0 & 1 \\ -4\dfrac{1 - x^2}{(1 + x^2)^2} & -2 \end{pmatrix},$$

$\tau = -2$, $\Delta = 4$, $\tau^2 - 4\Delta = -12$, and so $(0, 0)$ is a stable spiral point.

27. The corresponding plane autonomous system is

$$x' = y, \quad y' = -\frac{(\beta + \alpha^2 y^2)x}{1 + \alpha^2 x^2}$$

and the Jacobian matrix is

$$\mathbf{g}'(\mathbf{X}) = \begin{pmatrix} 0 & 1 \\ \dfrac{(\beta + \alpha y^2)(\alpha^2 x^2 - 1)}{(1 + \alpha^2 x^2)^2} & \dfrac{-2\alpha^2 yx}{1 + \alpha^2 x^2} \end{pmatrix}.$$

For $\mathbf{X} = (0, 0)$, $\tau = 0$ and $\Delta = \beta$. Since $\beta < 0$, we may conclude that $(0, 0)$ is a saddle point.

30. (a) The corresponding plane autonomous system is

$$x' = y, \quad y' = \epsilon\left(y - \frac{1}{3}y^3\right) - x$$

and so the only critical point is $(0, 0)$. Since the Jacobian matrix is

$$\mathbf{g}'(\mathbf{X}) = \begin{pmatrix} 0 & 1 \\ -1 & \epsilon(1 - y^2) \end{pmatrix},$$

$\tau = \epsilon$, $\Delta = 1$, and so $\tau^2 - 4\Delta = \epsilon^2 - 4$ at the critical point $(0, 0)$.

(b) When $\tau = \epsilon > 0$, $(0, 0)$ is an unstable critical point.

(c) When $\epsilon < 0$ and $\tau^2 - 4\Delta = \epsilon^2 - 4 < 0$, $(0, 0)$ is a stable spiral point. These two requirements can be written as $-2 < \epsilon < 0$.

(d) When $\epsilon = 0$, $x'' + x = 0$ and so $x = c_1 \cos t + c_2 \sin t$. Therefore all solutions are periodic (with period 2π) and so $(0, 0)$ is a center.

33. (a) $x' = 2xy = 0$ implies that either $x = 0$ or $y = 0$. If $x = 0$, then from $1 - x^2 + y^2 = 0$, $y^2 = -1$ and there are no real solutions. If $y = 0$, $1 - x^2 = 0$ and so $(1, 0)$ and $(-1, 0)$ are critical points. The Jacobian matrix is

$$\mathbf{g}'(\mathbf{X}) = \begin{pmatrix} 2y & 2x \\ -2x & 2y \end{pmatrix}$$

and so $\tau = 0$ and $\Delta = 4$ at either $\mathbf{X} = (1, 0)$ or $(-1, 0)$. We obtain no information about these critical points in this borderline case.

(b) $\dfrac{dy}{dx} = \dfrac{y'}{x'} = \dfrac{1 - x^2 + y^2}{2xy}$ or $2xy\dfrac{dy}{dx} = 1 - x^2 + y^2$. Letting $\mu = \dfrac{y^2}{x}$, it follows

that $\dfrac{d\mu}{dx} = \dfrac{1}{x^2} - 1$ and so $\mu = -\dfrac{1}{x} - x + 2c$. Therefore $\dfrac{y^2}{x} = -\dfrac{1}{x} - x + 2c$

which can be put in the form

$$(x - c)^2 + y^2 = c^2 - 1.$$

The solution curves are shown and so both $(1, 0)$ and $(-1, 0)$ are centers.

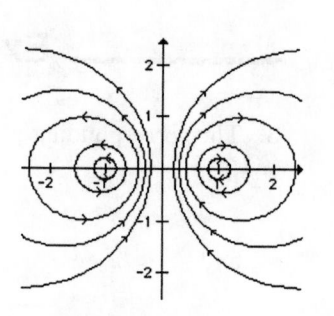

36. The corresponding plane autonomous system is

$$x' = y, \quad y' = \epsilon x^2 - x + 1$$

and so the critical points must satisfy $y = 0$ and

$$x = \frac{1 \pm \sqrt{1 - 4\epsilon}}{2\epsilon}.$$

Therefore we must require that $\epsilon \le \frac{1}{4}$ for real solutions to exist. We will use the Jacobian matrix

$$\mathbf{g}'(\mathbf{X}) = \begin{pmatrix} 0 & 1 \\ 2\epsilon x - 1 & 0 \end{pmatrix}$$

to attempt to classify $((1 \pm \sqrt{1 - 4\epsilon})/2\epsilon, 0)$ when $\epsilon \le 1/4$. Note that $\tau = 0$ and $\Delta = \mp\sqrt{1 - 4\epsilon}$. For $\mathbf{X} = ((1 + \sqrt{1 - 4\epsilon})/2\epsilon, 0)$ and $\epsilon < 1/4$, $\Delta < 0$ and so a saddle point occurs. For $\mathbf{X} = ((1 - \sqrt{1 - 4\epsilon})/2\epsilon, 0)$ $\Delta \ge 0$ and we are not able to classify this critical point using linearization.

39. (a) Letting $x = \theta$ and $y = x'$ we obtain the system $x' = y$ and $y' = 1/2 - \sin x$. Since $\sin \pi/6 = \sin 5\pi/6 = 1/2$ we see that $(\pi/6, 0)$ and $(5\pi/6, 0)$ are critical points of the system.

(b) The Jacobian matrix is

$$\mathbf{g}'(\mathbf{X}) = \begin{pmatrix} 0 & 1 \\ -\cos x & 0 \end{pmatrix}$$

and so

$$\mathbf{A}_1 = \mathbf{g}' = ((\pi/6, 0)) = \begin{pmatrix} 0 & 1 \\ -\sqrt{3}/2 & 0 \end{pmatrix} \quad \text{and} \quad \mathbf{A}_2 = \mathbf{g}' = ((5\pi/6, 0)) = \begin{pmatrix} 0 & 1 \\ \sqrt{3}/2 & 0 \end{pmatrix}.$$

Since $\det \mathbf{A}_1 > 0$ and the trace of $\mathbf{A}_1$ is 0, no conclusion can be drawn regarding the critical point $(\pi/6, 0)$. Since $\det \mathbf{A}_2 < 0$, we see that $(5\pi/6, 0)$ is a saddle point.

(c) From the system in part (a) we obtain the first-order differential equation

$$\frac{dy}{dx} = \frac{1/2 - \sin x}{y}.$$

Separating variables and integrating we obtain

$$\int y \, dy = \int \left(\frac{1}{2} - \sin x\right) dx$$

and

$$\frac{1}{2}y^2 = \frac{1}{2}x + \cos x + c_1$$

or

$$y^2 = x + 2\cos x + c_2.$$

For x_0 near $\pi/6$, if $\mathbf{X}(0) = (x_0, 0)$ then $c_2 = -x_0 - 2\cos x_0$ and $y^2 = x + 2\cos x - x_0 - 2\cos x_0$. Thus, there are two values of y for each x in a sufficiently small interval around $\pi/6$. Therefore $(\pi/6, 0)$ is a center.

Exercises 11.4

3. The corresponding plane autonomous system is

$$x' = y, \quad y' = -g\frac{f'(x)}{1 + [f'(x)]^2} - \frac{\beta}{m}y$$

and

$$\frac{\partial}{\partial x}\left(-g\frac{f'(x)}{1 + [f'(x)]^2} - \frac{\beta}{m}y\right) = -g\frac{(1 + [f'(x)]^2)f''(x) - f'(x)2f'(x)f''(x)}{(1 + [f'(x)]^2)^2}.$$

If $\mathbf{X}_1 = (x_1, y_1)$ is a critical point, $y_1 = 0$ and $f'(x_1) = 0$. The Jacobian at this critical point is therefore

$$\mathbf{g}'(\mathbf{X}_1) = \begin{pmatrix} 0 & 1 \\ -gf''(x_1) & -\frac{\beta}{m} \end{pmatrix}.$$

6. (a) If $f(x) = \cosh x$, $f'(x) = \sinh x$ and $[f'(x)]^2 + 1 = \sinh^2 x + 1 = \cosh^2 x$. Therefore

$$\frac{dy}{dx} = \frac{y'}{x'} = -g\frac{\sinh x}{\cosh^2 x}\frac{1}{y}.$$

We may separate variables to show that $y^2 = \frac{2g}{\cosh x} + c$. But $x(0) = x_0$ and $y(0) = x'(0) = v_0$. Therefore $c = v_0^2 - \frac{2g}{\cosh x_0}$ and so

$$y^2 = \frac{2g}{\cosh x} - \frac{2g}{\cosh x_0} + v_0^2.$$

Now

$$\frac{2g}{\cosh x} - \frac{2g}{\cosh x_0} + v_0^2 \geq 0 \quad \text{if and only if} \quad \cosh x \leq \frac{2g\cosh x_0}{2g - v_0^2\cosh x_0}$$

and the solution to this inequality is an interval $[-a, a]$. Therefore each x in $(-a, a)$ has two corresponding values of y and so the solution is periodic.

(b) Since $z = \cosh x$, the maximum height occurs at the largest value of x on the cycle. From (a), $x_{\max} = a$ where $\cosh a = \frac{2g\cosh x_0}{2g - v_0^2\cosh x_0}$. Therefore

$$z_{\max} = \frac{2g\cosh x_0}{2g - v_0^2\cosh x_0}.$$

9. (a) In the Lotka-Volterra Model the average number of predators is d/c and the average number of prey is a/b. But

$$x' = -ax + bxy - \epsilon_1 x = -(a + \epsilon_1)x + bxy$$

$$y' = -cxy + dy - \epsilon_2 y = -cxy + (d - \epsilon_2)y$$

and so the new critical point in the first quadrant is $(d/c - \epsilon_2/c, a/b + \epsilon_1/b)$.

(b) The average number of predators $d/c - \epsilon_2/c$ has decreased while the average number of prey $a/b + \epsilon_1/b$ has increased. The fishery science model is consistent with Volterra's principle.

12. $\Delta = r_1 r_2$, $\tau = r_1 + r_2$ and $\tau^2 - 4\Delta = (r_1 + r_2)^2 - 4r_1 r_2 = (r_1 - r_2)^2$. Therefore when $r_1 \neq r_2$, $(0,0)$ is an unstable node.

15. $\dfrac{K_1}{\alpha_{12}} < K_2 < K_1 \alpha_{21}$ and so $\alpha_{12}\alpha_{21} > 1$. Therefore $\Delta = (1 - \alpha_{12}\alpha_{21})\hat{x}\hat{y}\dfrac{r_1 r_2}{K_1 K_2} < 0$ and so $(\hat{x}, \hat{y})$ is a saddle point.

18. (a) The magnitude of the frictional force between the bead and the wire is $\mu(mg\cos\theta)$ for some $\mu > 0$. The component of this frictional force in the x-direction is

$$(\mu mg \cos\theta)\cos\theta = \mu mg \cos^2\theta.$$

But

$$\cos\theta = \frac{1}{\sqrt{1 + [f'(x)]^2}} \quad \text{and so} \quad \mu mg\cos^2\theta = \frac{\mu mg}{1 + [f'(x)]^2}.$$

It follows from Newton's Second Law that

$$mx'' = -mg\frac{f'(x)}{1 + [f'(x)]^2} - \beta x' + mg\frac{\mu}{1 + [f'(x)]^2}$$

and so

$$x'' = g\frac{\mu - f'(x)}{1 + [f'(x)]^2} - \frac{\beta}{m}x'.$$

(b) A critical point (x, y) must satisfy $y = 0$ and $f'(x) = \mu$. Therefore critical points occur at $(x_1, 0)$ where $f'(x_1) = \mu$. The Jacobian matrix of the plane autonomous system is

$$\mathbf{g}'(\mathbf{X}) = \begin{pmatrix} 0 & 1 \\ g\dfrac{(1 + [f'(x)]^2)(-f''(x)) - (\mu - f'(x))2f'(x)f''(x)}{(1 + [f'(x)]^2)^2} & -\dfrac{\beta}{m} \end{pmatrix}$$

and so at a critical point $\mathbf{X}_1$,

$$\mathbf{g}'(\mathbf{X}) = \begin{pmatrix} 0 & 1 \\ \dfrac{-gf''(x_1)}{1 + \mu^2} & -\dfrac{\beta}{m} \end{pmatrix}.$$

Therefore $\tau = -\dfrac{\beta}{m} < 0$ and $\Delta = \dfrac{gf''(x_1)}{1 + \mu^2}$. When $f''(x_1) < 0$, $\Delta < 0$ and so a saddle point occurs. When $f''(x_1) > 0$ and

$$\tau^2 - 4\Delta = \frac{\beta^2}{m^2} - 4g\frac{f''(x_1)}{1 + \mu^2} < 0,$$

$(x_1, 0)$ is a stable spiral point. This condition may also be written as

$$\beta^2 < 4gm^2\frac{f''(x_1)}{1 + \mu^2}.$$

21. The equation

$$x' = \alpha\frac{y}{1 + y}x - x = x\left(\frac{\alpha y}{1 + y} - 1\right) = 0$$

implies that $x = 0$ or $y = \dfrac{1}{\alpha - 1}$. When $\alpha > 0$, $\hat{y} = \dfrac{1}{\alpha - 1} > 0$. If $x = 0$, then from the differential equation for y', $y = \beta$. On the other hand, if $\hat{y} = \dfrac{1}{\alpha - 1}$, $\dfrac{\hat{y}}{1 + \hat{y}} = \dfrac{1}{\alpha}$ and so $\dfrac{1}{\alpha}\hat{x} - \dfrac{1}{\alpha - 1} + \beta = 0$. It follows that

$$\hat{x} = \alpha\left(\beta - \frac{1}{\alpha - 1}\right) = \frac{\alpha}{\alpha - 1}[(\alpha - 1)\beta - 1]$$

and if $\beta(\alpha - 1) > 1$, $\hat{x} > 0$. Therefore $(\hat{x}, \hat{y})$ is the unique critical point in the first quadrant. The Jacobian matrix is

$$\mathbf{g}'(\mathbf{X}) = \begin{pmatrix} \alpha \dfrac{y}{y+1} - 1 & \dfrac{\alpha x}{(1+y)^2} \\ -\dfrac{y}{1+y} & \dfrac{-x}{(1+y)^2} - 1 \end{pmatrix}$$

and for $\mathbf{X} = (\hat{x}, \hat{y})$, the Jacobian can be written in the form

$$\mathbf{g}'((\hat{x}, \hat{y})) = \begin{pmatrix} 0 & \dfrac{(\alpha - 1)^2}{\alpha} \hat{x} \\ -\dfrac{1}{\alpha} & -\dfrac{(\alpha - 1)^2}{\alpha^2} - 1 \end{pmatrix}.$$

It follows that

$$\tau = -\left[\frac{(\alpha - 1)^2}{\alpha^2} \hat{x} + 1 \right] < 0, \quad \Delta = \frac{(\alpha - 1)^2}{\alpha^2} \hat{x}$$

and so $\tau = -(\Delta + 1)$. Therefore $\tau^2 - 4\Delta = (\Delta + 1)^2 - 4\Delta = (\Delta - 1)^2 > 0$. Therefore $(\hat{x}, \hat{y})$ is a stable node.

Exercises 11.5

3. For $P = -x + y^2$ and $Q = x - y$, $\dfrac{\partial P}{\partial x} + \dfrac{\partial Q}{\partial y} = -2 < 0$. Therefore there are no periodic solutions by Theorem 11.5.

6. From $y' = xy - y = y(x - 1) = 0$ either $y = 0$ or $x = 1$. If $y = 0$, then from $2x + y^2 = 0$, $x = 0$. Likewise $x = 1$ implies that $2 + y^2 = 0$, which has no real solutions. Therefore $(0, 0)$ is the only critical point. But $\mathbf{g}'((0, 0))$ has determinant $\Delta = -2$. The single critical point is a saddle point and so, by the corollary to Theorem 11.4, there are no periodic solutions.

9. For $\delta(x, y) = e^{ax+by}$, $\dfrac{\partial}{\partial x}(\delta P) + \dfrac{\partial}{\partial y}(\delta Q)$ can be simplified to

$$e^{ax+by}[-bx^2 - 2ax + axy + (2b + 1)y].$$

Setting $a = 0$ and $b = -1/2$,

$$\frac{\partial}{\partial}(\delta P) + \frac{\partial}{\partial y}(\delta Q) = \frac{1}{2}x^2 e^{-\frac{1}{2}y}$$

which does not change signs. Therefore by Theorem 11.6 there are no periodic solutions.

12. The corresponding plane autonomous system is $x' = y$, $y' = g(x, y)$ and so

$$\frac{\partial P}{\partial x} + \frac{\partial Q}{\partial y} = \frac{\partial g}{\partial y} = \frac{\partial g}{\partial x'} \neq 0$$

in the region R. Therefore $\dfrac{\partial P}{\partial x} + \dfrac{\partial Q}{\partial y}$ cannot change signs and so there are no periodic solutions by Theorem 11.5.

15. If $\mathbf{n} = (-2x, -2y)$,

$$\mathbf{V} \cdot \mathbf{n} = 2x^2 - 4xy + 2y^2 + 2y^4 = 2(x - y)^2 + 2y^4 \geq 0.$$

Therefore $x^2 + y^2 \leq r$ serves as an invariant region for any $r > 0$ by Theorem 11.7.

18. The corresponding plane autonomous system is

$$x' = y, \quad y' = y(1 - 3x^2 - 2y^2) - x$$

and it is easy to see that $(0,0)$ is the only critical point. If $\mathbf{n} = (-2x, -2y)$ then

$$\mathbf{V} \cdot \mathbf{n} = -2xy - 2y^2(1 - 3x^2 - 2y^2) + 2xy = -2y^2(1 - 2r^2 - x^2).$$

If $r = \frac{1}{2}\sqrt{2}$, $2r^2 = 1$ and so $\mathbf{V} \cdot \mathbf{n} = 2x^2y^2 \geq 0$. Therefore $\frac{1}{4} \leq x^2 + y^2 \leq \frac{1}{2}$ serves as an invariant region. By Theorem 11.8(ii) there is at least one periodic solution.

21. (a) $\dfrac{\partial P}{\partial x} + \dfrac{\partial Q}{\partial y} = 2xy - 1 - x^2 \leq 2x - 1 - x^2 = -(x-1)^2 \leq 0$. Therefore there are no periodic solutions.

(b) If (x, y) is a critical point, $x^2y = \frac{1}{2}$ and so from $x' = x^2y - x + 1$, $\frac{1}{2} - x + 1 = 0$. Therefore $x = 3/2$ and so $y = 2/9$. For this critical point, $\tau = -31/12 < 0$, $\Delta = 9/4 > 0$, and $\tau^2 - 4\Delta < 0$. Therefore $(3/2, 2/9)$ is a stable spiral point and so, from Theorem 11.9(ii), $\lim_{x \to \infty} \mathbf{X}(t) = (3/2, 2/9)$.

——— Chapter 11 Review Exercises ———

3. A center or a saddle point

6. True

9. True

12. (a) If $\mathbf{X}(0) = \mathbf{X}_0$ lies on the line $y = -2x$, then $\mathbf{X}(t)$ approaches $(0,0)$ along this line. For all other initial conditions, $\mathbf{X}(t)$ approaches $(0,0)$ from the direction determined by the line $y = x$.

(b) If $\mathbf{X}(0) = \mathbf{X}_0$ lies on the line $y = -x$, then $\mathbf{X}(t)$ approaches $(0,0)$ along this line. For all other initial conditions, $\mathbf{X}(t)$ becomes unbounded and $y = 2x$ serves as an asymptote.

15. The corresponding plane autonomous system is $x' = y$, $y' = \mu(1 - x^2) - x$ and so the Jacobian at the critical point $(0,0)$ is

$$\mathbf{g}'((0,0)) = \begin{bmatrix} 0 & 1 \\ -1 & \mu \end{bmatrix}.$$

Therefore $\tau = \mu$, $\Delta = 1$ and $\tau^2 - 4\Delta = \mu^2 - 4$. Now $\mu^2 - 4 < 0$ if and only if $-2 < \mu < 2$. We may therefore conclude that $(0,0)$ is a stable node for $\mu < -2$, a stable spiral point for $-2 < \mu < 0$, an unstable spiral point for $0 < \mu < 2$, and an unstable node for $\mu > 2$.

18. The corresponding plane autonomous system is

$$x' = y, \quad y' = -\frac{\beta}{m}y - \frac{k}{m}(s + x)^3 + g$$

and so the Jacobian is

$$\mathbf{g}'(\mathbf{X}) = \begin{bmatrix} 0 & 1 \\ -\frac{3k}{m}(s + x)^2 & -\frac{\beta}{m} \end{bmatrix}.$$

For $\mathbf{X} = (0,0)$, $\tau = -\dfrac{\beta}{m} < 0$, $\Delta = \dfrac{3k}{m}s^2 > 0$. Therefore

$$\tau^2 - 4\Delta = \frac{\beta^2}{m^2} - \frac{12k}{m}s^2 = \frac{1}{m^2}(\beta^2 - 12kms^2).$$

Therefore $(0,0)$ is a stable node if $\beta^2 > 12kms^2$ and a stable spiral point provided $\beta^2 < 12kms^2$, where $ks^3 = mg$.

21. (a) If $x = \theta$ and $y = x' = \theta'$, the corresponding plane autonomous system is

$$x' = y, \quad y' = \omega^2 \sin x \cos x - \frac{g}{l}\sin x - \frac{\beta}{ml}y.$$

Therefore $\dfrac{\partial P}{\partial x} + \dfrac{\partial Q}{\partial y} = -\dfrac{\beta}{ml} < 0$ and so there are no periodic solutions.

(b) If (x, y) is a critical point, $y = 0$ and so $\sin x(\omega^2 \cos x - g/l) = 0$. Either $\sin x = 0$ (in which case $x = 0$) of $\cos x = g/\omega^2 l$. But if $\omega^2 < g/l$, $g/\omega^2 l > 1$ and so the latter equation has no real solutions. Therefore $(0, 0)$ is the only critical point if $\omega^2 < g/l$. The Jacobian matrix is

$$\mathbf{g}'(\mathbf{X}) = \begin{bmatrix} 0 & 1 \\ \omega^2 \cos 2x - \frac{g}{l}\cos x & -\frac{\beta}{ml} \end{bmatrix}$$

and so $\tau = -\beta/ml < 0$ and $\Delta = g/l - \omega^2 > 0$ for $\mathbf{X} = (0, 0)$. It follows that $(0, 0)$ is asymptotically stable and so after a small displacement, the pendulum will return to $\theta = 0$, $\theta' = 0$.

(c) If $\omega^2 > g/l$, $\cos x = g/\omega^2 l$ will have two solutions $x = \pm\hat{x}$ that satisfy $-\pi < x < \pi$. Therefore $(\pm\hat{x}, 0)$ are two additional critical points. If $\mathbf{X}_1 = (0, 0)$, $\Delta = g/l - \omega^2 < 0$ and so $(0, 0)$ is a saddle point. If $\mathbf{X}_1 = (\pm\hat{x}, 0)$, $\tau = -\beta/ml < 0$ and

$$\Delta = \frac{g}{l}\cos\hat{x} - \omega^2 \cos 2\hat{x} = \frac{g^2}{\omega^2 l^2} - \omega^2\left(2\frac{g^2}{\omega^4 l^2} - 1\right) = \omega^2 - \frac{g^2}{\omega^2 l^2} > 0.$$

Therefore $(\hat{x}, 0)$ and $(-\hat{x}, 0)$ are each stable. When $\theta(0) = \theta_0$, $\theta'(0) = 0$ and θ_0 is small we expect the pendulum to reach one of these two stable equilibrium positions.

(d) In (b), $(0, 0)$ is a stable spiral point provided

$$\tau^2 - 4\Delta = \frac{\beta^2}{m^2 l^2} - 4\left(\frac{g}{l} - \omega^2\right) < 0.$$

This condition is equivalent to $\beta < 2ml\sqrt{g/l - \omega^2}$. In (c), $(\pm\hat{x}, 0)$ are stable spiral points provided that

$$\tau^2 - 4\Delta = \frac{\beta^2}{m^2 l^2} - 4\left(\omega^2 - \frac{g^2}{\omega^2 l^2}\right) < 0.$$

This condition is equivalent to $\beta < 2ml\sqrt{\omega^2 - g^2/(\omega^2 l^2)}$.

12 Orthogonal Functions and Fourier Series

3. $\displaystyle\int_0^2 e^x(xe^{-x} - e^{-x})\,dx = \int_0^2 (x-1)\,dx = \left(\frac{1}{2}x^2 - x\right)\Big|_0^2 = 0$

6. $\displaystyle\int_{\pi/4}^{5\pi/4} e^x \sin x\,dx = \left(\frac{1}{2}e^x \sin x - \frac{1}{2}e^x \cos x\right)\Big|_{\pi/4}^{5\pi/4} = 0$

9. For $m \neq n$

$$\int_0^\pi \sin nx \sin mx\,dx = \frac{1}{2}\int_0^\pi [\cos(n-m)x - \cos(n+m)x]\,dx$$

$$= \frac{1}{2(n-m)}\sin(n-m)x\,\Big|_0^\pi - \frac{1}{2(n+m)}\sin 2(n+m)x\,\Big|_0^\pi$$

$$= 0.$$

For $m = n$

$$\int_0^\pi \sin^2 nx\,dx = \int_0^\pi \left[\frac{1}{2} - \frac{1}{2}\cos 2nx\right]\,dx = \frac{1}{2}x\,\Big|_0^\pi - \frac{1}{4n}\sin 2nx\,\Big|_0^\pi = \frac{\pi}{2}$$

so that

$$\|\sin nx\| = \sqrt{\frac{\pi}{2}}\,.$$

12. For $m \neq n$, we use Problems 11 and 10:

$$\int_{-p}^p \cos\frac{n\pi}{p}x \cos\frac{m\pi}{p}x\,dx = 2\int_0^p \cos\frac{n\pi}{p}x \cos\frac{m\pi}{p}x\,dx = 0$$

$$\int_{-p}^p \sin\frac{n\pi}{p}x \sin\frac{m\pi}{p}x\,dx = 2\int_0^p \sin\frac{n\pi}{p}x \sin\frac{m\pi}{p}x\,dx = 0.$$

Also

$$\int_{-p}^p \sin\frac{n\pi}{p}x \cos\frac{m\pi}{p}x\,dx = \frac{1}{2}\int_{-p}^p \left(\sin\frac{(n-m)\pi}{p}x + \sin\frac{(n+m)\pi}{p}x\right)\,dx = 0,$$

$$\int_{-p}^p 1 \cdot \cos\frac{n\pi}{p}x\,dx = \frac{p}{n\pi}\sin\frac{n\pi}{p}x\,\Big|_{-p}^p = 0,$$

$$\int_{-p}^p 1 \cdot \sin\frac{n\pi}{p}x\,dx = -\frac{p}{n\pi}\cos\frac{n\pi}{p}x\,\Big|_{-p}^p = 0,$$

and

$$\int_{-p}^p \sin\frac{n\pi}{p}x \cos\frac{n\pi}{p}x\,dx = \int_{-p}^p \frac{1}{2}\sin\frac{2n\pi}{p}x\,dx = -\frac{p}{4n\pi}\cos\frac{2n\pi}{p}x\,\Big|_{-p}^p = 0.$$

For $m = n$

$$\int_{-p}^p \cos^2\frac{n\pi}{p}x\,dx = \int_{-p}^p \left(\frac{1}{2} + \frac{1}{2}\cos\frac{2n\pi}{p}x\right)\,dx = p,$$

$$\int_{-p}^p \sin^2\frac{n\pi}{p}x\,dx = \int_{-p}^p \left(\frac{1}{2} - \frac{1}{2}\cos\frac{2n\pi}{p}x\right)\,dx = p,$$

and

$$\int_{-p}^{p} 1^2 dx = 2p$$

so that

$$\|1\| = \sqrt{2p}, \quad \left\|\cos\frac{n\pi}{p}x\right\| = \sqrt{p}, \quad \text{and} \quad \left\|\sin\frac{n\pi}{p}x\right\| = \sqrt{p}.$$

15. By orthogonality $\int_a^b \phi_0(x)\phi_n(x)dx = 0$ for $n = 1, 2, 3, \ldots$; that is, $\int_a^b \phi_n(x)dx = 0$ for $n = 1, 2, 3, \ldots$.

18. Setting

$$0 = \int_{-2}^{2} f_3(x)f_1(x)\, dx = \int_{-2}^{2} \left(x^2 + c_1 x^3 + c_2 x^4\right) dx = \frac{16}{3} + \frac{64}{5}c_2$$

and

$$0 = \int_{-2}^{2} f_3(x)f_2(x)\, dx = \int_{-2}^{2} \left(x^3 + c_1 x^4 + c_2 x^5\right) dx = \frac{64}{5}c_1$$

we obtain $c_1 = 0$ and $c_2 = -5/12$.

21. (a) The fundamental period is $2\pi/2\pi = 1$.

(b) The fundamental period is $2\pi/(4/L) = \frac{1}{2}\pi L$.

(c) The fundamental period of $\sin x + \sin 2x$ is 2π.

(d) The fundamental period of $\sin 2x + \cos 4x$ is $2\pi/2 = \pi$.

(e) The fundamental period of $\sin 3x + \cos 4x$ is 2π since the smallest integer multiples of $2\pi/3$ and $2\pi/4 = \pi/2$ that are equal are 3 and 4, respectively.

(f) The fundamental period of $f(x)$ is $2\pi/(n\pi/p) = 2p/n$.

Exercises 12.2

3. $a_0 = \int_{-1}^{1} f(x)\, dx = \int_{-1}^{0} 1\, dx + \int_{0}^{1} x\, dx = \frac{3}{2}$

$a_n = \int_{-1}^{1} f(x)\cos n\pi x\, dx = \int_{-1}^{0} \cos n\pi x\, dx + \int_{0}^{1} x\cos n\pi x\, dx = \frac{1}{n^2\pi^2}[(-1)^n - 1]$

$b_n = \int_{-1}^{1} f(x)\sin n\pi x\, dx = \int_{-1}^{0} \sin n\pi x\, dx + \int_{0}^{1} x\sin n\pi x\, dx = -\frac{1}{n\pi}$

$f(x) = \frac{3}{4} + \sum_{n=1}^{\infty}\left[\frac{(-1)^n - 1}{n^2\pi^2}\cos n\pi x - \frac{1}{n\pi}\sin n\pi x\right]$

6. $a_0 = \frac{1}{\pi}\int_{-\pi}^{\pi} f(x)\, dx = \frac{1}{\pi}\int_{-\pi}^{0} \pi^2\, dx + \frac{1}{\pi}\int_{0}^{\pi} \left(\pi^2 - x^2\right) dx = \frac{5}{3}\pi^2$

$a_n = \frac{1}{\pi}\int_{-\pi}^{\pi} f(x)\cos nx\, dx = \frac{1}{\pi}\int_{-\pi}^{0} \pi^2\cos nx\, dx + \frac{1}{\pi}\int_{0}^{\pi} \left(\pi^2 - x^2\right)\cos nx\, dx$

$= \frac{1}{\pi}\left(\frac{\pi^2 - x^2}{n}\sin nx\Big|_0^\pi + \frac{2}{n}\int_{0}^{\pi} x\sin nx\, dx\right) = \frac{2}{n^2}(-1)^{n+1}$

$b_n = \frac{1}{\pi}\int_{-\pi}^{\pi} f(x)\sin nx\, dx = \frac{1}{\pi}\int_{-\pi}^{0} \pi^2\sin nx\, dx + \frac{1}{\pi}\int_{0}^{\pi} \left(\pi^2 - x^2\right)\sin nx\, dx$

$= \frac{\pi}{n}[(-1)^n - 1] + \frac{1}{\pi}\left(\frac{x^2 - \pi^2}{n}\cos nx\Big|_0^\pi - \frac{2}{n}\int_{0}^{\pi} x\cos nx\, dx\right) = \frac{\pi}{n}(-1)^n + \frac{2}{n^3\pi}[1 - (-1)^n]$

$$f(x) = \frac{5\pi^2}{6} + \sum_{n=1}^{\infty} \left[\frac{2}{n^2}(-1)^{n+1}\cos nx + \left(\frac{\pi}{n}(-1)^n + \frac{2[1-(-1^n)]}{n^3\pi} \right) \sin nx \right]$$

9. $a_0 = \dfrac{1}{\pi} \displaystyle\int_{-\pi}^{\pi} f(x)\,dx = \dfrac{1}{\pi} \displaystyle\int_{0}^{\pi} \sin x\,dx = \dfrac{2}{\pi}$

$a_n = \dfrac{1}{\pi} \displaystyle\int_{-\pi}^{\pi} f(x)\cos nx\,dx = \dfrac{1}{\pi} \displaystyle\int_{0}^{\pi} \sin x \cos nx\,dx = \dfrac{1}{2\pi} \displaystyle\int_{0}^{\pi} [\sin(n+1)x + \sin(1-n)x]\,dx$

$\quad = \dfrac{1+(-1)^n}{\pi(1-n^2)} \quad$ for $n = 2, 3, 4, \ldots$

$a_1 = \dfrac{1}{2\pi} \displaystyle\int_{0}^{\pi} \sin 2x\,dx = 0$

$b_n = \dfrac{1}{\pi} \displaystyle\int_{-\pi}^{\pi} f(x)\sin nx\,dx = \dfrac{1}{\pi} \displaystyle\int_{0}^{\pi} \sin x \sin nx\,dx$

$\quad = \dfrac{1}{2\pi} \displaystyle\int_{0}^{\pi} [\cos(1-n)x - \cos(1+n)x]\,dx = 0 \quad$ for $n = 2, 3, 4, \ldots$

$b_1 = \dfrac{1}{2\pi} \displaystyle\int_{0}^{\pi} (1 - \cos 2x)\,dx = \dfrac{1}{2}$

$f(x) = \dfrac{1}{\pi} + \dfrac{1}{2}\sin x + \displaystyle\sum_{n=2}^{\infty} \dfrac{1+(-1)^n}{\pi(1-n^2)} \cos nx$

12. $a_0 = \dfrac{1}{2} \displaystyle\int_{-2}^{2} f(x)\,dx = \dfrac{1}{2} \left(\displaystyle\int_{0}^{1} x\,dx + \displaystyle\int_{1}^{2} 1\,dx \right) = \dfrac{3}{4}$

$a_n = \dfrac{1}{2} \displaystyle\int_{-2}^{2} f(x)\cos\dfrac{n\pi}{2}x\,dx = \dfrac{1}{2} \left(\displaystyle\int_{0}^{1} x\cos\dfrac{n\pi}{2}x\,dx + \displaystyle\int_{1}^{2} \cos\dfrac{n\pi}{2}x\,dx \right) = \dfrac{2}{n^2\pi^2} \left(\cos\dfrac{n\pi}{2} - 1 \right)$

$b_n = \dfrac{1}{2} \displaystyle\int_{-2}^{2} f(x)\sin\dfrac{n\pi}{2}x\,dx = \dfrac{1}{2} \left(\displaystyle\int_{0}^{1} x\sin\dfrac{n\pi}{2}x\,dx + \displaystyle\int_{1}^{2} \sin\dfrac{n\pi}{2}x\,dx \right) = \dfrac{2}{n^2\pi^2} \left(\sin\dfrac{n\pi}{2} + \dfrac{n\pi}{2}(-1)^{n+1} \right)$

$f(x) = \dfrac{3}{8} + \displaystyle\sum_{n=1}^{\infty} \left[\dfrac{2}{n^2\pi^2} \left(\cos\dfrac{n\pi}{2} - 1 \right) \cos\dfrac{n\pi}{2}x + \dfrac{2}{n^2\pi^2} \left(\sin\dfrac{n\pi}{2} + \dfrac{n\pi}{2}(-1)^{n+1} \right) \sin\dfrac{n\pi}{2}x \right]$

15. $a_0 = \dfrac{1}{\pi} \displaystyle\int_{-\pi}^{\pi} f(x)\,dx = \dfrac{1}{\pi} \displaystyle\int_{-\pi}^{\pi} e^x\,dx = \dfrac{1}{\pi}(e^\pi - e^{-\pi})$

$a_n = \dfrac{1}{\pi} \displaystyle\int_{-\pi}^{\pi} f(x)\cos nx\,dx = \dfrac{(-1)^n(e^\pi - e^{-\pi})}{\pi(1+n^2)}$

$b_n = \dfrac{1}{\pi} \displaystyle\int_{-\pi}^{\pi} f(x)\sin nx\,dx = \dfrac{1}{\pi} \displaystyle\int_{-\pi}^{\pi} e^x \sin nx\,dx = \dfrac{(-1)^n n(e^{-\pi} - e^\pi)}{\pi(1+n^2)}$

$f(x) = \dfrac{e^\pi - e^{-\pi}}{2\pi} + \displaystyle\sum_{n=1}^{\infty} \left[\dfrac{(-1)^n(e^\pi - e^{-\pi})}{\pi(1+n^2)} \cos nx + \dfrac{(-1)^n n(e^{-\pi} - e^\pi)}{\pi(1+n^2)} \sin nx \right]$

18. From Problem 17

$$\frac{\pi^2}{8} = \frac{1}{2}\left(\frac{\pi^2}{6} + \frac{\pi^2}{12} \right) = \frac{1}{2}\left(2 + \frac{2}{3^2} + \frac{2}{5^2} + \cdots \right) = 1 + \frac{1}{3^2} + \frac{1}{5^2} + \cdots.$$

21. Writing

$$f(x) = \frac{a_0}{2} + a_1\cos\frac{\pi}{p}x + \cdots + a_n\cos\frac{n\pi}{p}x + \cdots + b_1\sin\frac{\pi}{p}x + \cdots + b_n\sin\frac{n\pi}{p}x + \cdots$$

we see that $f^2(x)$ consists exclusively of squared terms of the form

$$\frac{a_0^2}{4}, \qquad a_n^2\cos^2\frac{n\pi}{p}x, \qquad b_n^2\sin^2\frac{n\pi}{p}x$$

185

and cross-product terms, with $m \neq n$, of the form

$$a_0 a_n \cos \frac{n\pi}{p} x, \qquad a_0 b_n \sin \frac{n\pi}{p} x, \qquad 2a_m a_n \cos \frac{m\pi}{p} x \cos \frac{n\pi}{p} x,$$

$$2a_m b_n \cos \frac{m\pi}{p} x \sin \frac{n\pi}{p} x, \qquad 2b_m b_n \sin \frac{m\pi}{p} x \sin \frac{n\pi}{p} x.$$

The integral of each cross-product term taken over the interval $(-p, p)$ is zero by orthogonality. For the squared terms we have

$$\frac{a_0^2}{4} \int_{-p}^{p} dx = \frac{a_0^2 p}{2}, \qquad a_n^2 \int_{-p}^{p} \cos^2 \frac{n\pi}{p} x \, dx = a_n^2 p, \qquad b_n^2 \int_{-p}^{p} \sin^2 \frac{n\pi}{p} x \, dx = b_n^2 p.$$

Thus

$$RMS(f) = \sqrt{\frac{1}{4} a_0^2 + \frac{1}{2} \sum_{n=1}^{\infty} (a_n^2 + b_n^2)}.$$

Exercises 12.3

3. Since $f(-x) = (-x)^2 - x = x^2 - x$, $f(x)$ is neither even nor odd.

6. Since $f(-x) = e^{-x} - e^x = -f(x)$, $f(x)$ is an odd function.

9. Since $f(x)$ is not defined for $x < 0$, it is neither even nor odd.

12. Since $f(x)$ is an even function, we expand in a cosine series:

$$a_0 = \int_{1}^{2} 1 \, dx = 1$$

$$a_n = \int_{1}^{2} \cos \frac{n\pi}{2} x \, dx = -\frac{2}{n\pi} \sin \frac{n\pi}{2}.$$

Thus

$$f(x) = \frac{1}{2} + \sum_{n=1}^{\infty} \frac{-2}{n\pi} \sin \frac{n\pi}{2} \cos \frac{n\pi}{2} x.$$

15. Since $f(x)$ is an even function, we expand in a cosine series:

$$a_0 = 2 \int_{0}^{1} x^2 \, dx = \frac{2}{3}$$

$$a_n = 2 \int_{0}^{1} x^2 \cos n\pi x \, dx = 2 \left(\frac{x^2}{n\pi} \sin n\pi x \Big|_{0}^{1} - \frac{2}{n\pi} \int_{0}^{1} x \sin n\pi x \, dx \right) = \frac{4}{n^2 \pi^2} (-1)^n.$$

Thus

$$f(x) = \frac{1}{3} + \sum_{n=1}^{\infty} \frac{4}{n^2 \pi^2} (-1)^n \cos n\pi x.$$

18. Since $f(x)$ is an odd function, we expand in a sine series:

$$b_n = \frac{2}{\pi} \int_{0}^{\pi} x^3 \sin nx \, dx = \frac{2}{\pi} \left(-\frac{x^3}{n} \cos nx \Big|_{0}^{\pi} + \frac{3}{n} \int_{0}^{\pi} x^2 \cos nx \, dx \right)$$

$$= \frac{2\pi^2}{n} (-1)^{n+1} - \frac{12}{n^2 \pi} \int_{0}^{\pi} x \sin nx \, dx$$

$$= \frac{2\pi^2}{n} (-1)^{n+1} - \frac{12}{n^2 \pi} \left(-\frac{x}{n} \cos nx \Big|_{0}^{\pi} + \frac{1}{n} \int_{0}^{\pi} \cos nx \, dx \right) = \frac{2\pi^2}{n} (-1)^{n+1} + \frac{12}{n^3} (-1)^n.$$

Thus

$$f(x) = \sum_{n=1}^{\infty} \left(\frac{2\pi^2}{n}(-1)^{n+1} + \frac{12}{n^3}(-1)^n \right) \sin nx.$$

21. Since $f(x)$ is an even function, we expand in a cosine series:

$$a_0 = \int_0^1 x\,dx + \int_1^2 1\,dx = \frac{3}{2}$$

$$a_n = \int_0^1 x \cos\frac{n\pi}{2}x\,dx + \int_1^2 \cos\frac{n\pi}{2}x\,dx = \frac{4}{n^2\pi^2}\left(\cos\frac{n\pi}{2} - 1\right).$$

Thus

$$f(x) = \frac{3}{4} + \sum_{n=1}^{\infty} \frac{4}{n^2\pi^2}\left(\cos\frac{n\pi}{2} - 1\right)\cos\frac{n\pi}{2}x.$$

24. Since $f(x)$ is an even function, we expand in a cosine series.

$$a_0 = \frac{2}{\pi/2}\int_0^{\pi/2}\cos x\,dx = \frac{4}{\pi}$$

$$a_n = \frac{2}{\pi/2}\int_0^{\pi/2}\cos x \cos\frac{n\pi}{\pi/2}x\,dx = \frac{4}{\pi}\int_0^{\pi/2}\cos x \cos 2nx\,dx$$

$$= \frac{2}{\pi}\int_0^{\pi/2}[\cos(2n-1)x + \cos(2n+1)x]\,dx = \frac{4(-1)^{n+1}}{\pi(4n^2-1)}$$

Thus

$$f(x) = \frac{2}{\pi} + \sum_{n=1}^{\infty} \frac{4(-1)^{n+1}}{\pi(4n^2-1)}\cos 2nx.$$

27. $a_0 = \frac{4}{\pi}\int_0^{\pi/2}\cos x\,dx = \frac{4}{\pi}$

$a_n = \frac{4}{\pi}\int_0^{\pi/2}\cos x \cos 2nx\,dx = \frac{2}{\pi}\int_0^{\pi/2}[\cos(2n+1)x + \cos(2n-1)x]\,dx = \frac{4(-1)^n}{\pi(1-4n^2)}$

$b_n = \frac{4}{\pi}\int_0^{\pi/2}\cos x \sin 2nx\,dx = \frac{2}{\pi}\int_0^{\pi/2}[\sin(2n+1)x + \sin(2n-1)x]\,dx = \frac{8n}{\pi(4n^2-1)}$

$f(x) = \frac{2}{\pi} + \sum_{n=1}^{\infty} \frac{4(-1)^n}{\pi(1-4n^2)}\cos 2nx$

$f(x) = \sum_{n=1}^{\infty} \frac{8n}{\pi(4n^2-1)}\sin 2nx$

30. $a_0 = \frac{1}{\pi}\int_\pi^{2\pi}(x-\pi)\,dx = \frac{\pi}{2}$

$a_n = \frac{1}{\pi}\int_\pi^{2\pi}(x-\pi)\cos\frac{n}{2}x\,dx = \frac{4}{n^2\pi}\left[(-1)^n - \cos\frac{n\pi}{2}\right]$

$b_n = \frac{1}{\pi}\int_\pi^{2\pi}(x-\pi)\sin\frac{n}{2}x\,dx = \frac{2}{n}(-1)^{n+1} - \frac{4}{n^2\pi}\sin\frac{n\pi}{2}$

$f(x) = \frac{\pi}{4} + \sum_{n=1}^{\infty} \frac{4}{n^2\pi}\left[(-1)^n - \cos\frac{n\pi}{2}\right]\cos\frac{n}{2}x$

$$f(x) = \sum_{n=1}^{\infty} \left(\frac{2}{n}(-1)^{n+1} - \frac{4}{n^2\pi} \sin \frac{n\pi}{2} \right) \sin \frac{n}{2} x$$

33. $a_0 = 2 \int_0^1 (x^2 + x)\, dx = \frac{5}{3}$

$$a_n = 2 \int_0^1 (x^2 + x) \cos n\pi x\, dx = \frac{2(x^2 + x)}{n\pi} \sin n\pi x \Big|_0^1 - \frac{2}{n\pi} \int_0^1 (2x + 1) \sin n\pi x\, dx = \frac{2}{n^2\pi^2}[3(-1)^n - 1]$$

$$b_n = 2 \int_0^1 (x^2 + x) \sin n\pi x\, dx = -\frac{2(x^2 + x)}{n\pi} \cos n\pi x \Big|_0^1 + \frac{2}{n\pi} \int_0^1 (2x + 1) \cos n\pi x\, dx$$

$$= \frac{4}{n\pi}(-1)^{n+1} + \frac{4}{n^3\pi^3}[(-1)^n - 1]$$

$$f(x) = \frac{5}{6} + \sum_{n=1}^{\infty} \frac{2}{n^2\pi^2}[3(-1)^n - 1] \cos n\pi x$$

$$f(x) = \sum_{n=1}^{\infty} \left(\frac{4}{n\pi}(-1)^{n+1} + \frac{4}{n^3\pi^3}[(-1)^n - 1] \right) \sin n\pi x$$

36. $a_0 = \frac{2}{\pi} \int_0^\pi x\, dx = \pi$

$$a_n = \frac{2}{\pi} \int_0^\pi x \cos 2nx\, dx = 0$$

$$b_n = \frac{2}{\pi} \int_0^\pi x \sin 2nx\, dx = -\frac{1}{n}$$

$$f(x) = \frac{\pi}{2} + \sum_{n=1}^{\infty} \left(-\frac{1}{n} \sin 2nx \right)$$

39. We have

$$b_n = \frac{2}{\pi} \int_0^\pi 5 \sin nt\, dt = \frac{10}{n\pi}[1 - (-1)^n]$$

so that

$$f(t) = \sum_{n=1}^{\infty} \frac{10[1 - (-1)^n]}{n\pi} \sin nt.$$

Substituting the assumption $x_p(t) = \displaystyle\sum_{n=1}^{\infty} B_n \sin nt$ into the differential equation then gives

$$x_p'' + 10x_p = \sum_{n=1}^{\infty} B_n(10 - n^2) \sin nt = \sum_{n=1}^{\infty} \frac{10[1 - (-1)^n]}{n\pi} \sin nt$$

and so $B_n = \dfrac{10[1 - (-1)^n]}{n\pi(10 - n^2)}$. Thus

$$x_p(t) = \frac{10}{\pi} \sum_{n=1}^{\infty} \frac{1 - (-1)^n}{n(10 - n^2)} \sin nt.$$

42. We have

$$a_0 = \frac{2}{(1/2)} \int_0^{1/2} t\, dt = \frac{1}{2}$$

$$a_n = \frac{2}{(1/2)} \int_0^{1/2} t \cos 2n\pi t\, dt = \frac{1}{n^2\pi^2}[(-1)^n - 1]$$

so that

$$f(t) = \frac{1}{4} + \sum_{n=1}^{\infty} \frac{(-1)^n - 1}{n^2 \pi^2} \cos 2n\pi t.$$

Substituting the assumption

$$x_p(t) = \frac{A_0}{2} + \sum_{n=1}^{\infty} A_n \cos 2n\pi t$$

into the differential equation then gives

$$\frac{1}{4} x_p'' + 12 x_p = 6 A_0 + \sum_{n=1}^{\infty} A_n (12 - n^2 \pi^2) \cos 2n\pi t = \frac{1}{4} + \sum_{n=1}^{\infty} \frac{(-1)^n - 1}{n^2 \pi^2} \cos 2n\pi t$$

and $A_0 = \dfrac{1}{24}$, $A_n = \dfrac{(-1)^n - 1}{n^2 \pi^2 (12 - n^2 \pi^2)}$. Thus

$$x_p(t) = \frac{1}{48} + \frac{1}{\pi^2} \sum_{n=1}^{\infty} \frac{(-1)^n - 1}{n^2 (12 - n^2 \pi^2)} \cos 2n\pi t.$$

45. (a) We have

$$b_n = \frac{2}{L} \int_0^L \frac{w_0 x}{L} \sin \frac{n\pi}{L} x \, dx = \frac{2 w_0}{n\pi} (-1)^{n+1}$$

so that

$$w(x) = \sum_{n=1}^{\infty} \frac{2 w_0}{n\pi} (-1)^{n+1} \sin \frac{n\pi}{L} x.$$

(b) If we assume $y(x) = \displaystyle\sum_{n=1}^{\infty} B_n \sin \frac{n\pi}{L} x$ then

$$y^{(4)} = \sum_{n=1}^{\infty} \frac{n^4 \pi^4}{L^4} B_n \sin \frac{n\pi}{L} x$$

and so the differential equation $EI y^{(4)} = w(x)$ gives

$$B_n = \frac{2 w_0 (-1)^{n+1} L^4}{EI n^5 \pi^5}.$$

Thus

$$y(x) = \frac{2 w_0 L^4}{EI \pi^5} \sum_{n=1}^{\infty} \frac{(-1)^{n+1}}{n^5} \sin \frac{n\pi}{L} x.$$

Exercises 12.4

In this section we make use of the following identities due to Euler's formula:

$$e^{in\pi} = e^{-in\pi} = (-1)^n, \qquad e^{-2in\pi} = 1, \qquad e^{-in\pi/2} = (-i)^n.$$

3. Identifying $p = 1/2$ we have

$$c_n = \int_{-1/2}^{1/2} f(x)e^{-2in\pi x}dx = \int_0^{1/4} e^{-2in\pi x}dx = -\frac{1}{2in\pi}e^{-2in\pi x}\Big|_0^{1/4}$$

$$= -\frac{1}{2in\pi}\left[e^{-in\pi/2} - 1\right] = -\frac{1}{2in\pi}\left[(-i)^n - 1\right] = \frac{i}{2n\pi}\left[(-i)^n - 1\right]$$

and

$$c_0 = \int_0^{1/4} dx = \frac{1}{4}.$$

Thus

$$f(x) = \frac{1}{4} + \frac{i}{2\pi}\sum_{\substack{n=-\infty \\ n \neq 0}}^{\infty}\frac{(-i)^n - 1}{n}e^{2in\pi x}.$$

6. Identifying $p = 1$ we have

$$c_n = \frac{1}{2}\int_{-1}^1 f(x)e^{-in\pi x}dx = \frac{1}{2}\left[\int_{-1}^0 e^x e^{-in\pi x}dx + \int_0^1 e^{-x}e^{-in\pi x}dx\right]$$

$$= \frac{1}{2}\left[-\frac{1}{1 - in\pi}e^{(1-in\pi)x}\Big|_{-1}^0 - \frac{1}{1 + in\pi}e^{-(1+in\pi)x}\Big|_0^1\right]$$

$$= \frac{e - (-1)^n}{e(1 - in\pi)} + \frac{1 - e^{-1}(-1)^n}{1 + in\pi} = \frac{2[e - (-1)^n]}{e(1 + n^2\pi^2)}.$$

Thus

$$f(x) = \sum_{n=-\infty}^{\infty}\frac{2[e - (-1)^n]}{e(1 + n^2\pi^2)}e^{in\pi x}.$$

9. Identifying $2p = \pi$ or $p = \pi/2$, and using $\sin x = (e^{ix} - e^{-ix})/2i$, we have

$$c_n = \frac{1}{\pi}\int_0^{\pi} f(x)e^{-2inx/\pi}dx$$

$$= \frac{1}{\pi}\int_0^{\pi}(\sin x)e^{-2inx/\pi}dx$$

$$= \frac{1}{\pi}\int_0^{\pi}\frac{1}{2i}(e^{ix} - e^{-ix})e^{-2inx/\pi}dx$$

$$= \frac{1}{2\pi i}\int_0^{\pi}\left(e^{(1-2n/\pi)ix} - e^{-(1+2n/\pi)ix}\right)dx$$

$$= \frac{1}{2\pi i}\left[\frac{1}{i(1 - 2n/\pi)}e^{(1-2n/\pi)ix}\right.$$

$$\left. + \frac{1}{i(1 + 2n/\pi)}e^{-(1+2n/\pi)ix}\right]_0^{\pi}$$

$$= \frac{\pi(1 + e^{-2in})}{\pi^2 - 4n^2}.$$

The fundamental period is $T = \pi$, so $\omega = 2\pi/\pi = 2$ and the values of $n\omega$ are $0, \pm 2, \pm 4, \pm 6, \ldots$. Values of $|c_n|$ for $n = 0, \pm 1, \pm 2, \pm 3, \pm 4$, and ± 5 are shown in the table. The bottom graph is a portion of the frequency spectrum.

n	-5	-4	-3	-2	-1	0	1	2	3	4	5
c_n	0.0198	0.0759	0.2380	0.4265	0.5784	0.6366	0.5784	0.4265	0.2380	0.0759	0.0198

12. From part (a) of Problem 11 and the fact that f is odd, $c_n + c_{-n} = a_n = 0$, so $c_{-n} = -c_n$. Then $b_n = i(c_n - c_{-n}) = 2ic_n$. From Problem 1, $b_n = 2i[1-(-1)^n]/n\pi i = 2[1-(-1)^n]/n\pi$, and the Fourier sine series of f is

$$f(x) = \sum_{i=1}^{\infty} \frac{2[1-(-1)^n]}{n\pi} \sin \frac{n\pi x}{2}.$$

Exercises 12.5

3. For $\lambda = 0$ the solution of $y'' = 0$ is $y = c_1 x + c_2$. The condition $y'(0) = 0$ implies $c_1 = 0$, so $\lambda = 0$ is an eigenvalue with corresponding eigenfunction 1.

For $\lambda < 0$ we have $y = c_1 \cosh\sqrt{-\lambda}\,x + c_2 \sin\sqrt{-\lambda}\,x$ and $y' = c_1\sqrt{-\lambda}\sinh\sqrt{-\lambda}\,x + c_2\sqrt{-\lambda}\cosh\sqrt{-\lambda}\,x$. The condition $y'(0) = 0$ implies $c_2 = 0$ and so $y = c_1\cosh\sqrt{-\lambda}\,x$. Now the condition $y'(L) = 0$ implies $c_1 = 0$. Thus $y = 0$ and there are no negative eigenvalues.

For $\lambda > 0$ we have $y = c_1\cos\sqrt{\lambda}\,x + c_2\sin\sqrt{\lambda}\,x$ and $y' = -c_1\sqrt{\lambda}\sin\sqrt{\lambda}\,x + c_2\sqrt{\lambda}\cos\sqrt{\lambda}\,x$. The condition $y'(0) = 0$ implies $c_2 = 0$ and so $y = c_1\cos\sqrt{\lambda}\,x$. Now the condition $y'(L) = 0$ implies $-c_1\sqrt{\lambda}\sin\sqrt{\lambda}\,L = 0$. For $c_1 \neq 0$ this condition will hold when $\sqrt{\lambda}\,L = n\pi$ or $\lambda = n^2\pi^2/L^2$, where $n = 1, 2, 3, \ldots$. These are the positive eigenvalues with corresponding eigenfunctions $\cos(n\pi/L)x$, $n = 1, 2, 3, \ldots$.

6. The eigenfunctions are $\sin\sqrt{\lambda_n}\,x$ where $\tan\sqrt{\lambda_n} = -\lambda_n$. Thus

$$\|\sin\sqrt{\lambda_n}\,x\|^2 = \int_0^1 \sin^2\sqrt{\lambda_n}\,x\,dx = \frac{1}{2}\int_0^1\left(1-\cos 2\sqrt{\lambda_n}\,x\right)dx$$

$$= \frac{1}{2}\left(x - \frac{1}{2\sqrt{\lambda_n}}\sin 2\sqrt{\lambda_n}\,x\right)\Big|_0^1 = \frac{1}{2}\left(1 - \frac{1}{2\sqrt{\lambda_n}}\sin 2\sqrt{\lambda_n}\right)$$

$$= \frac{1}{2}\left[1 - \frac{1}{2\sqrt{\lambda_n}}\left(2\sin\sqrt{\lambda_n}\cos\sqrt{\lambda_n}\right)\right]$$

$$= \frac{1}{2}\left[1 - \frac{1}{\sqrt{\lambda_n}}\tan\sqrt{\lambda_n}\cos\sqrt{\lambda_n}\cos\sqrt{\lambda_n}\right]$$

$$= \frac{1}{2}\left[1 - \frac{1}{\sqrt{\lambda_n}}\left(-\sqrt{\lambda_n}\cos^2\sqrt{\lambda_n}\right)\right] = \frac{1}{2}\left(1 + \cos^2\sqrt{\lambda_n}\right).$$

9. (a) An orthogonality relation is

$$\int_0^1 \cos x_m x \cos x_n x = 0$$

where $x_m \neq x_n$ are positive solutions of $\cot x = x$.

(b) The first two eigenfunctions in Problem 1 are $\cos 0.8603x$ and $\cos 3.4256x$. We use a CAS to compute

$$\int_0^1 (\cos 0.8603x)(\cos 3.4256x)\,dx = -1.8771 \times 10^{-6}.$$

12. To obtain the self-adjoint form we note that an integrating factor is $e^{\int -2x\,dx} = e^{-x^2}$. Thus, the differential equation is

$$e^{-x^2}y'' - 2xe^{-x^2}y' + 2ne^{-x^2}y = 0$$

and the self-adjoint form is

$$\frac{d}{dx}\left[e^{-x^2}y'\right] + 2ne^{-x^2}y = 0.$$

Identifying the weight function $p(x) = 2e^{-x^2}$ and noting that since $r(x) = e^{-x^2}$, $\lim_{x\to-\infty} r(x) = \lim_{x\to\infty} r(x) = 0$, we have the orthogonality relation

$$\int_{\infty}^{\infty} 2e^{-x^2} H_n(x)H_m(x)\, dx = 0, \ m \neq n.$$

Exercises 12.6

3. The boundary condition indicates that we use (15) and (16) in the text. With $b = 2$ we obtain

$$c_i = \frac{2}{4J_1^2(2\lambda_i)}\int_0^2 xJ_0(\lambda_i x)\, dx$$

$\boxed{t = \lambda_i x \qquad dt = \lambda_i\, dx}$

$$= \frac{1}{2J_1^2(2\lambda_i)} \cdot \frac{1}{\lambda_i^2}\int_0^{2\lambda_i} tJ_0(t)\, dt$$

$$= \frac{1}{2\lambda_i^2 J_1^2(2\lambda_i)}\int_0^{2\lambda_i} \frac{d}{dt}[tJ_1(t)]\, dt \qquad \text{[From (4) in the text]}$$

$$= \frac{1}{2\lambda_i^2 J_1^2(2\lambda_i)}tJ_1(t)\Big|_0^{2\lambda_i}$$

$$= \frac{1}{\lambda_i J_1(2\lambda_i)}.$$

Thus

$$f(x) = \sum_{i=1}^{\infty} \frac{1}{\lambda_i J_1(2\lambda_i)} J_0(\lambda_i x).$$

6. Writing the boundary condition in the form

$$2J_0(2\lambda) + 2\lambda J_0'(2\lambda) = 0$$

we identify $b = 2$ and $h = 2$. Using (17) and (18) in the text we obtain

$$c_i = \frac{2\lambda_i^2}{(4\lambda_i^2 + 4)J_0^2(2\lambda_i)}\int_0^2 xJ_0(\lambda_i x)\, dx$$

$\boxed{t = \lambda_i x \qquad dt = \lambda_i\, dx}$

$$= \frac{\lambda_i^2}{2(\lambda_i^2 + 1)J_0^2(2\lambda_i)} \cdot \frac{1}{\lambda_i^2}\int_0^{2\lambda_i} tJ_0(t)\, dt$$

$$= \frac{1}{2(\lambda_i^2 + 1)J_0^2(2\lambda_i)}\int_0^{2\lambda_i} \frac{d}{dt}[tJ_1(t)]\, dt \qquad \text{[From (4) in the text]}$$

$$= \frac{1}{2(\lambda_i^2 + 1)J_0^2(2\lambda_i)}tJ_1(t)\Big|_0^{2\lambda_i}$$

$$= \frac{\lambda_i J_1(2\lambda_i)}{(\lambda_i^2 + 1)J_0^2(2\lambda_i)}.$$

Thus

$$f(x) = \sum_{i=1}^{\infty} \frac{\lambda_i J_1(2\lambda_i)}{(\lambda_i^2 + 1) J_0^2(2\lambda_i)} J_0(\lambda_i x).$$

9. The boundary condition indicates that we use (19) and (20) in the text. With $b = 3$ we obtain

$$c_1 = \frac{2}{9} \int_0^3 x x^2 \, dx = \frac{2}{9} \frac{x^4}{4} \Big|_0^3 = \frac{9}{2},$$

$$c_i = \frac{2}{9 J_0^2(3\lambda_i)} \int_0^3 x J_0(\lambda_i x) x^2 \, dx$$

$$\boxed{t = \lambda_i x \qquad dt = \lambda_i \, dx}$$

$$= \frac{2}{9 J_0^2(3\lambda_i)} \cdot \frac{1}{\lambda_i^4} \int_0^{3\lambda_i} t^3 J_0(t) \, dt$$

$$= \frac{2}{9 \lambda_i^4 J_0^2(3\lambda_i)} \int_0^{3\lambda_i} t^2 \frac{d}{dt} \left[t J_1(t) \right] dt$$

$$\boxed{\begin{array}{ll} u = t^2 & dv = \frac{d}{dt}[t J_1(t)] \, dt \\ du = 2t \, dt & v = t J_1(t) \end{array}}$$

$$= \frac{2}{9 \lambda_i^4 J_0^2(3\lambda_i)} \left(t^3 J_1(t) \Big|_0^{3\lambda_i} - 2 \int_0^{3\lambda_i} t^2 J_1(t) \, dt \right)$$

With $n = 0$ in equation (5) in the text we have $J_0'(x) = -J_1(x)$, so the boundary condition $J_0'(3\lambda_i) = 0$ implies $J_1(3\lambda_i) = 0$. Then

$$c_i = \frac{2}{9 \lambda_i^4 J_0^2(3\lambda_i)} \left(-2 \int_0^{3\lambda_i} \frac{d}{dt} \left[t^2 J_2(t) \right] dt \right) = \frac{2}{9 \lambda_i^4 J_0^2(3\lambda_i)} \left(-2 t^2 J_2(t) \Big|_0^{3\lambda_i} \right)$$

$$= \frac{2}{9 \lambda_i^4 J_0^2(3\lambda_i)} \left[-18 \lambda_i^2 J_2(3\lambda_i) \right] = \frac{-4 J_2(3\lambda_i)}{\lambda_i^2 J_0^2(3\lambda_i)}.$$

Thus

$$f(x) = \frac{9}{2} - 4 \sum_{i=1}^{\infty} \frac{J_2(3\lambda_i)}{\lambda_i^2 J_0^2(3\lambda_i)} J_0(\lambda_i x).$$

15. We compute

$$c_0 = \frac{1}{2} \int_0^1 x P_0(x) \, dx = \frac{1}{2} \int_0^1 x \, dx = \frac{1}{4}$$

$$c_1 = \frac{3}{2} \int_0^1 x P_1(x) \, dx = \frac{3}{2} \int_0^1 x^2 \, dx = \frac{1}{2}$$

$$c_2 = \frac{5}{2} \int_0^1 x P_2(x) \, dx = \frac{5}{2} \int_0^1 \frac{1}{2}(3x^3 - x) dx = \frac{5}{16}$$

$$c_3 = \frac{7}{2} \int_0^1 x P_3(x) \, dx = \frac{7}{2} \int_0^1 \frac{1}{2}(5x^4 - 3x^2) dx = 0$$

$$c_4 = \frac{9}{2} \int_0^1 x P_4(x) \, dx = \frac{9}{2} \int_0^1 \frac{1}{8}(35x^5 - 30x^3 + 3x) dx = -\frac{3}{32}$$

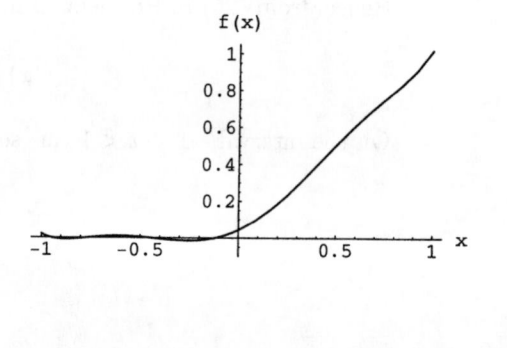

$$c_5 = \frac{11}{2} \int_0^1 x P_5(x) \, dx = \frac{11}{2} \int_0^1 \frac{1}{8}(63x^6 - 70x^4 + 15x^2) dx = 0$$

$$c_6 = \frac{13}{2} \int_0^1 x P_6(x) \, dx = \frac{13}{2} \int_0^1 \frac{1}{16}(231x^7 - 315x^5 + 105x^3 - 5x) dx = \frac{13}{256}.$$

Thus

$$f(x) = \frac{1}{4} P_0(x) + \frac{1}{2} P_1(x) + \frac{5}{16} P_2(x) - \frac{3}{32} P_4(x) + \frac{13}{256} P_6(x) + \cdots.$$

18. From Problem 17 we have

$$P_2(\cos\theta) = \frac{1}{4}(3\cos 2\theta + 1)$$

or

$$\cos 2\theta = \frac{4}{3} P_2(\cos\theta) - \frac{1}{3}.$$

Then, using $P_0(\cos\theta) = 1$,

$$F(\theta) = 1 - \cos 2\theta = 1 - \left[\frac{4}{3} P_2(\cos\theta) - \frac{1}{3}\right]$$

$$= \frac{4}{3} - \frac{4}{3} P_2(\cos\theta) = \frac{4}{3} P_0(\cos\theta) - \frac{4}{3} P_2(\cos\theta).$$

21. From (26) in Problem 19 in the text we find

$$c_0 = \int_0^1 x P_0(x) \, dx = \int_0^1 x \, dx = \frac{1}{2},$$

$$c_2 = 5 \int_0^1 x P_2(x) \, dx = 5 \int_0^1 \frac{1}{2}(3x^3 - x) dx = \frac{5}{8},$$

$$c_4 = 9 \int_0^1 x P_4(x) \, dx = 9 \int_0^1 \frac{1}{8}(35x^5 - 30x^3 + 3x) dx = -\frac{3}{16},$$

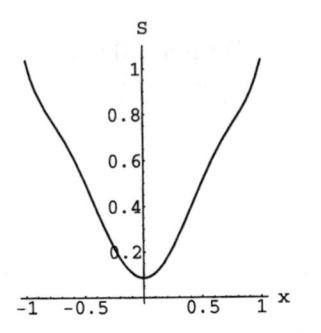

and

$$c_6 = 13 \int_0^1 x P_6(x) \, dx = 13 \int_0^1 \frac{1}{16}(231x^7 - 315x^5 + 105x^3 - 5x) dx = \frac{13}{128}.$$

Hence, from (25) in Problem 19 in the text,

$$f(x) = \frac{1}{2} P_0(x) + \frac{5}{8} P_2(x) - \frac{3}{16} P_4(x) + \frac{13}{128} P_6 + \cdots.$$

On the interval $-1 < x < 1$ this series represents the function $f(x) = |x|$.

Chapter 12 Review Exercises

3. Cosine, since f is even

6. 3/2, the average of 3 and 0

9. Since the coefficient of y in the differential equation is n^2, the weight function is the integrating factor

$$\frac{1}{a(x)}e^{\int (b/a)dx} = \frac{1}{1-x^2}e^{\int -\frac{x}{1-x^2}dx} = \frac{1}{1-x^2}e^{\frac{1}{2}\ln(1-x^2)} = \frac{\sqrt{1-x^2}}{1-x^2} = \frac{1}{\sqrt{1-x^2}}$$

on the interval $[-1, 1]$. The orthogonality relation is

$$\int_{-1}^{1} \frac{1}{\sqrt{1-x^2}} T_m(x)T_n(x)\,dx = 0, \quad m \neq n.$$

12. **(a)** For $m \neq n$

$$\int_0^L \sin\frac{(2n+1)\pi}{2L}x \sin\frac{(2m+1)\pi}{2L}x\,dx = \frac{1}{2}\int_0^L \left(\cos\frac{n-m}{L}\pi x - \cos\frac{n+m+\pi}{L}\pi x\right)dx = 0.$$

(b) From

$$\int_0^L \sin^2\frac{(2n+1)\pi}{2L}x\,dx = \int_0^L \left(\frac{1}{2} - \frac{1}{2}\cos\frac{(2n+1)\pi}{2L}x\right)dx = \frac{L}{2}$$

we see that

$$\left\|\sin\frac{(2n+1)\pi}{2L}x\right\| = \sqrt{\frac{L}{2}}.$$

15. Since

$$A_0 = 2\int_0^1 e^{-x}dx$$

and

$$A_n = 2\int_{-1}^1 e^{-x}\cos n\pi x\,dx = \frac{2}{1+n^2\pi^2}[(1-(-1)^n e^{-1}]$$

for $n = 1, 2, 3, \ldots$ we have

$$f(x) = 1 - e^{-1} + 2\sum_{n=1}^{\infty} \frac{1-(-1)^n e^{-1}}{1+n^2\pi^2}\cos n\pi x.$$

Since

$$B_n = 2\int_0^1 e^{-x}\sin n\pi x\,dx = \frac{2n\pi}{1+n^2\pi^2}[(1-(-1)^n e^{-1}]$$

for $n = 1, 2, 3, \ldots$ we have

$$f(x) = \sum_{n=1}^{\infty} \frac{2n\pi}{1+n^2\pi^2}[(1-(-1)^n e^{-1}]\sin n\pi x.$$

18. To obtain the self-adjoint form of the differential equation in Problem 17 we note that an integrating factor is $(1/x^2)e^{\int dx/x} = 1/x$. Thus the weight function is $9/x$ and an orthogonality relation is

$$\int_1^e \frac{9}{x}\cos\left(\frac{2n-1}{2}\pi\ln x\right)\cos\left(\frac{2m-1}{2}\pi\ln x\right)dx = 0, \quad m \neq n.$$

13 Boundary-Value Problems in Rectangular Coordinates

———————— **Exercises 13.1** ————————

3. If $u = XY$ then

$$u_x = X'Y,$$
$$u_y = XY',$$
$$X'Y = X(Y - Y'),$$

and

$$\frac{X'}{X} = \frac{Y - Y'}{Y} = \pm\lambda^2.$$

Then

$$X' \mp \lambda^2 X = 0 \quad \text{and} \quad Y' - (1 \mp \lambda^2)Y = 0$$

so that

$$X = A_1 e^{\pm\lambda^2 x},$$
$$Y = A_2 e^{(1 \mp \lambda^2)y},$$

and

$$u = XY = c_1 e^{y + c_2(x - y)}.$$

6. If $u = XY$ then

$$u_x = X'Y,$$
$$u_y = XY',$$
$$yX'Y = xXY',$$

and

$$\frac{X'}{xX} = \frac{Y'}{-yY} = \pm\lambda^2.$$

Then

$$X \mp \lambda^2 xX = 0 \quad \text{and} \quad Y' \pm \lambda^2 yY = 0$$

so that

$$X = A_1 e^{\pm\lambda^2 x^2/2},$$
$$Y = A_2 e^{\mp\lambda^2 y^2/2},$$

and

$$u = XY = c_1 e^{c_2(x^2 - y^2)}.$$

9. If $u = XT$ then

$$u_t = XT',$$
$$u_{xx} = X''T,$$
$$kX''T - XT = XT',$$

and we choose

196

$$\frac{T'}{T} = \frac{kX'' - X}{X} = -1 \pm k\lambda^2$$

so that

$$T' - (-1 \pm k\lambda^2)T = 0 \quad \text{and} \quad X'' - (\pm\lambda^2)X = 0.$$

For $\lambda^2 > 0$ we obtain

$$X = A_1 \cosh \lambda x + A_2 \sinh \lambda x \quad \text{and} \quad T = A_3 e^{(-1+k\lambda^2)t}$$

so that

$$u = XT = e^{(-1+k\lambda^2)t} \left(c_1 \cosh \lambda x + c_2 \sinh \lambda x \right).$$

For $-\lambda^2 < 0$ we obtain

$$X = A_1 \cos \lambda x + A_2 \sin \lambda x \quad \text{and} \quad T = A_3 e^{(-1-k\lambda^2)t}$$

so that

$$u = XT = e^{(-1-k\lambda^2)t} (c_3 \cos \lambda x + c_4 \sin \lambda x).$$

If $\lambda^2 = 0$ then

$$X'' = 0 \quad \text{and} \quad T' + T = 0,$$

and we obtain

$$X = A_1 x + A_2 \quad \text{and} \quad T = A_3 e^{-t}.$$

In this case

$$u = XT = e^{-t}(c_5 x + c_6)$$

12. If $u = XT$ then

$$u_t = XT',$$
$$u_{tt} = XT'',$$
$$u_{xx} = X''T,$$
$$a^2 X''T = XT'' + 2kXT',$$

and

$$\frac{X''}{X} = \frac{T'' + 2kT'}{a^2 T} = \pm\lambda^2$$

so that

$$X'' \mp \lambda^2 X = 0 \quad \text{and} \quad T'' + 2kT' \mp a^2\lambda^2 T = 0.$$

For $\lambda^2 > 0$ we obtain

$$X = A_1 e^{\lambda x} + A_2 e^{-\lambda x},$$
$$T = A_3 e^{(-k+\sqrt{k^2+a^2\lambda^2})t} + A_4 e^{(-k-\sqrt{k^2+a^2\lambda^2})t},$$

and

$$u = XT = \left(A_1 e^{\lambda x} + A_2 e^{-\lambda x} \right) \left(A_3 e^{(-k+\sqrt{k^2+a^2\lambda^2})t} + A_4 e^{(-k-\sqrt{k^2+a^2\lambda^2})t} \right).$$

For $-\lambda^2 < 0$ we obtain

$$X = A_1 \cos \lambda x + A_2 \sin \lambda x.$$

If $k^2 - a^2\lambda^2 > 0$ then

$$T = A_3 e^{(-k+\sqrt{k^2-a^2\lambda^2})t} + A_4 e^{(-k-\sqrt{k^2-a^2\lambda^2})t}.$$

If $k^2 - a^2\lambda^2 < 0$ then

$$T = e^{-kt}\left(A_3\cos\sqrt{a^2\lambda^2 - k^2}\,t + A_4\sin\sqrt{a^2\lambda^2 - k^2}\,t\right).$$

If $k^2 - a^2\lambda^2 = 0$ then

$$T = A_3 e^{-kt} + A_4 t e^{-kt}$$

so that

$$u = XT = (A_1\cos\lambda x + A_2\sin\lambda x)\left(A_3 e^{(-k+\sqrt{k^2-a^2\lambda^2})t} + A_4 e^{(-k-\sqrt{k^2-a^2\lambda^2})t}\right)$$

$$= (A_1\cos\lambda x + A_2\sin\lambda x)e^{-kt}\left(A_3\cos\sqrt{a^2\lambda^2 - k^2}\,t + A_4\sin\sqrt{a^2\lambda^2 - k^2}\,t\right)$$

$$= \left(A_1\cos\frac{k}{a}x + A_2\sin\frac{k}{a}x\right)\left(A_3 e^{-kt} + A_4 t e^{-kt}\right).$$

For $\lambda^2 = 0$ we obtain

$$X = A_1 x + A_2,$$
$$T = A_3 + A_4 e^{-2kt},$$

and

$$u = XT = (A_1 x + A_2)(A_3 + A_4 e^{-2kt}).$$

15. If $u = XY$ then

$$u_{xx} = X''Y,$$
$$u_{yy} = XY'',$$
$$X''Y + XY'' = XY,$$

and

$$\frac{X''}{X} = \frac{Y - Y''}{Y} = \pm\lambda^2$$

so that

$$X'' \mp \lambda^2 X = 0 \quad\text{and}\quad Y'' + (\pm\lambda^2 - 1)Y = 0.$$

For $\lambda^2 > 0$ we obtain

$$X = A_1 e^{\lambda x} + A_2 e^{-\lambda x}.$$

If $\lambda^2 - 1 > 0$ then

$$Y = A_3\cos\sqrt{\lambda^2 - 1}\,y + A_4\sin\sqrt{\lambda^2 - 1}\,y.$$

If $\lambda^2 - 1 < 0$ then

$$Y = A_3 e^{\sqrt{1-\lambda^2}\,y} + A_4 e^{-\sqrt{1-\lambda^2}\,y}.$$

If $\lambda^2 - 1 = 0$ then $Y = A_3 y + A_4$ so that

$$u = XY = \left(A_1 e^{\lambda x} + A_2 e^{-\lambda x}\right)\left(A_3\cos\sqrt{\lambda^2 - 1}\,y + A_4\sin\sqrt{\lambda^2 - 1}\,y\right),$$

$$= \left(A_1 e^{\lambda x} + A_2 e^{-\lambda x}\right)\left(A_3 e^{\sqrt{1-\lambda^2}\,y} + A_4 e^{-\sqrt{1-\lambda^2}\,y}\right)$$

$$= (A_1 e^x + A_2 e^{-x})(A_3 y + A_4).$$

For $-\lambda^2 < 0$ we obtain

$$X = A_1\cos\lambda x + A_2\sin\lambda x,$$
$$Y = A_3 e^{\sqrt{1+\lambda^2}\,y} + A_4 e^{-\sqrt{1+\lambda^2}\,y},$$

and

$$u = XY = (A_1 \cos \lambda x + A_2 \sin \lambda x)\left(A_3 e^{\sqrt{1+\lambda^2}\,y} + A_4 e^{-\sqrt{1+\lambda^2}\,y}\right).$$

For $\lambda^2 = 0$ we obtain

$$X = A_1 x + A_2,$$
$$Y = A_3 e^y + A_4 e^{-y},$$

and

$$u = XY = (A_1 x + A_2)(A_3 e^y + A_4 e^{-y}).$$

18. Identifying $A = 3$, $B = 5$, and $C = 1$, we compute $B^2 - 4AC = 13 > 0$. The equation is hyperbolic.

21. Identifying $A = 1$, $B = -9$, and $C = 0$, we compute $B^2 - 4AC = 81 > 0$. The equation is hyperbolic.

24. Identifying $A = 1$, $B = 0$, and $C = 1$, we compute $B^2 - 4AC = -4 < 0$. The equation is elliptic.

27. If $u = RT$ then

$$u_r = R'T,$$
$$u_{rr} = R''T,$$
$$u_t = RT',$$
$$RT' = k\left(R''T + \frac{1}{r}R'T\right),$$

and

$$\frac{r^2 R'' + rR'}{r^2 R} = \frac{T'}{kT} = \pm\lambda^2.$$

If we use $-\lambda^2 < 0$ then

$$r^2 R'' + rR' + \lambda^2 r^2 R = 0 \quad \text{and} \quad T'' \mp \lambda^2 kT = 0$$

so that

$$R = A_2 J_0(\lambda r) + A_3 Y_0(\lambda r),$$
$$T = A_1 e^{-k\lambda^2 t},$$

and

$$u = RT = e^{-k\lambda^2 t}[c_1 J_0(\lambda r) + c_2 Y_0(\lambda r)]$$

30. We identify $A = xy + 1$, $B = x + 2y$, and $C = 1$. Then $B^2 - 4AC = x^2 + 4y^2 - 4$. The equation $x^2 + 4y^2 = 4$ defines an ellipse. The partial differential equation is hyperbolic outside the ellipse, parabolic on the ellipse, and elliptic inside the ellipse.

Exercises 13.2

3. $k\dfrac{\partial^2 u}{\partial x^2} = \dfrac{\partial u}{\partial t}, \quad 0 < x < L,\ t > 0$

$u(0,t) = 100, \quad \left.\dfrac{\partial u}{\partial x}\right|_{x=L} = -hu(L,t), \quad t > 0$

$u(x,0) = f(x), \quad 0 < x < L$

6. $a^2 \dfrac{\partial^2 u}{\partial x^2} = \dfrac{\partial^2 u}{\partial t^2}$, $\quad 0 < x < L,\ t > 0$

$u(0,t) = 0,\quad u(L,t) = 0, \quad t > 0$

$u(x,0) = 0,\quad \left.\dfrac{\partial u}{\partial x}\right|_{t=0} = \sin\dfrac{\pi x}{L}, \quad 0 < x < L$

9. $\dfrac{\partial^2 u}{\partial x^2} + \dfrac{\partial^2 u}{\partial y^2} = 0, \quad 0 < x < 4,\ 0 < y < 2$

$\left.\dfrac{\partial u}{\partial x}\right|_{x=0} = 0,\quad u(4,y) = f(y), \quad 0 < y < 2$

$\left.\dfrac{\partial u}{\partial y}\right|_{y=0} = 0,\quad u(x,2) = 0, \quad 0 < x < 4$

Exercises 13.3

3. Using $u = XT$ and $-\lambda^2$ as a separation constant leads to

$$X'' + \lambda^2 X = 0,$$
$$X'(0) = 0,$$
$$X'(L) = 0,$$

and

$$T' + k\lambda^2 T = 0.$$

Then

$$X = c_1 \cos\frac{n\pi}{L}x \quad \text{and} \quad T = c_2 e^{-\frac{kn^2\pi^2}{L^2}t}$$

for $n = 0, 1, 2, \ldots$ so that

$$u = \sum_{n=0}^{\infty} A_n e^{-\frac{kn^2\pi^2}{L^2}t} \cos\frac{n\pi}{L}x.$$

Imposing

$$u(x,0) = f(x) = \sum_{n=0}^{\infty} A_n \cos\frac{n\pi}{L}x$$

gives

$$u(x,t) = \frac{1}{L}\int_0^L f(x)\,dx + \frac{2}{L}\sum_{n=1}^{\infty}\left(\int_0^L f(x)\cos\frac{n\pi}{L}x\,dx\right) e^{-\frac{kn^2\pi^2}{L^2}t} \cos\frac{n\pi}{L}x.$$

6. Using $u = XT$ and $-\lambda^2$ as a separation constant leads to

$$X'' + \lambda^2 X = 0,$$
$$X(0) = 0,$$
$$X(L) = 0,$$

and

$$T' + (h + k\lambda^2)T = 0.$$

Then

$$X = c_1 \sin\frac{n\pi}{L}x \quad \text{and} \quad T = c_2 e^{-\left(h + \frac{kn^2\pi^2}{L^2}\right)t}$$

for $n = 0, 1, 2, \ldots$ so that

$$u = \sum_{n=0}^{\infty} A_n e^{-\left(h + \frac{kn^2\pi^2}{L^2}\right)t} \sin \frac{n\pi}{L} x.$$

Imposing

$$u(x,0) = f(x) = \sum_{n=0}^{\infty} A_n \sin \frac{n\pi}{L} x$$

gives

$$u = \frac{2}{L} \sum_{n=1}^{\infty} \left(\int_0^L f(x) \sin \frac{n\pi}{L} x \, dx \right) e^{-\left(h + \frac{kn^2\pi^2}{L^2}\right)t} \sin \frac{n\pi}{L} x.$$

Exercises 13.4

3. Using $u = XT$ and $-\lambda^2$ as a separation constant leads to

$$X'' + \lambda^2 X = 0,$$
$$X(0) = 0,$$
$$X(L) = 0,$$

and

$$T'' + \lambda^2 a^2 T = 0.$$

Then

$$X = c_1 \sin \frac{n\pi}{L} x \quad \text{and} \quad T = c_2 \cos \frac{n\pi a}{L} t + c_3 \sin \frac{n\pi a}{L} t$$

for $n = 1, 2, 3, \ldots$ so that

$$u = \sum_{n=1}^{\infty} \left(A_n \cos \frac{n\pi a}{L} t + B_n \sin \frac{n\pi a}{L} t \right) \sin \frac{n\pi}{L} x.$$

Imposing

$$u(x,0) = \sum_{n=1}^{\infty} A_n \sin \frac{n\pi}{L} x$$

gives

$$A_n = \frac{2}{L} \left(\int_0^{L/3} \frac{3}{L} x \sin \frac{n\pi}{L} x \, dx + \int_{L/3}^{2L/3} \sin \frac{n\pi}{L} x \, dx + \int_{2L/3}^L \left(3 - \frac{3}{L} x \right) \sin \frac{n\pi}{L} x \, dx \right)$$

so that

$$A_1 = \frac{6\sqrt{3}}{\pi^2},$$
$$A_2 = A_3 = A_4 = 0,$$
$$A_5 = -\frac{6\sqrt{3}}{5^2 \pi^2},$$
$$A_6 = 0,$$
$$a_7 = \frac{6\sqrt{3}}{7^2 \pi^2} \ldots.$$

Imposing

$$u_t(x,0) = 0 = \sum_{n=1}^{\infty} B_n \frac{n\pi a}{L} \sin \frac{n\pi}{L} x$$

gives $B_n = 0$ for $n = 1, 2, 3, \ldots$ so that

$$u(x,t) = \frac{6\sqrt{3}}{\pi^2} \left(\cos \frac{\pi a}{L} t \sin \frac{\pi}{L} x - \frac{1}{5^2} \cos \frac{5\pi a}{L} t \sin \frac{5\pi}{L} x + \frac{1}{7^2} \cos \frac{7\pi a}{L} t \sin \frac{7\pi}{L} x - \cdots \right).$$

6. Using $u = XT$ and $-\lambda^2$ as a separation constant leads to

$$X'' + \lambda^2 X = 0,$$
$$X(0) = 0,$$
$$X(1) = 0,$$

and

$$T'' + \lambda^2 a^2 T = 0.$$

Then

$$X = c_1 \sin n\pi x \quad \text{and} \quad T = c_2 \cos n\pi a t + c_3 \sin n\pi a t$$

for $n = 1, 2, 3, \ldots$ so that

$$u = \sum_{n=1}^{\infty} (A_n \cos n\pi a t + B_n \sin n\pi a t) \sin n\pi x.$$

Imposing

$$u(x,0) = 0.01 \sin 3\pi x = \sum_{n=1}^{\infty} A_n \sin n\pi x$$

and

$$u_t(x,0) = 0 = \sum_{n=1}^{\infty} B_n n\pi a \sin n\pi x$$

gives $B_n = 0$ for $n = 1, 2, 3, \ldots$, $A_3 = 0.01$, and $A_n = 0$ for $n = 1, 2, 4, 5, 6, \ldots$ so that

$$u(x,t) = 0.01 \sin 3\pi x \cos 3\pi a t.$$

9. Using $u = XT$ and $-\lambda^2$ as a separation constant leads to

$$X'' + \lambda^2 X = 0,$$
$$X(0) = 0,$$
$$X(\pi) = 0,$$

and

$$T'' + 2\beta T' + \lambda^2 T = 0.$$

Then

$$X = c_1 \sin nx \quad \text{and} \quad T = e^{-\beta t} \left(c_2 \cos \sqrt{n^2 - \beta^2}\, t + c_3 \sin \sqrt{n^2 - \beta^2}\, t \right)$$

so that

$$u = \sum_{n=1}^{\infty} e^{-\beta t} \left(A_n \cos \sqrt{n^2 - \beta^2}\, t + B_n \sin \sqrt{n^2 - \beta^2}\, t \right) \sin nx.$$

Imposing

$$u(x,0) = f(x) = \sum_{n=1}^{\infty} A_n \sin nx$$

and

$$u_t(x,0) = 0 = \sum_{n=1}^{\infty} \left(B_n \sqrt{n^2 - \beta^2} - \beta A_n \right) \sin nx$$

gives

$$u(x,t) = e^{-\beta t} \sum_{n=1}^{\infty} A_n \left(\cos \sqrt{n^2 - \beta^2}\, t + \frac{\beta}{\sqrt{n^2 - \beta^2}} \sin \sqrt{n^2 - \beta^2}\, t \right) \sin nx,$$

where

$$A_n = \frac{2}{\pi} \int_0^{\pi} f(x) \sin nx \, dx.$$

12. (a) Using $X = c_1 \cosh \lambda x + c_2 \sinh \lambda x + c_3 \cos \lambda x + c_4 \sin \lambda x$ and $X(0) = 0$, $X'(0) = 0$ we find, in turn, $c_3 = -c_1$ and $c_4 = -c_2$. The conditions $X(L) = 0$ and $X'(L) = 0$ then yield the system of equations for c_1 and c_2:

$$c_1(\cosh \lambda L - \cos \lambda L) + c_2(\sinh \lambda L - \sin \lambda L) = 0$$

$$c_1(\lambda \sinh \lambda L + \lambda \sin \lambda L) + c_2(\lambda \cosh \lambda L - \lambda \cos \lambda L) = 0.$$

In order that this system have nontrivial solutions the determinant of the coefficients must be zero:

$$\lambda(\cosh \lambda L - \cos \lambda L)^2 - \lambda(\sinh^2 \lambda L - \sin^2 \lambda L) = 0.$$

$\lambda = 0$ is not an eigenvalue since this leads to $X = 0$. Thus the last equation simplifies to $\cosh \lambda L \cos \lambda L = 1$ or $\cosh x \cos x = 1$, where $x = \lambda L$.

(b) The equation $\cosh x \cos x = 1$ is the same as $\cos x = \operatorname{sech} x$. The figure indicates that the equation has an infinite number of roots.

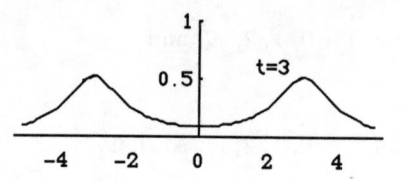

(c) Using a CAS we find the first four positive roots to be $x_1 = 4.7300$, $x_2 = 7.8532$, $x_3 = 10.9956$, and $x_4 = 14.1372$. Thus the first four eigenvalues are $\lambda_1 = x_1/L = 4.7300/L$, $\lambda_2 = x_2/L = 7.8532/L$, $\lambda_3 = x_3/L = 10.9956/L$, and $\lambda_4 = 14.1372/L$.

15. $u(x,t) = \dfrac{1}{2}[\sin(x + at) + \sin(x - at)] + \dfrac{1}{2a} \int_{x-at}^{x+at} ds$

$$= \frac{1}{2}[\sin x \cos at + \cos x \sin at + \sin x \cos at - \cos x \sin at] + \frac{1}{2a} s \Big|_{x-at}^{x+at} = \sin x \cos at + t$$

18.

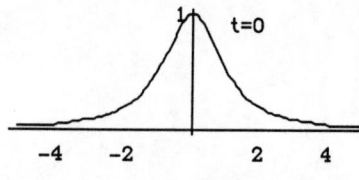

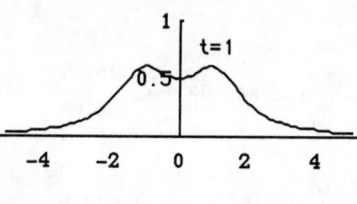

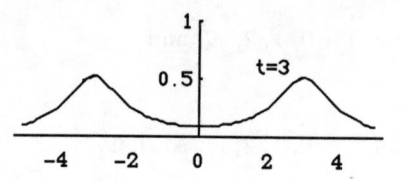

Exercises 13.5

3. Using $u = XY$ and $-\lambda^2$ as a separation constant leads to

$$X'' + \lambda^2 X = 0,$$

$$X(0) = 0,$$

$$X(a) = 0,$$

and

$$Y'' - \lambda^2 Y = 0,$$

$$Y(b) = 0.$$

Then

$$X = c_1 \sin \frac{n\pi}{a} x \quad \text{and} \quad Y = c_2 \cosh \frac{n\pi}{a} y - c_2 \frac{\cosh \frac{n\pi b}{a}}{\sinh \frac{n\pi b}{a}} \sinh \frac{n\pi}{a} y$$

for $n = 1, 2, 3, \ldots$ so that

$$u = \sum_{n=1}^{\infty} A_n \sin \frac{n\pi}{a} x \left(\cosh \frac{n\pi}{a} y - \frac{\cosh \frac{n\pi b}{a}}{\sinh \frac{n\pi b}{a}} \sinh \frac{n\pi}{a} y \right).$$

Imposing

$$u(x, 0) = f(x) = \sum_{n=1}^{\infty} A_n \sin \frac{n\pi}{a} x$$

gives

$$A_n = \frac{2}{a} \int_0^a f(x) \sin \frac{n\pi}{a} x \, dx$$

so that

$$u(x, y) = \frac{2}{a} \sum_{n=1}^{\infty} \left(\int_0^a f(x) \sin \frac{n\pi}{a} x \, dx \right) \sin \frac{n\pi}{a} x \left(\cosh \frac{n\pi}{a} y - \frac{\cosh \frac{n\pi b}{a}}{\sinh \frac{n\pi b}{a}} \sinh \frac{n\pi}{a} y \right).$$

6. Using $u = XY$ and λ^2 as a separation constant leads to

$$X'' - \lambda^2 X = 0,$$
$$X'(1) = 0,$$

and

$$Y'' + \lambda^2 Y = 0,$$
$$Y'(0) = 0,$$
$$Y'(\pi) = 0.$$

Then

$$Y = c_1 \cos ny$$

for $n = 0, 1, 2, \ldots$ and

$$X = c_2 \cosh nx - c_2 \frac{\sinh n}{\cosh n} \sinh nx$$

for $n = 0, 1, 2, \ldots$ so that

$$u = A_0 + \sum_{n=1}^{\infty} A_n \left(\cosh nx - \frac{\sinh n}{\cosh n} \sinh nx \right) \cos ny.$$

Imposing

$$u(0, y) = g(y) = A_0 + \sum_{n=1}^{\infty} A_n \cos ny$$

gives

$$A_0 = \frac{1}{\pi} \int_0^{\pi} g(y) \, dy \quad \text{and} \quad A_n = \frac{2}{\pi} \int_0^{\pi} g(y) \cos ny \, dy$$

for $n = 1, 2, 3, \ldots$ so that

$$u(x, y) = \frac{1}{\pi} \int_0^{\pi} g(y) \, dy + \sum_{n=1}^{\infty} \left(\frac{2}{\pi} \int_0^{\pi} g(y) \cos ny \, dy \right) \left(\cosh nx - \frac{\sinh n}{\cosh n} \sinh nx \right) \cos ny.$$

9. This boundary-value problem has the form of Problem 1 on page 707 in this section, with $a = b = 1$, $f(x) = 100$, and $g(x) = 200$. The solution, then, is

$$u(x, y) = \sum_{n=1}^{\infty} (A_n \cosh n\pi y + B_n \sinh n\pi y) \sin n\pi x,$$

where

$$A_n = 2 \int_0^1 100 \sin n\pi x \, dx = 200\left(\frac{1 - (-1)^n}{n\pi}\right)$$

and

$$B_n = \frac{1}{\sinh n\pi}\left[2 \int_0^1 200 \sin n\pi x \, dx - A_n \cosh n\pi\right]$$

$$= \frac{1}{\sinh n\pi}\left[400\left(\frac{1 - (-1)^n}{n\pi}\right) - 200\left(\frac{1 - (-1)^n}{n\pi}\right) \cosh n\pi\right]$$

$$= 200\left[\frac{1 - (-1)^n}{n\pi}\right][2 \operatorname{csch} n\pi - \coth n\pi].$$

12. Using $u = XY$ and $-\lambda^2$ as a separation constant leads to

$$X'' + \lambda^2 X = 0,$$
$$X'(0) = 0,$$
$$X'(\pi) = 0,$$

and

$$Y'' - \lambda^2 Y = 0.$$

By the boundedness of u as $y \to \infty$ we obtain $Y = c_1 e^{-ny}$ for $n = 1, 2, 3, \ldots$ or $Y = c_1$ and $X = c_2 \cos nx$ for $n = 0, 1, 2, \ldots$ so that

$$u = A_0 + \sum_{n=1}^{\infty} A_n e^{-ny} \cos nx.$$

Imposing

$$u(x, 0) = f(x) = A_0 + \sum_{n=1}^{\infty} A_n \cos nx$$

gives

$$A_0 = \frac{1}{\pi} \int_0^{\pi} f(x) \, dx \quad \text{and} \quad A_n = \frac{2}{\pi} \int_0^{\pi} f(x) \cos nx \, dx$$

so that

$$u(x, y) = \frac{1}{\pi} \int_0^{\pi} f(x) \, dx + \sum_{n=1}^{\infty} \left(\frac{2}{\pi} \int_0^{\pi} f(x) \cos nx \, dx\right) e^{-ny} \cos nx.$$

15. Referring to the discussion in the text under the heading **Superposition Principle**, we identify $a = b = \pi$, $f(x) = 0$, $g(x) = 1$, $F(y) = 1$, and $G(y) = 1$. Then $A_n = 0$ and

$$u_1(x, y) = \sum_{n=1}^{\infty} B_n \sinh ny \sin nx$$

where

$$B_n = \frac{2}{\pi \sinh n\pi} \int_0^{\pi} \sin nx \, dx = \frac{2[1 - (-1)^n]}{n\pi \sinh n\pi}.$$

Next

$$u_2(x,y) = \sum_{n=1}^{\infty} \left(A_n \cosh nx + B_n \sinh nx \right) \sin ny$$

where

$$A_n = \frac{2}{\pi} \int_0^{\pi} \sin ny \, dy = \frac{2[1-(-1)^n]}{n\pi}$$

and

$$B_n = \frac{1}{\sinh n\pi} \left(\frac{2}{\pi} \int_0^{\pi} \sin ny \, dy - A_n \cosh n\pi \right)$$

$$= \frac{1}{\sinh n\pi} \left(\frac{2[1-(-1)^n]}{n\pi} - \frac{2[1-(-1)^n]}{n\pi} \cosh n\pi \right)$$

$$= \frac{2[1-(-1)^n]}{n\pi \sinh n\pi}(1 - \cosh n\pi).$$

Now

$$A_n \cosh nx + B_n \sinh nx = \frac{2[1-(-1)^n]}{n\pi} \left[\cosh nx + \frac{\sinh nx}{\sinh n\pi}(1 - \cosh n\pi) \right]$$

$$= \frac{2[1-(-1)^n]}{n\pi \sinh n\pi}[\cosh nx \sinh n\pi + \sinh nx - \sinh nx \cosh n\pi]$$

$$= \frac{2[1-(-1)^n]}{n\pi \sinh n\pi}[\sinh nx + \sinh n(\pi - x)]$$

and

$$u(x,y) = u_1 + u_2 = \frac{2}{\pi} \sum_{n=1}^{\infty} \frac{1-(-1)^n}{n \sinh n\pi} \sinh ny \sin nx$$

$$+ \frac{2}{\pi} \sum_{n=1}^{\infty} \frac{[1-(-1)^n][\sinh nx + \sinh n(\pi - x)]}{n \sinh n\pi} \sin ny.$$

Exercises 13.6

3. If we let $u(x,t) = v(x,t) + \psi(x)$, then we obtain as in Example 1 in the text

$$k\psi'' + r = 0$$

or

$$\psi(x) = -\frac{r}{2k}x^2 + c_1 x + c_2.$$

The boundary conditions become

$$u(0,t) = v(0,t) + \psi(0) = u_0$$
$$u(1,t) = v(1,t) + \psi(1) = u_0.$$

Letting $\psi(0) = \psi(1) = u_0$ we obtain homogeneous boundary conditions in v:

$$v(0,t) = 0 \quad \text{and} \quad v(1,t) = 0.$$

Now $\psi(0) = \psi(1) = u_0$ implies $c_2 = u_0$ and $c_1 = r/2k$. Thus

$$\psi(x) = -\frac{r}{2k}x^2 + \frac{r}{2k}x + u_0 = u_0 - \frac{r}{2k}x(x-1).$$

To determine $v(x,t)$ we solve

$$k\frac{\partial^2 v}{\partial x^2} = \frac{\partial v}{dt}, \quad 0 < x < 1, \ t > 0$$

$$v(0,t) = 0, \quad v(1,t) = 0,$$

$$v(x,0) = \frac{r}{2k}x(x-1) - u_0.$$

Separating variables, we find

$$v(x,t) = \sum_{n=1}^{\infty} A_n e^{-kn^2\pi^2 t} \sin n\pi x,$$

where

$$A_n = 2\int_0^1 \left[\frac{r}{2k}x(x-1) - u_0\right]\sin n\pi x \, dx = 2\left[\frac{u_0}{n\pi} + \frac{r}{kn^3\pi^3}\right]\left[(-1)^n - 1\right]. \tag{1}$$

Hence, a solution of the original problem is

$$u(x,t) = \psi(x) + v(x,t)$$

$$= u_0 - \frac{r}{2k}x(x-1) + \sum_{n=1}^{\infty} A_n e^{-kn^2\pi^2 t} \sin n\pi x,$$

where A_n is defined in (1).

6. Substituting $u(x,t) = v(x,t) + \psi(x)$ into the partial differential equation gives

$$k\frac{\partial^2 v}{\partial x^2} + k\psi'' - hv - h\psi = \frac{\partial v}{\partial t}.$$

This equation will be homogeneous provided ψ satisfies

$$k\psi'' - h\psi = 0.$$

Since k and h are positive, the general solution of this latter equation is

$$\psi(x) = c_1 \cosh\sqrt{\frac{h}{k}}\, x + c_2 \sinh\sqrt{\frac{h}{k}}\, x.$$

From $\psi(0) = 0$ and $\psi(\pi) = u_0$ we find $c_1 = 0$ and $c_2 = u_0/\sinh\sqrt{h/k}\,\pi$. Hence

$$\psi(x) = u_0 \frac{\sinh\sqrt{h/k}\, x}{\sinh\sqrt{h/k}\, \pi}.$$

Now the new problem is

$$k\frac{\partial^2 v}{\partial x^2} - hv = \frac{\partial v}{\partial t}, \quad 0 < x < \pi, \ t > 0$$

$$v(0,t) = 0, \quad v(\pi,t) = 0, \quad t > 0$$

$$v(x,0) = -\psi(x), \quad 0 < x < \pi.$$

If we let $v = XT$ then

$$\frac{X''}{X} = \frac{T' + hT}{kT} = -\lambda^2$$

gives the separated differential equations

$$X'' + \lambda^2 X = 0 \quad \text{and} \quad T' + \left(h + k\lambda^2\right)T = 0.$$

The respective solutions are

$$X(x) = c_3 \cos \lambda x + c_4 \sin \lambda x$$

$$T(t) = c_5 e^{-(h+k\lambda^2)t}.$$

From $X(0) = 0$ we get $c_3 = 0$ and from $X(\pi) = 0$ we find $\lambda = n$ for $n = 1, 2, 3, \ldots$. Consequently, it follows that

$$v(x, t) = \sum_{n=1}^{\infty} A_n e^{-(h+kn^2)t} \sin nx$$

where

$$A_n = -\frac{2}{\pi} \int_0^{\pi} \psi(x) \sin nx \, dx.$$

Hence a solution of the original problem is

$$u(x, t) = u_0 \frac{\sinh \sqrt{h/k}\, x}{\sinh \sqrt{h/k}\, \pi} + e^{-ht} \sum_{n=1}^{\infty} A_n e^{-kn^2 t} \sin nx$$

where

$$A_n = -\frac{2}{\pi} \int_0^{\pi} u_0 \frac{\sinh \sqrt{h/k}\, x}{\sinh \sqrt{h/k}\, \pi} \sin nx \, dx.$$

Using the exponential definition of the hyperbolic sine and integration by parts we find

$$A_n = \frac{2u_0 nk(-1)^n}{\pi(h + kn^2)}.$$

9. Substituting $u(x, t) = v(x, t) + \psi(x)$ into the partial differential equation gives

$$a^2 \frac{\partial^2 v}{\partial x^2} + a^2 \psi'' + Ax = \frac{\partial^2 v}{\partial t^2}.$$

This equation will be homogeneous provided ψ satisfies

$$a^2 \psi'' + Ax = 0.$$

The general solution of this differential equation is

$$\psi(x) = -\frac{A}{6a^2} x^3 + c_1 x + c_2.$$

From $\psi(0) = 0$ we obtain $c_2 = 0$, and from $\psi(1) = 0$ we obtain $c_1 = A/6a^2$. Hence

$$\psi(x) = \frac{A}{6a^2}(x - x^3).$$

Now the new problem is

$$a^2 \frac{\partial^2 v}{\partial x^2} = \frac{\partial^2 v}{\partial t^2}$$

$$v(0, t) = 0, \quad v(1, t) = 0, \quad t > 0,$$

$$v(x, 0) = -\psi(x), \quad v_t(x, 0) = 0, \quad 0 < x < 1.$$

Identifying this as the wave equation solved in Section 13.4 in the text with $L = 1$, $f(x) = -\psi(x)$, and $g(x) = 0$ we obtain

$$v(x, t) = \sum_{n=1}^{\infty} A_n \cos n\pi a t \sin n\pi x$$

where

$$A_n = 2 \int_0^1 [-\psi(x)] \sin n\pi x \, dx = \frac{A}{3a^2} \int_0^1 (x^3 - x) \sin n\pi x \, dx = \frac{2A(-1)^n}{a^2 \pi^3 n^3}.$$

208

Thus

$$u(x,t) = \frac{A}{6a^2}(x - x^3) + \frac{2A}{a^2\pi^3}\sum_{n=1}^{\infty}\frac{(-1)^n}{n^3}\cos n\pi at \sin n\pi x.$$

12. Substituting $u(x,y) = v(x,y) + \psi(x)$ into Poisson's equation we obtain

$$\frac{\partial^2 v}{\partial x^2} + \psi''(x) + h + \frac{\partial^2 v}{\partial y^2} = 0.$$

The equation will be homogeneous provided ψ satisfies $\psi''(x) + h = 0$ or $\psi(x) = -\frac{h}{2}x^2 + c_1 x + c_2$. From $\psi(0) = 0$ we obtain $c_2 = 0$. From $\psi(\pi) = 1$ we obtain

$$c_1 = \frac{1}{\pi} + \frac{h\pi}{2}.$$

Then

$$\psi(x) = \left(\frac{1}{\pi} + \frac{h\pi}{2}\right)x - \frac{h}{2}x^2.$$

The new boundary-value problem is

$$\frac{\partial^2 v}{\partial x^2} + \frac{\partial^2 v}{\partial y^2} = 0$$

$$v(0,y) = 0, \quad v(\pi,y) = 0,$$

$$v(x,0) = -\psi(x), \quad 0 < x < \pi.$$

This is Problem 11 in Section 13.5. The solution is

$$v(x,y) = \sum_{n=1}^{\infty} A_n e^{-ny}\sin nx$$

where

$$A_n = \frac{2}{\pi}\int_0^{\pi}[-\psi(x)\sin nx]\,dx$$

$$= \frac{2(-1)^n}{m}\left(\frac{1}{\pi} + \frac{h\pi}{2}\right) - h(-1)^n\left(\frac{\pi}{n} + \frac{2}{n^2}\right).$$

Thus

$$u(x,y) = v(x,y) + \psi(x) = \left(\frac{1}{\pi} + \frac{h\pi}{2}\right)x - \frac{h}{2}x^2 + \sum_{n=1}^{\infty}A_n e^{-ny}\sin nx.$$

Exercises 13.7

3. Separating variables in Laplace's equation gives

$$X'' + \lambda^2 X = 0$$

$$Y'' - \lambda^2 Y = 0$$

and

$$X(x) = c_1 \cos\lambda x + c_2 \sin\lambda x$$

$$Y(y) = c_3 \cosh\lambda y + c_4 \sinh\lambda y.$$

From $u(0,y) = 0$ we obtain $X(0) = 0$ and $c_1 = 0$. From $u_x(a,y) = -hu(a,y)$ we obtain $X'(a) = -hX(a)$ and

$$\lambda\cos\lambda a = -h\sin\lambda a \quad \text{or} \quad \tan\lambda a = -\frac{\lambda}{h}.$$

Let λ_n, where $n = 1, 2, 3, \ldots$, be the consecutive positive roots of this equation. From $u(x, 0) = 0$ we obtain $Y(0) = 0$ and $c_3 = 0$. Thus

$$u(x, y) = \sum_{n=1}^{\infty} A_n \sinh \lambda_n y \sin \lambda_n x.$$

Now

$$f(x) = \sum_{n=1}^{\infty} A_n \sinh \lambda_n b \sin \lambda_n x$$

and

$$A_n \sinh \lambda_n b = \frac{\int_0^a f(x) \sin \lambda_n x \, dx}{\int_0^a \sin^2 \lambda_n x \, dx}.$$

Since

$$\int_0^a \sin^2 \lambda_n x \, dx = \frac{1}{2} \left[a - \frac{1}{2\lambda_n} \sin 2\lambda_n a \right] = \frac{1}{2} \left[a - \frac{1}{\lambda_n} \sin \lambda_n a \cos \lambda_n a \right]$$

$$= \frac{1}{2} \left[a - \frac{1}{h\lambda_n} (h \sin \lambda_n a) \cos \lambda_n a \right]$$

$$= \frac{1}{2} \left[a - \frac{1}{h\lambda_n} (-\lambda_n \cos \lambda_n a) \cos \lambda_n a \right] = \frac{1}{2h} \left[ah + \cos^2 \lambda_n a \right],$$

we have

$$A_n = \frac{2h}{\sinh \lambda_n b [ah + \cos^2 \lambda_n a]} \int_0^a f(x) \sin \lambda_n x \, dx.$$

6. Substituting $u(x, t) = v(x, t) + \psi(x)$ into the partial differential equation gives

$$a^2 \frac{\partial^2 v}{\partial x^2} + \psi''(x) = \frac{\partial^2 v}{\partial t^2}.$$

This equation will be homogeneous if $\psi''(x) = 0$ or $\psi(x) = c_1 x + c_2$. The boundary condition $u(0, t) = 0$ implies $\psi(0) = 0$ which implies $c_2 = 0$. Thus $\psi(x) = c_1 x$. Using the second boundary condition, we obtain

$$E \left(\frac{\partial v}{\partial x} + \psi' \right) \bigg|_{x=L} = F_0,$$

which will be homogeneous when

$$E\psi'(L) = F_0.$$

Since $\psi'(x) = c_1$ we conclude that $c_1 = F_0/E$ and

$$\psi(x) = \frac{F_0}{E} x.$$

The new boundary-value problem is

$$a^2 \frac{\partial^2 v}{\partial x^2} = \frac{\partial^2 v}{\partial t^2}, \quad 0 < x < L, \quad t > 0$$

$$v(0, t) = 0, \quad \frac{\partial v}{\partial x} \bigg|_{x=L} = 0, \quad t > 0,$$

$$v(x, 0) = -\frac{F_0}{E} x, \quad \frac{\partial v}{\partial t} \bigg|_{t=0} = 0, \quad 0 < x < L.$$

Referring to Example 2 in the text we see that

$$v(x, t) = \sum_{n=1}^{\infty} A_n \cos a \left(\frac{2n-1}{2L} \right) \pi t \sin \left(\frac{2n-1}{2L} \right) \pi x$$

where

$$-\frac{F_0}{E}x = \sum_{n=1}^{\infty} A_n \sin\left(\frac{2n-1}{2L}\right)\pi x$$

and

$$A_n = \frac{-F_0 \int_0^L x \sin\left(\frac{2n-1}{2L}\right)\pi x\,dx}{E \int_0^L \sin^2\left(\frac{2n-1}{2L}\right)\pi x\,dx} = \frac{8F_0 L(-1)^n}{E\pi^2(2n-1)^2}.$$

Thus

$$u(x,t) = v(x,t) + \psi(x)$$

$$= \frac{F_0}{E}x + \frac{8F_0 L}{E\pi^2}\sum_{n=1}^{\infty}\frac{(-1)^n}{(2n-1)^2}\cos a\left(\frac{2n-1}{2L}\right)\pi t \sin\left(\frac{2n-1}{2L}\right)\pi x.$$

9. (a) Using $u = XT$ and separation constant λ^4 we find

$$X^{(4)} - \lambda^4 X = 0$$

and

$$X(x) = c_1 \cos\lambda x + c_2 \sin\lambda x + c_3 \cosh\lambda x + c_4 \sinh\lambda x.$$

Since $u = XT$ the boundary conditions become

$$X(0) = 0, \quad X'(0) = 0, \quad X''(1) = 0, \quad X'''(1) = 0.$$

Now $X(0) = 0$ implies $c_1 + c_3 = 0$, while $X'(0) = 0$ implies $c_2 + c_4 = 0$. Thus

$$X(x) = c_1 \cos\lambda x + c_2 \sin\lambda x - c_1 \cosh\lambda x - c_2 \sinh\lambda x.$$

The boundary condition $X''(1) = 0$ implies

$$-c_1 \cos\lambda - c_2 \sin\lambda - c_1 \cosh\lambda - c_2 \sinh\lambda = 0$$

while the boundary condition $X'''(1) = 0$ implies

$$c_1 \sin\lambda - c_2 \cos\lambda - c_1 \sinh\lambda - c_2 \cosh\lambda = 0.$$

We then have the system of two equations in two unknowns

$$(\cos\lambda + \cosh\lambda)c_1 + (\sin\lambda + \sinh\lambda)c_2 = 0$$

$$(\sin\lambda - \sinh\lambda)c_1 - (\cos\lambda + \cosh\lambda)c_2 = 0.$$

This homogeneous system will have nontrivial solutions for c_1 and c_2 provided

$$\begin{vmatrix} \cos\lambda + \cosh\lambda & \sin\lambda + \sinh\lambda \\ \sin\lambda - \sinh\lambda & -\cos\lambda - \cosh\lambda \end{vmatrix} = 0$$

or

$$-2 - 2\cos\lambda\cosh\lambda = 0.$$

Thus, the eigenvalues are determined by the equation $\cos\lambda\cosh\lambda = -1$.

(b) Using a computer to graph $\cosh\lambda$ and $-1/\cos\lambda = -\sec\lambda$ we see that the first two positive eigenvalues occur near 1.9 and 4.7. Applying Newton's method with these initial values we find that the eigenvalues are $\lambda_1 = 1.8751$ and $\lambda_2 = 4.6941$.

3. We need to solve the partial differential equation

$$a^2 \left(\frac{\partial^2 u}{\partial x^2} + \frac{\partial^2 u}{\partial y^2} \right) = \frac{\partial^2 u}{\partial t^2}.$$

To separate this equation we try $u(x, y, t) = X(x)Y(y)T(t)$:

$$a^2 (X''YT + XY''T) = XYT''$$

$$\frac{X''}{X} = -\frac{Y''}{Y} + \frac{T''}{a^2 T} = -\lambda^2.$$

Then

$$X'' + \lambda^2 X = 0 \tag{1}$$

$$\frac{Y''}{Y} = \frac{T''}{a^2 T} + \lambda^2 = -\mu^2$$

$$Y'' + \mu^2 Y = 0 \tag{2}$$

$$T'' + a^2 \left(\lambda^2 + \mu^2 \right) T = 0. \tag{3}$$

The general solutions of equations (1), (2), and (3) are, respectively,

$$X(x) = c_1 \cos \lambda x + c_2 \sin \lambda x$$

$$Y(y) = c_3 \cos \mu y + c_4 \sin \mu y$$

$$T(t) = c_5 \cos a\sqrt{\lambda^2 + \mu^2}\, t + c_6 \sin a\sqrt{\lambda^2 + \mu^2}\, t.$$

The conditions $X(0) = 0$ and $Y(0) = 0$ give $c_1 = 0$ and $c_3 = 0$. The conditions $X(\pi) = 0$ and $Y(\pi) = 0$ yield two sets of eigenvalues:

$$\lambda = m, \ m = 1, 2, 3, \dots \quad \text{and} \quad \mu = n, \ n = 1, 2, 3, \dots.$$

A product solution of the partial differential equation that satisfies the boundary conditions is

$$u_{mn}(x, y, t) = \left(A_{mn} \cos a\sqrt{m^2 + n^2}\, t + B_{mn} \sin a\sqrt{m^2 + n^2}\, t \right) \sin mx \sin ny.$$

To satisfy the initial conditions we use the superposition principle:

$$u(x, y, t) = \sum_{m=1}^{\infty} \sum_{n=1}^{\infty} \left(A_{mn} \cos a\sqrt{m^2 + n^2}\, t + B_{mn} \sin a\sqrt{m^2 + n^2}\, t \right) \sin mx \sin ny.$$

The initial condition $u_t(x, y, 0) = 0$ implies $B_{mn} = 0$ and

$$u(x, y, t) = \sum_{m=1}^{\infty} \sum_{n=1}^{\infty} A_{mn} \cos a\sqrt{m^2 + n^2}\, t \sin mx \sin ny.$$

At $t = 0$ we have

$$xy(x - \pi)(y - \pi) = \sum_{m=1}^{\infty} \sum_{n=1}^{\infty} A_{mn} \sin mx \sin ny.$$

Using (11) and (12) in the text, it follows that

$$A_{mn} = \frac{4}{\pi^2} \int_0^\pi \int_0^\pi xy(x - \pi)(y - \pi) \sin mx \sin ny \, dx \, dy$$

$$= \frac{4}{\pi^2} \int_0^\pi x(x - \pi) \sin mx \, dx \int_0^\pi y(y - \pi) \sin ny \, dy$$

$$= \frac{16}{m^3 n^3 \pi^2} [(-1)^m - 1][(-1)^n - 1].$$

6. To separate Laplace's equation in three dimensions we try $u(x, y, z) = X(x)Y(y)Z(z)$:

$$X''YZ + XY''Z + XYZ'' = 0$$
$$\frac{X''}{X} = -\frac{Y''}{Y} - \frac{Z''}{Z} = -\lambda^2.$$

Then

$$X'' + \lambda^2 X = 0 \tag{4}$$

$$\frac{Y''}{Y} = -\frac{Z''}{Z} + \lambda^2 = -\mu^2$$

$$Y'' + \mu^2 Y = 0 \tag{5}$$

$$Z'' - (\lambda^2 + \mu^2)Z = 0. \tag{6}$$

The general solutions of equations (4), (5), and (6) are, respectively

$$X(x) = c_1 \cos \lambda x + c_2 \sin \lambda x$$
$$Y(y) = c_3 \cos \mu y + c_4 \sin \mu y$$
$$Z(z) = c_5 \cosh \sqrt{\lambda^2 + \mu^2}\, z + c_6 \sinh \sqrt{\lambda^2 + \mu^2}\, z.$$

The boundary and initial conditions are

$$u(0, y, z) = 0, \qquad\qquad u(a, y, z) = 0,$$
$$u(x, 0, z) = 0, \qquad\qquad u(x, b, z) = 0,$$
$$u(x, y, 0) = f(x, y), \qquad u(x, y, c) = 0.$$

The conditions $X(0) = Y(0) = 0$ give $c_1 = c_3 = 0$. The conditions $X(a) = Y(b) = 0$ yield two sets of eigenvalues:

$$\lambda = \frac{m\pi}{a}, \quad m = 1, 2, 3, \ldots \quad \text{and} \quad \mu = \frac{n\pi}{b}, \quad n = 1, 2, 3, \ldots.$$

Let

$$\omega_{mn}^2 = \frac{m^2 \pi^2}{a^2} + \frac{n^2 \pi^2}{b^2}.$$

Then the boundary condition $Z(c) = 0$ gives

$$c_5 \cosh c\omega_{mn} + c_6 \sinh c\omega_{mn} = 0$$

from which we obtain

$$Z(z) = c_5 \left(\cosh \omega_{mn} z - \frac{\cosh c\omega_{mn}}{\sinh c\omega_{mn}} \sinh \omega z \right)$$

$$= \frac{c_5}{\sinh c\omega_{mn}} (\sinh c\omega_{mn} \cosh \omega_{mn} z - \cosh c\omega_{mn} \sinh \omega_{mn} z)$$

$$= c_{mn} \sinh \omega_{mn}(c - z).$$

By the superposition principle

$$u(x, y, t) = \sum_{m=1}^{\infty} \sum_{n=1}^{\infty} A_{mn} \sinh \omega_{mn}(c - z) \sin \frac{m\pi}{a} x \sin \frac{n\pi}{b} y$$

where

$$A_{mn} = \frac{4}{ab \sinh c\omega_{mn}} \int_0^b \int_0^a f(x, y) \sin \frac{m\pi}{a} x \sin \frac{n\pi}{b} y \, dx \, dy.$$

Chapter 13 Review Exercises

3. Substituting $u(x,t) = v(x,t) + \psi(x)$ into the partial differential equation we obtain

$$k\frac{\partial^2 v}{\partial x^2} + k\psi''(x) = \frac{\partial v}{\partial t}.$$

This equation will be homogeneous provided ψ satisfies

$$k\psi'' = 0 \quad \text{or} \quad \psi = c_1 x + c_2.$$

Considering

$$u(0,t) = v(0,t) + \psi(0) = u_0$$

we set $\psi(0) = u_0$ so that $\psi(x) = c_1 x + u_0$. Now

$$-\frac{\partial u}{\partial x}\bigg|_{x=\pi} = -\frac{\partial v}{\partial x}\bigg|_{x=\pi} - \psi'(x) = v(\pi,t) + \psi(\pi) - u_1$$

is equivalent to

$$\frac{\partial v}{\partial x}\bigg|_{x=\pi} + v(\pi,t) = u_1 - \psi'(x) - \psi(\pi) = u_1 - c_1 - (c_1\pi + u_0),$$

which will be homogeneous when

$$u_1 - c_1 - c_1\pi - u_0 = 0 \quad \text{or} \quad c_1 = \frac{u_1 - u_0}{1 + \pi}.$$

The steady-state solution is

$$\psi(x) = \left(\frac{u_1 - u_0}{1 + \pi}\right) x + u_0.$$

6. The boundary-value problem is

$$\frac{\partial^2 u}{\partial x^2} + x^2 = \frac{\partial^2 u}{\partial t^2}, \quad 0 < x < 1, \quad t > 0,$$

$$u(0,t) = 1, \quad u(1,t) = 0, \quad t > 0,$$

$$u(x,0) = f(x), \quad u_t(x,0) = 0, \quad 0 < x < 1.$$

Substituting $u(x,t) = v(x,t) + \psi(x)$ into the partial differential equation gives

$$\frac{\partial^2 v}{\partial x^2} + \psi''(x) + x^2 = \frac{\partial^2 v}{\partial t^2}.$$

This equation will be homogeneous provided $\psi''(x) + x^2 = 0$ or

$$\psi(x) = -\frac{1}{12}x^4 + c_1 x + c_2.$$

From $\psi(0) = 1$ and $\psi(1) = 0$ we obtain $c_1 = -11/12$ and $c_2 = 1$. The new problem is

$$\frac{\partial^2 v}{\partial x^2} = \frac{\partial^2 v}{\partial t^2}, \quad 0 < x < 1, \quad t > 0,$$

$$v(0,t) = 0, \quad v(1,t) = 0, \quad t > 0,$$

$$v(x,0) = f(x) - \psi(x), \quad v_t(x,0) = 0, \quad 0 < x < 1.$$

From Section 13.4 in the text we see that $B_n = 0$,

$$A_n = 2\int_0^1 [f(x) - \psi(x)]\sin n\pi x\, dx = 2\int_0^1 \left[f(x) + \frac{1}{12}x^4 + \frac{11}{12}x - 1\right]\sin n\pi x\, dx,$$

and

$$v(x,t) = \sum_{n=1}^{\infty} A_n \cos n\pi t \sin n\pi x.$$

Thus

$$u(x,t) = v(x,t) + \psi(x) = -\frac{1}{12}x^4 - \frac{11}{12}x + 1 + \sum_{n=1}^{\infty} A_n \cos n\pi t \sin n\pi x.$$

9. Using $u = XY$ and λ^2 as a separation constant leads to

$$X'' - \lambda^2 X = 0,$$

and

$$Y'' + \lambda^2 Y = 0,$$

$$Y(0) = 0,$$

$$Y(\pi) = 0.$$

Then

$$Y = c_1 \sin ny \quad \text{and} \quad X = c_2 e^{-nx}$$

for $n = 1, 2, 3, \ldots$ (since u must be bounded as $x \to \infty$) so that

$$u = \sum_{n=1}^{\infty} A_n e^{-nx} \sin ny.$$

Imposing

$$u(0, y) = 50 = \sum_{n=1}^{\infty} A_n \sin ny$$

gives

$$A_n = \frac{2}{\pi} \int_0^{\pi} 50 \sin ny \, dy = \frac{100}{n\pi}[1 - (-1)^n]$$

so that

$$u(x,y) = \sum_{n=1}^{\infty} \frac{100}{n\pi}[1 - (-1)^n] e^{-nx} \sin ny.$$

12. Substituting $u(x,t) = v(x,t) + \psi(x)$ into the partial differential equation gives

$$k\frac{\partial^2 v}{\partial x^2} + k\psi'' + \sin 2\pi x = \frac{\partial v}{\partial t}.$$

This equation will be homogeneous provided ψ satisfies

$$k\psi'' + \sin 2\pi x = 0.$$

The general solution of this equation is

$$\psi(x) = \frac{1}{4k\pi^2} \sin 2\pi x + c_1 x + c_2.$$

From $\psi(0) = \psi(1) = 0$ we find that $c_1 = c_2 = 0$ and

$$\psi(x) = \frac{1}{4k\pi^2} \sin 2\pi x.$$

Now the new problem is

$$k\frac{\partial^2 v}{\partial x^2} = \frac{\partial v}{\partial t}, \quad 0 < x < 1, \quad t > 0$$

$$v(0,t) = 0, \quad v(1,t) = 0, \quad t > 0$$

$$v(x,0) = \sin \pi x - \psi(x), \quad 0 < x < 1.$$

If we let $v = XT$ then

$$\frac{X''}{X} = \frac{T'}{kT} = -\lambda^2$$

gives the separated differential equations

$$X'' + \lambda^2 X = 0 \quad \text{and} \quad T' + k\lambda^2 T = 0.$$

The respective solutions are

$$X(x) = c_3 \cos \lambda x + c_4 \sin \lambda x$$

$$T(t) = c_5 e^{-k\lambda^2 t}.$$

From $X(0) = 0$ we get $c_3 = 0$ and from $X(1) = 0$ we find $\lambda = n\pi$ for $n = 1, 2, 3, \ldots$. Consequently, it follows that

$$v(x, t) = \sum_{n=1}^{\infty} A_n e^{-kn^2\pi^2 t} \sin n\pi x$$

where

$$v(x, 0) = \sin \pi x - \frac{1}{4k\pi^2} \sin 2\pi x = 0$$

implies

$$A_n = 2 \int_0^1 \left(\sin \pi x - \frac{1}{4k\pi^2} \sin 2\pi x \right) \sin n\pi x \, dx.$$

By orthogonality $A_n = 0$ for $n = 3, 4, 5, \ldots$, and only A_1 and A_2 can be nonzero. We have

$$A_1 = 2 \left[\int_0^1 \sin^2 \pi x \, dx - \frac{1}{4k\pi^2} \int_0^1 \sin 2\pi x \sin \pi x \, dx \right] = 2 \int_0^1 \frac{1}{2}(1 - \cos 2\pi x) \, dx = 1$$

and

$$A_2 = 2 \left[\int_0^1 \sin \pi x \sin 2\pi x \, dx - \frac{1}{4k\pi^2} \int_0^1 \sin^2 2\pi x \, dx \right]$$

$$= -\frac{1}{2k\pi^2} \int_0^1 \frac{1}{2}(1 - \cos 4\pi x) \, dx = -\frac{1}{4k\pi^2}.$$

Therefore

$$v(x, t) = A_1 e^{-k\pi^2 t} \sin \pi x + A_2 e^{-k4\pi^2 t} \sin 2\pi x$$

$$= e^{-k\pi^2 t} \sin \pi x - \frac{1}{4k\pi^2} e^{-4k\pi^2 t} \sin 2\pi x$$

and

$$u(x, t) = v(x, t) + \psi(x) = e^{-k\pi^2 t} \sin \pi x + \frac{1}{4k\pi^2}(1 - e^{-4k\pi^2 t}) \sin 2\pi x.$$

14 Boundary-Value Problems in Other Coordinate Systems

_____ **Exercises 14.1** _____

3. We have

$$A_0 = \frac{1}{2\pi} \int_0^{2\pi} (2\pi\theta - \theta^2)\,d\theta = \frac{2\pi^2}{3}$$

$$A_n = \frac{1}{\pi} \int_0^{2\pi} (2\pi\theta - \theta^2) \cos n\theta\,d\theta = -\frac{4}{n^2}$$

$$B_n = \frac{1}{\pi} \int_0^{2\pi} (2\pi\theta - \theta^2) \sin n\theta\,d\theta = 0$$

and so

$$u(r,\theta) = \frac{2\pi^2}{3} - 4\sum_{n=1}^{\infty} \frac{r^n}{n^2} \cos n\theta.$$

6. We solve

$$\frac{\partial^2 u}{\partial r^2} + \frac{1}{r}\frac{\partial u}{\partial r} + \frac{1}{r^2}\frac{\partial^2 u}{\partial \theta^2} = 0, \quad 0 < \theta < \frac{\pi}{2}, \quad 0 < r < c,$$

$$u(c,\theta) = f(\theta), \quad 0 < \theta < \frac{\pi}{2},$$

$$u(r,0) = 0, \quad u(r,\pi/2) = 0, \quad 0 < r < c.$$

Proceeding as in Example 1 in the text we obtain the separated differential equations

$$r^2 R'' + rR' - \lambda^2 R = 0$$

$$\Theta'' + \lambda^2 \Theta = 0$$

with solutions

$$\Theta(\theta) = c_1 \cos \lambda\theta + c_2 \sin \lambda\theta$$

$$R(r) = c_3 r^\lambda + c_4 r^{-\lambda}.$$

Since we want $R(r)$ to be bounded as $r \to 0$ we require $c_4 = 0$. Applying the boundary conditions $\Theta(0) = 0$ and $\Theta(\pi/2) = 0$ we find that $c_1 = 0$ and $\lambda = 2n$ for $n = 1, 2, 3, \ldots$. Therefore

$$u(r,\theta) = \sum_{n=1}^{\infty} A_n r^{2n} \sin 2n\theta.$$

From

$$u(c,\theta) = f(\theta) = \sum_{n=1}^{\infty} A_n c^n \sin 2n\theta$$

we find

$$A_n = \frac{4}{\pi c^{2n}} \int_0^{\pi/2} f(\theta) \sin 2n\theta\,d\theta.$$

9. Proceeding as in Example 1 in the text and again using the periodicity of $u(r,\theta)$, we have

$$\Theta(\theta) = c_1 \cos \lambda\theta + c_2 \sin \lambda\theta$$

217

where $\lambda = n$ for $n = 0, 1, 2, \ldots$. Then

$$R(r) = c_3 r^n + c_4 r^{-n}.$$

[We do not have $c_4 = 0$ in this case since $0 < a \le r$.] Since $u(b, \theta) = 0$ we have

$$u(r, \theta) = A_0 \ln \frac{r}{b} + \sum_{n=1}^{\infty} \left[\left(\frac{b}{r} \right)^n - \left(\frac{r}{b} \right)^n \right] [A_n \cos n\theta + B_n \sin n\theta].$$

From

$$u(a, \theta) = f(\theta) = A_0 \ln \frac{a}{b} + \sum_{n=1}^{\infty} \left[\left(\frac{b}{a} \right)^n - \left(\frac{a}{b} \right)^n \right] [A_n \cos n\theta + B_n \sin n\theta]$$

we find

$$A_0 \ln \frac{a}{b} = \frac{1}{2\pi} \int_0^{2\pi} f(\theta)\, d\theta,$$

$$\left[\left(\frac{b}{a} \right)^n - \left(\frac{a}{b} \right)^n \right] A_n = \frac{1}{\pi} \int_0^{2\pi} f(\theta) \cos n\theta\, d\theta,$$

and

$$\left[\left(\frac{b}{a} \right)^n - \left(\frac{a}{b} \right)^n \right] B_n = \frac{1}{\pi} \int_0^{2\pi} f(\theta) \sin n\theta\, d\theta.$$

12. Letting $u(r, \theta) = v(r, \theta) + \psi(\theta)$ we obtain $\psi''(\theta) = 0$ and so $\psi(\theta) = c_1 \theta + c_2$. From $\psi(0) = 0$ and $\psi(\pi) = u_0$ we find, in turn, $c_2 = 0$ and $c_1 = u_0/\pi$. Therefore $\psi(\theta) = \frac{u_0}{\pi} \theta$. Now $u(1, \theta) = v(1, \theta) + \psi(\theta)$ so that $v(1, \theta) = u_0 - \frac{u_0}{\pi} \theta$. From

$$v(r, \theta) = \sum_{n=1}^{\infty} A_n r^n \sin n\theta \quad \text{and} \quad v(1, \theta) = \sum_{n=1}^{\infty} A_n \sin n\theta$$

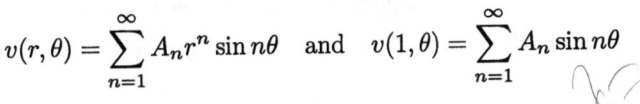

we obtain

$$A_n = \frac{2}{\pi} \int_0^{\pi} \left(u_0 - \frac{u_0}{\pi} \theta \right) \sin n\theta\, d\theta = \frac{2u_0}{\pi n}.$$

Thus

$$u(r, \theta) = \frac{u_0}{\pi} \theta + \frac{2u_0}{\pi} \sum_{n=1}^{\infty} \frac{r^n}{n} \sin n\theta.$$

Exercises 14.2

3. Referring to Example 2 in the text we have

$$R(r) = c_1 J_0(\lambda r) + c_2 Y_0(\lambda r)$$

$$Z(z) = c_3 \cosh \lambda z + c_4 \sinh \lambda z$$

where $c_2 = 0$ and $J_0(2\lambda) = 0$ defines the positive eigenvalues λ_n. From $Z(4) = 0$ we obtain

$$c_3 \cosh 4\lambda_n + c_4 \sinh 4\lambda_n = 0 \quad \text{or} \quad c_4 = -c_3 \frac{\cosh 4\lambda_n}{\sinh 4\lambda_n}.$$

Then

$$Z(z) = c_3 \left[\cosh \lambda_n z - \frac{\cosh 4\lambda_n}{\sinh 4\lambda_n} \sinh \lambda_n z \right] = c_3 \frac{\sinh 4\lambda_n \cosh \lambda_n z - \cosh 4\lambda_n \sinh \lambda_n z}{\sinh 4\lambda_n}$$

$$= c_3 \frac{\sinh \lambda_n (4 - z)}{\sinh 4\lambda_n}$$

and

$$u(r,z) = \sum_{n=1}^{\infty} A_n \frac{\sinh \lambda_n (4-z)}{\sinh 4\lambda_n} J_0(\lambda_n r).$$

From

$$u(r,0) = u_0 = \sum_{n=1}^{\infty} A_n J_0(\lambda_n r)$$

we obtain

$$A_n = \frac{2u_0}{4J_1^2(2\lambda_n)} \int_0^2 r J_0(\lambda_n r)\, dr = \frac{u_0}{\lambda_n J_1(2\lambda_n)}.$$

Thus the temperature in the cylinder is

$$u(r,z) = u_0 \sum_{n=1}^{\infty} \frac{\sinh \lambda_n (4-z) J_0(\lambda_n r)}{\lambda_n \sinh 4\lambda_n J_1(2\lambda_n)}.$$

6. Letting $u(r,t) = R(r)T(t)$ and separating variables we obtain

$$\frac{R'' + \frac{1}{r}R'}{R} = \frac{T'}{kT} = \mu \quad \text{and} \quad R'' + \frac{1}{r}R' - \mu R = 0, \quad T' - \mu k T = 0.$$

From the second equation we find $T(t) = e^{\mu k t}$. If $\mu > 0$, $T(t)$ increases without bound as $t \to \infty$. Thus we assume $\mu = -\lambda^2 \le 0$. Now

$$R'' + \frac{1}{r}R' + \lambda^2 R = 0$$

is a parametric Bessel equation with solution

$$R(r) = c_1 J_0(\lambda r) + c_2 Y_0(\lambda r).$$

Since Y_0 is unbounded as $r \to 0$ we take $c_2 = 0$, in which case $R(r) = c_1 J_0(\lambda r)$. If the edge $r = c$ is insulated we have the boundary condition $u_r(c,t) = 0$. Then

$$R'(c) = \lambda c_1 J_0'(\lambda c) = 0,$$

which defines an eigenvalue $\lambda = 0$ and positive eigenvalues λ_n. Thus

$$u(r,t) = A_0 + \sum_{n=1}^{\infty} A_n J_0(\lambda_n r) e^{-\lambda_n^2 k t}.$$

From

$$u(r,0) = f(r) = A_0 + \sum_{n=1}^{\infty} A_n J_0(\lambda_n r)$$

we find

$$A_0 = \frac{2}{c^2} \int_0^c r f(r)\, dr$$

$$A_n = \frac{2}{c^2 J_0^2(\lambda_n c)} \int_0^c r J_0(\lambda_n r) f(r)\, dr.$$

9. Substituting $u(r,t) = v(r,t) + \psi(r)$ into the partial differential equation gives

$$\frac{\partial^2 v}{\partial r^2} + \frac{1}{r}\frac{\partial v}{\partial r} + \psi'' + \frac{1}{r}\psi' = \frac{\partial v}{\partial t}.$$

This equation will be homogeneous provided $\psi'' + \frac{1}{r}\psi' = 0$ or $\psi(r) = c_1 \ln r + c_2$. Since $\ln r$ is unbounded as $r \to 0$ we take $c_1 = 0$. Then $\psi(r) = c_2$ and using $u(2,t) = v(2,t) + \psi(2) = 100$ we set $c_2 = \psi(r) = 100$. Referring to Problem 5 above, the solution of the boundary-value problem

$$\frac{\partial^2 v}{\partial r^2} + \frac{1}{r}\frac{\partial v}{\partial r} = \frac{\partial v}{\partial t}, \quad 0 < r < 2, \ t > 0,$$

$$v(2,t) = 0, \quad t > 0,$$

$$v(r,0) = u(r,0) - \psi(r)$$

is

$$v(r,t) = \sum_{n=1}^{\infty} A_n J_0(\lambda_n r) e^{-\lambda_n^2 t}$$

where

$$A_n = \frac{2}{2^2 J_1^2(2\lambda_n)} \int_0^2 r J_0(\lambda_n r)[u(r,0) - \psi(r)]\, dr$$

$$= \frac{1}{2 J_1^2(2\lambda_n)} \left[\int_0^1 r J_0(\lambda_n r)[200 - 100]\, dr + \int_1^2 r J_0(\lambda_n r)[100 - 100]\, dr \right]$$

$$= \frac{50}{J_1^2(2\lambda_n)} \int_0^1 r J_0(\lambda_n r)\, dr \qquad \boxed{x = \lambda_n r, \ dx = \lambda_n\, dr}$$

$$= \frac{50}{J_1^2(2\lambda_n)} \int_0^{\lambda_n} \frac{1}{\lambda_n^2} x J_0(x)\, dx$$

$$= \frac{50}{\lambda_n^2 J_1^2(2\lambda_n)} \int_0^{\lambda_n} \frac{d}{dx}[x J_1(x)]\, dx \qquad \boxed{\text{see (4) of Section 12.6 in text}}$$

$$= \frac{50}{\lambda_n^2 J_1^2(2\lambda_n)} (x J_1(x)) \Big|_0^{\lambda_n} = \frac{50 J_1(\lambda_n)}{\lambda_n J_1^2(2\lambda_n)}.$$

Thus

$$u(r,t) = v(r,t) + \psi(r) = 100 + 50 \sum_{n=1}^{\infty} \frac{J_1(\lambda_n) J_0(\lambda_n r)}{\lambda_n J_1^2(2\lambda_n)} e^{-\lambda^2 t}.$$

Exercises 14.3

3. The coefficients are given by

$$A_n = \frac{2n+1}{2c^n} \int_0^\pi \cos\theta P_n(\cos\theta) \sin\theta\, d\theta = \frac{2n+1}{2c^n} \int_0^\pi P_1(\cos\theta) P_n(\cos\theta) \sin\theta\, d\theta$$

$$\boxed{x = \cos\theta, \ dx = -\sin\theta\, d\theta}$$

$$= \frac{2n+1}{2c^n} \int_{-1}^1 P_1(x) P_n(x)\, dx.$$

Since $P_n(x)$ and $P_m(x)$ are orthogonal for $m \neq n$, $A_n = 0$ for $n \neq 1$ and

$$A_1 = \frac{2(1)+1}{2c^1} \int_{-1}^1 P_1(x) P_1(x)\, dx = \frac{3}{2c} \int_{-1}^1 x^2\, dx = \frac{1}{c}.$$

Thus

$$u(r,\theta) = \frac{r}{c} P_1(\cos\theta) = \frac{r}{c} \cos\theta.$$

6. Referring to Example 1 in the text we have

$$R(r) = c_1 r^n \quad \text{and} \quad \Theta(\theta) = P_n(\cos\theta).$$

Now $\Theta(\pi/2) = 0$ implies that n is odd, so

$$u(r,\theta) = \sum_{n=0}^{\infty} A_{2n+1} r^{2n+1} P_{2n+1}(\cos\theta).$$

From

$$u(c,\theta) = f(\theta) = \sum_{n=0}^{\infty} A_{2n+1} c^{2n+1} P_{2n+1}(\cos\theta)$$

we see that

$$A_{2n+1} c^{2n+1} = (4n+3) \int_0^{\pi/2} f(\theta) \sin\theta\, P_{2n+1}(\cos\theta)\, d\theta.$$

Thus

$$u(r,\theta) = \sum_{n=0}^{\infty} A_{2n+1} r^{2n+1} P_{2n+1}(\cos\theta)$$

where

$$A_{2n+1} = \frac{4n+3}{c^{2n+1}} \int_0^{\pi/2} f(\theta) \sin\theta\, P_{2n+1}(\cos\theta)\, d\theta.$$

9. Checking the hint, we find

$$\frac{1}{r}\frac{\partial^2}{\partial r^2}(ru) = \frac{1}{r}\frac{\partial}{\partial r}\left[r\frac{\partial u}{\partial r} + u\right] = \frac{1}{r}\left[r\frac{\partial^2 u}{\partial r^2} + \frac{\partial u}{\partial r} + \frac{\partial u}{\partial r}\right] = \frac{\partial^2 u}{\partial r^2} + \frac{2}{r}\frac{\partial u}{\partial r}.$$

The partial differential equation then becomes

$$\frac{\partial^2}{\partial r^2}(ru) = r\frac{\partial u}{\partial t}.$$

Now, letting $ru(r,t) = v(r,t) + \psi(r)$, since the boundary condition is nonhomogeneous, we obtain

$$\frac{\partial^2}{\partial r^2}[v(r,t) + \psi(r)] = r\frac{\partial}{\partial t}\left[\frac{1}{r}v(r,t) + \psi(r)\right]$$

or

$$\frac{\partial^2 v}{\partial r^2} + \psi''(r) = \frac{\partial v}{\partial t}.$$

This differential equation will be homogeneous if $\psi''(r) = 0$ or $\psi(r) = c_1 r + c_2$. Now

$$u(r,t) = \frac{1}{r}v(r,t) + \frac{1}{r}\psi(r) \quad \text{and} \quad \frac{1}{r}\psi(r) = c_1 + \frac{c_2}{r}.$$

Since we want $u(r,t)$ to be bounded as r approaches 0, we require $c_2 = 0$. Then $\psi(r) = c_1 r$. When $r = 1$

$$u(1,t) = v(1,t) + \psi(1) = v(1,t) + c_1 = 100,$$

and we will have the homogeneous boundary condition $v(1,t) = 0$ when $c_1 = 100$. Consequently, $\psi(r) = 100r$. The initial condition

$$u(r,0) = \frac{1}{r}v(r,0) + \frac{1}{r}\psi(r) = \frac{1}{r}v(r,0) + 100 = 0$$

implies $v(r,0) = -100r$. We are thus led to solve the new boundary-value problem

$$\frac{\partial^2 v}{\partial r^2} = \frac{\partial v}{\partial t}, \quad 0 < r < 1, \quad t > 0,$$

$$v(1,t) = 0, \quad \lim_{r\to 0}\frac{1}{r}v(r,t) < \infty,$$

$$v(r,0) = -100r.$$

221

Letting $v(r,t) = R(r)T(t)$ and separating variables leads to

$$R'' + \lambda^2 R = 0 \quad \text{and} \quad T' + \lambda^2 T = 0$$

with solutions

$$R(r) = c_3 \cos \lambda r + c_4 \sin \lambda r \quad \text{and} \quad T(t) = c_5 e^{-\lambda^2 t}.$$

The boundary conditions are equivalent to $R(1) = 0$ and $\lim\limits_{r \to 0} \dfrac{1}{r} R(r) < \infty$. Since

$$\lim_{r \to 0} \frac{1}{r} R(r) = \lim_{r \to 0} \frac{c_3 \cos \lambda r}{r} + \lim_{r \to 0} \frac{c_4 \sin \lambda r}{r} = \lim_{r \to 0} \frac{c_3 \cos \lambda r}{r} + c_4 \lambda < \infty$$

we must have $c_3 = 0$. Then $R(r) = c_4 \sin \lambda r$, and $R(1) = 0$ implies $\lambda = n\pi$ for $n = 1, 2, 3, \ldots$. Thus

$$v_n(r,t) = A_n e^{-n^2\pi^2 t} \sin n\pi r$$

for $n = 1, 2, 3, \ldots$. Using the condition $\lim_{r \to 0} \frac{1}{r} R(r) < \infty$ it is easily shown that there are no eigenvalues for $\lambda = 0$, nor does setting the common constant to $+\lambda^2$ when separating variables lead to any solutions. Now, by the superposition principle,

$$v(r,t) = \sum_{n=1}^{\infty} A_n e^{-n^2\pi^2 t} \sin n\pi r.$$

The initial condition $v(r,0) = -100r$ implies

$$-100r = \sum_{n=1}^{\infty} A_n \sin n\pi r.$$

This is a Fourier sine series and so

$$A_n = 2 \int_0^1 (-100r \sin n\pi r)\, dr = -200 \left[-\frac{r}{n\pi} \cos n\pi r \Big|_0^1 + \int_0^1 \frac{1}{n\pi} \cos n\pi r\, dr \right]$$

$$= -200 \left[-\frac{\cos n\pi}{n\pi} + \frac{1}{n^2\pi^2} \sin n\pi r \Big|_0^1 \right] = -200 \left[-\frac{(-1)^n}{n\pi} \right] = \frac{(-1)^n 200}{n\pi}.$$

A solution of the problem is thus

$$u(r,t) = \frac{1}{r} v(r,t) + \frac{1}{r} \psi(r) = \frac{1}{r} \sum_{n=1}^{\infty} (-1)^n \frac{20}{n\pi} e^{-n^2\pi^2 t} \sin n\pi r + \frac{1}{r}(100r)$$

$$= \frac{200}{\pi r} \sum_{n=1}^{\infty} \frac{(-1)^n}{n} e^{-n^2\pi^2 t} \sin n\pi r + 100.$$

12. Proceeding as in Example 1 we obtain

$$\Theta(\theta) = P_n(\cos \theta) \quad \text{and} \quad R(r) = c_1 r^n + c_2 r^{-(n+1)}$$

so that

$$u(r,\theta) = \sum_{n=0}^{\infty} (A_n r^n + B_n r^{-(n+1)}) P_n(\cos \theta).$$

To satisfy $\lim_{r \to \infty} u(r,\theta) = -Er\cos\theta$ we must have $A_n = 0$ for $n = 2, 3, 4, \ldots$. Then

$$\lim_{r \to \infty} u(r,\theta) = -Er\cos\theta = A_0 \cdot 1 + A_1 r \cos\theta,$$

so $A_0 = 0$ and $A_1 = -E$. Thus

$$u(r,\theta) = -Er\cos\theta + \sum_{n=0}^{\infty} B_n r^{-(n+1)} P_n(\cos\theta).$$

Now

$$u(c, \theta) = 0 = -Ec \cos \theta + \sum_{n=0}^{\infty} B_n c^{-(n+1)} P_n(\cos \theta)$$

so

$$\sum_{n=0}^{\infty} B_n c^{-(n+1)} P_n(\cos \theta) = Ec \cos \theta$$

and

$$B_n c^{-(n+1)} = \frac{2n+1}{2} \int_0^{\pi} Ec \cos \theta \, P_n(\cos \theta) \sin \theta \, d\theta.$$

Now $\cos \theta = P_1(\cos \theta)$ so, for $n \neq 1$,

$$\int_0^{\pi} \cos \theta \, P_n(\cos \theta) \sin \theta \, d\theta = 0$$

by orthogonality. Thus $B_n = 0$ for $n \neq 1$ and

$$B_1 = \frac{3}{2} Ec^3 \int_0^{\pi} \cos^2 \theta \sin \theta \, d\theta = Ec^3.$$

Therefore,

$$u(r, \theta) = -Er \cos \theta + Ec^3 r^{-2} \cos \theta.$$

Chapter 14 Review Exercises

3. The conditions $\Theta(0) = 0$ and $\Theta(\pi) = 0$ applied to $\Theta = c_1 \cos \lambda \theta + c_2 \sin \lambda \theta$ give $c_1 = 0$ and $\lambda = n$, $n = 1, 2, 3, \ldots$, respectively. Thus we have the Fourier sine-series coefficients

$$A_n = \frac{2}{\pi} \int_0^{\pi} u_0(\pi \theta - \theta^2) \sin n\theta \, d\theta = \frac{4u_0}{n^3 \pi}[1 - (-1)^n].$$

Thus

$$u(r, \theta) = \frac{4u_0}{\pi} \sum_{n=1}^{\infty} \frac{1 - (-1)^n}{n^3} r^n \sin n\theta.$$

6. We solve

$$\frac{\partial^2 u}{\partial r^2} + \frac{1}{r} \frac{\partial u}{\partial r} + \frac{1}{r^2} \frac{\partial^2 u}{\partial \theta^2} = 0, \quad r > 1, \quad 0 < \theta < \pi,$$

$$u(r, 0) = 0, \quad u(r, \pi) = 0, \quad r > 1,$$

$$u(1, \theta) = f(\theta), \quad 0 < \theta < \pi.$$

Separating variables we obtain

$$\Theta(\theta) = c_1 \cos \lambda \theta + c_2 \sin \lambda \theta$$

$$R(r) = c_3 r^{\lambda} + c_4 r^{-\lambda}.$$

Applying the boundary conditions $\Theta(0) = 0$, and $\Theta(\pi) = 0$ gives $c_1 = 0$ and $\lambda = n$ for $n = 1, 2, 3, \ldots$. Assuming $f(\theta)$ to be bounded, we expect the solution $u(r, \theta)$ to also be bounded as $r \to \infty$. This requires that $c_3 = 0$. Therefore

$$u(r, \theta) = \sum_{n=1}^{\infty} A_n r^{-n} \sin n\theta.$$

From

$$u(1, \theta) = f(\theta) = \sum_{n=1}^{\infty} A_n \sin n\theta$$

223

we obtain

$$A_n = \frac{2}{\pi} \int_0^\pi f(\theta) \sin n\theta \, d\theta.$$

9. Referring to Example 2 in Section 14.2 we have

$$R(r) = c_1 J_0(\lambda r) + c_2 Y_0(\lambda r)$$

$$Z(z) = c_3 \cosh \lambda z + c_4 \sinh \lambda z$$

where $c_2 = 0$ and $J_0(2\lambda) = 0$ defines the positive eigenvalues λ_n. From $Z'(0) = 0$ we obtain $c_4 = 0$. Then

$$u(r, z) = \sum_{n=1}^\infty A_n \cosh \lambda_n z J_0(\lambda_n r).$$

From

$$u(r, 4) = 50 = \sum_{n=1}^\infty A_n \cosh 4\lambda_n J_0(\lambda_n r)$$

we obtain (as in Example 1 of Section 14.1)

$$A_n \cosh 4\lambda_n = \frac{2(50)}{4 J_1^2(2\lambda_n)} \int_0^2 r J_0(\lambda_n r) \, dr = \frac{50}{\lambda_n J_1(2\lambda_n)} .$$

Thus the temperature in the cylinder is

$$u(r, z) = 50 \sum_{n=1}^\infty \frac{\cosh \lambda_n z J_0(\lambda_n r)}{\lambda_n \cosh 4\lambda_n J_1(2\lambda_n)} .$$

12. Since

$$\frac{1}{r} \frac{\partial^2}{\partial r^2} (ru) = \frac{1}{r} \frac{\partial}{\partial r} \left[r \frac{\partial u}{\partial r} + u \right] = \frac{1}{r} \left[r \frac{\partial^2 u}{\partial r^2} + \frac{\partial u}{\partial r} + \frac{\partial u}{\partial r} \right] = \frac{\partial^2 u}{\partial r^2} + \frac{2}{r} \frac{\partial u}{\partial r}$$

the differential equation becomes

$$\frac{1}{r} \frac{\partial^2}{\partial r^2} (ru) = \frac{\partial^2 u}{\partial t^2} \quad \text{or} \quad \frac{\partial^2}{\partial r^2} (ru) = r \frac{\partial^2 u}{\partial t^2} .$$

Letting $v(r, t) = ru(r, t)$ we obtain the boundary-value problem

$$\frac{\partial^2 v}{\partial r^2} = \frac{\partial^2 v}{\partial t^2} , \quad 0 < r < 1, \quad t > 0$$

$$\left. \frac{\partial v}{\partial r} \right|_{r=1} - v(1, t) = 0, \quad t > 0$$

$$v(r, 0) = rf(r), \quad \left. \frac{\partial v}{\partial t} \right|_{t=0} = rg(r), \quad 0 < r < 1.$$

If we separate variables using $v(r, t) = R(r)T(t)$ then we obtain

$$R(r) = c_1 \cos \lambda r + c_2 \sin \lambda r$$

$$T(t) = c_3 \cos \lambda t + c_4 \sin \lambda t.$$

Since $u(r, t) = v(r, t)/r$, in order to insure boundedness at $r = 0$ we define $c_1 = 0$. Then $R(r) = c_2 \sin \lambda r$. Now the boundary condition $R'(1) - R(1) = 0$ implies $\lambda \cos \lambda - \sin \lambda = 0$. Thus, the eigenvalues λ_n are the positive solutions of $\tan \lambda = \lambda$. We now have

$$v_n(r, t) = (A_n \cos \lambda_n t + B_n \sin \lambda_n t) \sin \lambda_n r.$$

For the eigenvalue $\lambda = 0$,

$$R(r) = c_1 r + c_2 \quad \text{and} \quad T(t) = c_3 t + c_4,$$

and boundedness at $r = 0$ implies $c_2 = 0$. We then take

$$v_0(r, t) = A_0 tr + B_0 r$$

so that

$$v(r, t) = A_0 tr + B_0 r + \sum_{n=1}^{\infty} (a_n \cos \lambda_n t + B_n \sin \lambda_n t) \sin \lambda_n r.$$

Now

$$v(r, 0) = rf(r) = B_0 r + \sum_{n=1}^{\infty} A_n \sin \lambda_n r.$$

Since $\{r, \sin \lambda_n r\}$ is an orthogonal set on $[0, 1]$,

$$\int_0^1 r \sin \lambda_n r \, dr = 0 \quad \text{and} \quad \int_0^1 \sin \lambda_n r \sin \lambda_n r \, dr = 0$$

for $m \neq n$. Therefore

$$\int_0^1 r^2 f(r) \, dr = B_0 \int_0^1 r^2 \, dr = \frac{1}{3} B_0$$

and

$$B_0 = 3 \int_0^1 r^2 f(r) \, dr.$$

Also

$$\int_0^1 rf(r) \sin \lambda_n r \, dr = A_n \int_0^1 \sin^2 \lambda_n r \, dr$$

and

$$A_n = \frac{\int_0^1 rf(r) \sin \lambda_n r \, dr}{\int_0^1 \sin^2 \lambda_n r \, dr}.$$

Now

$$\int_0^1 \sin^2 \lambda_n r \, dr = \frac{1}{2} \int_0^1 (1 - \cos 2\lambda_n r) \, dr = \frac{1}{2} \left[1 - \frac{\sin 2\lambda_n}{2\lambda_n} \right] = \frac{1}{2} [1 - \cos^2 \lambda_n].$$

Since $\tan \lambda_n = \lambda_n$,

$$1 + \lambda_n^2 = 1 + \tan^2 \lambda_n = \sec^2 \lambda_n = \frac{1}{\cos^2 \lambda_n}$$

and

$$\cos^2 \lambda_n = \frac{1}{1 + \lambda_n^2}.$$

Then

$$\int_0^1 \sin^2 \lambda_n r \, dr = \frac{1}{2} \left[1 - \frac{1}{1 + \lambda_n^2} \right] = \frac{\lambda_n^2}{2(1 + \lambda_n^2)}$$

and

$$A_n = \frac{2(1 + \lambda_n^2)}{\lambda_n^2} \int_0^1 rf(r) \sin \lambda_n r \, dr.$$

Similarly, setting

$$\frac{\partial v}{\partial t} \bigg|_{t=0} = rg(r) = A_0 r + \sum_{n=1}^{\infty} B_n \lambda_n \sin \lambda_n r$$

we obtain

$$A_0 = 3 \int_0^1 r^2 g(r) \, dr$$

and

$$B_n = \frac{2(1 + \lambda_n^2)}{\lambda_n^3} \int_0^1 rg(r) \sin \lambda_n r \, dr.$$

Therefore, since $v(r, t) = ru(r, t)$ we have

$$u(r, t) = A_0 t + B_0 + \sum_{n=1}^{\infty} (A_n \cos \lambda_n t + B_n \sin \lambda_n t) \frac{\sin \lambda_n r}{r},$$

where the λ_n are solutions of $\tan \lambda = \lambda$ and

$$A_0 = 3 \int_0^1 r^2 g(r) \, dr$$

$$B_0 = 3 \int_0^1 r^2 f(r) \, dr$$

$$A_n = \frac{2(1 + \lambda_n^2)}{\lambda_n^2} \int_0^1 rf(r) \sin \lambda_n r \, dr$$

$$B_n = \frac{2(1 + \lambda_n^2)}{\lambda_n^3} \int_0^1 rg(r) \sin \lambda_n r \, dr$$

for $n = 1, 2, 3, \ldots$.

15 Integral Transform Method

--- **Exercises 15.1** ---

3. By the first translation theorem,

$$\mathscr{L}\left\{e^t \operatorname{erf}(\sqrt{t}\,)\right\} = \mathscr{L}\left\{\operatorname{erf}(\sqrt{t}\,)\right\}\bigg|_{s\to s-1} = \frac{1}{s\sqrt{s+1}}\bigg|_{s\to s-1} = \frac{1}{\sqrt{s}\,(s-1)}.$$

6. We first compute

$$\frac{\sinh a\sqrt{s}}{s\sinh\sqrt{s}} = \frac{e^{a\sqrt{s}} - e^{-a\sqrt{s}}}{s(e^{\sqrt{s}} - e^{-\sqrt{s}})} = \frac{e^{(a-1)\sqrt{s}} - e^{-(a+1)\sqrt{s}}}{s(1 - e^{-2\sqrt{s}})}$$

$$= \frac{e^{(a-1)\sqrt{s}}}{s}\left[1 + e^{-2\sqrt{s}} + e^{-4\sqrt{s}} + \cdots\right] - \frac{e^{-(a+1)\sqrt{s}}}{s}\left[1 + e^{-2\sqrt{s}} + e^{-4\sqrt{s}} + \cdots\right]$$

$$= \left[\frac{e^{-(1-a)\sqrt{s}}}{s} + \frac{e^{-(3-a)\sqrt{s}}}{s} + \frac{e^{-(5-a)\sqrt{s}}}{s} + \cdots\right]$$

$$\qquad - \left[\frac{e^{-(1+a)\sqrt{s}}}{s} + \frac{e^{-(3+a)\sqrt{s}}}{s} + \frac{e^{-(5+a)\sqrt{s}}}{s} + \cdots\right]$$

$$= \sum_{n=0}^{\infty}\left[\frac{e^{-(2n+1-a)\sqrt{s}}}{s} - \frac{e^{-(2n+1+a)\sqrt{s}}}{s}\right].$$

Then

$$\mathscr{L}\left\{\frac{\sinh a\sqrt{s}}{s\sinh\sqrt{s}}\right\} = \sum_{n=0}^{\infty}\left[\mathscr{L}\left\{\frac{e^{-(2n+1-a)\sqrt{s}}}{s}\right\} - \mathscr{L}\left\{-\frac{e^{-(2n+1+a)\sqrt{s}}}{s}\right\}\right]$$

$$= \sum_{n=0}^{\infty}\left[\operatorname{erfc}\left(\frac{2n+1-a}{2\sqrt{t}}\right) - \operatorname{erfc}\left(\frac{2n+1+a}{2\sqrt{t}}\right)\right]$$

$$= \sum_{n=0}^{\infty}\left(\left[1 - \operatorname{erf}\left(\frac{2n+1-a}{2\sqrt{t}}\right)\right] - \left[1 - \operatorname{erf}\left(\frac{2n+1+a}{2\sqrt{t}}\right)\right]\right)$$

$$= \sum_{n=0}^{\infty}\left[\operatorname{erf}\left(\frac{2n+1+a}{2\sqrt{t}}\right) - \operatorname{erf}\left(\frac{2n+1-a}{2\sqrt{t}}\right)\right].$$

9. $\displaystyle\int_a^b e^{-u^2}\,du = \int_a^0 e^{-u^2}\,du + \int_0^b e^{-u^2}\,du = \int_0^b e^{-u^2}\,du - \int_0^a e^{-u^2}\,du$

$$= \frac{\sqrt{\pi}}{2}\operatorname{erf}(b) - \frac{\sqrt{\pi}}{2}\operatorname{erf}(a) = \frac{\sqrt{\pi}}{2}[\operatorname{erf}(b) - \operatorname{erf}(a)]$$

3. The solution of

$$a^2 \frac{d^2 U}{dx^2} - s^2 U = 0$$

is in this case

$$U(x, s) = c_1 e^{-(x/a)s} + c_2 e^{(x/a)s}.$$

Since $\lim_{x \to \infty} u(x, t) = 0$ we have $\lim_{x \to \infty} U(x, s) = 0$. Thus $c_2 = 0$ and

$$U(x, s) = c_1 e^{-(x/a)s}.$$

If $\mathscr{L}\{u(0, t)\} = \mathscr{L}\{f(t)\} = F(s)$ then $U(0, s) = F(s)$. From this we have $c_1 = F(s)$ and

$$U(x, s) = F(s) e^{-(x/a)s}.$$

Hence, by the second translation theorem,

$$u(x, t) = f\left(t - \frac{x}{a}\right) \mathscr{U}\left(t - \frac{x}{a}\right).$$

6. Transforming the partial differential equation gives

$$\frac{d^2 U}{dx^2} - s^2 U = -\frac{\omega}{s^2 + \omega^2} \sin \pi x.$$

Using undetermined coefficients we obtain

$$U(x, s) = c_1 \cosh sx + c_2 \sinh sx + \frac{\omega}{(s^2 + \pi^2)(s^2 + \omega^2)} \sin \pi x.$$

The transformed boundary conditions $U(0, s) = 0$ and $U(1, s) = 0$ give, in turn, $c_1 = 0$ and $c_2 = 0$. Therefore

$$U(x, s) = \frac{\omega}{(s^2 + \pi^2)(s^2 + \omega^2)} \sin \pi x$$

and

$$u(x, t) = \omega \sin \pi x \, \mathscr{L}^{-1}\left\{\frac{1}{(s^2 + \pi^2)(s^2 + \omega^2)}\right\}$$

$$= \frac{\omega}{\omega^2 - \pi^2} \sin \pi x \, \mathscr{L}^{-1}\left\{\frac{1}{\pi}\frac{\pi}{s^2 + \pi^2} - \frac{1}{\omega}\frac{\omega}{s^2 + \omega^2}\right\}$$

$$= \frac{\omega}{\pi(\omega^2 - \pi^2)} \sin \pi t \sin \pi x - \frac{1}{\omega^2 - \pi^2} \sin \omega t \sin \pi x.$$

9. Transforming the partial differential equation gives

$$\frac{d^2 U}{dx^2} - s^2 U = -sx e^{-x}.$$

Using undetermined coefficients we obtain

$$U(x, s) = c_1 e^{-sx} + c_2 e^{sx} - \frac{2s}{(s^2 - 1)^2} e^{-x} + \frac{s}{s^2 - 1} x e^{-x}.$$

The transformed boundary conditions $\lim_{x \to \infty} U(x, s) = 0$ and $U(0, s) = 0$ give, in turn, $c_2 = 0$ and $c_1 = 2s/(s^2 - 1)^2$. Therefore

$$U(x, s) = \frac{2s}{(s^2 - 1)^2} e^{-sx} - \frac{2s}{(s^2 - 1)^2} e^{-x} + \frac{s}{s^2 - 1} x e^{-x}.$$

From entries (13) and (26) in the table of Laplace transforms we obtain

$$u(x,t) = \mathscr{L}^{-1}\left\{\frac{2s}{(s^2-1)^2}\,e^{-sx} - \frac{2s}{(s^2-1)^2}\,e^{-x} + \frac{s}{s^2-1}\,xe^{-x}\right\}$$

$$= 2(t-x)\sinh(t-x)\,\mathscr{U}(t-x) - te^{-x}\sinh t + xe^{-x}\cosh t.$$

15. We use

$$U(x,s) = c_1 e^{-\sqrt{s}\,x} + c_2 e^{\sqrt{s}\,x} + \frac{u_0}{s}\,.$$

The condition $\lim_{x\to\infty} u(x,t) = u_0$ implies $\lim_{x\to\infty} U(x,s) = u_0/s$, so we define $c_2 = 0$. Then

$$U(x,s) = c_1 e^{-\sqrt{s}\,x} + \frac{u_0}{s}\,.$$

The transform of the remaining boundary conditions gives

$$\left.\frac{dU}{dx}\right|_{x=0} = U(0,s).$$

This condition yields $c_1 = -u_0/s(\sqrt{s}+1)$. Thus

$$U(x,s) = -u_0\,\frac{e^{-\sqrt{s}\,x}}{s(\sqrt{s}+1)} + \frac{u_0}{s}$$

and

$$u(x,t) = -u_0\,\mathscr{L}^{-1}\left\{\frac{e^{-x\sqrt{s}}}{s(\sqrt{s}+1)}\right\} + u_0\,\mathscr{L}^{-1}\left\{\frac{1}{s}\right\}$$

$$= u_0 e^{x+t}\operatorname{erfc}\left(\sqrt{t} + \frac{x}{2\sqrt{t}}\right) - u_0\operatorname{erfc}\left(\frac{x}{2\sqrt{t}}\right) + u_0 \qquad \boxed{\text{By (5) in the table in 15.1.}}$$

18. We use

$$U(x,s) = c_1 e^{-\sqrt{s}\,x} + c_2 e^{\sqrt{s}\,x}.$$

The condition $\lim_{x\to\infty} u(x,t) = 0$ implies $\lim_{x\to\infty} U(x,s) = 0$, so we define $c_2 = 0$. Then $U(x,s) = c_1 e^{-\sqrt{s}\,x}$. The transform of the remaining boundary condition gives

$$\left.\frac{dU}{dx}\right|_{x=0} = -F(s)$$

where $F(s) = \mathscr{L}\{f(t)\}$. This condition yields $c_1 = F(s)/\sqrt{s}$. Thus

$$U(x,s) = F(s)\,\frac{e^{-\sqrt{s}\,x}}{\sqrt{s}}\,.$$

Using entry (44) of the table of Laplace transforms and the convolution theorem we obtain

$$u(x,t) = \mathscr{L}^{-1}\left\{F(s)\cdot\frac{e^{-\sqrt{s}\,x}}{\sqrt{s}}\right\} = \frac{1}{\sqrt{\pi}}\int_0^t f(\tau)\,\frac{e^{-x^2/4(t-\tau)}}{\sqrt{t-\tau}}\,d\tau.$$

21. Transforming the partial differential equation gives

$$\frac{d^2 U}{dx^2} - sU = 0$$

and so

$$U(x,s) = c_1 e^{-\sqrt{s}\,x} + c_2 e^{\sqrt{s}\,x}.$$

The condition $\lim_{x \to -\infty} u(x, t) = 0$ implies $\lim_{x \to -\infty} U(x, s) = 0$, so we define $c_1 = 0$. The transform of the remaining boundary condition gives

$$\frac{dU}{dx}\bigg|_{x=1} = \frac{100}{s} - U(1, s).$$

This condition yields

$$c_2 \sqrt{s}\, e^{\sqrt{s}} = \frac{100}{s} - c_2 e^{\sqrt{s}}$$

from which it follows that

$$c_2 = \frac{100}{s(\sqrt{s}+1)}\, e^{-\sqrt{s}}.$$

Thus

$$U(x, s) = 100\, \frac{e^{-(1-x)\sqrt{s}}}{s(\sqrt{s}+1)}.$$

Using entry (49) of the table of Laplace transforms we obtain

$$u(x, t) = 100\, \mathscr{L}^{-1}\left\{ \frac{e^{-(1-x)\sqrt{s}}}{s(\sqrt{s}+1)} \right\} = 100\left[-e^{1-x+t}\, \mathrm{erfc}\left(\sqrt{t} + \frac{1-x}{\sqrt{t}}\right) + \mathrm{erfc}\left(\frac{1-x}{2\sqrt{t}}\right) \right].$$

24. The transform of the partial differential equation is

$$k\frac{d^2U}{dx^2} - hU + h\frac{u_m}{s} = sU - u_0$$

or

$$k\frac{d^2U}{dx^2} - (h+s)U = -h\frac{u_m}{s} - u_0.$$

By undetermined coefficients we find

$$U(x, s) = c_1 e^{\sqrt{(h+s)/k}\, x} + c_2 e^{-\sqrt{(h+s)/k}\, x} + \frac{hu_m + u_0 s}{s(s+h)}.$$

The transformed boundary conditions are $U'(0, s) = 0$ and $U'(L, s) = 0$. These conditions imply $c_1 = 0$ and $c_2 = 0$. By partial fractions we then get

$$U(x, s) = \frac{hu_m + u_0 s}{s(s+h)} = \frac{u_m}{s} - \frac{u_m}{s+h} + \frac{u_0}{s+h}.$$

Therefore,

$$u(x, t) = u_m \mathscr{L}^{-1}\left\{\frac{1}{s}\right\} - u_m \mathscr{L}^{-1}\left\{\frac{1}{s+h}\right\} + u_0 \mathscr{L}^{-1}\left\{\frac{1}{s+h}\right\} = u_m - u_m e^{-ht} + u_0 e^{-ht}.$$

27. We use

$$U(x, s) = c_1 e^{-\sqrt{RCs+RG}\, x} + c_2 e^{\sqrt{RCs+RG}\, x} + \frac{Cu_0}{Cs+G}.$$

The condition $\lim_{x \to \infty} \partial u/\partial x = 0$ implies $\lim_{x \to \infty} dU/dx = 0$, so we define $c_2 = 0$. Applying $U(0, s) = 0$ to

$$U(x, s) = c_1 e^{-\sqrt{RCsRG}\, x} + \frac{Cu_0}{Cs+G}$$

gives $c_1 = -Cu_0/(Cs+G)$. Therefore

$$U(x, s) = -Cu_0\, \frac{e^{-\sqrt{RCs+RG}\, x}}{Cs+G} + \frac{Cu_0}{Cs+G}$$

and

$$u(x,t) = u_0 \mathscr{L}^{-1}\left\{ \frac{1}{s+G/C} \right\} - u_0 \mathscr{L}^{-1}\left\{ \frac{e^{-x\sqrt{RC}\sqrt{s+G/C}}}{s+G/C} \right\}$$

$$= u_0 e^{-Gt/C} - u_0 e^{-Gt/C} \operatorname{erfc}\left(\frac{x\sqrt{RC}}{2\sqrt{t}} \right)$$

$$= u_0 e^{-Gt/C}\left[1 - \operatorname{erfc}\left(\frac{x}{2}\sqrt{\frac{RC}{t}} \right) \right]$$

$$= u_0 e^{-Gt/C} \operatorname{erf}\left(\frac{x}{2}\sqrt{\frac{RC}{t}} \right).$$

30. (a) We use

$$U(x,s) = c_1 e^{-(s/a)x} + c_2 e^{(s/a)x} + \frac{v_0^2 F_0}{(a^2 - v_0^2)s^2} e^{-(s/v_0)x}.$$

The condition $\lim_{x\to\infty} u(x,t) = 0$ implies $\lim_{x\to\infty} U(x,s) = 0$, so we must define $c_2 = 0$. Consequently

$$U(x,s) = c_1 e^{-(s/a)x} + \frac{v_0^2 F_0}{(a^2 - v_0^2)s^2} e^{-(s/v_0)x}.$$

The remaining boundary condition transforms into $U(0,s) = 0$. From this we find

$$c_1 = -v_0^2 F_0/(a^2 - v_0^2)s^2.$$

Therefore, by the second translation theorem

$$U(x,s) = -\frac{v_0^2 F_0}{(a^2 - v_0^2)s^2} e^{-(s/a)x} + \frac{v_0^2 F_0}{(a^2 - v_0^2)s^2} e^{-(s/v_0)x}$$

and

$$u(x,t) = \frac{v_0^2 F_0}{a^2 - v_0^2}\left[\mathscr{L}^{-1}\left\{ \frac{e^{-(x/v_0)s}}{s^2} \right\} - \mathscr{L}^{-1}\left\{ \frac{e^{-(x/a)s}}{s^2} \right\} \right]$$

$$= \frac{v_0^2 F_0}{a^2 - v_0^2}\left[\left(t - \frac{x}{v_0} \right) \mathscr{U}\left(t - \frac{x}{v_0} \right) - \left(t - \frac{x}{a} \right) \mathscr{U}\left(t - \frac{x}{a} \right) \right].$$

(b) In the case when $v_0 = a$ the solution of the transformed equation is

$$U(x,s) = c_1 e^{-(s/a)x} + c_2 e^{(s/a)x} - \frac{F_0}{2as} x e^{-(s/a)x}.$$

The usual analysis then leads to $c_1 = 0$ and $c_2 = 0$. Therefore

$$U(x,s) = -\frac{F_0}{2as} x e^{-(s/a)x}$$

and

$$u(x,t) = -\frac{x F_0}{2a} \mathscr{L}^{-1}\left\{ \frac{e^{-(x/a)s}}{s} \right\} = -\frac{x F_0}{2a} \mathscr{U}\left(t - \frac{x}{a} \right).$$

3. From formulas (5) and (6) in the text,

$$A(\alpha) = \int_0^3 x\cos\alpha x\, dx = \frac{x\sin\alpha x}{\alpha}\bigg|_0^3 - \frac{1}{\alpha}\int_0^3 \sin\alpha x\, dx$$

$$= \frac{3\sin 3\alpha}{\alpha} + \frac{\cos\alpha x}{\alpha^2}\bigg|_0^3 = \frac{3\alpha\sin 3\alpha + \cos 3\alpha - 1}{\alpha^2}$$

and

$$B(\alpha) = \int_0^3 x\sin\alpha x\, dx = -\frac{x\cos\alpha x}{\alpha}\bigg|_0^3 + \frac{1}{\alpha}\int_0^3 \cos\alpha x\, dx$$

$$= -\frac{3\cos 3\alpha}{\alpha} + \frac{\sin\alpha x}{\alpha^2}\bigg|_0^3 = \frac{\sin 3\alpha - 3\alpha\cos 3\alpha}{\alpha^2}.$$

Hence

$$f(x) = \frac{1}{\pi}\int_0^\infty \frac{(3\alpha\sin 3\alpha + \cos 3\alpha - 1)\cos\alpha x + (\sin 3\alpha - 3\alpha\cos 3\alpha)\sin\alpha x}{\alpha^2}\, d\alpha$$

$$= \frac{1}{\pi}\int_0^\infty \frac{3\alpha(\sin 3\alpha\cos\alpha x - \cos 3\alpha\sin\alpha x) + \cos 3\alpha\cos\alpha x + \sin 3\alpha\sin\alpha x - \cos\alpha x}{\alpha^2}\, d\alpha$$

$$= \frac{1}{\pi}\int_0^\infty \frac{3\alpha\sin\alpha(3 - x) + \cos\alpha(3 - x) - \cos\alpha x}{\alpha^2}\, d\alpha.$$

6. From formulas (5) and (6) in the text,

$$A(\alpha) = \int_{-1}^1 e^x\cos\alpha x\, dx$$

$$= \frac{e(\cos\alpha + \alpha\sin\alpha) - e^{-1}(\cos\alpha - \alpha\sin\alpha)}{1 + \alpha^2}$$

$$= \frac{2(\sinh 1)\cos\alpha - 2\alpha(\cosh 1)\sin\alpha}{1 + \alpha^2}$$

and

$$B(\alpha) = \int_{-1}^1 e^x\sin\alpha x\, dx$$

$$= \frac{e(\sin\alpha - \alpha\cos\alpha) - e^{-1}(-\sin\alpha - \alpha\cos\alpha)}{1 + \alpha^2}$$

$$= \frac{2(\cosh 1)\sin\alpha - 2\alpha(\sinh 1)\cos\alpha}{1 + \alpha^2}.$$

Hence

$$f(x) = \frac{1}{\pi}\int_0^\infty [A(\alpha)\cos\alpha x + B(\alpha)\sin\alpha x]\, d\alpha.$$

9. The function is even. Thus from formula (9) in the text

$$A(\alpha) = \int_0^\pi x\cos\alpha x\, dx = \frac{x\sin\alpha x}{\alpha}\bigg|_0^\pi - \frac{1}{\alpha}\int_0^\pi \sin\alpha x\, dx$$

$$= \frac{\pi\alpha\sin\pi\alpha}{\alpha} + \frac{1}{\alpha^2}\cos\alpha x\bigg|_0^\pi = \frac{\pi\alpha\sin\pi\alpha + \cos\pi\alpha - 1}{\alpha^2}.$$

Hence from formula (8) in the text

$$f(x) = \frac{2}{\pi} \int_0^\infty \frac{(\pi\alpha\sin\pi\alpha + \cos\pi\alpha - 1)\cos\alpha x}{\alpha^2}\, d\alpha.$$

12. The function is odd. Thus from formula (11) in the text

$$B(\alpha) = \int_0^\infty xe^{-x}\sin\alpha x\, dx.$$

Now recall

$$\mathcal{L}\{t\sin kt\} = -\frac{d}{ds}\mathcal{L}\{\sin kt\} = 2ks/(s^2+k^2)^2.$$

If we set $s = 1$ and $k = \alpha$ we obtain

$$B(\alpha) = \frac{2\alpha}{(1+\alpha^2)^2}.$$

Hence from formula (10) in the text

$$f(x) = \frac{4}{\pi} \int_0^\infty \frac{\alpha\sin\alpha x}{(1+\alpha^2)^2}\, d\alpha.$$

15. For the cosine integral,

$$A(\alpha) = \int_0^\infty xe^{-2x}\cos\alpha x\, dx.$$

But we know

$$\mathcal{L}\{t\cos kt\} = -\frac{d}{ds}\frac{s}{(s^2+k^2)} = \frac{(s^2-k^2)}{(s^2+k^2)^2}.$$

If we set $s = 2$ and $k = \alpha$ we obtain

$$A(\alpha) = \frac{4-\alpha^2}{(4+\alpha^2)^2}.$$

Hence

$$f(x) = \frac{2}{\pi} \int_0^\infty \frac{(4-\alpha^2)\cos\alpha x}{(4+\alpha^2)^2}\, d\alpha.$$

For the sine integral,

$$B(\alpha) = \int_0^\infty xe^{-2x}\sin\alpha x\, dx.$$

From Problem 12, we know

$$\mathcal{L}\{t\sin kt\} = \frac{2ks}{(s^2+k^2)^2}.$$

If we set $s = 2$ and $k = \alpha$ we obtain

$$B(\alpha) = \frac{4\alpha}{(4+\alpha^2)^2}.$$

Hence

$$f(x) = \frac{8}{\pi} \int_0^\infty \frac{\alpha\sin\alpha x}{(4+\alpha^2)^2}\, d\alpha.$$

18. From the formula for sine integral of $f(x)$ we have

$$f(x) = \frac{2}{\pi} \int_0^\infty \left(\int_0^\infty f(x)\sin\alpha x\, dx \right) \sin\alpha x\, dx$$

$$= \frac{2}{\pi} \left[\int_0^1 1\cdot\sin\alpha x\, d\alpha + \int_1^\infty 0\cdot\sin\alpha x\, d\alpha \right]$$

$$= \frac{2}{\pi} \frac{(-\cos\alpha x)}{x}\bigg|_0^1 = \frac{2}{\pi}\frac{1-\cos x}{x}.$$

233

—————— **Exercises 15.4** ——————————————————

For the boundary-value problems in this section it is sometimes useful to note that the identities

$$e^{i\alpha} = \cos\alpha + i\sin\alpha \quad \text{and} \quad e^{-i\alpha} = \cos\alpha - i\sin\alpha$$

imply

$$e^{i\alpha} + e^{-i\alpha} = 2\cos\alpha \quad \text{and} \quad e^{i\alpha} - e^{-i\alpha} = 2i\sin\alpha.$$

3. Using the Fourier transform, the partial differential equation equation becomes

$$\frac{dU}{dt} + k\alpha^2 U = 0 \quad \text{and so} \quad U(\alpha, t) = ce^{-k\alpha^2 t}.$$

Now

$$\mathscr{F}\{u(x,0)\} = U(\alpha, 0) = \sqrt{\pi}\, e^{-\alpha^2/4}$$

by the given result. This gives $c = \sqrt{\pi}\, e^{-\alpha^2/4}$ and so

$$U(\alpha, t) = \sqrt{\pi}\, e^{-(\frac{1}{4}+kt)\alpha^2}.$$

Using the given Fourier transform again we obtain

$$u(x,t) = \sqrt{\pi}\,\mathscr{F}^{-1}\{e^{-(1+4kt)\alpha^2/4}\} = \frac{1}{\sqrt{1+4kt}}\, e^{-x^2/(1+4kt)}.$$

6. (a) Using the Fourier sine transform, the partial differential equation becomes

$$\frac{dU}{dt} + k\alpha^2 U = k\alpha u_0.$$

The general solution of this linear equation is

$$U(\alpha, t) = ce^{-k\alpha^2 t} + \frac{u_0}{\alpha}.$$

But $U(\alpha, 0) = 0$ implies $c = -u_0/\alpha$ and so

$$U(\alpha, t) = u_0 \frac{1 - e^{-k\alpha^2 t}}{\alpha}$$

and

$$u(x,t) = \frac{2u_0}{\pi}\int_0^\infty \frac{1 - e^{-k\alpha^2 t}}{\alpha}\sin\alpha x\, d\alpha.$$

(b) The solution of part (a) can be written

$$u(x,t) = \frac{2u_0}{\pi}\int_0^\infty \frac{\sin\alpha x}{\alpha}\, d\alpha - \frac{2u_0}{\pi}\int_0^\infty \frac{\sin\alpha x}{\alpha}\, e^{-k\alpha^2 t}\, d\alpha.$$

Using $\displaystyle\int_0^\infty \frac{\sin\alpha x}{\alpha}\, d\alpha = \pi/2$ the last line becomes

$$u(x,t) = u_0 - \frac{2u_0}{\pi}\int_0^\infty \frac{\sin\alpha x}{\alpha}\, e^{-k\alpha^2 t}\, d\alpha.$$

9. Using the Fourier cosine transform we find

$$U(\alpha, t) = ce^{-k\alpha^2 t}.$$

Now

$$\mathscr{F}_C\{u(x,0)\} = \int_0^1 \cos\alpha x\, dx = \frac{\sin\alpha}{\alpha} = U(\alpha,0).$$

From this we obtain $c = (\sin\alpha)/\alpha$ and so

$$U(\alpha,t) = \frac{\sin\alpha}{\alpha} e^{-k\alpha^2 t}$$

and

$$u(x,t) = \frac{2}{\pi}\int_0^\infty \frac{\sin\alpha}{\alpha} e^{-k\alpha^2 t}\cos\alpha x\, d\alpha.$$

12. Using the Fourier sine transform we obtain

$$U(\alpha,t) = c_1\cos\alpha at + c_2\sin\alpha at.$$

Now

$$\mathscr{F}_S\{u(x,0)\} = \mathscr{F}\{xe^{-x}\} = \int_0^\infty xe^{-x}\sin\alpha x\, dx = \frac{2\alpha}{(1+\alpha^2)^2} = U(\alpha,0).$$

Also,

$$\mathscr{F}_S\{u_t(x,0)\} = \frac{dU}{dt}\bigg|_{t=0} = 0.$$

This last condition gives $c_2 = 0$. Then $U(\alpha,0) = 2\alpha/(1+\alpha^2)^2$ yields $c_1 = 2\alpha/(1+\alpha^2)^2$. Therefore

$$U(\alpha,t) = \frac{2\alpha}{(1+\alpha^2)^2}\cos\alpha at$$

and

$$u(x,t) = \frac{4}{\pi}\int_0^\infty \frac{\alpha\cos\alpha at}{(1+\alpha^2)^2}\sin\alpha x\, d\alpha.$$

15. Using the Fourier cosine transform with respect to x gives

$$U(\alpha,y) = c_1 e^{-\alpha y} + c_2 e^{\alpha y}.$$

Since we expect $u(x,y)$ to be bounded as $y \to \infty$ we define $c_2 = 0$. Thus

$$U(\alpha,y) = c_1 e^{-\alpha y}.$$

Now

$$\mathscr{F}_C\{u(x,0)\} = \int_0^1 50\cos\alpha x\, dx = 50\frac{\sin\alpha}{\alpha}$$

and so

$$U(\alpha,y) = 50\frac{\sin\alpha}{\alpha} e^{-\alpha y}$$

and

$$u(x,y) = \frac{100}{\pi}\int_0^\infty \frac{\sin\alpha}{\alpha} e^{-\alpha y}\cos\alpha x\, d\alpha.$$

18. The domain of y and the boundary condition at $y = 0$ suggest that we use a Fourier cosine transform. The transformed equation is

$$\frac{d^2 U}{dx^2} - \alpha^2 U - u_y(x,0) = 0 \quad\text{or}\quad \frac{d^2 U}{dx^2} - \alpha^2 U = 0.$$

Because the domain of the variable x is a finite interval we choose to write the general solution of the latter equation as

$$U(x,\alpha) = c_1\cosh\alpha x + c_2\sinh\alpha x.$$

Now $U(0, \alpha) = F(\alpha)$, where $F(\alpha)$ is the Fourier cosine transform of $f(y)$, and $U'(\pi, \alpha) = 0$ imply $c_1 = F(\alpha)$ and $c_2 = -F(\alpha) \sinh \alpha\pi / \cosh \alpha\pi$. Thus

$$U(x, \alpha) = F(\alpha) \cosh \alpha x - F(\alpha) \frac{\sinh \alpha\pi}{\cosh \alpha\pi} \sinh \alpha x = F(\alpha) \frac{\cosh \alpha(\pi - x)}{\cosh \alpha\pi}.$$

Using the inverse transform we find that a solution to the problem is

$$u(x, y) = \frac{2}{\pi} \int_0^\infty F(\alpha) \frac{\cosh \alpha(\pi - x)}{\cosh \alpha\pi} \cos \alpha y \, d\alpha.$$

21. Using the Fourier transform with respect to x gives

$$U(\alpha, y) = c_1 \cosh \alpha y + c_2 \sinh \alpha y.$$

The transform of the boundary condition $\dfrac{\partial u}{\partial y}\bigg|_{y=0} = 0$ is $\dfrac{dU}{dy}\bigg|_{y=0} = 0$. This condition gives $c_2 = 0$. Hence

$$U(\alpha, y) = c_1 \cosh \alpha y.$$

Now by the given information the transform of the boundary condition $u(x, 1) = e^{-x^2}$ is $U(\alpha, 1) = \sqrt{\pi} \, e^{-\alpha^2/4}$. This condition then gives $c_1 = \sqrt{\pi} \, e^{-\alpha^2/4} \cosh \alpha$. Therefore

$$U(\alpha, y) = \sqrt{\pi} \, \frac{e^{-\alpha^2/4} \cosh \alpha y}{\cosh \alpha}$$

and

$$U(x, y) = \frac{1}{2\sqrt{\pi}} \int_{-\infty}^\infty \frac{e^{-\alpha^2/4} \cosh \alpha y}{\cosh \alpha} e^{-i\alpha x} d\alpha$$

$$= \frac{1}{2\sqrt{\pi}} \int_{-\infty}^\infty \frac{e^{-\alpha^2/4} \cosh \alpha y}{\cosh \alpha} \cos \alpha x \, d\alpha$$

$$= \frac{1}{\sqrt{\pi}} \int_0^\infty \frac{e^{-\alpha^2/4} \cosh \alpha y}{\cosh \alpha} \cos \alpha x \, d\alpha.$$

Exercises 15.5

3. By the sifting property,

$$\mathscr{F}\{\delta(x)\} = \int_{-\infty}^\infty \delta(x) e^{i\alpha x} dx = e^{i\alpha 0} = 1.$$

6. Using a CAS we find

$$\mathscr{F}\{g(x)\} = \frac{1}{2}[\text{sign}(A - \alpha) + \text{sign}(A + \alpha)]$$

where $\text{sign}(t) = 1$ if $t > 0$ and $\text{sign} \, t = -1$ if $t < 0$. Thus

$$\mathscr{F}\{g(x)\} = \begin{cases} 1, & -A < \alpha < A \\ 0, & \text{elsewhere.} \end{cases}$$

Chapter 15 Review Exercises

3. The Laplace transform gives

$$U(x, s) = c_1 e^{-\sqrt{s+h}\, x} + c_2 e^{\sqrt{s+h}\, x} + \frac{u_0}{s+h}.$$

The condition $\lim_{x \to \infty} \partial u / \partial x = 0$ implies $\lim_{x \to \infty} dU/dx = 0$ and so we define $c_2 = 0$. Thus

$$U(x, s) = c_1 e^{-\sqrt{s+h}\, x} + \frac{u_0}{s+h}.$$

The condition $U(0, s) = 0$ then gives $c_1 = -u_0/(s+h)$ and so

$$U(x, s) = \frac{u_0}{s+h} - u_0 \frac{e^{-\sqrt{s+h}\, x}}{s+h}.$$

With the help of the first translation theorem we then obtain

$$u(x, t) = u_0 \mathscr{L}^{-1}\left\{\frac{1}{s+h}\right\} - u_0 \mathscr{L}^{-1}\left\{\frac{e^{-\sqrt{s+h}\, x}}{s+h}\right\} = u_0 e^{-ht} - u_0 e^{-ht} \operatorname{erfc}\left(\frac{x}{2\sqrt{t}}\right)$$

$$= u_0 e^{-ht}\left[1 - \operatorname{erfc}\left(\frac{x}{2\sqrt{t}}\right)\right] = u_0 e^{-ht} \operatorname{erf}\left(\frac{x}{2\sqrt{t}}\right).$$

6. The Laplace transform and undetermined coefficients gives

$$U(x, s) = c_1 \cosh sx + c_2 \sinh sx + \frac{s-1}{s^2 + \pi^2} \sin \pi x.$$

The conditions $U(0, s) = 0$ and $U(1, s) = 0$ give, in turn, $c_1 = 0$ and $c_2 = 0$. Thus

$$U(x, s) = \frac{s-1}{s^2 + \pi^2} \sin \pi x$$

and

$$u(x, t) = \sin \pi x \, \mathscr{L}^{-1}\left\{\frac{s}{s^2 + \pi^2}\right\} - \frac{1}{\pi} \sin \pi x \, \mathscr{L}^{-1}\left\{\frac{\pi}{s^2 + \pi^2}\right\}$$

$$= (\sin \pi x) \cos \pi t - \frac{1}{\pi}(\sin \pi x) \sin \pi t.$$

9. We solve the two problems

$$\frac{\partial^2 u_1}{\partial x^2} + \frac{\partial^2 u_1}{\partial y^2} = 0, \quad x > 0, \quad y > 0,$$

$$u_1(0, y) = 0, \quad y > 0,$$

$$u_1(x, 0) = \begin{cases} 100, & 0 < x < 1 \\ 0, & x > 1 \end{cases}$$

and

$$\frac{\partial^2 u_2}{\partial x^2} + \frac{\partial^2 u_2}{\partial y^2} = 0, \quad x > 0, \quad y > 0,$$

$$u_2(0, y) = \begin{cases} 50, & 0 < y < 1 \\ 0, & y > 1 \end{cases}$$

$$u_2(x, 0) = 0.$$

Using the Fourier sine transform with respect to x we find

$$u_1(x, y) = \frac{200}{\pi} \int_0^\infty \left(\frac{1 - \cos \alpha}{\alpha}\right) e^{-\alpha y} \sin \alpha x \, d\alpha.$$

Using the Fourier sine transform with respect to y we find

$$u_2(x, y) = \frac{100}{\pi} \int_0^\infty \left(\frac{1 - \cos \alpha}{\alpha} \right) e^{-\alpha x} \sin \alpha y \, d\alpha.$$

The solution of the problem is then

$$u(x, y) = u_1(x, y) + u_2(x, y).$$

12. Using the Laplace transform gives

$$U(x, s) = c_1 \cosh \sqrt{s}\, x + c_2 \sinh \sqrt{s}\, x.$$

The condition $u(0, t) = u_0$ transforms into $U(0, s) = u_0/s$. This gives $c_1 = u_0/s$. The condition $u(1, t) = u_0$ transforms into $U(1, s) = u_0/s$. This implies that $c_2 = u_0(1 - \cosh \sqrt{s})/s \sinh \sqrt{s}$. Hence

$$U(x, s) = \frac{u_0}{s} \cosh \sqrt{s}\, x + u_0 \left[\frac{1 - \cosh \sqrt{s}}{s \sinh \sqrt{s}} \right] \sinh \sqrt{s}\, x$$

$$= u_0 \left[\frac{\sinh \sqrt{s} \cosh \sqrt{s}\, x - \cosh \sinh \sqrt{s} \sinh \sqrt{s}\, x + \sinh \sqrt{s}\, x}{s \sinh \sqrt{s}} \right]$$

$$= u_0 \left[\frac{\sinh \sqrt{s}\,(1 - x) + \sinh \sqrt{s}\, x}{s \sinh \sqrt{s}} \right]$$

$$= u_0 \left[\frac{\sinh \sqrt{s}\,(1 - x)}{s \sinh \sqrt{s}} + \frac{\sinh \sqrt{s}\, x}{s \sinh \sqrt{s}} \right]$$

and

$$u(x, t) = u_0 \left[\mathscr{L}^{-1} \left\{ \frac{\sinh \sqrt{s}\,(1 - x)}{s \sinh \sqrt{s}} \right\} + \mathscr{L}^{-1} \left\{ \frac{\sinh \sqrt{s}\, x}{s \sinh \sqrt{s}} \right\} \right]$$

$$= u_0 \sum_{n=0}^\infty \left[\mathrm{erf} \left(\frac{2n + 2 - x}{2\sqrt{t}} \right) - \mathrm{erf} \left(\frac{2n + x}{2\sqrt{t}} \right) \right]$$

$$+ u_0 \sum_{n=0}^\infty \left[\mathrm{erf} \left(\frac{2n + 1 + x}{2\sqrt{t}} \right) - \mathrm{erf} \left(\frac{2n + 1 - x}{2\sqrt{t}} \right) \right].$$

16 Numerical Solutions of Partial Differential Equations

3. The figure shows the values of $u(x, y)$ along the boundary. We need to determine u_{11}, u_{21}, u_{12}, and u_{22}. By symmetry $u_{11} = u_{21}$ and $u_{12} = u_{22}$. The system is

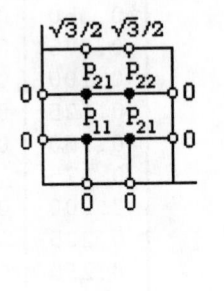

$$u_{21} + u_{12} + 0 + 0 - 4u_{11} = 0$$
$$0 + u_{22} + u_{11} + 0 - 4u_{21} = 0$$
$$u_{22} + \sqrt{3}/2 + 0 + u_{11} - 4u_{12} = 0$$
$$0 + \sqrt{3}/2 + u_{12} + u_{21} - 4u_{22} = 0$$

or

$$3u_{11} + u_{12} = 0$$
$$u_{11} - 3u_{12} = -\frac{\sqrt{3}}{2}.$$

Solving we obtain $u_{11} = u_{21} = \sqrt{3}/16$ and $u_{12} = u_{22} = 3\sqrt{3}/16$.

6. For Gauss-Seidel the coefficients of the unknowns u_{11}, u_{21}, u_{31}, u_{12}, u_{22}, u_{32}, u_{13}, u_{23}, u_{33} are shown in the matrix

$$\begin{bmatrix} 0 & .25 & 0 & .25 & 0 & 0 & 0 & 0 & 0 \\ .25 & 0 & .25 & 0 & .25 & 0 & 0 & 0 & 0 \\ 0 & .25 & 0 & 0 & 0 & .25 & 0 & 0 & 0 \\ .25 & 0 & 0 & 0 & .25 & 0 & .25 & 0 & 0 \\ 0 & .25 & 0 & .25 & 0 & .25 & 0 & .25 & 0 \\ 0 & 0 & .25 & 0 & .25 & 0 & 0 & 0 & .25 \\ 0 & 0 & 0 & .25 & 0 & 0 & 0 & .25 & 0 \\ 0 & 0 & 0 & 0 & .25 & 0 & .25 & 0 & .25 \\ 0 & 0 & 0 & 0 & 0 & .25 & 0 & .25 & 0 \end{bmatrix}$$

The constant terms are 7.5, 5, 20, 10, 0, 15, 17.5, 5, 27.5. We use 32.5 as the initial guess for each variable. Then $u_{11} = 21.92$, $u_{21} = 28.30$, $u_{31} = 38.17$, $u_{12} = 29.38$, $u_{22} = 33.13$, $u_{32} = 44.38$, $u_{13} = 22.46$, $u_{23} = 30.45$, and $u_{33} = 46.21$.

━━━━━━━━━━ **Exercises 16.2** ━━━━━━━━━━━━━━━━━━━━━━━━━━━━━━━

3. We identify $c = 1$, $a = 2$, $T = 1$, $n = 8$, and $m = 40$. Then $h = 2/8 = 0.25$, $k = 1/40 = 0.025$, and $\lambda = 2/5 = 0.4$.

TIME	X=0.25	X=0.50	X=0.75	X=1.00	X=1.25	X=1.50	X=1.75
0.000	1.0000	1.0000	1.0000	1.0000	0.0000	0.0000	0.0000
0.025	0.7074	0.9520	0.9566	0.7444	0.2545	0.0371	0.0053
0.050	0.5606	0.8499	0.8685	0.6633	0.3303	0.1034	0.0223
0.075	0.4684	0.7473	0.7836	0.6191	0.3614	0.1529	0.0462
0.100	0.4015	0.6577	0.7084	0.5837	0.3753	0.1871	0.0684
0.125	0.3492	0.5821	0.6428	0.5510	0.3797	0.2101	0.0861
0.150	0.3069	0.5187	0.5857	0.5199	0.3778	0.2247	0.0990
0.175	0.2721	0.4652	0.5359	0.4901	0.3716	0.2329	0.1078
0.200	0.2430	0.4198	0.4921	0.4617	0.3622	0.2362	0.1132
0.225	0.2186	0.3809	0.4533	0.4348	0.3507	0.2358	0.1160
0.250	0.1977	0.3473	0.4189	0.4093	0.3378	0.2327	0.1166
0.275	0.1798	0.3181	0.3881	0.3853	0.3240	0.2275	0.1157
0.300	0.1643	0.2924	0.3604	0.3626	0.3097	0.2208	0.1136
0.325	0.1507	0.2697	0.3353	0.3412	0.2953	0.2131	0.1107
0.350	0.1387	0.2495	0.3125	0.3211	0.2808	0.2047	0.1071
0.375	0.1281	0.2313	0.2916	0.3021	0.2666	0.1960	0.1032
0.400	0.1187	0.2150	0.2725	0.2843	0.2528	0.1871	0.0989
0.425	0.1102	0.2002	0.2549	0.2675	0.2393	0.1781	0.0946
0.450	0.1025	0.1867	0.2387	0.2517	0.2263	0.1692	0.0902
0.475	0.0955	0.1743	0.2236	0.2368	0.2139	0.1606	0.0858
0.500	0.0891	0.1630	0.2097	0.2228	0.2020	0.1521	0.0814
0.525	0.0833	0.1525	0.1967	0.2096	0.1906	0.1439	0.0772
0.550	0.0779	0.1429	0.1846	0.1973	0.1798	0.1361	0.0731
0.575	0.0729	0.1339	0.1734	0.1856	0.1696	0.1285	0.0691
0.600	0.0683	0.1256	0.1628	0.1746	0.1598	0.1214	0.0653
0.625	0.0641	0.1179	0.1530	0.1643	0.1506	0.1145	0.0617
0.650	0.0601	0.1106	0.1438	0.1546	0.1419	0.1080	0.0582
0.675	0.0564	0.1039	0.1351	0.1455	0.1336	0.1018	0.0549
0.700	0.0530	0.0976	0.1270	0.1369	0.1259	0.0959	0.0518
0.725	0.0497	0.0917	0.1194	0.1288	0.1185	0.0904	0.0488
0.750	0.0467	0.0862	0.1123	0.1212	0.1116	0.0852	0.0460
0.775	0.0439	0.0810	0.1056	0.1140	0.1050	0.0802	0.0433
0.800	0.0413	0.0762	0.0993	0.1073	0.0989	0.0755	0.0408
0.825	0.0388	0.0716	0.0934	0.1009	0.0931	0.0711	0.0384
0.850	0.0365	0.0674	0.0879	0.0950	0.0876	0.0669	0.0362
0.875	0.0343	0.0633	0.0827	0.0894	0.0824	0.0630	0.0341
0.900	0.0323	0.0596	0.0778	0.0841	0.0776	0.0593	0.0321
0.925	0.0303	0.0560	0.0732	0.0791	0.0730	0.0558	0.0302
0.950	0.0285	0.0527	0.0688	0.0744	0.0687	0.0526	0.0284
0.975	0.0268	0.0496	0.0647	0.0700	0.0647	0.0495	0.0268
1.000	0.0253	0.0466	0.0609	0.0659	0.0608	0.0465	0.0252

(x,y)	exact	approx	abs error
(0.25,0.1)	0.3794	0.4015	0.0221
(1,0.5)	0.1854	0.2228	0.0374
(1.5,0.8)	0.0623	0.0755	0.0132

6. **(a)** We identify $c = 15/88 \approx 0.1705$, $a = 20$, $T = 10$, $n = 10$, and $m = 10$. Then $h = 2$, $k = 1$, and $\lambda = 15/352 \approx 0.0426$.

TIME	X=2	X=4	X=6	X=8	X=10	X=12	X=14	X=16	X=18
0	30.0000	30.0000	30.0000	30.0000	30.0000	30.0000	30.0000	30.0000	30.0000
1	28.7216	30.0000	30.0000	30.0000	30.0000	30.0000	30.0000	30.0000	28.7216
2	27.5521	29.9455	30.0000	30.0000	30.0000	30.0000	30.0000	29.9455	27.5521
3	26.4800	29.8459	29.9977	30.0000	30.0000	30.0000	29.9977	29.8459	26.4800
4	25.4951	29.7089	29.9913	29.9999	30.0000	29.9999	29.9913	29.7089	25.4951
5	24.5882	29.5414	29.9796	29.9995	30.0000	29.9995	29.9796	29.5414	24.5882
6	23.7515	29.3490	29.9618	29.9987	30.0000	29.9987	29.9618	29.3490	23.7515
7	22.9779	29.1365	29.9373	29.9972	29.9998	29.9972	29.9373	29.1365	22.9779
8	22.2611	28.9082	29.9057	29.9948	29.9996	29.9948	29.9057	28.9082	22.2611
9	21.5958	28.6675	29.8670	29.9912	29.9992	29.9912	29.8670	28.6675	21.5958
10	20.9768	28.4172	29.8212	29.9862	29.9985	29.9862	29.8212	28.4172	20.9768

(b) We identify $c = 15/88 \approx 0.1705$, $a = 50$, $T = 10$, $n = 10$, and $m = 10$. Then $h = 5$, $k = 1$, and $\lambda = 3/440 \approx 0.0068$.

TIME	X=5	X=10	X=15	X=20	X=25	X=30	X=35	X=40	X=45
0	30.0000	30.0000	30.0000	30.0000	30.0000	30.0000	30.0000	30.0000	30.0000
1	29.7955	30.0000	30.0000	30.0000	30.0000	30.0000	30.0000	30.0000	29.7955
2	29.5937	29.9986	30.0000	30.0000	30.0000	30.0000	30.0000	29.9986	29.5937
3	29.3947	29.9959	30.0000	30.0000	30.0000	30.0000	30.0000	29.9959	29.3947
4	29.1984	29.9918	30.0000	30.0000	30.0000	30.0000	30.0000	29.9918	29.1984
5	29.0047	29.9864	29.9999	30.0000	30.0000	30.0000	29.9999	29.9864	29.0047
6	28.8136	29.9798	29.9998	30.0000	30.0000	30.0000	29.9998	29.9798	28.8136
7	28.6251	29.9720	29.9997	30.0000	30.0000	30.0000	29.9997	29.9720	28.6251
8	28.4391	29.9630	29.9995	30.0000	30.0000	30.0000	29.9995	29.9630	28.4391
9	28.2556	29.9529	29.9992	30.0000	30.0000	30.0000	29.9992	29.9529	28.2556
10	28.0745	29.9416	29.9989	30.0000	30.0000	30.0000	29.9989	29.9416	28.0745

(c) We identify $c = 50/27 \approx 1.8519$, $a = 20$, $T = 10$, $n = 10$, and $m = 10$. Then $h = 2$, $k = 1$, and $\lambda = 25/54 \approx 0.4630$.

TIME	X=2	X=4	X=6	X=8	X=10	X=12	X=14	X=16	X=18
0	18.0000	32.0000	42.0000	48.0000	50.0000	48.0000	42.0000	32.0000	18.0000
1	16.1481	30.1481	40.1481	46.1481	48.1481	46.1481	40.1481	30.1481	16.1481
2	15.1536	28.2963	38.2963	44.2963	46.2963	44.2963	38.2963	28.2963	15.1536
3	14.2226	26.8414	36.4444	42.4444	44.4444	42.4444	36.4444	26.8414	14.2226
4	13.4801	25.4452	34.7764	40.5926	42.5926	40.5926	34.7764	25.4452	13.4801
5	12.7787	24.2258	33.1491	38.8258	40.7407	38.8258	33.1491	24.2258	12.7787
6	12.1622	23.0574	31.6460	37.0842	38.9677	37.0842	31.6460	23.0574	12.1622
7	11.5756	21.9895	30.1875	35.4385	37.2238	35.4385	30.1875	21.9895	11.5756
8	11.0378	20.9636	28.8232	33.8340	35.5707	33.8340	28.8232	20.9636	11.0378
9	10.5230	20.0070	27.5043	32.3182	33.9626	32.3182	27.5043	20.0070	10.5230
10	10.0420	19.0872	26.2620	30.8509	32.4400	30.8509	26.2620	19.0872	10.0420

(d) We identify $c = 260/159 \approx 1.6352$, $a = 100$, $T = 10$, $n = 10$, and $m = 10$. Then $h = 10$, $k = 1$, and $\lambda = 13/795 \approx 0.0164$.

TIME	X=10	X=20	X=30	X=40	X=50	X=60	X=70	X=80	X=90
0	8.0000	16.0000	24.0000	32.0000	40.0000	32.0000	24.0000	16.0000	8.0000
1	8.0000	16.0000	23.6075	31.3459	39.2151	31.6075	23.7384	15.8692	8.0000
2	8.0000	15.9936	23.2279	30.7068	38.4452	31.2151	23.4789	15.7384	7.9979
3	7.9999	15.9812	22.8606	30.0824	37.6900	30.8229	23.2214	15.6076	7.9937
4	7.9996	15.9631	22.5050	29.4724	36.9492	30.4312	22.9660	15.4769	7.9874
5	7.9990	15.9399	22.1606	28.8765	36.2228	30.0401	22.7125	15.3463	7.9793
6	7.9981	15.9118	21.8270	28.2945	35.5103	29.6500	22.4610	15.2158	7.9693
7	7.9967	15.8791	21.5037	27.7261	34.8117	29.2610	22.2112	15.0854	7.9575
8	7.9948	15.8422	21.1902	27.1709	34.1266	28.8733	21.9633	14.9553	7.9439
9	7.9924	15.8013	20.8861	26.6288	33.4548	28.4870	21.7172	14.8253	7.9287
10	7.9894	15.7568	20.5911	26.0995	32.7961	28.1024	21.4727	14.6956	7.9118

9. (a) We identify $c = 15/88 \approx 0.1705$, $a = 20$, $T = 10$, $n = 10$, and $m = 10$. Then $h = 2$, $k = 1$, and $\lambda = 15/352 \approx 0.0426$.

TIME	X=2.00	X=4.00	X=6.00	X=8.00	X=10.00	X=12.00	X=14.00	X=16.00	X=18.00
0.00	30.0000	30.0000	30.0000	30.0000	30.0000	30.0000	30.0000	30.0000	30.0000
1.00	28.7733	29.9749	29.9995	30.0000	30.0000	30.0000	29.9998	29.9916	29.5911
2.00	27.6450	29.9037	29.9970	29.9999	30.0000	30.0000	29.9990	29.9679	29.2150
3.00	26.6051	29.7938	29.9911	29.9997	30.0000	29.9999	29.9970	29.9313	28.8684
4.00	25.6452	29.6517	29.9805	29.9991	30.0000	29.9997	29.9935	29.8839	28.5484
5.00	24.7573	29.4829	29.9643	29.9981	29.9999	29.9994	29.9881	29.8276	28.2524
6.00	23.9347	29.2922	29.9421	29.9963	29.9997	29.9988	29.9807	29.7641	27.9782
7.00	23.1711	29.0836	29.9134	29.9936	29.9995	29.9979	29.9711	29.6945	27.7237
8.00	22.4612	28.8606	29.8782	29.9899	29.9991	29.9966	29.9594	29.6202	27.4870
9.00	21.7999	28.6263	29.8362	29.9848	29.9985	29.9949	29.9454	29.5421	27.2666
10.00	21.1829	28.3831	29.7878	29.9783	29.9976	29.9927	29.9293	29.4610	27.0610

(b) We identify $c = 15/88 \approx 0.1705$, $a = 50$, $T = 10$, $n = 10$, and $m = 10$. Then $h = 5$, $k = 1$, and $\lambda = 3/440 \approx 0.0068$.

TIME	X=5.00	X=10.00	X=15.00	X=20.00	X=25.00	X=30.00	X=35.00	X=40.00	X=45.00
0.00	30.0000	30.0000	30.0000	30.0000	30.0000	30.0000	30.0000	30.0000	30.0000
1.00	29.7968	29.9993	30.0000	30.0000	30.0000	30.0000	30.0000	29.9998	29.9323
2.00	29.5964	29.9973	30.0000	30.0000	30.0000	30.0000	30.0000	29.9991	29.8655
3.00	29.3987	29.9939	30.0000	30.0000	30.0000	30.0000	30.0000	29.9980	29.7996
4.00	29.2036	29.9893	29.9999	30.0000	30.0000	30.0000	30.0000	29.9964	29.7345
5.00	29.0112	29.9834	29.9998	30.0000	30.0000	30.0000	29.9999	29.9945	29.6704
6.00	28.8212	29.9762	29.9997	30.0000	30.0000	30.0000	29.9999	29.9921	29.6071
7.00	28.6339	29.9679	29.9995	30.0000	30.0000	30.0000	29.9998	29.9893	29.5446
8.00	28.4490	29.9585	29.9992	30.0000	30.0000	30.0000	29.9997	29.9862	29.4830
9:00	28.2665	29.9479	29.9989	30.0000	30.0000	30.0000	29.9996	29.9827	29.4222
10.00	28.0864	29.9363	29.9986	30.0000	30.0000	30.0000	29.9995	29.9788	29.3621

(c) We identify $c = 50/27 \approx 1.8519$, $a = 20$, $T = 10$, $n = 10$, and $m = 10$. Then $h = 2$, $k = 1$, and $\lambda = 25/54 \approx 0.4630$.

TIME	X=2.00	X=4.00	X=6.00	X=8.00	X=10.00	X=12.00	X=14.00	X=16.00	X=18.00
0.00	18.0000	32.0000	42.0000	48.0000	50.0000	48.0000	42.0000	32.0000	18.0000
1.00	16.4489	30.1970	40.1562	46.1502	48.1531	46.1773	40.3274	31.2520	22.9449
2.00	15.3312	28.5350	38.3477	44.3130	46.3327	44.4671	39.0872	31.5755	24.6930
3.00	14.4219	27.0429	36.6090	42.5113	44.5759	42.9362	38.1976	31.7478	25.4131
4.00	13.6381	25.6913	34.9606	40.7728	42.9127	41.5716	37.4340	31.7086	25.6986
5.00	12.9409	24.4545	33.4091	39.1182	41.3519	40.3240	36.7033	31.5136	25.7663
6.00	12.3088	23.3146	31.9546	37.5566	39.8880	39.1565	35.9745	31.2134	25.7128
7.00	11.7294	22.2589	30.5939	36.0884	38.5109	38.0470	35.2407	30.8434	25.5871
8.00	11.1946	21.2785	29.3217	34.7092	37.2109	36.9834	34.5032	30.4279	25.4167
9.00	10.6987	20.3660	28.1318	33.4130	35.9801	35.9591	33.7660	29.9836	25.2181
10.00	10.2377	19.5150	27.0178	32.1929	34.8117	34.9710	33.0338	29.5224	25.0019

(d) We identify $c = 260/159 \approx 1.6352$, $a = 100$, $T = 10$, $n = 10$, and $m = 10$. Then $h = 10$, $k = 1$, and $\lambda = 13/795 \approx 0.0164$.

TIME	X=10.00	X=20.00	X=30.00	X=40.00	X=50.00	X=60.00	X=70.00	X=80.00	X=90.00
0.00	8.0000	16.0000	24.0000	32.0000	40.0000	32.0000	24.0000	16.0000	8.0000
1.00	8.0000	16.0000	24.0000	31.9979	39.7425	31.9979	24.0000	16.0026	8.3218
2.00	8.0000	16.0000	23.9999	31.9918	39.4932	31.9918	24.0000	16.0102	8.6333
3.00	8.0000	16.0000	23.9997	31.9820	39.2517	31.9820	24.0001	16.0225	8.9350
4.00	8.0000	16.0000	23.9993	31.9687	39.0176	31.9687	24.0002	16.0392	9.2272
5.00	8.0000	16.0000	23.9987	31.9520	38.7905	31.9521	24.0003	16.0599	9.5103
6.00	8.0000	15.9999	23.9978	31.9323	38.5701	31.9324	24.0005	16.0845	9.7846
7.00	8.0000	15.9999	23.9966	31.9097	38.3561	31.9098	24.0008	16.1126	10.0506
8.00	8.0000	15.9998	23.9951	31.8844	38.1483	31.8846	24.0012	16.1441	10.3084
9.00	8.0000	15.9997	23.9931	31.8566	37.9463	31.8569	24.0017	16.1786	10.5585
10.00	8.0000	15.9996	23.9908	31.8265	37.7499	31.8270	24.0023	16.2160	10.8012

12. We identify $c = 1$, $a = 1$, $T = 1$, $n = 5$, and $m = 20$. Then $h = 0.2$, $k = 0.04$, and $\lambda = 1$. The values below were obtained using *Excel*, which carries more than 12 significant digits. In order to see evidence of instability use $0 \leq t \leq 2$.

TIME	X=0.2	X=0.4	X=0.6	X=0.8	TIME	X=0.2	X=0.4	X=0.6	X=0.8
0.00	0.5878	0.9511	0.9511	0.5878	1.04	0.0000	0.0000	0.0000	0.0000
0.04	0.3633	0.5878	0.5878	0.3633	1.08	0.0000	0.0000	0.0000	0.0000
0.08	0.2245	0.3633	0.3633	0.2245	1.12	0.0000	0.0000	0.0000	0.0000
0.12	0.1388	0.2245	0.2245	0.1388	1.16	0.0000	0.0000	0.0000	0.0000
0.16	0.0858	0.1388	0.1388	0.0858	1.20	-0.0001	0.0001	-0.0001	0.0001
0.20	0.0530	0.0858	0.0858	0.0530	1.24	0.0001	-0.0002	0.0002	-0.0001
0.24	0.0328	0.0530	0.0530	0.0328	1.28	-0.0004	0.0006	-0.0006	0.0004
0.28	0.0202	0.0328	0.0328	0.0202	1.32	0.0010	-0.0015	0.0015	-0.0010
0.32	0.0125	0.0202	0.0202	0.0125	1.36	-0.0025	0.0040	-0.0040	0.0025
0.36	0.0077	0.0125	0.0125	0.0077	1.40	0.0065	-0.0106	0.0106	-0.0065
0.40	0.0048	0.0077	0.0077	0.0048	1.44	-0.0171	0.0277	-0.0277	0.0171
0.44	0.0030	0.0048	0.0048	0.0030	1.48	0.0448	-0.0724	0.0724	-0.0448
0.48	0.0018	0.0030	0.0030	0.0018	1.52	-0.1172	0.1897	-0.1897	0.1172
0.52	0.0011	0.0018	0.0018	0.0011	1.56	0.3069	-0.4965	0.4965	-0.3069
0.56	0.0007	0.0011	0.0011	0.0007	1.60	-0.8034	1.2999	-1.2999	0.8034
0.60	0.0004	0.0007	0.0007	0.0004	1.64	2.1033	-3.4032	3.4032	-2.1033
0.64	0.0003	0.0004	0.0004	0.0003	1.68	-5.5064	8.9096	-8.9096	5.5064
0.68	0.0002	0.0003	0.0003	0.0002	1.72	14.416	-23.326	23.326	-14.416
0.72	0.0001	0.0002	0.0002	0.0001	1.76	-37.742	61.067	-61.067	37.742
0.76	0.0001	0.0001	0.0001	0.0001	1.80	98.809	-159.88	159.88	-98.809
0.80	0.0000	0.0001	0.0001	0.0000	1.84	-258.68	418.56	-418.56	258.685
0.84	0.0000	0.0000	0.0000	0.0000	1.88	677.24	-1095.8	1095.8	-677.245
0.88	0.0000	0.0000	0.0000	0.0000	1.92	-1773.1	2868.9	-2868.9	1773.1
0.92	0.0000	0.0000	0.0000	0.0000	1.96	4641.9	-7510.8	7510.8	-4641.9
0.96	0.0000	0.0000	0.0000	0.0000	2.00	-12153	19663	-19663	12153
1.00	0.0000	0.0000	0.0000	0.0000					

Exercises 16.3

3. (a) Identifying $h = 1/5$ and $k = 0.5/10 = 0.05$ we see that $\lambda = 0.25$.

TIME	X=0.2	X=0.4	X=0.6	X=0.8
0.00	0.5878	0.9511	0.9511	0.5878
0.05	0.5808	0.9397	0.9397	0.5808
0.10	0.5599	0.9059	0.9059	0.5599
0.15	0.5256	0.8505	0.8505	0.5256
0.20	0.4788	0.7748	0.7748	0.4788
0.25	0.4206	0.6806	0.6806	0.4206
0.30	0.3524	0.5701	0.5701	0.3524
0.35	0.2757	0.4460	0.4460	0.2757
0.40	0.1924	0.3113	0.3113	0.1924
0.45	0.1046	0.1692	0.1692	0.1046
0.50	0.0142	0.0230	0.0230	0.0142

(b) Identifying $h = 1/5$ and $k = 0.5/20 = 0.025$ we see that $\lambda = 0.125$.

TIME	X=0.2	X=0.4	X=0.6	X=0.8
0.00	0.5878	0.9511	0.9511	0.5878
0.03	0.5860	0.9482	0.9482	0.5860
0.05	0.5808	0.9397	0.9397	0.5808
0.08	0.5721	0.9256	0.9256	0.5721
0.10	0.5599	0.9060	0.9060	0.5599
0.13	0.5445	0.8809	0.8809	0.5445
0.15	0.5257	0.8507	0.8507	0.5257
0.18	0.5039	0.8153	0.8153	0.5039
0.20	0.4790	0.7750	0.7750	0.4790
0.23	0.4513	0.7302	0.7302	0.4513
0.25	0.4209	0.6810	0.6810	0.4209
0.28	0.3879	0.6277	0.6277	0.3879
0.30	0.3527	0.5706	0.5706	0.3527
0.33	0.3153	0.5102	0.5102	0.3153
0.35	0.2761	0.4467	0.4467	0.2761
0.38	0.2352	0.3806	0.3806	0.2352
0.40	0.1929	0.3122	0.3122	0.1929
0.43	0.1495	0.2419	0.2419	0.1495
0.45	0.1052	0.1701	0.1701	0.1052
0.48	0.0602	0.0974	0.0974	0.0602
0.50	0.0149	0.0241	0.0241	0.0149

6. We identify $c = 24944.4$, $k = 0.00010022$ seconds $= 0.10022$ milliseconds, and $\lambda = 0.25$. Time in the table is expressed in milliseconds.

TIME	X=10	X=20	X=30	X=40	X=50
0.00000	0.2000	0.2667	0.2000	0.1333	0.0667
0.10022	0.1958	0.2625	0.2000	0.1333	0.0667
0.20045	0.1836	0.2503	0.1997	0.1333	0.0667
0.30067	0.1640	0.2307	0.1985	0.1333	0.0667
0.40089	0.1384	0.2050	0.1952	0.1332	0.0667
0.50111	0.1083	0.1744	0.1886	0.1328	0.0667
0.60134	0.0755	0.1407	0.1777	0.1318	0.0666
0.70156	0.0421	0.1052	0.1615	0.1295	0.0665
0.80178	0.0100	0.0692	0.1399	0.1253	0.0661
0.90201	-0.0190	0.0340	0.1129	0.1184	0.0654
1.00223	-0.0435	0.0004	0.0813	0.1077	0.0638
1.10245	-0.0626	-0.0309	0.0464	0.0927	0.0610
1.20268	-0.0758	-0.0593	0.0095	0.0728	0.0564
1.30290	-0.0832	-0.0845	-0.0278	0.0479	0.0493
1.40312	-0.0855	-0.1060	-0.0639	0.0184	0.0390
1.50334	-0.0837	-0.1237	-0.0974	-0.0150	0.0250
1.60357	-0.0792	-0.1371	-0.1275	-0.0511	0.0069
1.70379	-0.0734	-0.1464	-0.1533	-0.0882	-0.0152
1.80401	-0.0675	-0.1515	-0.1747	-0.1249	-0.0410
1.90424	-0.0627	-0.1528	-0.1915	-0.1595	-0.0694
2.00446	-0.0596	-0.1509	-0.2039	-0.1904	-0.0991
2.10468	-0.0585	-0.1467	-0.2122	-0.2165	-0.1283
2.20491	-0.0592	-0.1410	-0.2166	-0.2368	-0.1551
2.30513	-0.0614	-0.1349	-0.2175	-0.2507	-0.1772
2.40535	-0.0643	-0.1294	-0.2154	-0.2579	-0.1929
2.50557	-0.0672	-0.1251	-0.2105	-0.2585	-0.2005
2.60580	-0.0696	-0.1227	-0.2033	-0.2524	-0.1993
2.70602	-0.0709	-0.1219	-0.1942	-0.2399	-0.1889
2.80624	-0.0710	-0.1225	-0.1833	-0.2214	-0.1699
2.90647	-0.0699	-0.1236	-0.1711	-0.1972	-0.1435
3.00669	-0.0678	-0.1244	-0.1575	-0.1681	-0.1115
3.10691	-0.0649	-0.1237	-0.1425	-0.1348	-0.0761
3.20713	-0.0617	-0.1205	-0.1258	-0.0983	-0.0395
3.30736	-0.0583	-0.1139	-0.1071	-0.0598	-0.0042
3.40758	-0.0547	-0.1035	-0.0859	-0.0209	0.0279
3.50780	-0.0508	-0.0889	-0.0617	0.0171	0.0552
3.60803	-0.0460	-0.0702	-0.0343	0.0525	0.0767
3.70825	-0.0399	-0.0478	-0.0037	0.0840	0.0919
3.80847	-0.0318	-0.0221	0.0297	0.1106	0.1008
3.90870	-0.0211	0.0062	0.0648	0.1314	0.1041
4.00892	-0.0074	0.0365	0.1005	0.1464	0.1025
4.10914	0.0095	0.0680	0.1350	0.1558	0.0973
4.20936	0.0295	0.1000	0.1666	0.1602	0.0897
4.30959	0.0521	0.1318	0.1937	0.1606	0.0808
4.40981	0.0764	0.1625	0.2148	0.1581	0.0719
4.51003	0.1013	0.1911	0.2291	0.1538	0.0639
4.61026	0.1254	0.2164	0.2364	0.1485	0.0575
4.71048	0.1475	0.2373	0.2369	0.1431	0.0532
4.81070	0.1659	0.2526	0.2315	0.1379	0.0512
4.91093	0.1794	0.2611	0.2217	0.1331	0.0514
5.01115	0.1867	0.2620	0.2087	0.1288	0.0535

Chapter 16 Review Exercises

3. (a)

TIME	X=0.0	X=0.2	X=0.4	X=0.6	X=0.8	X=1.0
0.00	0.0000	0.2000	0.4000	0.6000	0.8000	0.0000
0.01	0.0000	0.2000	0.4000	0.6000	0.5500	0.0000
0.02	0.0000	0.2000	0.4000	0.5375	0.4250	0.0000
0.03	0.0000	0.2000	0.3844	0.4750	0.3469	0.0000
0.04	0.0000	0.1961	0.3609	0.4203	0.2922	0.0000
0.05	0.0000	0.1883	0.3346	0.3734	0.2512	0.0000

(b)

TIME	X=0.0	X=0.2	X=0.4	X=0.6	X=0.8	X=1.0
0.00	0.0000	0.2000	0.4000	0.6000	0.8000	0.0000
0.01	0.0000	0.2000	0.4000	0.6000	0.8000	0.0000
0.02	0.0000	0.2000	0.4000	0.6000	0.5500	0.0000
0.03	0.0000	0.2000	0.4000	0.5375	0.4250	0.0000
0.04	0.0000	0.2000	0.3844	0.4750	0.3469	0.0000
0.05	0.0000	0.1961	0.3609	0.4203	0.2922	0.0000

(c) The table in part (b) is the same as the table in part (a) shifted downward one row.

17 Functions of a Complex Variable

━━━━━━━━━━ **Exercises 17.1** ━━━━━━━━━━

3. $i^8 = (i^2)^4 = (-1)^4 = 1$

6. $-3 - 9i$

9. $11 - 10i$

12. $-2 - 2i$

15. $\dfrac{2-4i}{3+5i} \cdot \dfrac{3-5i}{3-5i} = \dfrac{-14-22i}{34} = -\dfrac{7}{17} - \dfrac{11}{17}i$

18. $\dfrac{3-i}{11-2i} \cdot \dfrac{11+2i}{11+2i} = \dfrac{35-5i}{125} = \dfrac{7}{25} - \dfrac{1}{25}i$

21. $(1+i)(10+10i) = 10(1+i)^2 = 20i$

24. $(2+3i)(-i)^2 = -2 - 3i$

27. $\dfrac{x}{x^2+y^2}$

30. 0

33. $2x + 2yi = -9 + 2i$ implies $2x = -9$ and $2y = 2$. Hence $z = -\frac{9}{2} + i$.

36. $x^2 - y^2 - 4x + (-2xy - 4y)i = 0 + 0i$ implies $x^2 - y^2 - 4x = 0$ and $y(-2x - 4) = 0$. If $y = 0$ then $x(x - 4) = 0$ and so $z = 0$ and $z = 4$. If $-2x - 4 = 0$ or $x = -2$ then $12 - y^2 = 0$ or $y = \pm 2\sqrt{3}$. This gives $z = -2 + 2\sqrt{3}\,i$ and $z = -2 - 2\sqrt{3}\,i$.

39. $|z_1 - z_2| = |(x_1 - x_2) + i(y_1 - y_2)| = \sqrt{(x_1 - x_2)^2 + (y_1 - y_2)^2}$ which is the distance formula in the plane.

━━━━━━━━━━ **Exercises 17.2** ━━━━━━━━━━

3. $3\left(\cos\dfrac{3\pi}{2} + i\sin\dfrac{3\pi}{2}\right)$

6. $5\sqrt{2}\left(\cos\dfrac{7\pi}{4} + i\sin\dfrac{7\pi}{4}\right)$

9. $\dfrac{3\sqrt{2}}{2}\left(\cos\dfrac{5\pi}{4} + i\sin\dfrac{5\pi}{4}\right)$

12. $z = -8 + 8i$

15. $z_1 z_2 = 8\left[\cos\left(\dfrac{\pi}{8} + \dfrac{3\pi}{8}\right) + i\sin\left(\dfrac{\pi}{8} + \dfrac{3\pi}{8}\right)\right] = 8i; \quad \dfrac{z_1}{z_2} = \dfrac{1}{2}\left[\cos\left(\dfrac{\pi}{8} - \dfrac{3\pi}{8}\right) + i\sin\left(\dfrac{\pi}{8} - \dfrac{3\pi}{8}\right)\right] = \dfrac{\sqrt{2}}{4} - \dfrac{\sqrt{2}}{4}i$

18. $\left[4\sqrt{2}\left(\cos\dfrac{\pi}{4} + i\sin\dfrac{\pi}{4}\right)\right]\left[\sqrt{2}\left(\cos\dfrac{3\pi}{4} + i\sin\dfrac{3\pi}{4}\right)\right] = 8\left[\cos\left(\dfrac{\pi}{4} + \dfrac{3\pi}{4}\right) + i\sin\left(\dfrac{\pi}{4} + \dfrac{3\pi}{4}\right)\right] = -8$

21. $2^9\left[\cos\dfrac{9\pi}{3} + i\sin\dfrac{9\pi}{3}\right] = -512$

24. $(2\sqrt{2})^4\left[\cos\dfrac{8\pi}{3} + i\sin\dfrac{8\pi}{3}\right] = -32 + 32\sqrt{3}\,i$

27. $8^{1/3} = 2\left[\cos\dfrac{2k\pi}{3} + i\sin\dfrac{2k\pi}{3}\right], \quad k = 0, 1, 2$

$w_0 = 2[\cos 0 + i\sin 0] = 2; \quad w_1 = 2\left[\cos\dfrac{2\pi}{3} + i\sin\dfrac{2\pi}{3}\right] = -1 + \sqrt{3}\,i$

$w_2 = 2\left[\cos\dfrac{4\pi}{3} + i\sin\dfrac{4\pi}{3}\right] = -1 - \sqrt{3}\,i$

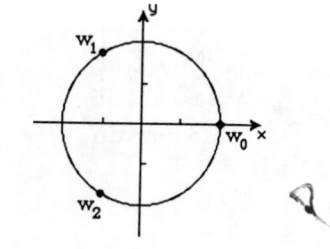

30. $(-1+i)^{1/3} = 2^{1/6} \left[\cos\left(\dfrac{\pi}{4} + \dfrac{2k\pi}{3}\right) + i\sin\left(\dfrac{\pi}{4} + \dfrac{2k\pi}{3}\right) \right], \quad k = 0, 1, 2$

$w_0 = 2^{1/6} \left[\cos\dfrac{\pi}{4} + i\sin\dfrac{\pi}{4} \right] = \dfrac{1}{\sqrt[3]{2}} + \dfrac{1}{\sqrt[3]{2}} i = 0.7937 + 0.7937i$

$w_1 = 2^{1/6} \left[\cos\dfrac{11\pi}{12} + i\sin\dfrac{11\pi}{12} \right] = -1.0842 + 0.2905i$

$w_2 = 2^{1/6} \left[\cos\dfrac{19\pi}{12} + i\sin\dfrac{19\pi}{12} \right] = 0.2905 - 1.0842i$

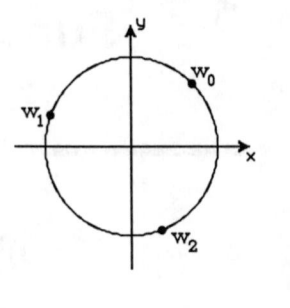

33. The solutions are the four fourth roots of -1;

$$w_k = \cos\dfrac{\pi + 2k\pi}{4} + i\sin\dfrac{\pi + 2k\pi}{4}, \quad k = 0, 1, 2, 3.$$

We have

$w_1 = \cos\dfrac{\pi}{4} + i\sin\dfrac{\pi}{4} = \dfrac{\sqrt{2}}{2} + \dfrac{\sqrt{2}}{2} i$ $\qquad$ $w_3 = \cos\dfrac{5\pi}{4} + i\sin\dfrac{5\pi}{4} = -\dfrac{\sqrt{2}}{2} - \dfrac{\sqrt{2}}{2} i$

$w_2 = \cos\dfrac{3\pi}{4} + i\sin\dfrac{3\pi}{4} = -\dfrac{\sqrt{2}}{2} + \dfrac{\sqrt{2}}{2} i$ $\qquad$ $w_4 = \cos\dfrac{7\pi}{4} + i\sin\dfrac{7\pi}{4} = \dfrac{\sqrt{2}}{2} - \dfrac{\sqrt{2}}{2} i.$

36. $\dfrac{\left[8\left(\cos\dfrac{3\pi}{8} + i\sin\dfrac{3\pi}{8} \right) \right]^3}{\left[2\left(\cos\dfrac{\pi}{16} + i\sin\dfrac{\pi}{16} \right) \right]^{10}} = \dfrac{2^9}{2^{10}} \left[\cos\left(\dfrac{9\pi}{8} - \dfrac{10\pi}{16} \right) + i\left(\dfrac{9\pi}{8} - \dfrac{10\pi}{16} \right) \right] = \dfrac{1}{2}\left(\cos\dfrac{\pi}{2} + i\sin\dfrac{\pi}{2} \right) = \dfrac{1}{2} i$

39. **(a)** $\mathrm{Arg}(z_1) = \pi, \quad \mathrm{Arg}(z_2) = \dfrac{\pi}{2}, \quad \mathrm{Arg}(z_1 z_2) = -\dfrac{\pi}{2}, \quad \mathrm{Arg}(z_1) + \mathrm{Arg}(z_2) = \dfrac{3\pi}{2} \neq \mathrm{Arg}(z_1 z_2)$

(b) $\mathrm{Arg}(z_1/z_2) = -\dfrac{\pi}{2}, \quad \mathrm{Arg}(z_1) - \mathrm{Arg}(z_2) = \pi - \dfrac{\pi}{2} = \dfrac{\pi}{2} \neq \mathrm{Arg}(z_1/z_2)$

Exercises 17.3

3.

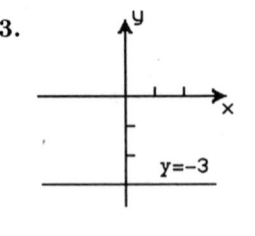

6. $|z+1/2|=2$

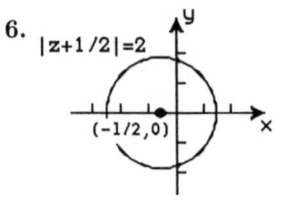

9.

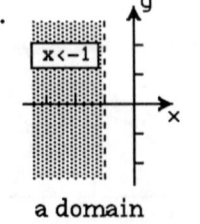

a domain

12.

a domain

15.

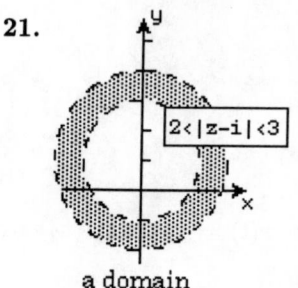

$x^2 - y^2 > 0$

not a domain

18.

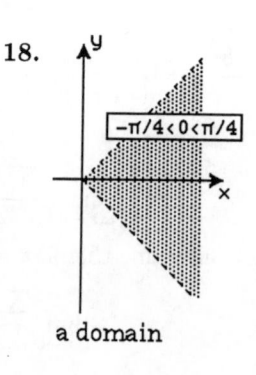

$-\pi/4 < \theta < \pi/4$

a domain

21.

$2 < |z - i| < 3$

a domain

24. $|\mathrm{Re}(z)| = |x|$ is the same as $\sqrt{x^2}$ and $|z| = \sqrt{x^2 + y^2}$. Since $y^2 \geq 0$ the inequality $\sqrt{x^2} \leq \sqrt{x^2 + y^2}$ is true for all complex numbers.

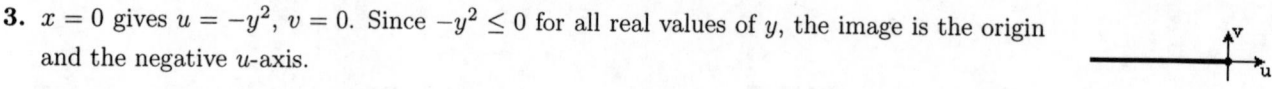

Exercises 17.4

3. $x = 0$ gives $u = -y^2$, $v = 0$. Since $-y^2 \leq 0$ for all real values of y, the image is the origin and the negative u-axis.

6. $y = -x$ gives $u = 0$, $v = -2x^2$. Since $-x^2 \leq 0$ for all real values of x, the image is the origin and the negative v-axis.

9. $f(z) = (x^2 - y^2 - 3x) + i(2xy - 3y + 4)$

12. $f(z) = (x^4 - 6x^2y^2 + y^4) + i(4x^3y - 4xy^3)$

15. (a) $f(0 + 2i) = -4 + i$ **(b)** $f(2 - i) = 3 - 9i$ **(c)** $f(5 + 3i) = 1 + 86i$

18. (a) $f(0 + \frac{\pi}{4}i) = \frac{\sqrt{2}}{2} + \frac{\sqrt{2}}{2}i$ **(b)** $f(-1 - \pi i) = -e^{-1}$ **(c)** $f(3 + \frac{\pi}{3}i) = \frac{1}{2}e^3 + \frac{\sqrt{3}}{2}e^3 i$

21. $\lim\limits_{z \to i} \dfrac{z^4 - 1}{z - i} = \lim\limits_{z \to i} \dfrac{(z^2 - 1)(z - i)(z + i)}{z - i} = -4i$

24. Along the line $x = 1$, $\lim\limits_{z \to 1} \dfrac{x + y - 1}{z - 1} = \lim\limits_{y \to 0} \dfrac{y}{iy} = \dfrac{1}{i} = -i$, whereas along the x-axis,

$\lim\limits_{z \to 1} \dfrac{x + y - 1}{z - 1} = \lim\limits_{x \to 1} \dfrac{x - 1}{x - 1} = 1$.

27. $f'(z) = 12z^2 - (6 + 2i)z - 5$

30. $f'(z) = (z^5 + 3iz^3)(4z^2 + 3iz^2 + 4z - 6i) + (z^4 + iz^3 + 2z^2 - 6iz)(5z^4 + 9iz^2)$

33. $f'(z) = \dfrac{(2z+i)3 - (3z-4+8i)2}{(2z+1)^2} = \dfrac{8-13i}{(2z+i)^2}$

36. $0,\ 2-5i$

39. We have
$$\lim_{\Delta z \to 0} \frac{\overline{z+\Delta z} - \overline{z}}{\Delta z} = \lim_{\Delta z \to 0} \frac{\overline{\Delta z}}{\Delta z}.$$

If we let $\Delta z \to 0$ along a horizontal line then $\Delta z = \Delta x$, $\overline{\Delta z} = \Delta x$, and
$$\lim_{\Delta z \to 0} \frac{\overline{\Delta z}}{\Delta z} = \lim_{\Delta x \to 0} \frac{\Delta x}{\Delta x} = 1.$$

If we let $\Delta z \to 0$ along a vertical line then $\Delta z = i\Delta y$, $\overline{\Delta z} = -i\Delta y$, and
$$\lim_{\Delta z \to 0} \frac{\overline{\Delta z}}{\Delta z} = \lim_{\Delta y \to 0} \frac{-i\Delta y}{i\Delta y} = -1.$$

Since these two limits are not equal, $f(z) = \overline{z}$ cannot be differentiable at any z.

42. The system $\dfrac{dx}{dt} = -y$, $\dfrac{dy}{dt} = x$ can be solved as in Section 3.11. We obtain $x(t) = c_1 \cos t + c_2 \sin t$, $y(t) = c_1 \sin t - c_2 \cos t$.

45. If $y = \frac{1}{2}x^2$ the equations $u = x^2 - y^2$, $v = 2xy$ give $u = x^2 - \frac{1}{4}x^4$, $v = x^3$. With the aid of a computer, the graph of these parametric equations is shown.

Exercises 17.5

3. $u = x$, $v = 0$; $\dfrac{\partial u}{\partial x} = 1$, $\dfrac{\partial v}{\partial y} = 0$. Since $1 \neq 0$, f is not analytic at any point.

6. $u = x^2 - y^2$, $v = -2xy$; $\dfrac{\partial u}{\partial x} = 2x$, $\dfrac{\partial v}{\partial y} = -2x$; $\dfrac{\partial u}{\partial y} = -2y$, $-\dfrac{\partial v}{\partial x} = 2y$

The Cauchy-Riemann equations hold only at $(0,0)$. Since there is no neighborhood about $z = 0$ within which f is differentiable we conclude f is nowhere analytic.

9. $u = e^x \cos y$, $v = e^x \sin y$; $\dfrac{\partial u}{\partial x} = e^x \cos y = \dfrac{\partial v}{\partial y}$; $\dfrac{\partial u}{\partial y} = -e^x \sin y = -\dfrac{\partial v}{\partial x}$. f is analytic for all z.

12. $u = 4x^2 + 5x - 4y^2 + 9$, $v = 8xy + 5y - 1$; $\dfrac{\partial u}{\partial x} = 8x + 5 = \dfrac{\partial v}{\partial y}$, $\dfrac{\partial u}{\partial y} = -8y = -\dfrac{\partial v}{\partial x}$. f is analytic for all z.

15. $\dfrac{\partial u}{\partial x} = 3 = b = \dfrac{\partial v}{\partial y}$; $\dfrac{\partial u}{\partial y} = -1 = -a = -\dfrac{\partial v}{\partial x}$. f is analytic for all z when $b = 3$, $a = 1$.

18. $u = 3x^2y^2$, $v = -6x^2y^2$; $\dfrac{\partial u}{\partial x} = 6xy^2$, $\dfrac{\partial v}{\partial y} = -12x^2y$; $\dfrac{\partial u}{\partial y} = 6x^2y$, $-\dfrac{\partial v}{\partial x} = 12xy^2$

u and v are continuous and have continuous first partial derivatives. The Cauchy-Riemann equations are satisfied whenever $6xy(y + 2x) = 0$ and $6xy(x - 2y) = 0$. The point satisfying $y + 2x = 0$ and $x - 2y = 0$ is $z = 0$. The points that satisfy $6xy = 0$ are the points along the y-axis ($x = 0$) or along the x-axis ($y = 0$). The function f is differentiable but not analytic on either axis; there is no neighborhood about any point $z = x$ or $z = iy$ within which f is differentiable.

21. Since f is entire,

$$f'(z) = \frac{\partial u}{\partial x} + i\frac{\partial v}{\partial x} = e^x \cos y + ie^x \sin y = f(z).$$

24. $\frac{\partial^2 u}{\partial x^2} = 0$, $\frac{\partial^2 u}{\partial y^2} = 0$ gives $\frac{\partial^2 u}{\partial x^2} + \frac{\partial^2 u}{\partial y^2} = 0$. Thus u is harmonic. Now $\frac{\partial u}{\partial x} = 2 - 2y = \frac{\partial v}{\partial y}$ implies $v = 2y - y^2 + h(x)$,

$\frac{\partial u}{\partial y} = -2x = -\frac{\partial v}{\partial x} = -h'(x)$ implies $h'(x) = 2x$ or $h(x) = x^2 + C$. Therefore $f(z) = 2x - 2xy + i(2y - y^2 + x^2 + C)$.

27. $\frac{\partial^2 u}{\partial x^2} = \frac{2y^2 - 2x^2}{(x^2 + y^2)^2}$, $\frac{\partial^2 u}{\partial y^2} = \frac{2x^2 - 2y^2}{(x^2 + y^2)^2}$ gives $\frac{\partial^2 u}{\partial x^2} + \frac{\partial^2 u}{\partial y^2} = 0$. Thus u is harmonic. Now $\frac{\partial u}{\partial x} = \frac{2x}{x^2 + y^2} = \frac{\partial v}{\partial y}$

implies $v = 2\tan^{-1}\frac{y}{x} + h(x)$, $\frac{\partial u}{\partial y} = \frac{2y}{x^2 + y^2} = -\frac{\partial v}{\partial x} = \frac{2y}{x^2 + y^2} - h'(x)$ implies $h'(x) = 0$ or $h(x) = C$.

Therefore $f(z) = \log_e(x^2 + y^2) + i\left(\tan^{-1}\frac{y}{x} + C\right)$, $z \neq 0$.

30. $f(x) = \frac{x}{x^2 + y^2} - i\frac{y}{x^2 + y^2}$. The level curves $u(x, y) = c_1$ and $v(x, y) = c_2$ are the family of circles $x = c_1(x^2 + y^2)$

and $-y = c_2(x^2 + y^2)$, with the exception that $(0, 0)$ is not on the circumference of any circle.

Exercises 17.6

3. $e^{-1 + \frac{\pi}{4}i} = e^{-1}\cos\frac{\pi}{4} + ie^{-1}\sin\frac{\pi}{4} = e^{-1}\left(\frac{\sqrt{2}}{2} + \frac{\sqrt{2}}{2}i\right)$

6. $e^{-\pi + \frac{3\pi}{2}i} = e^{-\pi}\cos\frac{3\pi}{2} + ie^{-\pi}\sin\frac{3\pi}{2} = -e^{-\pi}i$

9. $e^{5i} = \cos 5 + i\sin 5 = 0.2837 - 0.9589i$

12. $e^{5 + \frac{5\pi}{2}i} = e^5\cos\frac{5\pi}{2} + ie^5\sin\frac{5\pi}{2} = e^5 i$

15. $e^{z^2} = e^{x^2 - y^2 + 2xyi} = e^{x^2 - y^2}\cos 2xy + ie^{x^2 - y^2}\sin 2xy$

18. $\frac{e^{z_1}}{e^{z_2}} = \frac{e^{x_1}\cos y_1 + ie^{x_1}\sin y_1}{e^{x_2}\cos y_2 + ie^{x_2}\sin y_2} = \frac{(e^{x_1}\cos y_1 + ie^{x_1}\sin y_1)(e^{x_2}\cos y_2 - ie^{x_2}\sin y_2)}{e^{2x_2}}$

$= e^{x_1 - x_2}[(\cos y_1\cos y_2 + \sin y_1\sin y_2) + i(\sin y_1\cos y_2 - \cos y_1\sin y_2)]$

$= e^{x_1 - x_2}[\cos(y_1 - y_2) + i\sin(y_1 - y_2)] = e^{x_1 - x_2 + i(y_1 - y_2)} = e^{(x_1 + iy_1) - (x_2 + iy_2)} = e^{z_1 - z_2}$

21. $u = e^x\cos y$, $v = -e^x\sin y$; $\frac{\partial u}{\partial x} = e^x\cos y$, $\frac{\partial v}{\partial y} = -e^x\cos y$; $\frac{\partial u}{\partial y} = -e^x\sin y$, $-\frac{\partial v}{\partial x} = e^x\sin y$

Since the Cauchy-Riemann equations are not satisfied at any point, f is nowhere analytic.

24. $\ln(-ei) = \log_e e + i\left(-\frac{\pi}{2} + 2n\pi\right) = 1 + \left(-\frac{\pi}{2} + 2n\pi\right)i$

27. $\ln(\sqrt{2} + \sqrt{6}i) = \log_e 2\sqrt{2} + i\left(\frac{\pi}{3} + 2n\pi\right) = 1.0397 + \left(\frac{\pi}{3} + 2n\pi\right)i$

30. $\text{Ln}(-e^3) = \log_e e^3 + \pi i = 3 + \pi i$

33. $\text{Ln}(1 + \sqrt{3}i)^5 = \text{Ln}(16 - 16\sqrt{3}i) = \log_e 32 - \frac{\pi}{3}i = 3.4657 - \frac{\pi}{3}i$

36. $\frac{1}{z} = \ln(-1) = \log_e 1 + i(\pi + 2n\pi) = (2n + 1)\pi i$ and so $z = -\frac{i}{(2n + 1)\pi}$.

39. $(-i)^{4i} = e^{4i\ln(-i)} = e^{4i[\log_e 1 + i(-\frac{\pi}{2} + 2n\pi)]} = e^{(2 - 8n)\pi}$

42. $(1 - i)^{2i} = e^{2i\ln(1 - i)} = e^{2i[\log_e \sqrt{2} + i(-\frac{\pi}{4} + 2n\pi)]} = e^{\frac{\pi}{2} - 4n\pi}[\cos(\log_e 2) + i\sin(\log_e 2)] = e^{-4n\pi}[3.7004 + 3.0737i]$

45. If $z_1 = i$ and $z_2 = -1 + i$ then

$$\text{Ln}(z_1 z_2) = \text{Ln}(-1 - i) = \log_e \sqrt{2} - \frac{3\pi}{4} i,$$

whereas

$$\text{Ln} z_1 + \text{Ln} z_2 = \frac{\pi}{2} i + \left(\log_e \sqrt{2} + \frac{3\pi}{4} i \right) = \log_e \sqrt{2} + \frac{5\pi}{4} i.$$

48. (a) $(i^i)^2 = (e^{i \ln i})^2 = [e^{-(\frac{\pi}{2} + 2n\pi)}]^2 = e^{-(\pi + 4n\pi)}$ and $i^{2i} = e^{2i \ln i} = e^{-(\pi + 4n\pi)}$

(b) $(i^2)^i = (-1)^i = e^{i \ln(-1)} = e^{-(\pi + 2n\pi)}$, whereas $i^{2i} = e^{-(\pi + 4n\pi)}$

Exercises 17.7

3. $\sin\left(\frac{\pi}{4} + i\right) = \sin\frac{\pi}{4} \cosh(1) + i \cos\frac{\pi}{4} \sinh(1) = 1.0911 + 0.8310i$

6. $\cot\left(\frac{\pi}{2} + 3i\right) = \dfrac{\cos(\frac{\pi}{2} + 3i)}{\sin(\frac{\pi}{2} + 3i)} = \dfrac{-i \sinh(3)}{\cosh(3)} = -0.9951i$

9. $\cosh(\pi i) = \cos(i(\pi i)) = \cos(-\pi) = \cos \pi = -1$

12. $\cosh(2 + 3i) = \cosh(2) \cos(3) + i \sinh(2) \sin(3) = -3.7245 + 0.5118i$

15. $\dfrac{e^{iz} - e^{-iz}}{2i} = 2$ gives $e^{2(iz)} - 4i e^{iz} - 1 = 0$. By the quadratic formula, $e^{iz} = 2i \pm \sqrt{3}\, i$ and so

$$iz = \ln[(2 \pm \sqrt{3})i]$$

$$z = -i\left[\log_e(2 \pm \sqrt{3}) + \left(\frac{\pi}{2} + 2n\pi\right) i \right] = \frac{\pi}{2} + 2n\pi - i\log_e(2 \pm \sqrt{3}), \quad n = 0, \pm1., \pm 2, \ldots.$$

18. $\dfrac{e^z - e^{-z}}{2} = -1$ gives $e^{2z} + 2e^z - 1 = 0$. By the quadratic formula, $e^z = -1 \pm \sqrt{2}$, and so

$$z = \ln(-1 \pm \sqrt{2})$$

$$z = \log_e(\sqrt{2} - 1) + 2n\pi i \quad \text{or} \quad z = \log_e(\sqrt{2} + 1) + (\pi + 2n\pi)i,$$

$n = 0, \pm1, \pm2, \ldots .$

21. $\cos z = \cosh 2$ implies $\cos x \cosh y - i \sin x \sinh y = \cosh 2 + 0i$ and so we must have $\cos x \cosh y = \cosh 2$ and $\sin x \sinh y = 0$. The last equation has solutions $x = n\pi$, $n = 0, \pm1, \pm2, \ldots$, or $y = 0$. For $y = 0$ the first equation becomes $\cos x = \cosh 2$. Since $\cosh 2 > 1$ this equation has no solutions. For $x = n\pi$ the first equation becomes $(-1)^n \cosh y = \cosh 2$. Since $\cosh y > 0$ we see n must be even, say, $n = 2k$, $k = 0, \pm1, \pm2, \ldots$. Now $\cosh y = \cosh 2$ implies $y = \pm 2$. Solutions of the original equation are then

$$z = 2k\pi \pm 2i, \quad k = 0, \pm1, \pm2, \ldots .$$

24. $\sinh z = \dfrac{e^{x+iy} - e^{-x-iy}}{2} = \frac{1}{2}(e^x e^{iy} - e^{-x} e^{-iy}) = \frac{1}{2}[e^x(\cos y + i \sin y) - e^{-x}(\cos y - i \sin y)]$

$$= \left(\frac{e^x - e^{-x}}{2} \right) \cos y + i \left(\frac{e^x + e^{-x}}{2} \right) \sin y = \sinh x \cos y + i \cosh x \sin y$$

27. $|\cosh z|^2 = \cosh^2 x \cos^2 y + \sinh^2 x \sin^2 y = (1 + \sinh^2 x) \cos^2 y + \sinh^2 x \sin^2 y$

$$= \cos^2 y + \sinh^2 x(\cos^2 y + \sin^2 y) = \cos^2 y + \sinh^2 x$$

30. $\tan z = \dfrac{\sin z}{\cos z} = \dfrac{\sin z \cos z}{|\cos z|^2} = \dfrac{[\sin x \cosh y + i \cos x \sinh y][\cos x \cosh y + i \sin x \sinh y]}{\cos^2 x + \sinh^2 y}$

$\qquad = \dfrac{(\sin x \cos x \cosh^2 y - \sin x \cos x \sinh^2 y)}{\cos^2 x + \sinh^2 y} + i \,\dfrac{\cos^2 x \sinh y \cosh y + \sin^2 x \sinh y \cosh y}{\cos^2 x + \sinh^2 y}$

$\qquad = \dfrac{\sin x \cos x (\cosh^2 y - \sinh^2 y)}{\cos^2 x + \sinh^2 y} + i \,\dfrac{\sin y \cosh y (\cos^2 x + \sin^2 x)}{\cos^2 x + \sinh^2 y}$

$\qquad = \dfrac{\sin x \cos x}{\cos^2 x + \sinh^2 y} + i \,\dfrac{\sinh y \cosh y}{\cos^2 x + \sinh^2 y} = \dfrac{\sin 2x}{2(\cos^2 x + \sinh^2 y)} + i \,\dfrac{\sinh 2y}{2(\cos^2 x + \sinh^2 y)}$

But

$$2\cos^2 x + 2\sinh^2 y = (2\cos^2 x - 1) + (2\sinh^2 y + 1) = \cos 2x + \cosh 2y.$$

Therefore $\tan z = u + iz$ where

$$u = \frac{\sin 2x}{\cos 2x + \cosh 2y}, \qquad v = \frac{\sinh 2y}{\cos 2x + \cosh 2y}.$$

Exercises 17.8

3. $\sin^{-1} 0 = -i \ln(\pm 1) = \begin{cases} 2n\pi + i \log_e 1 \\ (2n+1)\pi + i \log_e 1 \end{cases} = \begin{cases} 2n\pi \\ (2n+1)\pi \end{cases} = n\pi, \quad n = 0, \pm 1, \pm 2, \ldots$

6. $\cos^{-1} 2i = -i \ln[(2 \pm \sqrt{5}\,)i] = \begin{cases} 2n\pi - \frac{\pi}{2} + i \log_e(2 + \sqrt{5}\,) \\ 2n\pi + \frac{\pi}{2} - i \log_e(2 + \sqrt{5}\,) \end{cases}, \quad n = 0, \pm 1, \pm 2, \ldots$

9. $\tan^{-1} 1 = \dfrac{i}{2} \ln \dfrac{i+1}{i-1} = \dfrac{i}{2} \ln(-i) = -n\pi + \dfrac{\pi}{4} + \dfrac{i}{2} \log_e 1 = \dfrac{\pi}{4} - n\pi, \quad n = 0, \pm 1, \pm 2, \ldots$

Note that this can also be written as $\tan^{-1} 1 = \frac{\pi}{4} + n\pi, \quad n = 0, \pm 1, \pm 2, \ldots$.

12. $\cosh^{-1} i = \ln[(1 + \pm\sqrt{2}\,)i] = \begin{cases} \log_e(1 + \sqrt{2}\,) + (\frac{\pi}{2} + 2n\pi)i \\ \log_e(\sqrt{2} - 1) + (-\frac{\pi}{2} + 2n\pi)i \end{cases}, \quad n = 0, \pm 1, \pm 2, \ldots$

Chapter 17 Review Exercises

3. $-7/25$

6. The closed annular region between the circles $|z + 2| = 1$ and $|z + 2| = 3$. These circles have center at $z = -2$.

9. $z = \ln(2i) = \log_e 2 + i\left(\dfrac{\pi}{2} + 2n\pi\right), \quad n = 0, \pm 1, \pm 2, \ldots$

12. $f(-1 + i) = -33 + 26i$

15. $\mathrm{Ln}(-ie^3) = \log_e e^3 + \left(-\dfrac{\pi}{2}\right)i = 3 - \dfrac{\pi}{2} i$

18. $-\dfrac{1}{13} - \dfrac{17}{13} i$

21. The region satisfying $xy \leq 1$ is shown in the figure.

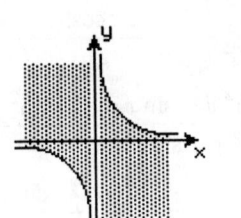

24. The region satisfying $y < x$ is shown in the figure.

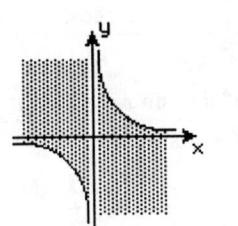

27. The four fourth roots of $1 - i$ are given by

$$w_R = 2^{1/8} \left[\cos\left(-\frac{\pi}{16} + \frac{k\pi}{2}\right) + i\sin\left(-\frac{\pi}{16} + \frac{k\pi}{2}\right) \right], \quad n = 0,\ 1,\ 2,\ 3$$

$$w_0 = 2^{1/8} \left[\cos\left(-\frac{\pi}{16}\right) + i\sin\left(-\frac{\pi}{16}\right) \right] = 1.0696 - 0.2127i$$

$$w_1 = 2^{1/8} \left[\cos\frac{7\pi}{16} + i\sin\frac{7\pi}{16} \right] = 0.2127 + 1.0696i$$

$$w_2 = 2^{1/8} \left[\cos\frac{15\pi}{16} + i\sin\frac{15\pi}{16} \right] = -1.0696 + 0.2127i$$

$$w_3 = 2^{1/8} \left[\cos\frac{23\pi}{16} + i\sin\frac{23\pi}{16} \right] = -0.2127 - 1.0696i.$$

30. $\text{Im}(z - 3\bar{z}) = 4y$, $z\text{Re}(z^2) = (x^3 - xy^2) + i(x^2y - y^3)$. Thus,

$$f(z) = (4y + x^3 - xy^2 - 5x) + i(x^2y - y^3 - 5y).$$

33. $z = z^{-1}$ gives $z^2 = 1$ or $(z - 1)(z + 1) = 0$. Thus $z = \pm 1$.

36. $z^2 = \bar{z}^2$ gives $xy = -xy$ or $xy = 0$. This implies $x = 0$ or $y = 0$. All real numbers ($y = 0$) and all pure imaginary numbers ($x = 0$) satisfy the equation.

39. $\text{Ln}(1 + i)(1 - i) = \text{Ln}(2) = \log_e 2$; $\quad \text{Ln}(1 + i) = \log_e \sqrt{2} + \frac{\pi}{4} i$; $\quad \text{Ln}(1 - i) = \log_e \sqrt{2} - \frac{\pi}{4} i$. Therefore

$$\text{Ln}(1 + i) + \text{Ln}(1 - i) = 2\log_e \sqrt{2} = \log_e 2 = \text{Ln}(1 + i)(1 - i).$$

18 Integration in the Complex Plane

3. $\displaystyle\int_C z^2\,dz = (3+2i)^3 \int_{-2}^{2} t^2\,dt = \frac{16}{3}(3+2i)^3 = -48 + \frac{736}{3}\,i$

6. $\displaystyle\int_C |z|^2\,dz = \int_1^2 \left(2t^5 + \frac{2}{t}\right)dt - i\int_1^2 \left(t^2 + \frac{1}{t^4}\right)dt = 21 + \ln 4 - \frac{21}{8}\,i$

9. Using $y = -x + 1$, $0 \le x \le 1$, $z = x + (-x+1)i$, $dz = (1-i)\,dx$,

$$\int_C (x^2 + iy^3)\,dz = (1-i)\int_1^0 [x^2 + (1-x)^3 i]\,dx = -\frac{7}{12} + \frac{1}{12}\,i.$$

12. $\displaystyle\int_C \sin z\,dz = \int_{C_1} \sin z\,dz + \int_{C_2} \sin z\,dz$ where C_1 and C_2 are the line segments $y = 0$, $0 \le x \le 1$, and $x = 1$, $0 \le y \le 1$, respectively. Now

$$\int_{C_1} \sin z\,dz = \int_0^1 \sin x\,dx = 1 - \cos 1$$

$$\int_{C_2} \sin z\,dz = i\int_0^1 \sin(1 + iy)\,dy = \cos 1 - \cos(1 + i).$$

Thus

$$\int_C \sin z\,dz = (1 - \cos 1) + (\cos 1 - \cos(1+i)) = 1 - \cos(1+i) = (1 - \cos 1 \cosh 1) + i\sin 1 \sinh 1 = 0.1663 + 0.9889i.$$

15. We have

$$\oint_C ze^z\,dz = \int_{C_1} ze^z\,dz + \int_{C_2} ze^z\,dz + \int_{C_3} ze^z\,dz + \int_{C_4} ze^z\,dz.$$

On C_1, $y = 0$, $0 \le x \le 1$, $z = x$, $dz = dx$,

$$\int_{C_1} ze^z\,dz = \int_0^1 xe^x\,dx = xe^x - e^x \Big|_0^1 = 1.$$

On C_2, $x = 1$, $0 \le y \le 1$, $z = 1 + iy$, $dz = i\,dy$,

$$\int_{C_2} ze^z\,dz = i\int_0^1 (1 + iy)e^{1+iy}\,dy = ie^{i+1}.$$

On C_3, $y = 1$, $0 \le x \le 1$, $z = x + i$, $dz = dx$,

$$\int_{C_3} ze^z\,dz = \int_1^0 (x + i)e^{x+i}\,dx = (i-1)e^i - ie^{1+i}.$$

On C_4, $x = 0$, $0 \le y \le 1$, $z = iy$, $dz = i\,dy$,

$$\int_{C_4} ze^z\,dz = -\int_1^0 ye^{iy}\,dy = (1 - i)e^i - 1.$$

Thus

$$\oint_C ze^z\,dz = 1 + ie^{i+1} + (i-1)e^i - ie^{1+i} + (1-i)e^i - 1 = 0.$$

18. We have

$$\oint_C (2z-1)\,dz = \int_{C_1} (2z-1)\,dz + \int_{C_2} (2z-1)\,dz + \int_{C_3} (2z-1)\,dz.$$

On C_1, $y=0$, $0 \le x \le 1$, $z=x$, $dz=dx$,

$$\int_{C_1} (2z-1)\,dz = \int_0^1 (2x-1)\,dx = 0.$$

On C_2, $x=1$, $0 \le y \le 1$, $z=1+iy$, $dz = i\,dy$,

$$\int_{C_2} (2z-1)\,dz = -2\int_0^1 y\,dy + i\int_0^1 dy = -1+i.$$

On C_3, $y=x$, $z=x+ix$, $dz=(1+i)\,dx$,

$$\int_{C_3} (2z-1)\,dz = (1+i)\int_1^0 (2x-1+2ix)\,dx = 1-i.$$

Thus

$$\oint_C (2z-1)\,dz = 0-1+i+1-i = 0.$$

21. On C, $y=-x+1$, $0 \le x \le 1$, $z=x+(-x+1)i$, $dz=(1-i)\,dx$,

$$\int_C (z^2-z+2)\,dz = (1-i)\int_0^1 [x^2-(1-x)^2-x+2+(3x-2x^2-1)i]\,dx = \frac{4}{3} - \frac{5}{3}i.$$

24. On C, $x=\sin t$, $y=\cos t$, $0 \le t \le \pi/2$ or $z=ie^{-it}$, $dz=e^{-it}\,dt$,

$$\int_C (z^2-z+2)\,dz = \int_0^{\pi/2} (-e^{-2it} - ie^{-it} + 2)e^{-it}\,dt = \int_0^{\pi/2} (-e^{-3it} - ie^{-2it} + 2e^{-it})\,dt$$

$$= -\frac{1}{3}ie^{-3\pi i/2} + \frac{1}{2}e^{-\pi i} + 2ie^{-\pi i/2} + \frac{1}{3}i - \frac{1}{2} - 2i = \frac{4}{3} - \frac{5}{3}i.$$

27. The length of the line segment from $z=0$ to $z=1+i$ is $\sqrt{2}$. In addition, on this line segment

$$|z^2+4| \le |z|^2+4 \le |1+i|^2+4 = 6.$$

Thus $\left| \int_C (z^2+4)\,dz \right| \le 6\sqrt{2}$.

30. With $z_k^* = z_k$,

$$\int_C z\,dz = \lim_{\|P\|\to 0} \sum_{k=1}^n z_k(z_k - z_{k-1})$$

$$= \lim_{\|P\|\to 0} [(z_1^2 - z_1 z_0) + (z_2^2 - z_2 z_1) + \cdots + (z_n^2 - z_n z_{n-1})]. \qquad (1)$$

With $z_k^* = z_{k-1}$,

$$\int_C z\,dz = \lim_{\|P\|\to 0} \sum_{k=1}^n z_{k-1}(z_k - z_{k-1})$$

$$= \lim_{\|P\|\to 0} [(z_0 z_1 - z_0^2) + (z_1 z_2 - z_1^2) + \cdots + (z_{n-1} z_n - z_{n-1}^2)]. \qquad (2)$$

Adding (1) and (2) gives

$$2\int_C z\,dz = \lim_{\|P\|\to 0} (z_n^2 - z_0^2) \quad \text{or} \quad \int_C z\,dz = \frac{1}{2}(z_n^2 - z_0^2).$$

33. For $f(z) = 2z$, $\overline{f(z)} = 2\bar{z}$, so on $z = e^{it}$, $\bar{z} = e^{-it}$, $dz = ie^{it}\,dt$, and

$$\oint_C \overline{f(z)}\,dz = \int_0^{2\pi} (e^{-it})(ie^{it}\,dt) = 2i\int_0^{2\pi} dt = 4\pi i.$$

Thus circulation $= \mathrm{Re}\left(\oint_C \overline{f(z)}\,dz\right) = 0$, net flux $= \mathrm{Im}\left(\oint_C \overline{f(z)}\,dz\right) = 4\pi$.

Exercises 18.2

3. $f(z) = \dfrac{z}{2z+3}$ is discontinuous at $z = -3/2$ but is analytic within and on the circle $|z| = 1$.

6. $f(z) = \dfrac{e^z}{2z^2 + 11z + 15}$ is discontinuous at $z = -5/2$ and at $z = -3$ but is analytic within and on the circle $|z| = 1$.

9. By the principle of deformation of contours we can choose the more convenient circular contour C_1 defined by $|z| = 1$. Thus

$$\oint_C \frac{1}{z}\,dz = \oint_{C_1} \frac{1}{z}\,dz = 2\pi i$$

by (4) of Section 18.2.

12. By Theorem 18.4 and (4) of Section 18.2,

$$\oint_C \left(z + \frac{1}{z^2}\right) dz = \oint_C \frac{1}{z}\,dz + \oint_C \frac{1}{z^2}\,dz = 0 + 0 = 0.$$

15. By partial fractions, $\displaystyle\oint_C \frac{2z+1}{z(z+1)}\,dz = \oint_C \frac{1}{z}\,dz + \oint_C \frac{1}{z+1}\,dz.$

(a) By Theorem 18.4 and (4) of Section 18.2,

$$\oint_C \frac{1}{z}\,dz + \oint_C \frac{1}{z+1}\,dz = 2\pi i + 0 = 2\pi i.$$

(b) By writing $\displaystyle\oint_C = \oint_{C_1} + \oint_{C_2}$ where C_1 and C_2 are the circles $|z| = 1/2$ and $|z+1| = 1/2$, respectively, we have by Theorem 18.4 and (4) of Section 18.2,

$$\oint_C \frac{1}{z}\,dz + \oint_C \frac{1}{z+1}\,dz = \oint_{C_1} \frac{1}{z}\,dz + \oint_{C_1} \frac{1}{z+1}\,dz + \oint_{C_2} \frac{1}{z}\,dz + \oint_{C_2} \frac{1}{z+1}\,dz$$

$$= 2\pi i + 0 + 0 + 2\pi i = 4\pi i.$$

(c) Since $f(z) = \dfrac{2z+1}{z(z+1)}$ is analytic within and on C it follows from Theorem 18.4 that

$$\oint_C \frac{2z+1}{z^2+z}\,dz = 0.$$

18. (a) By writing $\displaystyle\oint_C = \oint_{C_1} + \oint_{C_2}$ where C_1 and C_2 are the circles $|z+2| = 1$ and $|z-2i| = 1$, respectively, we have by Theorem 18.4 and (4) of Section 18.2,

$$\oint_C \left(\frac{3}{z+2} - \frac{1}{z-2i}\right) dz = \oint_{C_1} \frac{3}{z+2}\,dz - \oint_{C_1} \frac{1}{z-2i}\,dz + \oint_{C_2} \frac{3}{z+2}\,dz - \oint_{C_2} \frac{1}{z-2i}\,dz$$

$$= 3(2\pi i) - 0 + 0 - 2\pi i = 4\pi i.$$

(b) Since $|z - 2i| = 1$ does not contain -2 and does contain $2i$, the integral is $0 - 2\pi i$.

21. We have

$$\oint_C \frac{8z - 3}{z^2 - z} \, dz = \oint_{C_1} \frac{8z - 3}{z^2 - z} \, dz - \oint_{C_2} \frac{8z - 3}{z^2 - z} \, dz$$

where C_1 and C_2 are the closed portions of the curve C enclosing $z = 0$ and $z = 1$, respectively. By partial fractions, Theorem 18.4, and (4) of Section 18.2,

$$\oint_{C_1} \frac{8z - 3}{z^2 - z} \, dz = 5 \oint_{C_1} \frac{1}{z - 1} \, dz + 3 \oint_{C_1} \frac{1}{z} \, dz = 5(0) + 3(2\pi i) = 6\pi i$$

$$\oint_{C_1} \frac{8z - 3}{z^2 - z} \, dz = 5 \oint_{C_2} \frac{1}{z - 1} \, dz + 3 \oint_{C_2} \frac{1}{z} \, dz = 5(2\pi i) + 3(0) = 10\pi i.$$

Thus

$$\oint_C \frac{8z - 3}{z^2 - z} \, dz = 6\pi i - 10\pi i = -4\pi i.$$

24. Write

$$\oint_C (z^2 + z + \mathrm{Re}(z)) \, dz = \oint_C (z^2 + z) \, dz + \oint_C \mathrm{Re}(z) \, dz.$$

By Theorem 18.4, $\oint_C (z^2 + z) \, dz = 0$. However, since $\mathrm{Re}(z) = x$ is not analytic,

$$\oint_C x \, dz = \oint_{C_1} x \, dz + \oint_{C_2} x \, dz + \oint_{C_3} x \, dz$$

where C_1 is $y = 0$, $0 \le x \le 1$, C_2 is $x = 1$, $0 \le y \le 2$, and C_3 is $y = 2x$, $0 \le x \le 1$. Thus,

$$\oint_C x \, dz = \int_0^1 x \, dx + i \int_0^2 dy + (1 + 2i) \int_1^0 x \, dx = \frac{1}{2} + 2i - \frac{1}{2}(1 + 2i) = i.$$

Exercises 18.3

3. The given integral is independent of the path. Thus

$$\int_C 2z \, dz = \int_{-2+7i}^{2-i} 2z \, dz = z^2 \Big|_{-2+7i}^{2-i} = 48 + 24i.$$

6. $\displaystyle \int_{-2i}^1 (3z^2 - 4z + 5i) \, dz = z^3 - 2z^2 + 5iz \Big|_{-2i}^1 = -19 - 3i$

9. $\displaystyle \int_{-i/2}^{1-i} (2z + 1)^2 \, dz = \frac{1}{6}(2z + 1)^3 \Big|_{-i/2}^{1-i} = -\frac{7}{6} - \frac{22}{3}i$

12. $\displaystyle \int_{1-i}^{1+2i} ze^{z^2} \, dz = \frac{1}{2}e^{z^2} \Big|_{1-i}^{1+2i} = \frac{1}{2}[e^{-3+4i} - e^{-2i}] = \frac{1}{2}(e^{-3}\cos 4 - \cos 2) + \frac{1}{2}(e^{-3}\sin 4 + \sin 2)i = 0.1918 + 0.4358i$

15. $\displaystyle \int_{\pi i}^{2\pi i} \cosh z \, dz = \sinh z \Big|_{\pi i}^{2\pi i} = \sinh 2\pi i - \sinh \pi i = i \sin 2\pi - i \sin \pi = 0$

18. $\displaystyle \int_{1+i}^{4+4i} \frac{1}{z} \, dz = \mathrm{Ln}\, z \Big|_{1+i}^{4+4i} = \mathrm{Ln}(4 + 4i) - \mathrm{Ln}(1 + i) = \log_e 4\sqrt{2} + \frac{\pi}{4}i - \left(\log_e \sqrt{2} + \frac{\pi}{4}i\right) = \log_e 4 = 1.3863$

21. Integration by parts gives

$$\int e^z \cos z \, dz = \frac{1}{2}e^z(\cos z + \sin z) + C$$

and so

$$\int_\pi^i e^z \cos z\, dz = \frac{1}{2} e^z (\cos z + \sin z) \Big|_\pi^i = \frac{1}{2}[e^i(\cos i + \sin i) - e^\pi(\cos \pi + \sin \pi)]$$

$$= \frac{1}{2}[(\cos 1 \cosh 1 - \sin 1 \sinh 1 + e^\pi) + i(\cos 1 \sinh 1 + \sin 1 \cosh 1) = 11.4928 + 0.9667i.$$

24. Integration by parts gives

$$\int z^2 e^z\, dz = z^2 e^z - 2ze^z + 2e^z + C$$

and so

$$\int_0^{\pi i} z^2 e^z\, dz = e^z(z^2 - 2z + 2) \Big|_0^{\pi i} = e^{\pi i}(-\pi^2 - 2\pi i + 2) - 2 = \pi^2 - 4 + 2\pi i.$$

--- **Exercises 18.4** ---

3. By Theorem 18.9 with $f(z) = e^z$,

$$\oint_C \frac{e^z}{z - \pi i}\, dz = 2\pi i e^{\pi i} = -2\pi i.$$

6. By Theorem 18.9 with $f(z) = \frac{1}{3} \cos z$,

$$\oint_C \frac{\frac{1}{3} \cos z}{z - \frac{\pi}{3}}\, dz = 2\pi i \left(\frac{1}{3} \cos \frac{\pi}{3}\right) = \frac{\pi}{3} i.$$

9. By Theorem 18.9 with $f(z) = \frac{z^2 + 4}{z - i}$,

$$\oint_C \frac{\frac{z^2 + 4}{z - i}}{z - 4i}\, dz = 2\pi i \left(-\frac{12}{3i}\right) = -8\pi.$$

12. By Theorem 18.10 with $f(z) = z$, $f'(z) = 1$, $f''(z) = 0$, and $f'''(z) = 0$,

$$\oint_C \frac{z}{(z - (-i))^4}\, dz = \frac{2\pi i}{3!}(0) = 0.$$

15. **(a)** By Theorem 18.9 with $f(z) = \frac{2z + 5}{z - 2}$,

$$\oint_C \frac{\frac{2z + 5}{z - 2}}{z}\, dz = 2\pi i \left(-\frac{5}{2}\right) = -5\pi i.$$

(b) Since the circle $|z - (-1)| = 2$ encloses only $z = 0$, the value of the integral is the same as in part **(a)**.

(c) From Theorem 18.9 with $f(z) = \frac{2z + 5}{z}$,

$$\oint_C \frac{\frac{2z + 5}{z}}{z - 2}\, dz = 2\pi i \left(\frac{9}{2}\right) = 9\pi i.$$

(d) Since the circle $|z - (-2i)| = 1$ encloses neither $z = 0$ nor $z = 2$ it follows from the Cauchy-Goursat Theorem, Theorem 18.4, that

$$\oint_C \frac{2z + 5}{z(z - 2)}\, dz = 0.$$

18. (a) By Theorem 18.10 with $f(z) = \dfrac{1}{z - 4}$, $f'(z) = -\dfrac{1}{(z - 4)^2}$, and $f''(z) = \dfrac{2}{(z - 4)^3}$,

$$\oint_C \frac{\dfrac{1}{z - 4}}{z^3}\, dz = \frac{2\pi i}{2!}\left(\frac{2}{-64}\right) = -\frac{\pi}{32}\, i.$$

(b) By the Cauchy-Goursat Theorem, Theorem 18.4,

$$\oint_C \frac{1}{z^3(z - 4)}\, dz = 0.$$

21. We have

$$\oint_C \frac{1}{z^3(z - 1)^2}\, dz = \oint_{C_1} \frac{\dfrac{1}{(z - 1)^2}}{z^3}\, dz + \oint_{C_2} \frac{\dfrac{1}{z^3}}{(z - 1)^2}\, dz,$$

where C_1 and C_2 are the circles $|z| = 1/3$ and $|z - 1| = 1/3$, respectively. By Theorem 18.10,

$$\oint_{C_1} \frac{\dfrac{1}{(z - 1)^2}}{z^3}\, dz = \frac{2\pi i}{2!}(6) = 6\pi i, \qquad \oint_{C_2} \frac{\dfrac{1}{z^3}}{(z - 1)^2}\, dz = \frac{2\pi i}{1!}(-3) = -6\pi i.$$

Thus

$$\oint_C \frac{1}{z^3(z - 1)^2}\, dz = 6\pi i - 6\pi i = 0.$$

24. We have

$$\oint_C \frac{e^{iz}}{(z^2 + 1)^2}\, dz = \oint_{C_1} \frac{\dfrac{e^{iz}}{(z + i)^2}}{(z - i)^2}\, dz - \oint_{C_2} \frac{\dfrac{e^{iz}}{(z - i)^2}}{(z - (-i))^2}\, dz,$$

where C_1 and C_2 are the closed portions of the curve C enclosing $z = i$ and $z = -i$, respectively. By Theorem 18.10,

$$\oint_{C_1} \frac{\dfrac{e^{iz}}{(z + i)^2}}{(z - i)^2}\, dz = \frac{2\pi i}{1!}\left(\frac{-4e^{-1}}{-8i}\right) = \pi e^{-1}, \qquad \oint_{C_2} \frac{\dfrac{e^{iz}}{(z - i)^2}}{(z - (-i))^2}\, dz = \frac{2\pi i}{1!}\left(\frac{0}{8i}\right) = 0.$$

Thus

$$\oint_C \frac{e^{iz}}{(z^2 + 1)^2}\, dz = \pi e^{-1}.$$

Chapter 18 Review Exercises

3. True

6. $\pi(-16 + 8i)$

9. True (Use partial fractions and write the given integral as two integrals.)

12. 12π

15. $\displaystyle\int_C |z^2|\, dz = \int_0^2 (t^4 + t^2)\, dt + 2i \int_0^2 (t^5 + t^3)\, dt = \frac{136}{15} + \frac{88}{3}\, i$

18. $\displaystyle\int_{3i}^{1-i} (4z - 6)\, dz = 2z^2 - 6z \Big|_{3i}^{1-i} = 12 + 20i$

21. On $|z| = 1$, let $z = e^{it}$, $dz = ie^{it}\, dt$, so that

$$\oint_C (z^{-2} + z^{-1} + z + z^2)\, dz = i \int_0^{2\pi} (e^{-2it} + e^{-it} + e^{it} + e^{2it}) e^{it}\, dt = -e^{-it} + it + \frac{1}{2}e^{2it} + \frac{1}{3}e^{3it} \Big|_0^{2\pi} = 2\pi i.$$

24. By Theorem 18.10 with $f(z) = \dfrac{\cos z}{z-1}$ and $f'(z) = \dfrac{\sin z - \cos z - z\sin z}{(z-1)^2}$,

$$\oint_C \frac{\dfrac{\cos z}{z-1}}{z^2}\, dz = \frac{2\pi i}{1!}\left(\frac{-1}{1}\right) = -2\pi i.$$

27. Using the principle of deformation of contours we choose C to be the more convenient circular contour $|z+i| = \frac{1}{4}$. On this circle $z = -i + \frac{1}{4}e^{it}$ and $dz = \frac{1}{4}ie^{it}\, dt$. Thus

$$\oint_C \frac{z}{z+i}\, dz = i \int_0^{2\pi} \left(\frac{1}{4}e^{it} - i\right) dt = 2\pi.$$

30. We have

$$\left| \int_C \mathrm{Ln}(z+1)\, dz \right| \le |\text{max of } \mathrm{Ln}(z+1) \text{ on } C| \cdot 2,$$

where 2 is the length of the line segment. Now

$$|\mathrm{Ln}(z+1)| \le |\log_e(z+1)| + |\mathrm{Arg}(z+1)|.$$

But $\max \mathrm{Arg}(z+1) = \pi/4$ when $z = i$ and $\max|z+1| = \sqrt{10}$ when $z = 2+i$. Thus,

$$\left| \int_C \mathrm{Ln}(z+1)\, dz \right| \le \left(\frac{1}{2}\log_e 10 + \frac{\pi}{4}\right) 2 = \log_e 10 + \frac{\pi}{2}.$$

19 Series and Residues

3. $0, 2, 0, 2, 0$

6. Converges. To see this write the general term as $\left(\dfrac{2}{5}\right)^n \dfrac{1+n2^{-n}i}{1+3n5^{-n}i}$.

9. Diverges. To see this write the general term as $\sqrt{n}\left(1+\dfrac{1}{\sqrt{n}}i^n\right)$.

12. Write $z_n = \left(\dfrac{1}{4}+\dfrac{1}{4}i\right)^n$ in polar form as $z_n = \left(\dfrac{\sqrt{2}}{4}\right)^n \cos n\theta + i\left(\dfrac{\sqrt{2}}{4}\right)^n \sin n\theta$. Now

$$\operatorname{Re}(z_n) = \left(\dfrac{\sqrt{2}}{4}\right)^n \cos n\theta \to 0 \text{ as } n \to \infty$$

and

$$\operatorname{Im}(z_n) = \left(\dfrac{\sqrt{2}}{4}\right)^n \sin n\theta \to 0 \text{ as } n \to \infty$$

since $\sqrt{2}/4 < 1$.

15. We identify $a = 1$ and $z = 1 - i$. Since $|z| = \sqrt{2} > 1$ the series is divergent.

18. We identify $a = 1/2$ and $z = i$. Since $|z| - 1$ the series is divergent.

21. From

$$\lim_{n\to\infty}\left|\dfrac{\dfrac{1}{(1-2i)^{n+2}}}{\dfrac{1}{(1-2i)^{n+1}}}\right| = \dfrac{1}{|1-2i|} = \dfrac{1}{\sqrt{5}}$$

we see that the radius of convergence is $R = \sqrt{5}$. The circle of convergence is $|z - 2i| = \sqrt{5}$.

24. From

$$\lim_{n\to\infty}\left|\dfrac{\dfrac{1}{(n+1)^2(3+4i)^{n+1}}}{\dfrac{1}{n^2(3+4i)^n}}\right| = \lim_{n\to\infty}\left(\dfrac{n}{n+1}\right)^2 \dfrac{1}{|3+4i|} = \dfrac{1}{5}$$

we see that the radius of convergence is $R = 5$. The circle of convergence is $|z + 3i| = 5$.

27. From

$$\lim_{n\to\infty}\sqrt[n]{\left|\dfrac{1}{5^{2n}}\right|} = \lim_{n\to\infty}\dfrac{1}{25} = \dfrac{1}{25}$$

we see that the radius of convergence is $R = 25$. The circle of convergence is $|z - 4 - 3i| = 25$.

30. (a) The circle of convergence is $|z| = 1$. Since the series of absolute values

$$\sum_{k=1}^{\infty}\left|\dfrac{z^k}{k^2}\right| = \sum_{k=1}^{\infty}\dfrac{|z|^k}{k^2} = \sum_{k=1}^{\infty}\dfrac{1}{k^2}$$

converges, the given series is absolutely convergent for every z on $|z| = 1$. Since absolute convergence implies convergence, the given series converges for all z on $|z| = 1$.

(b) The circle of convergence is $|z| = 1$. On the circle, $n|z|^n \to \infty$ as $n \to \infty$. This implies $nz^n \not\to 0$ as $n \to \infty$. Thus by Theorem 19.3 the series is divergent for every z on the circle $|z| = 1$.

Exercises 19.2

3. Differentiating $\dfrac{1}{1+2z} = 1 - 2z + 2^2 z^2 - 2^3 z^3 + \cdots$ gives $\dfrac{-2}{(1+2z)^2} = -2 + 2 \cdot 2^2 z - 3 \cdot 2^3 z^2 + \cdots$. Thus

$$\frac{1}{(1+2z)} = 1 - 2 \cdot (2z) + 3 \cdot (2z)^2 - \cdots = \sum_{k=1}^{\infty} (-1)^{k-1} k (2z)^{k-1}. \quad R = 1/2$$

6. Replacing z in $e^z = \displaystyle\sum_{k=0}^{\infty} \frac{z^k}{k!}$ by $-z^2$ and multiplying the result by z gives $ze^{-z^2} = \displaystyle\sum_{k=0}^{\infty} \frac{(-1)^k}{k!} z^{2k+1}$. $\quad R = \infty$

9. Replacing z in $\cos z = \displaystyle\sum_{k=0}^{\infty} (-1)^k \frac{z^{2k}}{(2k)!}$ by $z/2$ gives $\cos \dfrac{z}{2} = \displaystyle\sum_{k=0}^{\infty} \frac{(-1)^k}{(2k)!} \left(\frac{z}{2}\right)^{2k}$. $\quad R = \infty$

12. Using the identity $\cos z = \dfrac{1}{2}(1 + \cos 2z)$ and the series $\cos z = \displaystyle\sum_{k=0}^{\infty} (-1)^k \frac{z^{2k}}{(2k)!}$ gives

$$\cos^2 z = \frac{1}{2} + \frac{1}{2} \sum_{k=0}^{\infty} (-1)^k \frac{(2z)^{2k}}{(2k)!} = 1 + \sum_{k=1}^{\infty} (-1)^k \frac{2^{2k-1}}{(2k)!} z^{2k}. \quad R = \infty$$

15. Using (5) of Section 19.1,

$$\frac{1}{3-z} = \frac{1}{3 - 2i - (z - 2i)} = \frac{1}{3 - 2i} \cdot \frac{1}{1 - \dfrac{z - 2i}{3 - 2i}}$$

$$= \frac{1}{3 - 2i} \left[1 + \frac{z - 2i}{3 - 2i} + \frac{(z - 2i)^2}{(3 - 2i)^2} + \frac{(z - 2i)^3}{(3 - 2i)^3} + \cdots \right]$$

$$= \frac{1}{3 - 2i} + \frac{z - 2i}{(3 - 2i)^2} + \frac{(z - 2i)^2}{(3 - 2i)^3} + \frac{(z - 2i)^3}{(3 - 2i)^4} + \cdots = \sum_{k=0}^{\infty} \frac{(z - 2i)^k}{(3 - 2i)^{k+1}}. \quad R = \sqrt{13}$$

18. Using (5) of Section 19.1,

$$\frac{1+z}{1-z} = -1 + \frac{2}{1-z} = -1 + \frac{2}{1 - i - (z - i)} = -1 + \frac{2}{1 - i} \cdot \frac{1}{1 - \dfrac{z - i}{1 - i}}$$

$$= -1 + \frac{2}{1 - i} \left[1 + \frac{z - i}{1 - i} + \frac{(z - i)^2}{(1 - i)^2} + \frac{(z - i)^3}{(1 - i)^3} + \cdots \right]$$

$$= -1 + \frac{2}{1 - i} + \frac{2(z - i)}{(1 - i)^2} + \frac{2(z - i)^2}{(1 - i)^3} + \frac{2(z - i)^3}{(1 - i)^4} + \cdots = -1 + \sum_{k=0}^{\infty} \frac{2(z - i)^k}{(1 - i)^{k+1}}. \quad R = \sqrt{2}$$

21. Using $e^z = e^{3i} \cdot e^{z-3i}$ and (12) of Section 19.2, $e^z = e^{3i} \displaystyle\sum_{k=0}^{\infty} \frac{(z - 3i)^k}{k!}$. $\quad R = \infty$

24. Using (8) of Section 19.2, $e^{1/(1+z)} = e - ez + \dfrac{3e}{2} z^2 - \cdots$.

27. The distance from $2 + 5i$ to i is $|2 + 5i - i| = |2 + 4i| = 2\sqrt{5}$.

30. The series are

$$f(z) = \sum_{k=0}^{\infty} (-1)^k \frac{(z-3)^k}{3^{k+1}}, \quad R = 3$$

and

$$f(z) = \sum_{k=0}^{\infty} (-1)^k \frac{(z-1-i)^k}{(1+i)^{k+1}}, \quad R = \sqrt{2}.$$

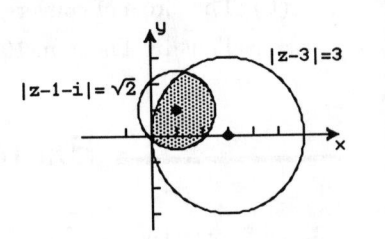

33. From $e^z \approx 1 + z + \dfrac{z^2}{2}$ we obtain

$$e^{(1+i)/10} \approx 1 + \frac{1+i}{10} + \frac{(1+i)^2}{100} = 1.1 + 0.12i.$$

36. $e^{iz} = \displaystyle\sum_{k=0}^{\infty} \frac{(iz)^k}{k!} = 1 + i\,\frac{z}{1!} - \frac{z^2}{2!} - i\,\frac{z^3}{3!} + \frac{z^4}{4!} + i\,\frac{z^5}{5!} - \frac{z^6}{6!} - i\,\frac{z^7}{7!} + \cdots$

$$= \left(1 - \frac{z^2}{2!} + \frac{z^4}{4!} - \frac{z^6}{6!} + \cdots\right) + i\left(\frac{z}{1!} - \frac{z^3}{3!} + \frac{z^5}{5!} - \frac{z^7}{7!} + \cdots\right) = \cos z + i \sin z$$

Exercises 19.3

3. $f(z) = 1 - \dfrac{1}{1!z^2} + \dfrac{1}{2!z^4} - \dfrac{1}{3!z^6} + \cdots$

6. $f(z) = z\left(1 - \dfrac{1}{2!z^2} + \dfrac{1}{4!z^4} - \dfrac{1}{6!z^6} + \cdots\right) = z - \dfrac{1}{2!z} + \dfrac{1}{4!z^3} - \dfrac{1}{6!z^5} + \cdots$

9. $f(z) = \dfrac{1}{z-3} \cdot \dfrac{1}{3+z-3} = \dfrac{1}{3(z-3)} \cdot \dfrac{1}{1 + \dfrac{z-3}{3}} = \dfrac{1}{3(z-3)}\left[1 - \dfrac{z-3}{3} + \dfrac{(z-3)^2}{3^2} - \dfrac{(z-3)^3}{3^3} + \cdots\right]$

$$= \frac{1}{3(z-3)} - \frac{1}{3^2} + \frac{z-3}{3^3} - \frac{(z-3)^2}{3^4} + \cdots$$

12. $f(z) = \dfrac{1}{3}\left[\dfrac{1}{z-3} - \dfrac{1}{z}\right] = \dfrac{1}{3}\left[\dfrac{1}{-4+z+1} - \dfrac{1}{z+1-1}\right] = \dfrac{1}{3}\left[-\dfrac{1}{4} \cdot \dfrac{1}{1 - \dfrac{z+1}{4}} - \dfrac{1}{z+1} \cdot \dfrac{1}{1 - \dfrac{1}{z+1}}\right]$

$$= \frac{1}{3}\left[-\frac{1}{4}\left(1 + \frac{z+1}{4} + \frac{(z+1)^2}{4^2} + \frac{(z+1)^3}{4^3} + \cdots\right) - \frac{1}{z+1}\left(1 + \frac{1}{z+1} + \frac{1}{(z+1)^2} + \frac{1}{(z+1)^3} + \cdots\right)\right]$$

$$= \cdots - \frac{1}{(z+1)^2} - \frac{1}{z+1} - \frac{1}{12} - \frac{z+1}{3 \cdot 4^2} - \frac{(z+1)^2}{3 \cdot 4^3} - \cdots$$

15. $f(z) = \dfrac{1}{z-1} \cdot \dfrac{-1}{1-(z-1)} = \dfrac{-1}{z-1}\left[1 + (z-1) + (z-1)^2 + (z-1)^3 + \cdots\right] = -\dfrac{1}{z-1} - 1 - (z-1) - (z-1)^2 - \cdots$

18. $f(z) = \dfrac{1}{3(z+1)} + \dfrac{2}{3} \cdot \dfrac{1}{(z+1)-3} = \dfrac{1}{3(z+1)} + \dfrac{2}{3(z+1)} \cdot \dfrac{1}{1 - \dfrac{3}{z+1}}$

$$= \frac{1}{3(z+1)} + \frac{2}{3(z+1)}\left(1 + \frac{3}{z+1} + \frac{3^2}{(z+1)^2} + \frac{3^3}{(z+1)^3} + \cdots\right)$$

$$= \frac{1}{z+1} + \frac{2}{(z+1)^2} + \frac{2 \cdot 3}{(z+1)^3} + \frac{2 \cdot 3^2}{(z+1)^4} + \cdots$$

21. $f(z) = \frac{1}{z}(1-z)^{-2} = \frac{1}{z}\left(1 + (-2)(-z) + \frac{(-2)(-3)}{z!}(-z)^2 + \frac{(-2)(-3)(-4)}{3!}(-z)^3 + \cdots\right) = \frac{1}{z} + 2 + 3z + 4z^2 + \cdots$

24. $f(z) = \frac{1}{(z-3)^3} \cdot \frac{-1}{1 - (z-1)} = \frac{-1}{(z-1)^3}[1 + (z-1) + (z-1)^2 + (z-1)^3 + \cdots]$

$$= -\frac{1}{(z-1)^3} - \frac{1}{(z-1)^2} - \frac{1}{z-1} - 1 - (z-1) - \cdots$$

27. $f(z) = z + \frac{2}{z-2} = 1 + (z-1) + \frac{2}{-1+z-1} = 1 + (z-1) + \frac{2}{z-1} \cdot \frac{1}{1 - \frac{1}{z-1}}$

$$= 1 + (z-1) + \frac{2}{z-1}\left(1 + \frac{1}{z-1} + \frac{1}{(z-1)^2} + \frac{1}{(z-1)^3} + \cdots\right) = \cdots + \frac{2}{(z-1)^2} + \frac{2}{z-1} + 1 + (z-1)$$

Exercises 19.4

3. Since $f(-2+i) = f'(-2+i) = 0$ and $f''(z) = 2$ for all z, $z = -2 + i$ is a zero of order two.

6. Write $f(z) = (z^2 + 9)/z = (z - 3i)(z + 3i)/z$ to see that $3i$ and $-3i$ are zeros of f. Now $f'(z) = 1 - 9/z^2$ and $f'(3i) = f'(-3i) = 2 \neq 0$. This indicates that each zero is of order one.

9. From

$$f(z) = z(1 - \cos z^2) = z\left(-\frac{z^4}{2!} + \frac{z^8}{4!} - \cdots\right) = z^5\left(-\frac{1}{2!} + \frac{z^4}{4!} - \cdots\right)$$

we see that $z = 0$ is a zero of order five.

12. From the series $e^z = -\sum_{k=0}^{\infty} \frac{(z - \pi i)^k}{k!}$ centered at πi and

$$f(z) = 1 - \pi i + z + e^z = 1 - \pi i + z + \left(-1 - \frac{z - \pi i}{1!} - \frac{(z - \pi i)^2}{2!} - \frac{(z - \pi i)^3}{3!} - \cdots\right)$$

$$= -\frac{(z - \pi i)^2}{2!} - \frac{(z - \pi i)^3}{3!} - \cdots = (z - \pi i)^2\left(-\frac{1}{2!} - \frac{z - \pi i}{3!} - \cdots\right)$$

we see that $z = \pi i$ is a zero of order two.

15. From $f(z) = \frac{1 + 4i}{(z + 2)(z + i)^4}$ and Theorem 19.11 we see that -2 is a simple pole and $-i$ is a pole of order four.

18. From $z^2 \sin \pi z = z^3\left(\pi - \frac{\pi^3 z^2}{3!} + \cdots\right)$ we see $z = 0$ is a zero of order three. From $f(z) = \frac{\cos \pi z}{z^2 \sin \pi z}$ and Theorem 19.11 we see 0 is a pole of order three. The numbers n, $n \pm 1$, ± 2, ... are simple poles.

21. From $1 - e^z = 1 - \left(1 + \frac{z}{1!} + \frac{z^2}{2!} + \cdots\right) = z\left(-1 - \frac{z}{2!} - \cdots\right)$ we see that $z = 0$ is a zero of order one. By periodicity of e^z it follows that $z = 2n\pi i$, $n = 0, \pm 1, \pm 2, \ldots$ are zeros of order one. From $f(z) = \frac{1}{1 - e^z}$ and Theorem 19.11 we see that the numbers $2n\pi i$, $n = 0, \pm 1, \pm 2, \ldots$ are simple poles.

24. From the Laurent series

$$f(z) = z^3\left[\frac{1}{z} - \frac{1}{3!}\left(\frac{1}{z}\right)^3 + \frac{1}{5!}\left(\frac{1}{z}\right)^5 - \frac{1}{7!}\left(\frac{1}{z}\right)^7 + \cdots\right] = z^2 - \frac{1}{3!} + \frac{1}{5!z^2} - \frac{1}{7!z^4} + \cdots, \quad 0 < |z|,$$

we see that the principal part contains an infinite number of nonzero terms. Hence $z = 0$ is an essential singularity.

Exercises 19.5

3. $f(z) = -\dfrac{3}{z} - \dfrac{1}{z-2} = -\dfrac{3}{z} + \dfrac{1}{2} \cdot \dfrac{1}{1 - \dfrac{z}{2}} = -\dfrac{3}{z} + \dfrac{1}{2}\left(1 + \dfrac{z}{2} + \dfrac{z^2}{2^2} + \dfrac{z^3}{2^3} + \cdots\right) = -\dfrac{3}{z} + \dfrac{1}{2} + \dfrac{z}{2^2} + \dfrac{z^2}{2^3} + \cdots$

$\operatorname{Res}(f(z), 0) = -3$

6. $f(z) = \dfrac{e^{-2}}{(z-2)^2} e^{-(z-2)} = \dfrac{e^{-2}}{(z-2)^2}\left(1 - \dfrac{z-2}{1!} + \dfrac{(z-2)^2}{2!} - \dfrac{(z-2)^3}{3!} + \cdots\right)$

$= \dfrac{e^{-2}}{(z-2)^2} - \dfrac{e^{-2}}{z-2} + \dfrac{e^{-2}}{2} - \dfrac{e^{-2}(z-2)}{3!} + \cdots$

$\operatorname{Res}(f(z), 2) = -e^{-2}$

9. $\operatorname{Res}(f(z), 1) = \lim\limits_{z \to 1}(z-1)\dfrac{1}{z^2(z+2)(z-1)} = \lim\limits_{z \to 1}\dfrac{1}{z^2(z+2)} = \dfrac{1}{3}$

$\operatorname{Res}(f(z), -2) = \lim\limits_{z \to -2}(z+2)\dfrac{1}{z^2(z+2)(z-1)} = \lim\limits_{z \to -2}\dfrac{1}{z^2(z-1)} = -\dfrac{1}{12}$

$\operatorname{Res}(F(z), 0) = \dfrac{1}{1!}\lim\limits_{z \to 0}\dfrac{d}{dz}\left[z^2 \cdot \dfrac{1}{z^2(z+2)(z-1)}\right] = \lim\limits_{z \to 0}\dfrac{-2z-1}{(z+2)^2(z-1)^2} = -\dfrac{1}{4}$

12. $\operatorname{Res}(f(z), -3) = \lim\limits_{z \to -3}(z+3)\cdot\dfrac{2z-1}{(z-1)^4(z+3)} = \lim\limits_{z \to -3}\dfrac{2z-1}{(z-1)^4} = -\dfrac{7}{256}$

$\operatorname{Res}(f(z), 1) = \dfrac{1}{3!}\lim\limits_{z \to 1}\dfrac{d^3}{dz^3}\left[(z-1)^4 \cdot \dfrac{2z-1}{(z-1)^4(z+3)}\right] = \dfrac{1}{6}\lim\limits_{z \to 1}\dfrac{-42}{(z+3)^4} = -\dfrac{7}{256}$

15. Using $\dfrac{d}{dz}\cos z = -\sin z$ and the result in (4),

$$\operatorname{Res}\left(f(z), (2n+1)\dfrac{\pi}{2}\right) = \dfrac{1}{-\sin z}\bigg|_{z=(2n+1)\frac{\pi}{2}} = \dfrac{1}{-\sin(2n+1)\frac{\pi}{2}} = (-1)^{n+1}.$$

18. (a) $\displaystyle\oint_C \dfrac{z+1}{z^2(z-2i)}\,dz = 2\pi i\,\operatorname{Res}(f(z), 0) = \pi\left(-1 + \dfrac{1}{2}i\right)$

(b) $\displaystyle\oint_C \dfrac{z+1}{z^2(z-2i)}\,dz = 2\pi i\,\operatorname{Res}(f(z), 2i) = \pi\left(1 - \dfrac{1}{2}i\right)$

(c) $\displaystyle\oint_C \dfrac{z+1}{z^2(z-2i)}\,dz = 2\pi i[\operatorname{Res}(f(z), 0) + \operatorname{Res}(f(z), 2i)] = 2\pi i\left[\dfrac{1}{4} + \dfrac{1}{2}i + \left(-\dfrac{1}{4} - \dfrac{1}{2}i\right)\right] = 0$

21. $\displaystyle\oint_C \dfrac{1}{z^2 + 4z + 13}\,dz = 2\pi i\,\operatorname{Res}(f(z), -2 + 3i) = \dfrac{\pi}{3}$

24. $\displaystyle\oint_C \dfrac{z}{(z+1)(z^2+1)}\,dz = 2\pi i[\operatorname{Res}(f(z), i) + \operatorname{Res}(f(z), -i)] = 2\pi i\left[\dfrac{1}{4} - \dfrac{1}{4}i + \dfrac{1}{4} + \dfrac{1}{4}i\right] = \pi i$

27. $\displaystyle\oint_C \dfrac{\tan z}{z}\,dz = 2\pi i\,\operatorname{Res}\left(f(z), \dfrac{\pi}{2}\right) = -4i$. Note: $z = 0$ is not a pole. See Example 1, Section 19.4.

30. $\displaystyle\oint_C \dfrac{2z-1}{z^2(z^3+1)}\,dz = 2\pi i\left[\operatorname{Res}(f(z), 0) + \operatorname{Res}(f(z), -1) + \operatorname{Res}\left(f(z), \dfrac{1}{2} + \dfrac{\sqrt{3}}{2}i\right)\right]$

$= 2\pi i\left[2 + (-1) + \left(-\dfrac{1}{2} - \dfrac{1}{6}\sqrt{3}i\right)\right] = \pi\left(\dfrac{\sqrt{3}}{3} + i\right)$

Exercises 19.6

3. $\displaystyle\int_0^{2\pi} \frac{\cos\theta}{3+\sin\theta}\,d\theta = \oint_C \frac{z^2+1}{z(z^2+6iz-1)}\,dz = 2\pi i[\mathrm{Res}(f(z),0) + \mathrm{Res}(f(z),-3+2\sqrt{2}\,i)] = 0$

6. $\displaystyle\int_0^{\pi} \frac{d\theta}{1+\sin^2\theta} = \frac{1}{2}\int_0^{2\pi} \frac{d\theta}{1+\sin^2\theta} = -\frac{2}{i}\oint_C \frac{z}{z^4-6z^2+1}\,dz$

$$= \left(-\frac{2}{i}\right) 2\pi i[\mathrm{Res}(f(z),\sqrt{3-2\sqrt{2}}) + \mathrm{Res}(f(z),-\sqrt{3-2\sqrt{2}})] = \frac{\pi}{\sqrt{2}}$$

9. We use $\cos 2\theta = (z^2+z^{-2})/2$.

$$\int_0^{2\pi} \frac{\cos 2\theta}{5-4\cos\theta}\,d\theta = \frac{i}{2}\oint_C \frac{z^4+1}{z^2(2z^2-5z+2)}\,dz = \left(\frac{i}{2}\right)2\pi i\left[\mathrm{Res}(f(z),0) + \mathrm{Res}\left(f(z),\frac{1}{2}\right)\right] = \frac{\pi}{6}$$

12. $\displaystyle\int_{-\infty}^{\infty} \frac{1}{x^2-2x+25}\,dx = 2\pi i\,\mathrm{Res}(f(z),1+2\sqrt{6}\,i) = \frac{\pi}{2\sqrt{6}}$

15. $\displaystyle\int_{-\infty}^{\infty} \frac{1}{(x^2+1)^3}\,dx = 2\pi i\,\mathrm{Res}(f(z),i) = \frac{3\pi}{8}$

18. $\displaystyle\int_{-\infty}^{\infty} \frac{dx}{(x^2+1)^2(x^2+9)} = 2\pi i[\mathrm{Res}(f(z),i) + \mathrm{Res}(f(z),3i)] = \frac{5\pi}{96}$

21. $\displaystyle\int_{-\infty}^{\infty} \frac{e^{ix}}{x^2+1}\,dx = 2\pi i\,\mathrm{Res}(f(z),i) = \pi e^{-1}$. Therefore, $\displaystyle\int_{-\infty}^{\infty} \frac{\cos x}{x^2+1}\,dx = \mathrm{Re}\left(\int_{-\infty}^{\infty} \frac{e^{ix}}{x^2+1}\,dx\right) = \pi e^{-1}$.

24. $\displaystyle\int_{-\infty}^{\infty} \frac{e^{ix}}{(x^2+4)^2}\,dx = 2\pi i\,\mathrm{Res}(f(z),2i) = \frac{3e^{-2}}{16}\pi$; $\displaystyle\int_{-\infty}^{\infty} \frac{\cos x}{(x^2+4)^2}\,dx = \mathrm{Re}\left(\int_{-\infty}^{\infty} \frac{e^{ix}}{(x^2+4)^2}\,dx\right) = \frac{3e^{-2}}{16}\pi$.

Therefore

$$\int_0^{\infty} \frac{\cos x}{(x^2+4)^2}\,dx = \frac{1}{2}\left(\frac{3e^{-2}}{16}\pi\right) = \frac{3e^{-2}}{32}\pi.$$

27. $\displaystyle\int_{-\infty}^{\infty} \frac{e^{2ix}}{x^4+1}\,dx = 2\pi i\left[\mathrm{Res}\left(f(z),\frac{1}{\sqrt{2}}+\frac{1}{\sqrt{2}}i\right) + \mathrm{Res}\left(f(z),-\frac{1}{\sqrt{2}}+\frac{1}{\sqrt{2}}i\right)\right]$

$$= 2\pi i\left[\left(-\frac{\sqrt{2}}{8}-\frac{\sqrt{2}}{8}i\right)e^{(-\sqrt{2}+\sqrt{2}\,i)} + \left(\frac{\sqrt{2}}{8}-\frac{\sqrt{2}}{8}i\right)e^{(-\sqrt{2}-\sqrt{2}\,i)}\right]$$

$$= \pi e^{-\sqrt{2}}\left[\frac{\sqrt{2}}{2}\cos\sqrt{2} + \frac{\sqrt{2}}{2}\sin\sqrt{2}\right]$$

$$\int_{-\infty}^{\infty} \frac{\cos 2x}{x^4+1}\,dx = \mathrm{Re}\left(\int_{-\infty}^{\infty} \frac{e^{2ix}}{x^4+1}\,dx\right) = \pi e^{-\sqrt{2}}\left[\frac{\sqrt{2}}{2}\cos\sqrt{2} + \frac{\sqrt{2}}{2}\sin\sqrt{2}\right]$$

Therefore

$$\int_0^{\infty} \frac{\cos 2x}{x^4+1}\,dx = \pi e^{-\sqrt{2}}\frac{\sqrt{2}}{4}(\cos\sqrt{2} + \sin\sqrt{2}).$$

30. $\displaystyle\int_{-\infty}^{\infty} \frac{xe^{ix}}{(x^2+1)(x^2+4)}\,dx = 2\pi i[\mathrm{Res}(f(z),i) + \mathrm{Res}(f(z),2i)] = 2\pi i\left[\frac{1}{6}e^{-1} - \frac{1}{6}e^{-2}\right] = \frac{\pi}{3}(e^{-1}-e^{-2})i$;

$$\int_{-\infty}^{\infty} \frac{x\sin x}{(x^2+1)(x^2+4)}\,dx = \mathrm{Im}\left(\int_{-\infty}^{\infty} \frac{xe^{ix}}{(x^2+1)(x^2+4)}\,dx\right) = \frac{\pi}{3}(e^{-1}-e^{-2}).$$

Therefore

$$\int_0^{\infty} \frac{x\sin x}{(x^2+1)(x^2+4)}\,dx = \frac{1}{2}\left[\frac{\pi}{3}(e^{-1}-e^{-2})\right] = \frac{\pi}{6}(e^{-1}-e^{-2}).$$

33. $\displaystyle\int_0^\pi \frac{d\theta}{(a+\cos\theta)^2} = \frac{1}{2}\int_0^{2\pi}\frac{d\theta}{(a+\cos\theta)^2} = \frac{2}{i}\oint_C \frac{z}{(z^2+2az+1)^2}\,dz$ $(C$ is $|z|=1) = \frac{2}{i}\oint_C \frac{z}{(z-r_1)^2(z-r_2)^2}\,dz$

where $r_1 = -a + \sqrt{a^2-1}$, $r_2 = -a - \sqrt{a^2-1}$. Now

$$\oint_C \frac{z}{(z-r_1)^2(z-r_2)^2}\,dz = 2\pi i\,\mathrm{Res}(f(z), r_1) = 2\pi i\frac{a}{4(\sqrt{a^2-1})^3} = \frac{a\pi}{2(\sqrt{a^2-1})^3}\,i.$$

Thus

$$\int_0^\pi \frac{d\theta}{(a+\cos\theta)^2} = \frac{2}{i}\cdot\frac{a\pi}{2(\sqrt{a^2-1})^3}\,i = \frac{a\pi}{(\sqrt{a^2-1})^3}.$$

When $a = 2$ we obtain

$$\int_0^\pi \frac{d\theta}{(2+\cos\theta)^2} = \frac{2\pi}{(\sqrt{3})^3} \quad\text{and so}\quad \int_0^{2\pi}\frac{d\theta}{(2+\cos\theta)^2} = \frac{4\pi}{3\sqrt{3}}.$$

36. Using the Fourier sine transform with respect to y the partial differential equation becomes $\dfrac{d^2U}{dx^2} - \alpha^2 U = 0$ and so

$$U(x,\alpha) = c_1\cosh\alpha x + c_2\sinh\alpha x.$$

The boundary condition $u(0,y)$ becomes $U(0,\alpha) = 0$ and so $c_1 = 0$. Thus $U(x,\alpha) = c_2\sinh\alpha x$. Now to evaluate

$$U(\pi,\alpha) = \int_0^\infty \frac{2y}{y^4+4}\sin\alpha y\,dy = \int_{-\infty}^\infty \frac{y}{y^4+4}\sin\alpha y\,dy$$

we use the contour integral $\displaystyle\int_C \frac{ze^{i\alpha z}}{z^4+4}\,dz$ and

$$\int_{-\infty}^\infty \frac{xe^{i\alpha x}}{x^4+4}\,dx = 2\pi i[\mathrm{Res}(f(z),1+i) + \mathrm{Res}(f(z),-1+i)] = 2\pi i\left[-\frac{1}{8}ie^{(-1+i)\alpha} + \frac{1}{8}ie^{(-1-i)\alpha}\right] = \frac{\pi}{2}(e^{-\alpha}\sin\alpha)i$$

$$\int_{-\infty}^\infty \frac{x\sin\alpha x}{x^4+4}\,dx = \mathrm{Im}\left(\int_{-\infty}^\infty \frac{xe^{i\alpha x}}{x^4+4}\,dx\right) = \frac{\pi}{2}e^{-\alpha}\sin\alpha.$$

Finally, $U(\pi,\alpha) = \dfrac{\pi}{2}e^{-\alpha}\sin\alpha = c_2\sinh\alpha\pi$ gives $c_2 = \dfrac{\pi}{2}\dfrac{e^{-\alpha}\sin\alpha}{\sinh\alpha\pi}$. Hence $U(x,\alpha) = \dfrac{\pi}{2}\dfrac{e^{-\alpha}\sin\alpha}{\sinh\alpha\pi}\sinh\alpha x$ and

$$u(x,y) = \int_0^\infty \frac{e^{-\alpha}\sin\alpha}{\sinh\alpha\pi}\sinh\alpha x\sin\alpha y\,d\alpha.$$

Chapter 19 Review Exercises

3. False   **6.** True   **9.** $1/\pi$ **12.** False

15. $f(z) = \dfrac{1}{z^4}\left[1 - \left(1 + \dfrac{iz}{1!} + \dfrac{i^2z^2}{2!} + \dfrac{i^3z^3}{3!} + \dfrac{i^4z^4}{4!} + \cdots\right)\right] = -\dfrac{i}{z^3} + \dfrac{1}{2!z^2} + \dfrac{i}{3!z} - \dfrac{1}{4!} - \dfrac{iz}{5!} + \cdots$

18. $\dfrac{1-\cos z^2}{z^5} = \dfrac{1}{z^5}\left[1 - \left(1 - \dfrac{z^4}{2!} + \dfrac{z^8}{4!} - \dfrac{z^{12}}{6!} + \dfrac{z^{16}}{8!} - \cdots\right)\right] = \dfrac{1}{2!z} - \dfrac{z^3}{4!} + \dfrac{z^7}{6!} - \dfrac{z^{11}}{8!} + \cdots$

21. $\displaystyle\oint_C \frac{2z+5}{z(z+2)(z-1)^4}\,dz = 2\pi i[\mathrm{Res}(f(z),0) + \mathrm{Res}(f(z),-2)] = \frac{404}{81}\pi i$

24. $\oint_C \dfrac{z+1}{\sinh z}\, dz = 2\pi i[\text{Res}(f(z),0) + \text{Res}(f(z),\pi i)] = 2\pi i[1 + (-\pi i - 1)] = 2\pi^2$

27. $\oint_C \dfrac{1}{z(e^z - 1)}\, dz = 2\pi i\, \text{Res}(f(z),0) = -\pi i.$ Note: $z = 0$ is a pole of order two, and so

$$\text{Res}(f(z),0) = \lim_{z \to 0} \frac{d}{dz}\, z^2 \cdot \frac{1}{z^2\left(1 + \dfrac{z}{2!} + \dfrac{z^2}{3!} + \cdots\right)} = \lim_{z \to 0} -\frac{\left(\dfrac{1}{2!} + \dfrac{2z}{3!} + \cdots\right)}{\left(1 + \dfrac{z}{2!} + \dfrac{z^2}{3!} + \cdots\right)^2} = -\frac{1}{2}$$

30. $\oint_C \csc \pi z\, dz = 2\pi i[\text{Res}(f(z),0) + \text{Res}(f(z),1) + \text{Res}(f(z),2)] = 2\pi i\left[\dfrac{1}{\pi} + \left(-\dfrac{1}{\pi}\right) + \dfrac{1}{\pi}\right] = 2i$

33. $\displaystyle\int_0^{2\pi} \dfrac{\cos^2 \theta}{2 + \sin \theta}\, d\theta = \dfrac{1}{2}\oint_C \dfrac{z^4 + 2z^2 + 1}{z^2(z^2 + 4iz - 1)}\, dz$ $(C$ is $|z| = 1)$ $= \pi i[\text{Res}(f(z),0) + \text{Res}(f(z),(-2 + \sqrt{3}\,)i)]$

$$= \pi i[-4i + 2\sqrt{3}\,i] = (4 - 2\sqrt{3}\,)\pi$$

[Note: The answer in the text is correct but not simplified.]

36. We have

$$\oint_C e^{-a^2 z^2} e^{ibz}\, dz = \int_{-r}^{r} + \int_{C_1} + \int_{C_2} + \int_{C_3} = 0$$

by the Cauchy-Goursat Theorem. Therefore

$$\int_{-r}^{r} = -\int_{C_1} - \int_{C_2} - \int_{C_3}.$$

Let C_1 and C_3 denote the vertical sides of the rectangle. By the ML-inequality, $\displaystyle\int_{C_1} \to 0$ and $\displaystyle\int_{C_3} \to 0$ as $r \to \infty$. On C_2, $z = x + \dfrac{b}{2a^2}\,i$, $-r \le x \le r$, $dz = dx$,

$$\int_{-\infty}^{\infty} e^{-ax^2} e^{ibx}\, dx = -\int_{\infty}^{-\infty} e^{-a^2\left(x + \frac{b}{2a^2}i\right)^2} e^{ib\left(x + \frac{b}{2a^2}i\right)}\, dx = \int_{-\infty}^{\infty} e^{-a^2 x^2} e^{-b^2/4a^2}\, dx$$

$$\int_{-\infty}^{\infty} e^{-ax^2}(\cos bx + i\sin bx)\, dx = e^{-b^2/4a^2}\int_{-\infty}^{\infty} e^{-a^2 x^2}\, dx.$$

Using the given value of $\displaystyle\int_{-\infty}^{\infty} e^{-a^2 x^2}\, dx$ and equating real and imaginary parts gives

$$\int_{-\infty}^{\infty} e^{-ax^2} \cos bx\, dx = \frac{\sqrt{\pi}}{a}\, e^{-b^2/4a^2} \qquad \text{and so} \qquad \int_0^{\infty} e^{-ax^2} \cos bx\, dx = \frac{\sqrt{\pi}}{2a}\, e^{-b^2/4a^2}.$$

20 Conformal Mappings and Applications

_____ **Exercises 20.1** _____

3. For $w = z^2$, $u = x^2 - y^2$ and $v = 2xy$. If $xy = 1$, $v = 2$ and so the hyperbola $xy = 1$ is mapped onto the line $v = 2$.

6. If $\theta = \pi/4$, then $v = \theta = \pi/4$. In addition $u = \log_e r$ will vary from $-\infty$ to ∞. The image is therefore the horizontal line $v = \pi/4$.

9. For $w = e^z$, $u = e^x \cos y$ and $v = e^x \sin y$. Therefore if $e^x \cos y = 1$, $u = 1$. The curve $e^x \cos y = 1$ is mapped into the line $u = 1$. Since $v = \dfrac{\sin y}{\cos y} = \tan y$, v varies from $-\infty$ to ∞ and the image is the line $u = 1$.

12. For $w = \dfrac{1}{z}$, $u = \dfrac{x}{x^2 + y^2}$ and $v = \dfrac{-y}{x^2 + y^2}$. The line $y = 0$ is mapped to the line $v = 0$, and, from Problem 2, the line $y = 1$ is mapped onto the circle $|w + \frac{1}{2} i| = \frac{1}{2}$. Since $f(\frac{1}{2} i) = -2i$, the region $0 \le y \le 1$ is mapped onto the points in the half-plane $v \le 0$ which are on or outside the circle $|w + \frac{1}{2} i| = \frac{1}{2}$. (The image does not include the point $w = 0$.)

15. The mapping $w = z + 4i$ is a translation which maps the circle $|z| = 1$ to a circle of radius $r = 1$ and with center $w = 4i$. This circle may be described by $|w - 4i| = 1$.

18. Since $w = (1 + i)z = \sqrt{2}\, e^{i\pi/4}z$, the mapping is the composite of a rotation through $45°$ and a magnification by $\alpha = \sqrt{2}$. The image of the first quadrant is therefore the angular wedge $\pi/4 \le \text{Arg}\, w \le 3\pi/4$.

21. We first let $z_1 = z - i$ to map the region $1 \le y \le 4$ to the region $0 \le y_1 \le 3$. We then let $w = e^{-i\pi/2}z_1$ to rotate this strip through $-90°$. Therefore $w = -i(z - i) = -iz - 1$ maps $1 \le y \le 4$ to the strip $0 \le u \le 3$.

24. The mapping $w = iz$ will rotate the strip $-1 \le x \le 1$ through $90°$ so that the strip $-1 \le v \le 1$ results.

27. By Example 1, Section 20.1, $z_1 = e^z$ maps the strip $0 \le y \le \pi$ onto the upper half-plane $y_1 \ge 0$, or $0 \le \text{Arg}\, z_1 \le \pi$. The power function $w = z_1^{3/2}$ changes the opening of this wedge by a factor of $3/2$ so the wedge $0 \le \text{Arg}\, w \le 3\pi/2$ results. The composite of these two mappings is $w = (e^z)^{3/2} = e^{3z/2}$.

30. The mapping $z_1 = -(z - \pi i)$ lowers R by π units in the vertical direction and then rotates the resulting region through $180°$. The image region R_1 is upper half-plane $y_1 \ge 0$. By Example 1, Section 20.1, $w = \text{Ln}\, z_1$ maps R_1 onto the strip $0 \le v \le \pi$. The composite of these two mappings is $w = \text{Ln}(\pi i - z)$.

_____ **Exercises 20.2** _____

3. $f'(z) = 1 + e^z$ and $1 + e^z = 0$ for $z = \pm i \pm 2n\pi i$. Therefore f is conformal except for $z = \pi i \pm 2n\pi i$.

6. The function $f(z) = \pi i - \frac{1}{2}[\text{Ln}(z + 1) + \text{Ln}(z - 1)]$ is analytic except on the branch cut $x - 1 \le 0$ or $x \le 1$, and

$$f'(z) = -\frac{1}{2}\left(\frac{1}{z + 1} + \frac{1}{z - 1}\right) = -\frac{z}{z^2 - 1}$$

is non-zero for $z \ne 0$, ± 1. Therefore f is conformal except for $z = x$, $x \le 1$.

9. $f(z) = (\sin z)^{1/4}$ is the composite of $z_1 = \sin z$ and $w = z_1^{1/4}$. The region $-\pi/2 \le x \le \pi/2$, $y \ge 0$ is mapped to the upper half-plane $y_2 \ge 0$ by $z_1 = \sin z$ (See Example 2) and the power function $w = z_1^{1/4}$ maps this upper

270

half-plane to the angular wedge $0 \le \operatorname{Arg} w \le \pi/4$. The real interval $[-\pi/2, \pi/2]$ is first mapped to $[-1, 1]$ and then to the union of the line segments from $e^{i\frac{\pi}{4}}$ to 0 and 0 to 1.

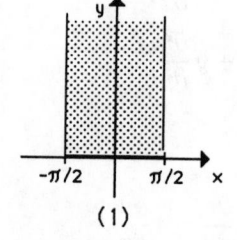 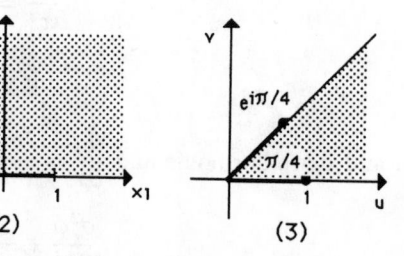

(1)   (2)   (3)

12. Using C-3 $w = e^z$ maps R onto the target region R'. The image of AB is shown in the figure.

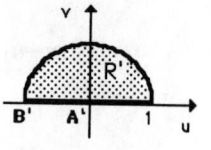

15. Using H-6, $z_1 = \dfrac{e^{\pi/z} + e^{-\pi/z}}{e^{\pi/z} - e^{-\pi/z}}$ maps R onto the upper half-plane $y_1 \ge 0$, and $w = z_1^{1/2}$ maps this half-plane onto the target region R'. Therefore

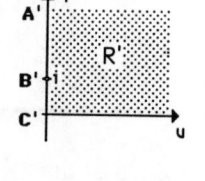

$$w = \left(\frac{e^{\pi/z} + e^{-\pi/z}}{e^{\pi/z} - e^{-\pi/z}} \right)^{1/2}$$

and the image of AB is shown in the figure.

18. Using E-9, $z_1 = \cosh z$ maps R onto the upper half-plane $y_1 \ge 0$. Using M-7, $w = z_1 + \operatorname{Ln} z_1 + 1$ maps this half-plane onto the target region R'. Therefore $w = \cosh z + \operatorname{Ln}(\cosh z) + 1$ and the image of AB is shown in the figure.

21. In this exercise we find a conformal mapping $w = f(z)$ that maps the given region R onto the upper half-plane $v \ge 0$ and transfers the boundary conditions so that the resulting Dirichlet problem is as shown in the figure. We have $f(z) = i\,\dfrac{1-z}{1+z}$, using H-1, and so $u = U(f(z)) = \dfrac{1}{\pi} \operatorname{Arg}\left(\dfrac{1-z}{1+z}\right)$. The

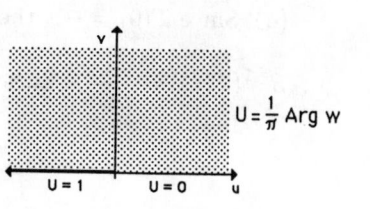

solution may also be written as $u(x, y) = \dfrac{1}{\pi} \tan^{-1}\left(\dfrac{1 - x^2 - y^2}{2y}\right)$.

24. The mapping $z_1 = z^2$ maps R onto the region R_1 defined by $y_1 \ge 0$, $|z_1| \ge 1$ and shown in H-3, and $w = \dfrac{1}{2}\left(z_1 + \dfrac{1}{z_1}\right)$ maps R_1 onto the upper half-plane $v \ge 0$. Letting $c_0 = 5$,

$$u = \frac{5}{\pi}\left[\operatorname{Arg}\left(\frac{1}{2}\left[z^2 + \frac{1}{z^2} \right] - 1 \right) - \operatorname{Arg}\left(\frac{1}{2}\left[z^2 + \frac{1}{z^2} \right] + 1 \right) \right].$$

27. (a) If $u = \dfrac{\partial^2 \phi}{\partial x^2} + \dfrac{\partial^2 \phi}{\partial y^2}$,

$$\frac{\partial^2 u}{\partial x^2} + \frac{\partial^2 u}{\partial y^2} = \frac{\partial^4 \phi}{\partial x^4} + 2\frac{\partial^4 \phi}{\partial x^2 \partial y^2} + \frac{\partial^4 \phi}{\partial y^4} = 0$$

since ϕ is assumed to be biharmonic.

(b) If $g = u + iv$, then $\phi = \text{Re}(\bar{z}g(z)) = xu + yv$.

$$\frac{\partial^2 \phi}{\partial x^2} = 3\frac{\partial u}{\partial x} + x\frac{\partial^2 u}{\partial x^2} + y\frac{\partial^2 v}{\partial x^2}$$

$$\frac{\partial^2 \phi}{\partial y^2} = 2\frac{\partial v}{\partial y} + x\frac{\partial^2 u}{\partial y^2} + y\frac{\partial^2 v}{\partial y^2}.$$

Since u and v are harmonic and $\dfrac{\partial u}{\partial x} = \dfrac{\partial v}{\partial y}$,

$$\frac{\partial^2 \phi}{\partial x^2} + \frac{\partial^2 \phi}{\partial y^2} = 2\frac{\partial u}{\partial x} + 2\frac{\partial v}{\partial y} = 4\frac{\partial u}{\partial x}.$$

Now $u_1 = \dfrac{\partial u}{\partial x}$ is also harmonic and so $\dfrac{\partial^2 u_1}{\partial x^2} + \dfrac{\partial^2 u_1}{\partial y^2} = 0$. But

$$\frac{\partial^2 u_1}{\partial x^2} + \frac{\partial^2 u_1}{\partial y^2} = \frac{1}{4}\left[\frac{\partial^4 \phi}{\partial x^4} + 2\frac{\partial^4 \phi}{\partial x^2 \partial y^2} + \frac{\partial^4 \phi}{\partial y^4}\right]$$

and so ϕ is biharmonic.

Exercises 20.3

3. (a) For $T(z) = \dfrac{z+1}{z-1}$, $T(0) = -1$, $T(1) = \infty$, and $T(\infty) = 1$.

(b) The circle $|z| = 1$ passes through the pole at $z = 1$ and so the image is a line. Since $T(-1) = 0$ and $T(i) = -i$, the image is the line $u = 0$. If $|z - 1| = 1$,

$$|w - 1| = \left|\frac{z+1}{z-1} - 1\right| = \frac{2}{|z-1|} = 2$$

and so the image is the circle $|w - 1| = 2$ in the w-plane.

(c) Since $T(0) = -1$, the image of the disk $|z| \leq 1$ is the half-plane $u \leq 0$.

6. $S^{-1}(T(z)) = \dfrac{az+b}{cz+d}$ where

$$\begin{bmatrix} a & b \\ c & d \end{bmatrix} = \text{adj}\left(\begin{bmatrix} 2 & 1 \\ 1 & 1 \end{bmatrix}\right)\begin{bmatrix} i & 0 \\ 1 & -2i \end{bmatrix} = \begin{bmatrix} -1+i & 2i \\ 2-i & -4i \end{bmatrix}.$$

Therefore

$$S^{-1}(T(z)) = \frac{(-1+i)z + 2i}{(2-i)z - 4i} \quad \text{and} \quad S^{-1}(w) = \frac{w-1}{-w+2}.$$

9. $T(z) = \dfrac{(z-z_1)(z_2-z_3)}{(z-z_3)(z_2-z_1)}$ maps z_1, z_2, z_3 to 0, 1, ∞. Therefore $T(z) = \dfrac{(z+1)(-2)}{(z-2)(1)} = -2\dfrac{z+1}{z-2}$ maps -1, 0,

2 to 0, 1, ∞.

12. $S(w) = z = \dfrac{(w-w_1)(w_2-w_3)}{(w-w_3)(w_2-w_1)}$ maps w_1, w_2, w_3 to 0, 1, ∞ and so $z = \dfrac{(w-1-i)(-1+i)}{(w-1+i)(-1-i)}$ maps $1+i$, 0,

$1-i$ to 0, 1, ∞. Solving for w, we find that $w = \dfrac{2z-2}{(1+i)z-1+i}$ maps 0, 1, ∞ to $1+i$, 0, $1-i$.

15. Using the cross-ratio formula (7),

$$S(w) = \frac{(w+1)(-3)}{(w-3)(1)} = \frac{(z-1)(2i)}{(z+i)(i-1)} = T(z).$$

We can solve for w to obtain

$$w = 3 \frac{(1+i)z + (1-i)}{(-3+5i)z - 3 - 5i}.$$

Alternatively we can apply the matrix method to compute $w = S^{-1}(T(z))$.

18. The mapping $T(z) = \frac{1}{2}\frac{z+1}{z}$ maps -1, 1, 0 to 0, 1, ∞ and maps each of the two circles in R to lines since both circles pass through the pole at $z = 0$. Since $T(\frac{1}{2} + \frac{1}{2}i) = 1 - i$ and $T(1) = 1$, the circle $|z - \frac{1}{2}| = \frac{1}{2}$ is mapped onto the line $u = 1$. Likewise, the circle $|z + \frac{1}{2}| = \frac{1}{2}$ is mapped onto the line $u = 0$. The transferred boundary conditions are shown in the figure and $U(u,v) = u$ is the solution. The solution to the Dirichlet problem in Figure 20.38 is

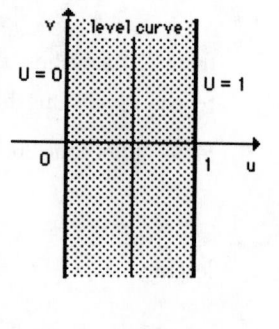

$$u = U(T(z)) = \text{Re}\left(\frac{1}{2}\frac{z+1}{z}\right) = \frac{1}{2} + \frac{1}{2}\frac{x}{x^2 + y^2}.$$

The level curves $u = c$ are the circles with centers on the x-axis which pass through the origin. The level curve $u = \frac{1}{2}$, however, is the vertical line $x = 0$.

21. $T_2(T_1(z)) = \dfrac{a_2 T_1(z) + b_2}{c_2 T_1(z) + d_2} = \dfrac{a_2 \dfrac{a_1 z + b_1}{c_1 z + d_1} + b_2}{c_2 \dfrac{a_1 z + b_1}{c_1 z + d_1} + d_2} = \dfrac{a_1 a_2 z + a_2 b_1 + b_2 c_1 z + b_2 d_1}{a_1 c_2 z + b_1 c_2 + c_1 d_2 z + d_1 d_2} = \dfrac{(a_1 a_2 + c_1 b_2)z + (b_1 a_2 + d_1 b_2)}{(a_1 c_2 + c_1 d_2)z + (b_1 c_2 + d_1 d_2)}$

Exercises 20.4

3. $\arg f'(t) = -\dfrac{1}{2}\text{Arg}(t+1) + \dfrac{1}{2}\text{Arg}(t-1) = \begin{cases} 0, & t < -1 \\ \pi/2, & -1 < t < 1 \\ 0, & t > 1 \end{cases}$

and $\alpha_1 = \pi/2$ and $\alpha_2 = 3\pi/2$. Since $f(-1) = 0$, the image of the upper half-plane is the region shown in the figure.

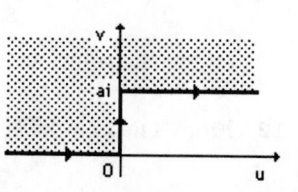

6. Since $\alpha_1 = \pi/3$ and $\alpha_2 = \pi/2$, $\alpha_1/\pi - 1 = -2/3$ and $\alpha_2/\pi - 1 = -1/2$ and so $f'(z) = A(z+1)^{-2/3}z^{-1/2}$ for some constant A.

9. Since $\alpha_1 = \alpha_2 = \pi/2$, $f'(z) = A(z+1)^{-1/2}(z-1)^{-1/2} = A/(z^2 - 1)^{1/2}$. Therefore $f(z) = A\cosh^{-1}z + B$. But $f(-1) = \pi i$ and $f(1) = 0$. Since $\cosh^{-1}1 = 0$, $B = 0$. Since $\cosh^{-1}(-1) = \pi i$, $\pi i = A(\pi i)$ and so $A = 1$. Hence $f(z) = \cosh^{-1}z$.

12. From (3), $f'(z) = Az^{-3/4}(z-1)^{\alpha_2/\pi - 1}$. But $\alpha_2 \to \pi$ as $\theta \to 0$. This suggests that we examine $f'(z) = Az^{-3/4}$. Therefore $f(z) = A_1 z^{1/4} + B_1$. But $f(0) = 0$ and $f(1) = 1$ so that $B_1 = 0$ and $A_1 = 1$. Hence $f(z) = z^{1/4}$ and we recognize that this power function maps the upper half-plane onto the wedge $0 \le \text{Arg}\, w \le \pi/4$.

Exercises 20.5

3. The harmonic function

$$u_1 = \frac{5}{\pi}[\pi - \text{Arg}(z-1)] = \begin{cases} 5, & x > 1 \\ 0, & x < 1 \end{cases}, \quad \text{and} \quad u_2 = -\frac{1}{\pi}\text{Arg}\left(\frac{z+1}{z+2}\right) + \frac{1}{\pi}\text{Arg}\left(\frac{z}{z+1}\right)$$

from (3) satisfies all boundary conditions except that $u_2 = 0$ for $x > 1$. Therefore $u = u_1 + u_2$ is the solution to the given Dirichlet problem.

6. From Theorem 20.5,

$$u(x,y) = \frac{y}{\pi}\int_{-\infty}^{\infty} \frac{\cos t}{(x-t)^2 + y^2}\,dt = \frac{y}{\pi}\int_{-\infty}^{\infty} \frac{\cos(x-s)}{s^2 + y^2}\,ds$$

letting $s = x - t$. But $\cos(x-s) = \cos x \cos x + \sin x \sin s$. It follows that

$$u(x,y) = \frac{y\cos x}{\pi}\int_{-\infty}^{\infty} \frac{\cos s}{s^2 + y^2}\,ds + \frac{y\sin x}{\pi}\int_{-\infty}^{\infty} \frac{\sin s}{s^2 + y^2}\,ds = \frac{y\cos x}{\pi}\left(\frac{\pi e^{-y}}{y}\right) = e^{-y}\cos x, \quad y > 0.$$

9. Using H-1, $f(z) = i\dfrac{1-z}{1+z}$ maps R onto the upper half-plane R'. The corresponding Dirichlet problem in R' is shown in the figure. From (3),

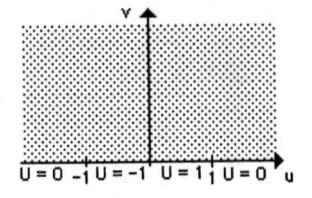

$$U = -\frac{1}{\pi}\text{Arg}\left(\frac{w}{w+1}\right) + \frac{1}{\pi}\text{Arg}\left(\frac{w-1}{w}\right) = \frac{1}{\pi}\left[\text{Arg}\left(\frac{w-1}{w}\right) - \text{Arg}\left(\frac{w}{w+1}\right)\right].$$

The harmonic function $u = U(f(z))$ may be simplified to

$$u = \frac{1}{\pi}\left[\text{Arg}\left(\frac{(1-i)z - (1+i)}{1-z}\right) - \text{Arg}\left(\frac{1-z}{-(1+i)z + 1 - i}\right)\right]$$

and is the solution to the original Dirichlet problem in R.

12. From Theorem 20.6, $u(x,y) = \dfrac{1}{2\pi}\displaystyle\int_{-\pi}^{\pi} e^{-|t|}\dfrac{1-|z|^2}{|e^{it} - z|^2}\,dt$. Therefore,

$$u(0,0) = \frac{1}{2\pi}\int_{-\pi}^{\pi} e^{-|t|}dt = \frac{1}{\pi}\int_{0}^{\pi} e^{-t}\,dt = \frac{1}{\pi}(1 - e^{-\pi}).$$

With the aid of Simpson's Rule, $u(0.5, 0) = 0.5128$ and $u(-0.5, 0) = 0.1623$.

15. For $u(e^{i\theta}) = \sin\theta + \cos\theta$, the Fourier series solution (6) reduces to
$$u(r,\theta) = r\sin\theta + r\cos\theta \quad \text{or} \quad u(x,y) = y + x.$$
The corresponding system of level curves is shown in the figure.

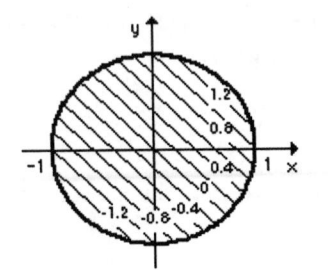

Exercises 20.6

3. $g(z) = \dfrac{x}{x^2 + y^2} - \dfrac{y}{x^2 + y^2} i = \dfrac{1}{z}$ is analytic for $z \neq 0$ and so div $\mathbf{F} = 0$ and curl $\mathbf{F} = 0$ by Theorem 20.7. A complex potential is $G(z) = \text{Ln}\,z$ and

$$\phi(x, y) = \text{Re}(G(z)) = \frac{1}{2} \log_e(x^2 + y^2).$$

The equipotential lines $\phi(x, y) = c$ are circles $x^2 + y^2 = e^{2c}$ and are shown in the figure.

6. The function $f(z) = \dfrac{1}{z}$ maps the original region R to the strip $-\frac{1}{2} \leq v \leq 0$ (see Example 2, Section 20.1). The boundary conditions transfer as shown in the figure. $U = -2v$ is the solution in the horizontal strip and so

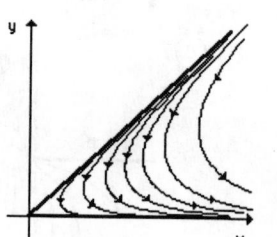

$$\phi(x, y) = -2\text{Im}\left(\frac{1}{z}\right) = \frac{2y}{x^2 + y^2}$$

is the potential in the original region R. The equipotential lines $\dfrac{2y}{x^2 + y^2} = c$ may be written as

$$x^2 + \left(y + \frac{1}{c}\right)^2 = \left(\frac{1}{c}\right)^2 \text{ for } c \neq 0 \text{ and are circles. If } c = 0, \text{ we obtain the line } y = 0. \text{ Note that}$$

$\phi(x, y) = \text{Re}\left(\dfrac{2i}{z}\right)$ and so $G(z) = \dfrac{2i}{z}$ is a complex potential. The corresponding vector field is

$$\mathbf{F} = \overline{G'(z)} = \frac{2i}{\bar{z}^2} = \left(\frac{-4xy}{(x^2 + y^2)^2}, \frac{2(x^2 - y^2)}{(x^2 + y^2)^2}\right).$$

9. (a) $\psi(x, y) = \text{Im}(z^4) = 4xy(x^2 - y^2)$ and so $\psi(x, y) = 0$ when $y = x$ and $y = 0$.

(b) $\mathbf{V} = \overline{G'(z)} = \overline{4z^3} = 4(x^3 - 3xy^2, y^3 - 3x^2y)$

(c) In polar coordinates $r^4 \sin 4\theta = c$ or $r = (c \csc 4\theta)^{1/4}$, for $0 < \theta < \pi/4$, are the streamlines. See the figure.

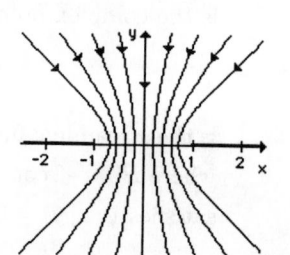

12. (a) The image of R under $w = i\sin^{-1} z$ is the horizontal strip (see E-6) $-\pi/2 \leq v \leq \pi/2$ and

$$\psi(x, y) = \text{Im}(i\sin^{-1} z) = \begin{cases} \pi/2, & x \geq 1 \\ -\pi/2, & x \leq -1 \end{cases}.$$

Each piece of boundary is therefore a streamline.

(b) $\mathbf{V} = \overline{G'(z)} = \overline{\dfrac{i}{(1 - z^2)^{1/2}}} = \dfrac{-i}{(1 - \bar{z}^2)^{1/2}}$

(c) The streamlines are the images of the lines $v = b$, $-\pi/2 < b < \pi/2$ under $z = -i\sin w$ and are therefore hyperbolas. See Example 2, Section 20.2, and the figure. Note that at $z = 0$, $v = -i$ and the flow is downward.

15. (a) For $f(z) = \pi i - \frac{1}{2}[\text{Ln}(z + 1) + \text{Ln}(z - 1)]$

$$f(t) = \pi i - \frac{1}{2}[\log_e |t + 1| + \log_e |t - 1| + i\text{Arg}(t + 1) + i\text{Arg}(t - 1)]$$

and so $\text{Im}(f(t)) = \begin{cases} 0, & t < -1 \\ \pi/2, & -1 < t < 1 \\ \pi, & t > 1 \end{cases}$ Hence $\text{Im}(G(z)) = \psi(x,y) = 0$ on the boundary of R.

(b) $x = -\dfrac{1}{2}[\log_e |t+1+ic| + \log_e |t-1+ic|], \quad y = \pi - \dfrac{1}{2}[\text{Arg}(t+1+ic) + \text{Arg}(t-1+ic)] \quad$ for $c > 0$

(c)

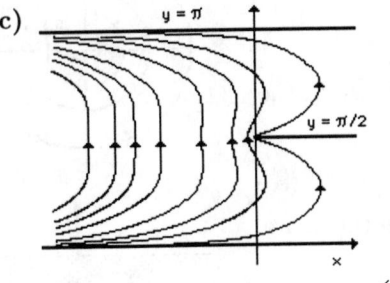

18. (a) For $f(z) = 2(z+1)^{1/2} + \text{Ln}\left(\dfrac{(z+1)^{1/2}-1}{(z+1)^{1/2}+1}\right)$,

$$f(t) = 2(t+1)^{1/2} + \text{Ln}\dfrac{(t+1)^{1/2}-1}{(t+1)^{1/2}+1}.$$

If we write $(t+1)^{1/2} = |t+1|^{1/2}e^{(i/2)\text{Arg}(t+1)}$, we may conclude that

$$\text{Im}(f(t)) = \begin{cases} 0, & t > 0 \\ \pi, & -1 < t < 0 \end{cases} \quad \text{and} \quad \text{Re}(f(t)) = 0 \text{ for } t < -1.$$

Therefore $\text{Im}(G(z)) = \psi(x,y) = 0$ on the boundary of R.

(b) $x = \text{Re}\left[2(t+ic+1)^{1/2} + \text{Ln}\dfrac{(t+ic+1)^{1/2}-1}{(t+ic+1)^{1/2}+1}\right]$

$y = \text{Im}\left[2(t+ic+1)^{1/2} + \text{Ln}\dfrac{(t+ic+1)^{1/2}-1}{(t+ic+1)^{1/2}+1}\right] \quad$ for $c > 0$

(c)

21. $f(z) = z^2$ maps the first quadrant onto the upper half-plane and $f(\xi_0) = f(1) = 1$. Therefore $G(z) = \text{Ln}(z^2-1)$ is the complex potential, and so

$$\psi(x,y) = \text{Arg}(z^2-1) = \tan^{-1}\left(\dfrac{2xy}{x^2-y^2-1}\right)$$

is the streamline function where $\tan^{-1}$ is chosen to be between 0 and π. If $\psi(x,y) = c$, then $x^2 + Bxy - y^2 - 1 = 0$ where $B = -2\cot c$. Each hyperbola in the family passes through $(1,0)$ and the boundary of the first quadrant satisfies $\psi(x,y) = 0$.

24. (a) $\mathbf{V} = \dfrac{a+ib}{\bar{z}} = \left(\dfrac{ax-by}{x^2+y^2}, \dfrac{bx+ay}{x^2+y^2}\right)$ and since $(x'(t), y'(t)) = \mathbf{V}$, the path of the particle satisfies

$$\dfrac{dx}{dt} = \dfrac{ax-by}{x^2+y^2}, \qquad \dfrac{dy}{dt} = \dfrac{bx+ay}{x^2+y^2}.$$

(b) Switching to polar coordinates,

$$\frac{dr}{dt} = \frac{1}{r}\left(x\frac{dx}{dt} + y\frac{dy}{dt}\right) = \frac{1}{r}\left(\frac{ax^2 - bxy}{r^2} + \frac{bxy + ay^2}{r^2}\right) = \frac{a}{r}$$

$$\frac{d\theta}{dt} = \frac{1}{r^2}\left(-y\frac{dx}{dt} + x\frac{dy}{dt}\right) = \frac{1}{r^2}\left(\frac{-axy + by^2}{r^2} + \frac{bx^2 + axy}{r^2}\right) = \frac{b}{r^2}.$$

Therefore $\dfrac{dr}{d\theta} = \dfrac{a}{b}r$ and so $r = ce^{a\theta/b}$.

(c) $\dfrac{dr}{dt} = \dfrac{a}{r} < 0$ if and only if $a < 0$, and in this case r is decreasing and the curve spirals inward. $\dfrac{d\theta}{dt} = \dfrac{b}{r^2} < 0$ if and only if $b < 0$, and in this case θ is decreasing and the curve is traversed clockwise.

Chapter 20 Review Exercises

3. The wedge $0 \le \operatorname{Arg} w \le 2\pi/3$. See figure 20.6 in the text.

6. A line, since $|z - 1| = 1$ passes through the pole at $z = 2$.

9. False. $\overline{g(z)} = P - iQ$ is analytic. See Theorem 20.7.

12. First use $z_1 = z^2$ to map the first quadrant onto the upper half-plane $y_1 \ge 0$, and segment AB to the negative real axis. We then use $w = \frac{1}{\pi}(z_1 + \operatorname{Ln} z_1 + 1)$ to map this half-plane onto the target region R'. The composite transformation is

$$w = \frac{1}{\pi}[z^2 + \operatorname{Ln}(z^2) + 1]$$

and the image of AB is the ray extending to the left from $w = i$ along the line $v = 1$.

15. The inversion $w = 1/z$ maps R onto the horizontal strip $-1 \le v \le -1/2$ and the transferred boundary conditions are shown in the figure. The solution in the strip is $U = 2v + 2$ and so

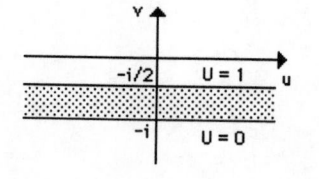

$$u = U\left(\frac{1}{z}\right) = 2\operatorname{Im}\left(\frac{1}{z}\right) + 2 = \frac{-2y}{x^2 + y^2} + 2$$

is the solution to the original Dirichlet problem in R.

18. (a) From Theorem 20.5,

$$u(x,y) = \frac{y}{\pi}\int_{-\infty}^{\infty}\frac{\sin t}{(x-t)^2 + y^2}\,dt = \frac{y}{\pi}\int_{-\infty}^{\infty}\frac{\sin(x-s)}{s^2 + y^2}]\,ds \quad (\text{letting } s = x - t).$$

But $\sin(x - s) = \sin x \cos s - \cos x \sin s$. We now proceed as in the solution to Problem 6, Section 20.5 to show that $u(x,y) = e^{-y}\sin x$.

(b) For $u(e^{i\theta}) = \sin\theta$, the Fourier Series solution (6) in Section 20.5 reduces to $u(r,\theta) = r\sin\theta$.

Appendix II Gamma Function

3. If $t = x^3$, then $dt = 3x^2\,dx$ and $x^4\,dx = \frac{1}{3}t^{2/3}\,dt$. Now

$$\int_0^\infty x^4 e^{-x^3}\,dx = \int_0^\infty \frac{1}{3}t^{2/3}e^{-t}\,dt = \frac{1}{3}\int_0^\infty t^{2/3}e^{-t}\,dt = \frac{1}{3}\Gamma\left(\frac{5}{3}\right) = \frac{1}{3}(0.89) \approx 0.297.$$

6. For $x > 0$

$$\Gamma(x+1) = \int_0^\infty t^x e^{-t}dt \qquad \boxed{u = t^x,\; du = xt^{x-1}\,dt;\;\; dv = e^{-t}\,dt,\; v = -e^{-t}}$$

$$= -t^x e^{-t}\Big|_0^\infty - \int_0^\infty xt^{x-1}(-e^{-t})\,dt = x\int_0^\infty t^{x-1}e^{-t}dt = x\Gamma(x).$$

Notes

Notes

Notes

Notes

Notes

Notes